Abstract Algebra

Abstract Algebra
A Concrete Introduction

Robert H. Redfield

Hamilton College

Addison
Wesley

Boston   San Francisco   New York
London   Toronto   Sydney   Tokyo   Singapore   Madrid
Mexico City   Munich   Paris   Cape Town   Hong Kong   Montreal

Sponsoring Editors: Carolyn Lee-Davis / Laurie Rosatone
Assistant Editor: RoseAnne Johnson
Managing Editor: Karen Guardino
Associate Production Supervisor: Cindy Cody
Cover Design: Dardani Gasc Design
Cover Art: One Wisdom Une Fen'tre, 1912, Robert Delaunay 1885–1941
Source: French Muse National d'Art de Moderne, Paris/Superstock
Design Supervisor: Barbara Atkinson
Manufacturing Buyer: Evelyn Beaton
Marketing Manager: Michael Boezi
Compositor: Nesbitt Graphics, Inc.

For permission to use copyrighted material, grateful acknowledgment is made to the copyright holders on p. 446, which is hereby made part of this copyright page.

Maple is a registered trademark of Waterloo Maple, Inc.

Library of Congress Cataloging-in-Publication Data

Redfield, Robert H.
    Abstract algebra: a concrete introduction / Robert H. Redfield.
        p. cm.
    Includes bibliographical references and index.
    ISBN 0-201-43721-X
        1.   Algebra, Abstract. I. Title.

    QA162.R43 2000
    512'.02—dc21   00-032753

Please visit our website at www.awl.com/redfield

ISBN 0-201-43721-X

1 2 3 4 5 6 7 8 9 10 CRW 03 02 01 00

To my wife Mary Javorski

and

my children Signe, Lisbeth, and Catherine

Preface

This book is intended as a text for a beginning course in abstract algebra. Its pedagogical object is to introduce the fundamental structures of modern abstract algebra (fields, rings, and groups); its method is to set a problem as a goal and then to introduce the structures and their properties as they become necessary for the solution of the problem.

Adopting such an approach allows the subject to be presented concretely and with motivation. The presentation starts in the familiar setting of the rational, real, and complex numbers, and from there onward, new concepts and results flow naturally and inevitably from the problem to be solved and the concrete examples that have been met in the many computational exercises as well as in the exposition of the previous chapters. This progression allows the subject to be introduced in a thorough but not intimidating way and provides the assurance of concrete examples as it introduces the power of abstraction.

Computation and Proof

Although general theorems are proved in their entirety and detailed proofs are required for many exercises, concrete examples and computational problems are not only abundant but also necessary for a complete understanding of the theorems and their proofs. Examples both motivate and illustrate the definitions and the theorems, and it would be an unusual student who could supply the proofs required in the exercises without first solving at least some of the computational problems.

Historical Notes

Most chapters conclude with a "Historical Note," usually a biography of a mathematician who has in some way contributed to the topic of the chapter. The main purpose of these notes is not the usual one of putting the detailed mathematics presented in the chapter in a broader mathematical and historical context, although this is certainly a secondary purpose. Rather, the historical notes are meant to show the flesh and blood behind the mathematics, the happiness and sadness, the successes and failures, and the disruptions of politics and war. They are meant to describe both some of the general milieu in which the mathematics was created and some of the particular incidents in the life of one (or more) of the individual creators.

Appendices

The appendices are primarily for reference. There is a Greek alphabet for anyone who has not already learned it by osmosis, and a review of proof techniques to provide a quick reference for those who might have forgotten some terminology or methods. The fundamental theorems of linear algebra, which are used in Part 2 of the body of the text, are proved in the general setting of vector spaces over arbitrary fields; and, finally, some of the results of Part 2 are used to solve two of the famous questions that baffled the Greek geometers: can an angle be trisected and can a cube be duplicated?

Instructor's Solutions Manual

Solutions to the odd-numbered computational exercises are included at the end of the book. There is also a solutions manual available for the instructor from the publisher. It contains detailed solutions to all the exercises, both those involving computations and those involving proofs. (ISBN: 0-201-70901-5)

Note to the Instructor: Organization

There are two ways to proceed through the text depending on the time allotted and the inclination of the instructor.

Parts 1 through 4 give the modern version of Galois's proof that there are fifth degree polynomial equations which are not solvable by radicals. This path introduces all three algebraic structures in an order (first fields, then rings, and finally groups) which progresses from the concrete and familiar (fields) to the more abstract and unusual (groups) and which also parallels the order in which these structures were recognized. A class of good students can cover this material in a fourteen week semester.

Parts 1, 3, and 5, on the other hand, form a course in elementary group theory whose object is to determine all finite groups with fewer than sixteen elements. This path is better suited to a short semester and is appropriate for an instructor whose primary interest lies in group theory. It is not even necessary to complete all of Part 5. The first four chapters of Part 5 culminate with the fundamental theorem of finite commutative groups (which in fact characterizes all such groups, not just the ones with a small number of elements). In whatever fashion Parts 1, 3, and 5 are covered, a traveller who has reached the end of this path can go back to the beginning of Part 2 and in another shorter semester complete the book.

Note that Chapter 2 is somewhat of an anomaly. Although it motivates Chapter 3 on the complex numbers and begins the study of the insolvability of quintic equations by showing how to solve cubic and quartic equations, it is not strictly necessary for either path. It can be skipped without detriment to any of the material in the subsequent chapters.

Note also that since Part 3 occurs in both paths, some of the material in its chapters may not be appropriate for those following the second, group theory, path. To specify the material which is necessary for the first path but not the second, paragraphs in Part 3 which involve material from Part 2 are marked by a line on the left, as are this paragraph and the next, and are therefore inappropriate for someone traveling along the second path.

That is, determining polynomial equations which are not solvable by radicals (Parts 1 through 4) involves <u>all of Part 3,</u> whereas determining all small groups (Parts 1, 3, and 5) involves only the paragraphs in Part 3 which are <u>unmarked.</u>

Note to the Instructor: Summary

For those who are already familiar with some of the material introduced in the course of the succeeding chapters, the flow of the book may be described as follows.

The object of Parts 1 through 4 is to show that exact solutions, similar to those given by the familiar quadratic formula, cannot be found in general for equations involving polynomials of degrees larger than four.

To do this, we need some fundamental information about the integers and the real, rational, and complex numbers. **Part 1** reviews these structures along with some of their important properties. It also introduces some less familiar structures: the integers mod n (finite structures whose elements are subsets of the integers) and the quaternions (a generalization of the complex numbers that Hamilton discovered in the nineteenth century). Part 1 also includes a description of the algorithms that solve cubic and quartic polynomial equations.

While **Part 2** introduces fields and rings in general, it is mostly concerned with subfields of the complex numbers and polynomials with coefficients in these subfields. These polynomial rings share many familiar properties with the integers and are crucial to characterizing simple extension fields of these subfields. Such extension fields can be viewed as vector spaces over their base fields, and we show how to use polynomials to construct vector space bases of these extension fields and thus how to calculate them. Galois's crucial observation was that sets of certain functions, called automorphisms, on these fields are intricately connected with the roots of the polynomials; these sets of automorphisms are introduced and it is shown how to calculate some of the simple ones. Functions from a set to itself can be composed with each other, and the sets of automorphisms determined in Part 2 behave nicely with respect to composition. In this way, they form examples of groups, the last of the three algebraic structures to be studied.

Part 3 introduces groups in general and derives many of their elementary properties. The set of all one-to-one and onto functions from a set to itself, the so-called permutations of the set, form a group with respect to composition and are used, along with the additive groups of integers mod n and a finite group derived from Hamilton's quaternions, to motivate and illustrate the general definitions and theorems found in Part 3. Many small groups are calculated and used as examples and counterexamples. Once we have derived the necessary properties of groups, we are in a position to find polynomial equations which are not solvable by radicals.

Part 4 determines conditions on a group of automorphisms that imply that the generating polynomial equation is not solvable by radicals and then shows how to determine polynomial equations which satisfy these conditions.

As indicated above, Parts 1, 3, and 5 comprise a beginning course in group theory whose goal is the determination of all groups with fewer than sixteen elements.

Part 5 takes up the theory of groups where Part 3 left off. It begins by showing how all finite commutative groups can be expressed in terms of the groups of integers mod n. This is the fundamental theorem of finite commutative groups mentioned above, and as noted there, it is possible to conclude the course at this point. Part 5 then turns to the study of noncommutative groups. Subsequent chapters include results (Cauchy's theorem and the Sylow theorems) that are fundamental to the study of finite groups and which are used to determine all small groups. Two of the groups with twelve elements are examples of general constructions (alternating groups and semidirect products) that are described in the concluding two chapters. Of course, working with small finite groups involves a substantial amount of computation to complement the proofs of the general theorems.

Acknowledgments

My students and my colleagues at Hamilton College allowed me to experiment until I found the correct approach to this material; I could not have written this book without them. The students put up with a multitude of revisions, and I am especially indebted to those students who pointed out inconsistencies and typographical errors in the various versions. And without the indulgence of my colleagues, I would never have had the opportunity to try so many different variations on the same theme.

I also wish to express my gratitude to the reviewers engaged by Addison Wesley Longman including: Larry Cummings, University of Waterloo; Ulrich Daepp, Bucknell University; Mark Gross, Cornell University; David Handel, Wayne State University; Robert Hunter, Penn State University; Joel Iiams, University of North

Dakota; Calvin Jongsma, Dordt College; Tom Jager, Calvin College; David Kenoyer, Plattsburgh State University; David Leep, University of Kentucky; Steven Levandosky, University of Texas, Austin; Edith Luchins, Rensselaer Polytechnic Institute; John Masterson, Michigan State University; Rennie Mirollo, Boston College; Robin Sue Sanders, Buffalo State College; Francis Sullivan, Clemson University; Mark Taylor, Acadia University; John Wilson, Centre College; Paul Wilson, Rochester Institute of Technology; and Cynthia Woodburn, Pittsburgh State University.

I am also greatly indebted to my editor, Carolyn Lee-Davis, of Addison Wesley Longman, who ensured that the manuscript was the best I could produce. As well, I am indebted to the team at Nesbitt Graphics (in particular, Janet Nuciforo) who are responsible for the overall look and feel of the book.

Lastly, I want to thank my family for putting up with my absences on those many nights and weekends that I spent working on the manuscript.

Robert H. Redfield

Contents

Introduction   xvii
Historical Note: Al-Khwarizmi   xix

PART 1   PRELIMINARIES   1

Chapter 1   Properties of the Integers   3
Historical Note: Augustus de Morgan   16

Chapter 2   Solving Cubic and Quartic Polynomial Equations   18
Historical Note: How the Cubic and Quartic Equations
    Were Solved   31

Chapter 3   Complex Numbers   34
Historical Note: Highlights in the Development of the
    Complex Numbers   48

Chapter 4   Some Other Examples   58
Historical Note: William Rowan Hamilton   66

PART 2   ALGEBRAIC EXTENSION FIELDS   69

Chapter 5   Fields   70
Chapter 6   Solvability by Radicals   81
Historical Note: Niels Henrik Abel   91

Chapter 7   Rings   95
Chapter 8   Ways in Which Polynomials Are Like the Integers   103
Historical Note: Julia Robinson   116

Chapter 9   **Principal Ideals**   **119**
                   Historical Note: Emmy Noether   127

Chapter 10   **Algebraic Elements**   **130**

Chapter 11   **Eisenstein's Irreducibility Criterion**   **140**
                   Historical Note: Gotthold Eisenstein   146

Chapter 12   **Extension Fields as Vector Spaces**   **148**

Chapter 13   **Automorphisms of Fields**   **159**
                   Historical Note: Evariste Galois   167

Chapter 14   **Counting Automorphisms**   **170**
                   Historical Note: Richard Dedekind   180

PART 3   ELEMENTARY GROUP THEORY   183

Chapter 15   **Groups**   **184**
                   Historical Note: Walther von Dyck   194

Chapter 16   **Permutation Groups**   **196**

Chapter 17   **Group Homomorphisms**   **203**
                   Historical Note: Arthur Cayley   209

Chapter 18   **Subgroups**   **211**

Chapter 19   **Subgroups Generated by Subsets**   **220**

Chapter 20   **Cosets**   **226**

Chapter 21   **Finite Groups and Lagrange's Theorem**   **233**
                   Historical Note: Joseph Louis Lagrange   241

Chapter 22   **Equivalence Relations and Cauchy's Theorem**   **243**
                   Historical Note: Augustin-Louis Cauchy   251

Chapter 23   **Normal Subgroups and Quotient Groups**   **254**
                   Historical Note: Otto Hölder   259

Chapter 24   **The Homomorphism Theorem for Groups**   **262**
                   Historical Note: B. L. van der Waerden   268

PART 4   POLYNOMIAL EQUATIONS NOT SOLVABLE BY RADICALS   271

Chapter 25   **Galois Groups of Radical Extensions**   **272**

Chapter 26   **Solvable Groups and Commutator Subgroups**   **280**
                   Historical Note: William Burnside   287

Chapter 27   **Solvable Galois Groups**   288
Chapter 28   **Polynomial Equations Not Solvable by Radicals**   296
              Historical Note: Paolo Ruffini   299

PART 5   FINITE GROUPS   301

Chapter 29   **Finite External Direct Products of Groups**   302
              Historical Note: J. H. M. Wedderburn   310
Chapter 30   **Finite Internal Direct Products of Groups**   312
Chapter 31   **Abelian Groups with Prime Power Order**   320
Chapter 32   **The Fundamental Theorem of Finite Abelian Groups**   329
              Historical Note: Leopold Kronecker   337
Chapter 33   **Dihedral Groups**   339
              Historical Note: Felix Klein   345
Chapter 34   **Cauchy's Theorem**   348
Chapter 35   **The Sylow Theorems**   359
              Historical Note: Peter Ludvig Sylow   370
Chapter 36   **Groups of Order Less Than 16**   372
Chapter 37   **Groups of Even Permutations**   380
              Historical Note: Camille Jordan   386
Chapter 38   **Semidirect Products**   388

Appendix A   **The Greek Alphabet**   396
Appendix B   **Proving Theorems**   397
              Historical Note: George Boole   410
Appendix C   **Vector Spaces Over Fields**   413
Appendix D   **Constructions with Straightedge and Compass**   418

              Answers to Odd-Numbered Computational Exercises   426
              Bibliography   444
              Photo Credits   446
              Notation Index   447
              Subject Index   448

Introduction

The word "algebra" comes from the title, *Al-jabr wa-al-muqabalah,* of a book written by Abu Ja'far Muhammad ibn Musa al-Khwarizmi circa A.D. 830. Al-Khwarizmi used *al-jabr* to refer to the adding of equal amounts to each side of an equation to eliminate negative terms and *al-muqabalah* to refer to the subtracting of equal amounts from each side of an equation to eliminate positive terms. (For instance, *al-jabr* reduces $3x = 10 - 2x$ to $5x = 10$ by adding $2x$ to each side, while *al-muqabalah* converts $7 + 2x = 5 + 4x$ to $2 + 2x = 4x$ by subtracting 5 from each side.) Using the English word "root" to refer to the solution of a polynomial equation also comes from Arabic: Latin translators of Greek mathematics use *latus* (the side of a square), while Latin translators of Arabic mathematics use *radix* (root).

In this way, algebra came to be synonymous with the study of equations, so much so that the mathematical origin of algebra is frequently attributed to Diophantos, who, around 250 B.C., discussed ways of solving mathematical problems (which he always stated in words) by using a special symbol that he called "the number" for the unknown quantity. For instance, if we denote "the number" by s, the statement and solution of Diophantos's problem 20 reads as follows.

"*Problem 20.* To find two numbers such that the square of either added to the other gives a square.

"Let the first number be s, the other $2s + 1$, then the square of the first plus the second gives a square. The square of the second plus the first is $4s^2 + 5s + 1$. This must be equal to a square. I form this square from $2s - 2$, then it is $4s^2 + 4 - 8s$, and s becomes $\frac{3}{13}$. The first number is $\frac{3}{13}$, the second $\frac{19}{13}$."[1]

The mathematics which the Muslims inherited from the Greeks, including that of Diophantos, made the divisions of estates very complicated, if not impossible, and thus prompted al-Khwarizmi to develop a more accurate, flexible, and comprehensive method. Similarly, when European trade with the Muslim world shifted from a barter system to a monetary system, such tools became necessary for commercial transactions, and in response, the works of the Muslim mathematicians were translated into Latin for the use of European merchants, the most prominent compendium being the *Liber abbaci* (1202) of Leonardo of Pisa (better known as Fibonacci).

Work on methods for solving higher order equations progressed slowly until the sixteenth century when Cardano published *Ars magna,* a book that included general methods for solving equations involving polynomials of the fourth degree or less. All attempts at finding a general method for solving equations of the fifth degree continued to fail, and finally at the beginning of the nineteenth century, Ruffini and Abel showed independently that such a general method could not be found. Of particular importance was the general approach to the problem, taken a few years later by Galois, which over time changed algebra fundamentally: his insights turned the subject matter of algebra from the study of equations into the study of algebraic structures.

Of course, modern computers can find decimal approximations to the solutions of many polynomial equations. Such approximations can be very useful, when approximate solutions are necessary, and acceptable, when proper caution is exercised, but their use can lead the unwary astray.

In particular, as we have indicated above and will show below, exact solutions of equations involving polynomials of degree greater than four cannot always be found. For this reason, the programmers who create computer algebra systems, knowing that they cannot find an algorithm which will solve such equations exactly, construct these systems to give exact solutions for polynomial equations only of degree four or less. For equations involving polynomials of higher degree, these systems give numerical solutions only as decimal approximations.

For instance, when the computer algebra system Maple™ is asked to solve $x^3 - 3x^2 - 3x - 3 = 0$, it can, and does, give the real root exactly as

$$r = \sqrt[3]{4 + 2\sqrt{2}} + \frac{2}{\sqrt[3]{4 + 2\sqrt{2}}} + 1,$$

and if it is asked to check the solution, it calculates $r^3 - 3r^2 - 3r - 3$ to be exactly 0. But when it is asked to solve $4x^5 - 10x^2 + 5 = 0$ numerically, it must respond with a decimal approximation. For instance, Maple gives $v = 1.164$ as a floating point solution of this equation correct to four digits. Of course, such a solution, correct to only four digits, is not especially accurate; when asked to check the solution, Maple returns the value $4v^5 - 10v^2 + 5 = -.0017$.

Although the accuracy of a Maple calculation can be many more than four digits, the time and memory that are required for solving an equation increase as the required accuracy increases. For example, Maple will immediately find a floating point solution, s_1, correct to one hundred digits, of the equation $x^{55} - 1234^{33} = 0$. However, on checking the solution, Maple finds that $s_1^{55} - 1234^{33} = -2000$. A floating point solution s_2, which is correct to one thousand digits, should be more accurate. It takes Maple thirty seconds to find such a solution and the result is indeed much more accurate: Maple computes $s_2^{55} - 1234^{33} = -1 \times 10^{-897}$.

The time is takes to find a reasonably accurate solution also increases if the equation is more difficult to solve. For instance, it takes thirty seconds for Maple to find a floating point solution t_1, correct to one hundred digits, of the equation $x^{55} - 1234^{333} = 0$. This solution is not at all accurate: Maple calculates $t_1^{55} - 1234^{333} = -1.1 \times 10^{931}$. A floating point solution correct to one thousand digits should be more accurate, but, after ten minutes, Maple still has not found one. Nonetheless, because this equation is very simple, we can quickly find a solution by using a different method; namely, we can compute t_2, the 55th root of 1234^{333}, correct to one thousand digits. But even this solution is far from accurate: Maple calculates $t_2^{55} - 1234^{333} = -6 \times 10^{30}$.

In general, more complicated equations lead to less accurate approximate solutions whose computation requires lengthier calculations. This trouble could of course be avoided if all polynomial equations could be solved by formulas similar to the quadratic formula. But we know from the work of Ruffini, Abel, and Galois that no such general solutions exist. So, because of their efforts, mathematicians who need numerical solutions to polynomial equations now know that instead of wasting time and energy trying to find exact solutions, they must turn their attention to finding approximate solutions that are as accurate as possible and to determining exactly when these approximate solutions can be used with confidence.

As well, the work of Galois in particular marks a watershed in the development of algebra in particular and of mathematics in general. He showed that the solutions of equations could be determined by studying the structures underlying the equations, and over the last two centuries, mathematicians have incorporated this outlook into all branches of the subject.

HISTORICAL NOTE

Al-Khwarizmi

Abu Ja'far Muhammad ibn Musa al-Khwarizmi was born before 800 and died after 847. The epithet "al-Khwarizmi" would indicate that he came from south of the Aral Sea in central Asia but sometimes an additional epithet "al-Qutrubbulli" was added and this would indicate that he came from a district near Baghdad instead. Quite possibly he was born near Baghdad, while his ancestors came from central Asia. He was most likely an orthodox Muslim, and under the Caliph al-Ma'mun (reigned 813–833), he became a member of an academy of science that Caliph Harun al-Rashid had earlier established in Baghdad. Al-Ma'mun took a great interest in the academy and it owed its preeminence primarily to him.

Caliph al-Wathiq, who reigned after al-Ma'mun, possibly sent al-Khwarizmi on missions to the Caucasus in 842 and later to the Byzantine empire. According to one source, al-Khwarizmi was among a group of astronomers who at the sickbed of al-Wathiq predicted on the basis of his horoscope that he would live another fifty years, only to have him die ten days later.

Al-Khwarizmi is most famous for his book *Algebra* (the full English title is *The Compendious Book on Calculation by al-jabr and al-muqabala)*, an elementary treatise which provided, in al-Khwarizmi's words, "what is easiest and most useful in arithmetic, such as people constantly require in cases of inheritance, legacies, partition, lawsuits, and trade, and in all their dealings with one another, or where the measuring of lands, the digging of canals, geometrical computations, and other objects of various sorts and kinds are concerned."[2] As mentioned above, the words *al-jabr* and *al-muqabala* refer respectively to the processes of adding equal terms to both sides of an equation to eliminate negative terms and of subtracting equal amounts from both sides of an equation to reduce positive terms. For instance, *al-jabr* converts the equation $x^2 = 40x - 4x^2$ to $5x^2 = 40x$ by adding $4x^2$ to both sides, and *al-muqabala* reduces the equation $50 + x^2 = 29 + 10x$ to $21 + x^2 = 10x$ by subtracting 29 from both sides. The word *al-jabr* became the common appellation of later books in Arabic on the same subject and their Latin translations gave rise to the English word "algebra."

Ironically, in view of the modern usage of "algebra," al-Khwarizmi's style was highly <u>non</u>symbolic. For instance, his solution of the equation $x^2 + 10x = 39$ reads as follows: "For instance: one square and ten roots of the same amount to thirty-nine dirhems; that is to say, what must be the square which, when increased by ten of its own roots, amounts to thirty-nine? The solution is: you halve the number of roots, which in the present instance yields five. This you multiply by itself; the product is twenty-five. Add this to thirty-nine; the sum is sixty-four. Now take the root of this, which is eight, and subtract from it half the number of roots, which is four. The remainder is three. This is the root of the square you sought for; the square itself is nine."[3] (In modern terminology and notation, he completes the square to transform the equation into $(x + 5)^2 = 39 + 25 = 64$; then $x + 5 = \sqrt{64} = 8$; and hence $x = 3$.)

To prove that his method of solving the problem is valid, he uses a geometrical construction reminiscent of Euclid. However, it seems most likely that he relied less on the Greek tradition than on either the Hindu tradition or a local Syriac-Persian-Jewish tradition. That he used material from the last tradition seems especially likely in view of his short but well-informed and accurate treatise on the Jewish calendar. While his scientific achievements never reached the level of those of either the Greeks or the Hindus, they were extremely influential.

References

1. van der Waerden, B. L. *Geometry and Algebra in Ancient Civilizations.* New York: Springer-Verlag, 1983, p.101.
2. Toomer, G. J. "Al-Khwarizmi, Abu Ja'far Muhammad Ibn Musa," *Dictionary of Scientific Biography* 16 vols., Gillispie, Charles Coulston, ed. in chief, New York: Charles Scribner's Sons, 1970, vol. VII, p. 359.
3. van der Waerden, B. L. *A History of Algebra.* New York: Springer-Verlag, 1985, pp. 7–8.

PART 1

Preliminaries

Properties of the Integers

Our objective is to show that the familiar formula for solving quadratic polynomial equations does not extend to all polynomial equations, and to attain this goal, we will need to investigate certain abstract algebraic structures in some detail. In general, algebraic structures are characterized by the finitistic nature of their operations, and for this reason, properties of the integers will underlie several of our general results and many of our examples. This preliminary chapter deals with those properties that we will need in the sequel. Most of these properties will probably be familiar, although perhaps in a form different from that given in the sequel.

We start with an axiom, the well-ordering principle, which we will need to derive the remaining results of the chapter. Note that after we have used this axiom to prove the results of this chapter, it will only rarely be necessary to invoke it again.

To state the axiom, we need to recall the following terminology.

> **DEFINITION**
>
> A subset S of the integers is **bounded below** if there is an integer n such that for all $s \in S$, $n \le s$; an element $m \in S$ is **minimal** if for all $s \in S$, $m \le s$.

Example 1.1 The set P of positive integers is bounded below by all negative integers, by 0, and by 1. Since none of the negative integers are in P, no negative integer can be a minimal element of P, and similarly, since $0 \notin P$, 0 cannot be a minimal element of P. However, the integer 1 is a lower bound and is in P, and hence 1 is a minimal element of P. ❖

The axiom we need is then the following.

Axiom **Well-Ordering Principle**

Every nonempty subset of the integers that is bounded below contains a minimal element.

This principle is stronger than it may at first appear. It differentiates the integers from related sets like the real numbers by using discreteness. For example, consider the following.

Example 1.2 While the set of real numbers whose square is strictly greater than 2 is bounded below (by 1, among other numbers), it does not <u>contain</u> a minimal element. On the other hand, the same subset of the integers, i.e., all those integers whose squares are strictly greater than 2, is also bounded below (also by 1), but in this case the subset does contain a minimal element, namely, 2. ❖

From the well-ordering principle, we can deduce two principles of mathematical induction and several elementary results about the integers. We start by deriving the division algorithm and some of its consequences.

The Division Algorithm

The most common method of dividing one integer into another is long division. When this method is expressed in terms of integer equations and inequalities, it is called the division algorithm. In the following statement of the algorithm, q corresponds to the quotient determined by long division and r to the remainder.

Theorem 1.3 **Division Algorithm**

Let m and n be integers and suppose that $n > 0$. Then there exist integers q and r such that

$$m = nq + r \quad and \quad 0 \le r < n.$$

Proof Let S denote the set of differences $m - nx$, where x is an integer and $m - nx \ge 0$. Since $n > 0, n \ge 1$ and hence $-n \le -1$. Then $m \ge -|m| \ge -n|m| = n(-|m|)$ so that $m - n(-|m|) \ge 0$, and thus, since $-|m|$ is an integer, we have $m - n(-|m|) \in S$, i.e., S is not empty. So since S is bounded below by 0, the well-ordering principle implies that S contains a minimal element r, and since $r \in S$, $0 \le r$ and $m = nq + r$ for some integer q. It remains to show that $r < n$. Suppose by way of contradiction that $r \ge n$. Then $r - n = m - nq - n = m - n(q + 1)$ and $r - n \ge 0$ so that $r - n \in S$. But, since $r - n < r$, this contradicts our assumption that r is minimal, and hence we may conclude that $r < n$. ∎

The division algorithm is extremely useful not only in its familiar form of long division, but also for deriving the existence and properties of greatest common divisors. For completeness, we recall the following definition.

> **DEFINITION**
>
> A **greatest common divisor** of integers m and n, not both zero, is an integer d with the following properties:
> - (*i*) d is positive;
> - (*ii*) d divides both m and n;
> - (*iii*) if k divides both m and n, then k divides d.

Note that condition (*ii*) of the definition says that d is a common divisor of m and n, while condition (*iii*) ensures that d is the greatest such divisor. Condition (*i*) is required to ensure that there is only one greatest common divisor.

It is the division algorithm, together with the well-ordering principle, that allows us to show that any two nonzero integers always have exactly one greatest common divisor. Thus, we may refer to <u>the</u> greatest common divisor of the two integers. The method of proof also yields the extremely useful observation that the greatest common divisor may be expressed as a "linear combination" of the integers. Specifically, we have the following.

Theorem 1.4   *Any two nonzero integers m and n have a unique greatest common divisor d. Furthermore, there exist integers μ and ν such that*

$$d = \mu m + \nu n.$$

Proof   Let S denote the set of positive integers of the form $xm + yn$, where x and y are integers. Since the square of any nonzero integer is always positive, $m^2 + n^2 \in S$, and thus S is not empty. But S is bounded below by 0, and hence the well-ordering principle implies that S contains a minimal element d. Since $d \in S$, d is positive and there exist integers μ and ν such that $d = \mu m + \nu n$. We will show (1) that d is a greatest common divisor of m and n and (2) that it is the only one.

(1) We have noted that d is positive. By the division algorithm, there exist integers q and r such that $m = dq + r$ and $0 \le r < d$. Then $r = m - dq = (1 - \mu q)m + (-\nu q)n$, and hence if $0 < r$, $r \in S$. Since $r < d$, this would contradict the minimality of d, and thus $r = 0$. Then $m = dq$, i.e., d divides m. Similarly, d divides n. If k divides both m and n, then there exist integers α and β such that $m = \alpha k$ and $n = \beta k$. But then

$$d = \mu m + \nu n = \mu \alpha k + \nu \beta k = (\mu \alpha + \nu \beta)k$$

and hence k divides d. We conclude that d is a greatest common divisor of m and n.

(2) Suppose that j is some other greatest common divisor of m and n. By (*iii*) in the definition of greatest common divisor, d divides j and j divides d. Since d and

j are both positive by (i), this implies that $d \leq j$ and $j \leq d$ and hence that $j = d$. ∎

Example 1.5 The greatest common divisor of 15 and 100 is certainly 5, and 5 may be written $5 = (7)(15) + (-1)(100)$. Note that although the greatest common divisor is unique, the integers μ and ν, given in Theorem 1.4, are not. For instance, 5 may also be written $5 = (-13)(15) + (2)(100)$. ❖

We can use the set S defined in the proof of Theorem 1.4 to create an algorithm (Example 1.7) that computes the greatest common divisor of two integers m and n and simultaneously finds appropriate integers μ and ν. The following characterization of S is the key to creating the algorithm.

Proposition 1.6 *Let m and n be nonzero integers and let S denote the set of positive integers of the form xm + yn, where x and y are integers. Then, for any integer d, the following statements are equivalent:*

(i)   *d is the greatest common divisor of m and n;*
(ii) *d is a common divisor of m and n and d ∈ S;*
(iii) *S = {kd | k is a positive integer}.*

Proof Let $P = \{kd \mid k$ is a positive integer$\}$, and note that to show that (i), (ii), and (iii) are equivalent, it suffices to show that (i) ⇒ (ii), (ii) ⇒ (iii), and (iii) ⇒ (i).

The definition of greatest common divisor together with Theorem 1.4 shows that (i) ⇒ (ii).

To see that (ii) ⇒ (iii), assume that (ii) holds. We must show that $P \subseteq S$ and $S \subseteq P$. Observe first that if $p \in P$, then $p = kd$ for some positive integer k, and since $d \in S$ by (ii), $d = xm + yn$ for integers x and y. Then $p = kd = (kx)m + (ky)n \in S$ so that $P \subseteq S$. Conversely, any $s \in S$ may be written in the form $s = am + bn$, and since d is a common divisor of m and n, m and n may be written $m = \alpha d$ and $n = \beta d$ for integers α and β. Then $s = (a\alpha + b\beta)d$, and since both s and d are positive, $\alpha a + \beta b$ is positive. It follows that $s \in P$, and hence that $S \subseteq P$.

To see that (iii) ⇒ (i), assume that (iii) holds. We need to show that the three defining properties of greatest common divisors hold for d. Note first that d is positive because $d = 1 \cdot d \in P$ and by hypothesis, $P = S$. As well, $|m| = 1|m| + 0|n| \in S$, and hence by hypothesis, $|m| \in P$. But d divides every element of P, and thus, since $m = |m|$ or $m = -|m|$, d divides m. Similarly, d divides n. Finally, suppose that c is a common divisor of m and n, i.e., $m = zc$ and $n = wc$. Since $d \in S$, $d = xm + yn$ for integers x and y, and hence $d = (xz + yw)c$, i.e., c divides d. It follows that d is a greatest common divisor of m and n. ∎

Example 1.7 **Euclidean Algorithm for integers** We will determine the greatest common divisor d of 585 and 104 and simultaneously find integers μ and ν such that $d = 585\mu + 104\nu$.

First apply the division algorithm to 585 and 104:

$$585 = (104)(5) + 65. \tag{1}$$

Then apply it to the divisor, 104, and the remainder, 65, of the resulting equation:

$$104 = (65)(1) + 39. \tag{2}$$

Continue in this fashion until obtaining a remainder of 0:

$$65 = (39)(1) + 26; \tag{3}$$
$$39 = (26)(1) + 13; \tag{4}$$
$$26 = (13)(2) + 0. \tag{5}$$

The last nonzero remainder is 13, the remainder of equation (4). By equation (5), this remainder divides the divisor of equation (4), viz., 26, and thus also divides the dividend of equation (4), viz., 39. Similar arguments show that the last nonzero remainder, 13, divides the divisor and dividend of each equation and hence divides 585 and 104. Furthermore, we may write this last nonzero remainder in terms of the original numbers, 585 and 104, by solving each of the equations for its remainder and then substituting the result into the previous equation.

$$
\begin{aligned}
13 &= 39 - 26 \cdot 1 \\
&= 39 - (65 - 39 \cdot 1) \cdot 1 = -65 + 39 \cdot 2 \\
&= -65 + (104 - 65 \cdot 1) \cdot 2 = 104 \cdot 2 - 65 \cdot 3 \\
&= 104 \cdot 2 - (585 - 104 \cdot 5) \cdot 3 = -585 \cdot 3 + 104 \cdot 17 \\
&= 585(-3) + 104(17).
\end{aligned}
$$

Since 13 divides both 585 and 104 and may be written in the form $13 = 585\mu + 104\nu$, Proposition 1.6 shows that 13 is the greatest common divisor of 585 and 104. ❖

In their turn, greatest common divisors can be used to gather information about prime numbers. First recall the definition of prime number. (Note that 1 is <u>not</u> a prime number.)

DEFINITION

An integer $p > 1$ is **prime** if its only positive divisors are 1 and p.

Since a greatest common divisor is in particular a divisor, the only possibilities for the greatest common divisor of a prime number and another number are the given prime and 1. Thus, if the prime does not divide the other number, then their greatest common divisor must be 1. We prove next that this property in fact characterizes the prime numbers.

Proposition 1.8   *An integer $p > 1$ is prime if and only if for all nonzero integers n, either p divides n or the greatest common divisor of p and n is 1.*

Proof   We observed previously that if p is prime and p does not divide n, then the greatest common divisor of p and n must be 1. On the other hand, suppose that for all $n \neq 0$, either p divides n or the greatest common divisor of p and n is 1, and let a be a positive divisor of p. Then $p = ab$ for some positive integer b. The greatest common divisor of a and p is certainly a itself, and hence if $a \neq p$, then the condition implies that $a = 1$. It follows that p is prime. ∎

Example 1.9   For instance, 13 is prime, and if 13 does not divide an integer $n \neq 0$, the greatest common divisor of 13 and n must be 1. On the other hand, the greatest common divisor of 15 and 10 is 5, and hence, since $1 \neq 5 \neq 15$, 15 is not prime. ❖

Using the characterization given in Proposition 1.8, we can prove the useful (and familiar) result that whenever a prime divides a product, it must divide at least one of the factors.

Proposition 1.10   *Let m and n be integers and let p be a prime. If p divides mn, then p divides m or p divides n.*

Proof   If $mn = 0$, then one of the two numbers is 0 and hence p divides it. If $mn \neq 0$, then neither m nor n is 0, and if p does not divide m, then by Proposition 1.8, the greatest common divisor of p and m must be 1. So by Theorem 1.4, there exist integers π and μ such that $1 = \pi p + \mu m$. Then $n = \pi p n + \mu m n$ and hence, since p divides both mn and $\pi p n$, p also divides n. ∎

Mathematical Induction

To prove that a statement is true for all integers greater than a given integer, the most straightforward strategy is to pick an arbitrary integer greater than the given integer and prove the statement. However, suppose we wish to prove that if a prime p divides a product of arbitrary length $m_1 \cdots m_n$, then p divides one of the factors $m_1, \ldots, m_n$. We could argue that by Proposition 1.10, p must divide either m_1 or $m_2 \cdots m_n$. And if p divides m_1, we are done, while otherwise, p divides $m_2 \cdots m_n$. In the latter case, we may apply Proposition 1.10 again, and thus, proceeding in a similar fashion, we see that p must divide one of the factors $m_1, \ldots, m_n$. This proof is correct but the instructions to proceed "in a similar fashion" are vague and might hide a mistake. One way to make proofs like this more convincing is to use the principle of mathematical induction.

This principle is a property of the integers which is one of the most powerful consequences of the well-ordering principle. It can be phrased in several equivalent ways, of which we give two, the most common phrasing (Theorem 1.11) and then a variant with a somewhat stronger induction hypothesis (Theorem 1.16). The latter formulation will be essential for a small number of proofs in the sequel.

Theorem 1.11   **First Principle of Mathematical Induction**

Let N be an integer and suppose that for each integer n ≥ N, P(n) is a statement about n which is either true or false. If

(*i*)   *P(N) is true, and*

(*ii*)   *for all k ≥ N, P(k + 1) is true whenever P(k) is true,*

then P(n) is true for all integers n ≥ N.

Proof   Let S denote the set of all integers s that are greater than or equal to N and for which $P(s)$ is false. It clearly suffices to show that S is empty. Suppose by way of contradiction that S is not empty. Then, since S is bounded below by N, the well-ordering principle implies that it must contain a minimal member, μ. By (*i*), $\mu \neq N$, and thus $\mu - 1 \geq N$. But, since μ is minimal in S, $\mu - 1 \notin S$ so that $P(\mu - 1)$ is true. Then by (*ii*), $P(\mu)$ is also true so that $\mu \notin S$, a contradiction. We conclude that S is empty and hence that $P(n)$ is true for all $n \geq N$. ∎

Induction may be used in the following way to remove the vagueness of the above proof that if a prime divides a product, then it divides one of the factors. This kind of use of induction is very common: induction is the most common method of extending results about binary combinations to arbitrary finite combinations.

Proposition 1.12   *Let $m_1, \ldots, m_n$ be integers and let p be a prime. If p divides $m_1 \cdots m_n$, then p divides m_i for some i.*

Proof   Assume that p is a prime, and for each integer $n \geq 1$, let $P(n)$ be the statement: if $m_1, \ldots, m_n$ are integers and p divides $m_1 \cdots m_n$, then p divides m_i for some i. We will prove the proposition by using induction to show that $P(n)$ is true for all $n \geq 1$.

(*i*)   For $n = 1$, $P(n)$ is the statement: if m_1 is an integer and p divides m_1, then p divides m_1. This is certainly true.

(*ii*)   Suppose that $k \geq 1$ and that $P(k)$ is true: if $m_1, \ldots, m_k$ are integers and p divides $m_1 \cdots m_k$, then p divides m_i for some i. We need to show that $P(k + 1)$ is true. So suppose further that $m_1, \ldots, m_{k+1}$ are integers and that p divides $m_1 \cdots m_{k+1}$. Then $m_1 \cdots m_k$ and m_{k+1} are both integers and p divides $(m_1 \cdots m_k)m_{k+1}$. Thus, by Proposition 1.10, p divides $m_1 \cdots m_k$ or p divides m_{k+1}. In the latter case, we are done; in the former case, we may apply the induction hypothesis (that $P(k)$ holds) and conclude that p divides m_i for some $i \leq k$. We have thus shown that $P(k + 1)$ is true, and hence we have shown by induction that $P(n)$ is true for all $n \geq 1$. ∎

The first step in constructing any proof by induction is of course, identifying the statement $P(n)$. Typically, as in the preceding proof, $P(n)$ is merely a restatement of the result to be proved. This is especially easy when the result to be proved is an equation, as in the following.

Proposition 1.13   *For any positive integer n,* $\sum_{i=1}^{n} i^2 = \frac{1}{6}n\,(n+1)(2n+1).$

Proof   For $n \geq 1$, let $P(n)$ be the equation: $\sum_{i=1}^{n} i^2 = \frac{1}{6}n\,(n+1)(2n+1)$. We will use induction to show that $P(n)$ is true for all $n \geq 1$.

(*i*)   For $n = 1$, $\sum_{i=1}^{n} i^2 = \sum_{i=1}^{1} i^2 = 1^2 = 1 = \frac{1}{6}(1)(1+1)(2\cdot 1+1)$, and thus $P(1)$ holds.

(*ii*) Suppose that $k \geq 1$ and that $P(k)$ is true, i.e., that

$$\sum_{i=1}^{k} i^2 = \frac{1}{6}k\,(k+1)(2k+1).$$

Then

$$\sum_{i=1}^{k+1} i^2 = \sum_{i=1}^{k} i^2 + (k+1)^2 = \frac{1}{6}k\,(k+1)(2k+1) + (k+1)^2$$
$$= \frac{1}{6}(k+1)[k(2k+1) + 6(k+1)]$$
$$= \frac{1}{6}(k+1)[2k^2 + 7k + 6]$$
$$= \frac{1}{6}(k+1)[k+2][2k+3]$$
$$= \frac{1}{6}(k+1)[(k+1)+1][2(k+1)+1].$$

So $P(k+1)$ holds, and thus by induction, $P(n)$ holds for all $n \geq 1$. ∎

Note the strategy that is employed in the induction step (*ii*) of the above proof. In this step, a succession of equalities links the left side, $\sum_{i=1}^{k+1} i^2$, of the equation $P(k+1)$ with its right side, $\frac{1}{6}(k+1)[(k+1)+1][2(k+1)+1]$. The first objective of these equalities is to rewrite the summation in such a way that the induction hypothesis, $P(k)$, can be applied. Once this is done, elementary arithmetic is used to rewrite the resulting expression in the desired form. In proofs like the one above, it is usually easiest not to expand each expression and then simplify, but rather to find as many common factors as possible.

In cases such as Propositions 1.12 and 1.13, the statement $P(n)$ is a literal restating of what is to be proved. Sometimes it is necessary to rephrase the statement of the proposition to ensure that it is in the proper form, i.e., that it is a statement about an integer n. For example, consider the following result.

Proposition 1.14   *A sum of two consecutive squares is odd.*

Proof   For any integer n, let $P(n)$ be the statement $n^2 + (n+1)^2$ is odd. We need to show that $P(n)$ is true for all integers n. Note first that if $k < 0$, then $k^2 + (k+1)^2 =$

$((-k - 1) + 1)^2 + (-k - 1)^2$, and hence $P(k)$ is true if and only if $P(-k - 1)$ is true. But since $k < 0$, $k \leq -1$, and thus $-k - 1 \geq 0$. So it suffices to show that $P(n)$ is true for all $n \geq 0$. We will do this by induction on n.

(i)   If $n = 0$, then $n^2 + (n + 1)^2 = 1$, which is odd. So $P(0)$ holds.

(ii)   Suppose that $n \geq 1$ and that $P(n)$ holds, i.e., that $n^2 + (n + 1)^2$ is odd. Then $(n + 1)^2 + (n + 2)^2 = (n + 1)^2 + (n^2 + 2n + 4) = ((n + 1)^2 + n^2) + (2n + 4)$. By the induction hypothesis, $(n + 1)^2 + n^2$ is odd and obviously $2n + 4$ is even. So, since the sum of an odd integer and an even integer is odd, $(n + 1)^2 + (n + 2)^2$ is odd. It follows by induction that $P(n)$ holds for all $n \geq 0$. ∎

We will not need Propositions 1.13 and 1.14 in the discussion that follows; they are included only as examples of proofs by induction. On the other hand, the fact that an integer may be written as a product of primes in at most one way is a result we will need and one whose proof requires the first principle of mathematical induction.

Proposition 1.15   *Suppose that a is an integer and that $p_1, \ldots, p_n$ are primes such that $p_1 \leq \cdots \leq p_n$ and $a = p_1 \cdots p_n$. If $q_1, \ldots, q_m$ are primes such that $q_1 \leq \cdots \leq q_m$ and $a = q_1 \cdots q_m$, then $m = n$ and $q_i = p_i$ for all $1 \leq i \leq n$.*

Proof   The proof is by induction with $P(n)$ being the statement: if a is an integer, if $p_1, \ldots, p_n$ are primes such that $p_1 \leq \cdots \leq p_n$ and $a = p_1 \cdots p_n$, and if $q_1, \ldots, q_m$ are primes such that $q_1 \leq \cdots \leq q_m$ and $a = q_1 \cdots q_m$, then $m = n$ and $q_i = p_i$ for all $1 \leq i \leq n$.

(i)   If a is an integer, if p_1 is a prime, and if $a = p_1$, then a is prime and hence can have only one prime factor, namely itself. So $P(1)$ holds.

(ii)   Suppose that $P(n)$ holds and that the hypotheses of $P(n + 1)$ hold. That is, suppose that $P(n)$ holds, that a is an integer, that $p_1, \ldots, p_{n+1}$ are primes such that $p_1 \leq \cdots \leq p_{n+1}$ and $a = p_1 \cdots p_{n+1}$, and that $q_1, \ldots, q_m$ are primes such that $q_1 \leq \cdots \leq q_m$ and $a = q_1 \cdots q_m$. We need to show that $m = n + 1$ and that $p_i = q_i$ for all $1 \leq i \leq n + 1$. Our hypotheses tell us that $p_1 \cdots p_{n+1} = a = q_1 \cdots q_m$ and hence that p_{n+1} divides $q_1 \cdots q_m$. So by Propositon 1.12, p_{n+1} divides q_i for some i, and therefore $p_{n+1} \leq q_i \leq q_m$. The same argument, with the roles of p_{n+1} and q_m reversed, shows that as well $q_m \leq p_{n+1}$ and hence that $p_{n+1} = q_m$. But then $p_1 \cdots p_n = q_1 \cdots q_{m-1}$, and by the induction hypothesis, $n = m - 1$ and $p_i = q_i$ for all $1 \leq i \leq n$. We conclude that $m = n + 1$ and that $p_i = q_i$ for all $1 \leq i \leq n + 1$, i.e., that $P(n + 1)$ holds. ∎

We mentioned earlier that the principle of mathematical induction may be phrased in several ways. For a few of the proofs in the sequel, we will need the following version of the principle.

Theorem 1.16   **Second Principle of Mathematical Induction**

Let N be an integer and suppose that for each integer $n \geq N$, $P(n)$ is a statement about n which is either true or false. If

> (*i*)   *$P(N)$ is true, and*
> (*ii*)   *for all integers $k \geq N$, $P(k)$ is true whenever $P(m)$ is true for all integers m such that $k > m \geq N$,*

then $P(n)$ is true for all integers $n \geq N$.

Proof   Use a proof similar to that of Theorem 1.11. ■

Note that since there are no integers m such that $N > m \geq N$, it follows from condition (*ii*) of Theorem 1.16 that $P(N)$ is true. For this reason, condition (*i*) is actually redundant. Note also that although the induction hypothesis of Theorem 1.16 (*ii*) may appear to be stronger than that of Theorem 1.11 (*ii*), the two results are in fact logically equivalent (i.e., each result can be derived from the other—see Exercises 52 and 53 at the end of the chapter).

We have an immediate use for the second principle of mathematical induction. We need it to show that any integer may be expressed as the product of primes in at most one way. Note that a product is allowed to have only one factor; i.e., in the same way that $2 \cdot 3 \cdot 5$ is a product with three factors, 2 is a product with one factor.

Proposition 1.17   **Fundamental Theorem of Arithmetic**

Any integer greater than 1 can be uniquely written as a product of primes.

Proof   By Proposition 1.15, it suffices merely to show that each integer greater than 1 can be written as a product of primes. We will prove this by using the second principle of mathematical induction with $P(n)$ being the following statement: n can be written as a product of primes.

> (*i*)   Since 2 is prime, $P(2)$ holds.
> (*ii*)   Suppose that $k \geq 2$ and that $P(m)$ holds for all integers m such that $k > m \geq 2$. If k is prime, then it is already written as a product of primes. If k is not prime, then $k = ab$ for integers a and b, both strictly between 1 and k. Thus, since both $P(a)$ and $P(b)$ hold by the induction hypothesis, a and b can each be written as a product of primes. Then juxtaposing these two products expresses k as a product of primes, i.e., $P(k)$ holds. It follows from Theorem 1.16 that $P(n)$ holds for all $n > 1$. ■

For many of the examples that follow, we will need to know that numbers like $\sqrt[3]{2}$ are not rational numbers. This result may be proved by using Proposition 1.12 or Proposition 1.17. To illustrate the use of Proposition 1.17, we choose the latter method and leave the former as an exercise (Exercise 48). The proofs are similar and are essentially what is given by Euclid.

Proposition 1.18 *If p is a prime number and n is an integer greater than one, then $\sqrt[n]{p}$ is not a rational number.*

Proof Suppose by way of contradiction that $\sqrt[n]{p}$ is rational, i.e., that there are positive integers a and b that have no common prime factor and which are such that $(\frac{a}{b})^n = p$. Then $a^n = pb^n$ and hence p divides a^n. By Proposition 1.17, a and b may be written as products of primes: $a = p_1 \cdots p_m$, $b = q_1 \cdots q_k$. Then $p_1{}^n \cdots p_m{}^n = a^n = pb^n = pq_1{}^n \cdots q_k{}^n$, and thus by Proposition 1.17, the same primes occur in the products $p_1{}^n \cdots p_m{}^n$ and $pq_1{}^n \cdots q_k{}^n$, i.e., $p_j = p$ for some j. Renumbering the p_i if necessary, we may assume that $j = 1$, and thus $p_1{}^{n-1} p_2{}^n \cdots p_m{}^n = q_1{}^n \cdots q_k{}^n$. Since $n > 1$, $n - 1 > 0$, and hence $p = p_1$ occurs as a factor of $p_1{}^{n-1} p_2{}^n \cdots p_m{}^n$. So p also occurs as a factor of $q_1{}^n \cdots q_k{}^n$, and thus, again by Proposition 1.17, $q_r = p$ for some r. But then a and b have the common factor p, a contradiction of our choice of a and b. We conclude that $\sqrt[n]{p}$ cannot be a rational number. ∎

Exercises

1. State the contrapositive of Proposition 1.10.
2. State the contrapositive of Proposition 1.12.

In Exercises 3–8, use the Euclidean algorithm described in Example 1.7 to find (1) the greatest common divisor d of m and n and (2) integers μ and ν such that $d = \mu m + \nu n$.

3. $m = 15, n = 35$ 4. $m = 15, n = 42$

5. $m = 105, n = 216$ 6. $m = 28, n = 35$

7. $m = 65, n = 143$ 8. $m = 130, n = 231$

In Exercises 9–14, use induction to prove that the given equation holds for all positive integers n.

9. $\sum_{i=1}^{n} i = \frac{1}{2}n(n + 1)$.

10. $\sum_{i=1}^{n} i^3 = \left[\frac{n(n + 1)}{2}\right]^2$.

11. $\sum_{i=1}^{n} \frac{1}{i(i + 1)} = \frac{n}{n + 1}$.

12. $\sum_{i=1}^{n} (2i - 1) = n^2$.

13. $\sum_{i=1}^{n} \frac{1}{(2i - 1)(2i + 1)} = \frac{n}{2n + 1}$

14. $\sum_{i=1}^{n} (6i - 5) = 3n^2 - 2n$.

Exercises 15–22 concern the Fibonacci numbers, which are defined recursively as follows:

$$F_1 = F_2 = 1, \qquad F_{n+2} = F_{n+1} + F_n \quad \text{for } n \geq 1.$$

In each exercise, use induction to prove the given statement.

15. For all $n \geq 1$, $F_{n+1}F_{n+2} - F_nF_{n+3} = (-1)^n$.

16. For all $n \geq 2$, $F_{n-1}F_{n+1} - (F_n)^2 = (-1)^n$.

17. For all $n \geq 1$, $F_1 + F_2 + \cdots + F_n = F_{n+2} - 1$.

18. For all $n \geq 1$, F_{3n-2} and F_{3n-1} are odd, and F_{3n} is even.

19. For all $n \geq 1$, F_n and F_{n+1} have greatest common divisor 1.

20. For all $m \geq 2$, $n \geq 1$, $F_{m+n} = F_{m-1}F_n + F_mF_{n+1}$.

21. For all $m, n \geq 1$, F_m divides F_n whenever m divides n. (*Hint:* Use induction on k, where $n = mk$; for the induction step, use Exercise 20 above.)

22. For all $n \geq 1$, $F_n = \dfrac{(1 + \sqrt{5})^n - (1 - \sqrt{5})^n}{2^n\sqrt{5}}$.

In Exercises 23–30, show that the given number is not rational.

23. $\sqrt{6}$ 24. $\sqrt{24}$ 25. $\sqrt[3]{72}$

26. $\sqrt{30}$ 27. $\sqrt{14}$ 28. $\sqrt[3]{36}$

29. $\sqrt{210}$ 30. $\sqrt[4]{216}$

31. State and prove a general theorem that encompasses Exercises 23–30.

In Exercises 32–35, prove the given statement or show that it is false by producing a counterexample.

32. Every finite nonempty subset of the real numbers contains a minimal element.

33. Every nonempty subset of the integers contains a minimal element.

34. Every nonempty subset of the positive integers contains a minimal element.

35. Every nonempty subset of the rational numbers that is bounded below contains a minimal element.

36. Use induction to prove that for all positive integers n, three divides $4^n - 1$.

37. Use induction to prove that three divides any sum of three consecutive cubes.

38. The formula for the sum of a geometric series $a + ar + ar^2 + \cdots$ is calculated by converting the partial sums $a + ar + \cdots + ar^{n-1}$ to their closed form $\dfrac{a(1 - r^n)}{1 - r}$ and then taking the limit. Use induction to prove that for real numbers a and r, $r \neq 1$,

$$a + ar + \cdots + ar^{n-1} = \frac{a(1 - r^n)}{1 - r}.$$

39. Suppose that m, n are integers with no common prime factors. Prove that if k is an integer such that m divides nk, then m divides k.

40. Suppose that m, n are integers with no common prime factors. Prove that if k is an integer that is divisible by both m and n, then mn divides k.

Exercises 41–45 concern the binomial coefficients $\binom{n}{r}$. Recall that these are defined as follows: $0! = 1$; for $n > 0$, $n! = 1 \cdot 2 \cdots n$; for integers $0 \le n$ and $0 \le r \le n$, $\binom{n}{r} = \dfrac{n!}{r!(n-r)!}$. Exercise 44 is the binomial theorem; Exercise 45 concerns the Fibonacci numbers defined previously.

41. Prove that for integers $0 < n$ and $0 < r \le n$, $\binom{n+1}{r} = \binom{n}{r-1} + \binom{n}{r}$.

42. Use induction to prove that for integers $0 \le n$ and $0 \le r \le n$, $\binom{n}{r}$ is an integer.

43. Prove that if p is prime and $0 < r < p$, then p divides $\binom{p}{r}$.

44. Use induction to prove the binomial theorem: if x and y are real numbers and n is a positive integer, then

$$(x + y)^n = \sum_{r=0}^{n} \binom{n}{r} x^{n-r} y^r.$$

45. For the Fibonacci numbers defined previously, use induction to show that for all $n \ge 1$, $F_n = \binom{n-1}{0} + \binom{n-2}{1} + \cdots + \binom{n-j}{j-1}$, where the summation ends with the largest j such that $2j \le n + 1$.

46. Give a detailed proof of the second principle of mathematical induction (Theorem 1.16). Let N be an integer and suppose that for each integer $n \ge N$, $P(n)$ is a statement that is either true or false. If, for all $k \ge N$, $P(k)$ is true whenever $P(m)$ is true for all $k > m \ge N$, show that $P(n)$ is true for all integers $n \ge N$.

47. Define the greatest common divisor of a finite sequence of nonzero integers $n_1, \ldots, n_k$. Then show that any sequence $n_1, \ldots, n_k$ of nonzero integers has a unique greatest common divisor d and that there exist integers $v_1, \ldots, v_k$ such that $d = v_1 n_1 + \cdots + v_k n_k$.

48. Use Propositon 1.12 to prove Proposition 1.18.

49. Explain why the second, rather than the first, principle of mathematical induction is used to prove the fundamental theorem of arithmetic.

50. Use the second principle of mathematical induction to prove that for all $n \ge 3$, $3^n > n^3$.

51. Give a general statement of the Euclidean algorithm used in Example 1.7. Then prove that this general algorithm does indeed find the greatest common divisor.

52. Derive the first principle of mathematical induction (Theorem 1.11) from the second (Theorem 1.16).

53. Derive the second principle of mathematical induction (Theorem 1.16) from the first (Theorem 1.11).

54. Consider the theorem: *all students in this class will receive the same final grade.* Explain what is wrong with the following "proof" of this theorem.

> ***Proof***   The proof is by induction on the number of students in the class.
>
> (*i*)   If there is only one student in the class, then all students in the class will certainly receive the same final grade.
>
> (*ii*)   Suppose that in any class with *k* students all students will receive the same final grade and consider a class with *k* + 1 students. Removing one student from this set leaves a class with *k* students, all of whom will receive the same final grade by the induction hypothesis. Replacing this student and removing another again leaves a class with *k* students, all of whom will also receive the same final grade. Repeating this process for all *k* + 1 students shows that all of them will receive the same final grade. ∎

55. Derive the well-ordering principle from the first principle of mathematical induction

HISTORICAL NOTE

Augustus De Morgan

Augustus De Morgan was born in Madura, Madras, India, in June 1806, and he died in London, England, on March 18, 1871. His father worked for the East India Company and his mother was the daughter of a pupil of Abraham de Moivre and the granddaughter of James Dodson, who computed a table of anti-logarithms. When Augustus was just seven months old, his family moved to England, and when he was ten years old, his father died. After attending a succession of private high schools, he entered Trinity College, Cambridge, in February 1823.

He was blind in one eye and never participated in sports. However, he played the flute exquisitely and was a prominent member of the university's musical clubs. In 1827, he graduated with a Bachelor of Arts degree, but to obtain a fellowship, he needed a Master of Arts degree and for this it was necessary to pass a theological examination. He refused to take the examination and was left with no choice but to pursue a new career in London. He read law but found he preferred teaching mathematics, and thus when London University (as distinct from the University of London) was founded on explicitly nonsectarian principles in 1828, he applied for, and was appointed to, the position of Professor of Mathematics. However, in 1831, when, in the view of De Morgan and several of his colleagues, the university dismissed a professor of anatomy without giving sufficient reasons, he resigned his chair, and only on the accidental death of his successor was he reappointed in 1836.

One of De Morgan's first London friends was William Frend, like De Morgan an arithmetician and actuary, and a man of similar religious principles. Frend's daughter, Sophia Elizabeth, also became a friend of De Morgan's, such a close friend that in 1837 they were married. They had several children and their son George became a mathematician, who, with his father, founded the London Mathematical Society. Augustus was its first president, George its first secretary.

In 1866, the university senate recommended that a Unitarian clergyman be appointed to a chair in mental philosophy. The university council appointed a layman instead and, feeling that the university had compromised its principle of religious neutrality, De Morgan resigned again. But this time he was older and had little to fall back on. His pupils managed to secure him a small pension, but then, in 1868, his son George died, followed quickly by one of his daughters. In 1871, he himself died of nervous prostration. He was survived by his wife, Sophia, who became a spiritualist and wrote his biography in 1882.

De Morgan was a prolific writer. The founders of London University also founded, at the same time, the Society for the Diffusion of Useful Knowledge and under its imprint published the *Penny Cyclopedia*. De Morgan contributed 850 articles (one-sixth of all that appeared) to that publication alone and regularly wrote for at least fifteen other periodicals. His mathematical work includes a book on calculus and contributions to logic and actuarial science. His formulation of logic contained the laws that still bear his name, and in the preface to *The Mathematical Analysis of Logic*, George Boole acknowledged his debt to De Morgan. Another notable contribution concerns induction: although mathematicians had been using induction, they had done so with little clarity; in 1838, De Morgan carefully defined it and gave it its name.

One of his most curious works was called *Budget of Paradoxes*. He explained the title as follows. "A great many individuals, ever since the rise of the mathematical method, have, each for himself, attacked its direct and indirect consequences. I shall call each of these persons a *paradoxer*, and his system a *paradox*. I use the word in the old sense: a paradox is something which is apart from general opinion, either in subject matter, method, or conclusion."[1] De Morgan had accumulated a large collection of paradoxical books written by angle trisectors, cube duplicators, gravitation subverters, etc. The *Budget* reviews them all. For instance, James Smith proved that $\pi = 3\frac{1}{8}$ by assuming that this value was correct and from that assumption concluding that π could have no other value. A modern reader interested in such material should consult Underwood Dudley's *Mathematical Cranks*.

Reference

1. MacFarlane, Alexander. *Ten British Mathematicians*. New York: John Wiley and Sons, 1916, p. 30.

2 Solving Cubic and Quartic Polynomial Equations

It is easy to solve a polynomial equation of the first degree: if $ax + b = 0, a \neq 0$, then $x = \frac{-b}{a}$. And how to solve a quadratic equation is known to every high school graduate. In both cases, the solutions are exact. While a computer or sufficiently sophisticated calculator can find approximate solutions of equations involving polynomials of higher degrees, such machines cannot in general find exact solutions. The objective of this chapter is to find such exact solutions for polynomial equations of the third and fourth degrees. We will devote Parts 2–4 of the book to showing that similar general methods can never succeed in finding exact solutions to polynomial equations of the fifth degree. Such a negative result helps those interested in numerical solutions only insofar as it ensures that no effort will be expended in a fruitless search for exact solutions and hence that time and effort will be directed toward finding good approximations as generally applicable as possible. For those interested in exact solutions, however, a negative result raises more questions than it answers, the most obvious being to determine those equations that do have exact solutions. Furthermore, the point of view that must be taken to prove this negative result, in retrospect, can be seen to have changed fundamentally the questions that algebra addresses and to have stimulated giant advances, in algebra in particular and in mathematics in general. The approach taken in the succeeding chapters will reflect this fundamental change.

For completeness, we begin by deriving the quadratic formula.

The Quadratic Formula

The usual method of solving the quadratic equation, $ax^2 + bx + c = 0, a \neq 0$, is to apply the quadratic formula,

$$x = \frac{-b \pm \sqrt{b^2 - 4ac}}{2a}.$$

One way to derive this formula is to observe that if $b = 0$, then it is easy to solve for x: $x = \pm\sqrt{\frac{-c}{a}}$. So one way of solving the general equation is to find a simple substitution that will yield a quadratic in which the coefficient of x is 0. The simplest substitution is one of the form $x = y - t$. Since we will have to divide by a at some point, we do that first and then substitute for x:

$$x^2 + \frac{b}{a}x + \frac{c}{a} = 0,$$

$$(y - t)^2 + \frac{b}{a}(y - t) + \frac{c}{a} = 0,$$

$$y^2 + \left(\frac{b}{a} - 2t\right)y + \left(t^2 - \frac{b}{a}t + \frac{c}{d}\right) = 0.$$

So letting $t = \frac{b}{2a}$ yields the equation $y^2 - \frac{b^2}{4a^2} + \frac{c}{a} = 0$, which is easily solved for y:

$$y = \pm\sqrt{\frac{b^2 - 4ac}{4a^2}}.$$

And it follows that the solutions of the original quadratic equation are

$$x = \frac{-b}{2a} \pm \frac{\sqrt{b^2 - 4ac}}{2a}.$$

The methods for solving cubic and quartic polynomial equations both begin in a way analogous to the derivation just shown, namely, by forming an equivalent equation which does not contain the next lowest power of x. In the case of the cubic, the first general solution to be found was that of the reduced equation.

Cubic Polynomial Equations

As described in the historical note at the end of this chapter, the first cubic equations to be solved were those of the form $x^3 + px + q = 0$. The key to solving this equation is to note that the apparently more complicated equation $w^6 + Bw^3 + C = 0$ is in fact easy to solve because it is a quadratic in w^3: $(w^3)^2 + Bw^3 + C = 0$. We would like to convert the original equation, $x^3 + px + q = 0$, into this easily solved form by using a simple substitution. This might appear to be impossible until we realize that the equation $w^6 + Bw^3 + C = 0$ is equivalent to the equation $w^3 + B + \frac{C}{w^3} = 0$ and hence that a substitution of the form $x = w + \frac{a}{w}$ might perform the conversion for us. And indeed when we substitute this expression into the polynomial, we find that

$$x^3 + px + q = \left(w + \frac{a}{w}\right)^3 + p\left(w + \frac{a}{w}\right) + q$$

$$= w^3 + (p + 3a)w + q + (3a^2 + ap)\frac{1}{w} + \frac{a^3}{w^3}.$$

So if we pick $a = -\frac{p}{3}$, then the coefficients of w and $\frac{1}{w}$ both vanish and the original cubic equation becomes $w^3 + q - \frac{p^3}{27w^3} = 0$. Multiplying both sides of this equation by w^3 yields the desired quadratic equation in w^3, $(w^3)^2 + qw^3 - \frac{p^3}{27} = 0$, which can be solved for w^3 by using the quadratic formula previously derived:

$$w^3 = -\frac{q}{2} \pm \sqrt{\frac{q^2}{4} + \frac{p^3}{27}}.$$

In particular, if

$$\alpha = \sqrt[3]{-\frac{q}{2} + \sqrt{\frac{q^2}{4} + \frac{p^3}{27}}},$$

then $x = \alpha - \frac{p}{3\alpha}$ solves the original equation, and if

$$\beta = \sqrt[3]{-\frac{q}{2} - \sqrt{\frac{q^2}{4} + \frac{p^3}{27}}},$$

then $\beta - \frac{p}{3\beta}$ also solves the original equation.

In fact, these solutions of the original equation are the same. For

$$\alpha\beta = \sqrt[3]{-\frac{q}{2} + \sqrt{\frac{q^2}{4} + \frac{p^3}{27}}} \sqrt[3]{-\frac{q}{2} - \sqrt{\frac{q^2}{4} + \frac{p^3}{27}}} = -\frac{p}{3}$$

and hence $x = \alpha - \frac{p}{3\beta} = \alpha + \beta = -\frac{p}{3\beta} + \beta$. So we may write a general solution of the cubic equation $x^3 + px + q = 0$ in the form

$$x = \alpha + \beta = \sqrt[3]{-\frac{q}{2} + \sqrt{\frac{q^2}{4} + \frac{p^3}{27}}} + \sqrt[3]{-\frac{q}{2} - \sqrt{\frac{q^2}{4} + \frac{p^3}{27}}}.$$

This formula was known to Cardano and is named after him.

Note that there are difficulties with this general formula which may not be immediately apparent. For instance, as we will see in the next chapter, there are always three distinct cube roots of a nonzero complex number (see Proposition 3.8), and it is not always obvious which of the three should be taken as "the" cube root. As well, it may happen that adding two cube roots describes a real solution as the sum of two nonreal numbers and that in this case it can be far from obvious how to simplify the sum (cf. Examples 2.1, 2.3, and 3.10). Anomalies such as these were known to Cardano and his contemporaries and helped to spur the study of complex numbers. We will see the results of these efforts in the next chapter.

Cardano's formula provides a single solution of the original equation. Can we find the other solutions? To this end, observe that the solution we have found is a simple combination of α and β: $\alpha + \beta = 1\alpha + 1\beta$. So assume the other solutions are similarly simple combinations of α and β, i.e., that they have the form $a\alpha + b\beta$. We observed that $\alpha\beta = -\frac{p}{3}$, and hence we have

$$(a\alpha + b\beta)^3 + p(a\alpha + b\beta) + q$$
$$= a^3\alpha^3 + 3a^2\alpha^2 b\beta + 3a\alpha b^2\beta^2 + b^3\beta^3 + pa\alpha + pb\beta + q$$
$$= a^3\alpha^3 + 3a^2\alpha b\left(-\frac{p}{3}\right) + 3ab^2\beta\left(-\frac{p}{3}\right) + b^3\beta^3 + pa\alpha + pb\beta + q$$
$$= (a^3\alpha^3 + b^3\beta^3 + q) + pa\alpha(-ab + 1) + pb\beta(-ab + 1).$$

So since

$$\alpha^3 + \beta^3 = \left(\sqrt[3]{-\frac{q}{2} + \sqrt{\frac{q^2}{4} + \frac{p^3}{27}}}\right)^3 + \left(\sqrt[3]{-\frac{q}{2} - \sqrt{\frac{q^2}{2} + \frac{p^3}{27}}}\right)^3 = -q,$$

this last expression will be 0 if $a^3 = 1$, $b^3 = 1$, and $ab = 1$. In particular, a and b must solve $z^3 - 1 = 0$. But $z^3 - 1 = (z - 1)(z^2 + z + 1)$, and thus the solutions of $z^3 - 1 = 0$ are 1 and, by the quadratic formula, $-\frac{1}{2} + \frac{\sqrt{-3}}{2}$ and $-\frac{1}{2} - \frac{\sqrt{-3}}{2}$. So there are three possibilities for a. If $a = 1$, then $b = ab = 1$ and we have the solution, $x = \alpha + \beta$, provided by Cardano's formula. For the other possibilities, note that since $ab = 1$, $b = a^{-1}$, and that since $\left(-\frac{1}{2} + \frac{\sqrt{-3}}{2}\right)\left(-\frac{1}{2} - \frac{\sqrt{-3}}{2}\right) = 1$, $\left(-\frac{1}{2} + \frac{\sqrt{-3}}{2}\right)^{-1} = \left(-\frac{1}{2} - \frac{\sqrt{-3}}{2}\right)$ and $\left(-\frac{1}{2} - \frac{\sqrt{-3}}{2}\right)^{-1} = \left(-\frac{1}{2} + \frac{\sqrt{-3}}{2}\right)$. So if $a = -\frac{1}{2} + \frac{\sqrt{-3}}{2}$, then $b = -\frac{1}{2} - \frac{\sqrt{-3}}{2}$, and if $a = -\frac{1}{2} - \frac{\sqrt{-3}}{2}$, then $b = -\frac{1}{2} + \frac{\sqrt{-3}}{2}$. We conclude that the complete set of solutions of the cubic equation $x^3 + px + q = 0$ is:

$$\alpha + \beta, \quad \text{and} \quad \left(-\frac{1}{2} + \frac{\sqrt{-3}}{2}\right)\alpha + \left(-\frac{1}{2} - \frac{\sqrt{-3}}{2}\right)\beta,$$

$$\text{and} \quad \left(-\frac{1}{2} - \frac{\sqrt{-3}}{2}\right)\alpha + \left(-\frac{1}{2} + \frac{\sqrt{-3}}{2}\right)\beta,$$

where $\alpha = \sqrt[3]{-\frac{q}{2} + \sqrt{\frac{q^2}{4} + \frac{p^3}{27}}}$ and $\beta = \sqrt[3]{-\frac{q}{2} - \sqrt{\frac{q^2}{4} + \frac{p^3}{27}}}$.

Example 2.1 Let $f(x) = x^3 - 3x + 1$. To solve $f(x) = 0$, we substitute $x = w - \frac{-3}{3w} = w + \frac{1}{w}$ to obtain

$$\left(w + \frac{1}{w}\right)^3 - 3\left(w + \frac{1}{w}\right) + 1 = w^3 + 1 + \frac{1}{w^3}.$$

Applying the quadratic formula to the equation $(w^3)^2 + w^3 + 1 = 0$, we see that

$$w^3 = -\frac{1}{2} \pm \frac{\sqrt{-3}}{2}$$

and hence that

$$x = \sqrt[3]{-\frac{1}{2} + \frac{\sqrt{-3}}{2}} + \frac{1}{\sqrt[3]{-\frac{1}{2} + \frac{\sqrt{-3}}{2}}} = \sqrt[3]{-\frac{1}{2} + \frac{\sqrt{-3}}{2}} + \sqrt[3]{-\frac{1}{2} - \frac{\sqrt{-3}}{2}}$$

solves $f(x) = 0$.

Now we have $\alpha = \sqrt[3]{-\frac{1}{2} + \frac{\sqrt{-3}}{2}}$ and $\beta = \sqrt[3]{-\frac{1}{2} - \frac{\sqrt{-3}}{2}}$, and thus we know how to find the other two solutions. Specifically, we combine α and β with $\left(-\frac{1}{2} + \frac{\sqrt{-3}}{2}\right)$ and $\left(-\frac{1}{2} - \frac{\sqrt{-3}}{2}\right)$ in the following ways. A second solution is

$$x = \left(-\frac{1}{2} + \frac{\sqrt{-3}}{2}\right)\sqrt[3]{-\frac{1}{2} + \frac{\sqrt{-3}}{2}} + \left(-\frac{1}{2} - \frac{\sqrt{-3}}{2}\right)\sqrt[3]{-\frac{1}{2} - \frac{\sqrt{-3}}{2}}$$

$$= \left(-\frac{1}{2} + \frac{\sqrt{-3}}{2}\right)^{4/3} + \left(-\frac{1}{2} - \frac{\sqrt{-3}}{2}\right)^{4/3}$$

and a third is

$$x = \left(-\frac{1}{2} - \frac{\sqrt{-3}}{2}\right)\sqrt[3]{-\frac{1}{2} + \frac{\sqrt{-3}}{2}} + \left(-\frac{1}{2} + \frac{\sqrt{-3}}{2}\right)\sqrt[3]{-\frac{1}{2} - \frac{\sqrt{-3}}{2}}$$

$$= \left(-\frac{1}{2} + \frac{\sqrt{-3}}{2}\right)^{7/3} + \left(-\frac{1}{2} - \frac{\sqrt{-3}}{2}\right)^{7/3}.$$

FIGURE 2.1

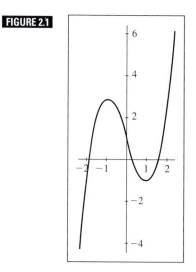

Written in this way, the solutions of $f(x) = 0$ do not appear to be real. However, if we use the techniques of single variable calculus to draw the graph of the function $f(x) = x^3 - 3x + 1$, we obtain the graph shown in Figure 2.1, from which we can immediately conclude that the equation $f(x) = 0$ has three distinct <u>real</u> solutions.

Even without the graph, the intermediate value theorem allows us to come to the same conclusion. For since $f(-3) = -17 < 0, f(0) = 1 > 0, f(1) = -1 < 0$, and $f(2) = 3 > 0$, we know from this theorem that there are three distinct real values of $x, x = r_1, r_2, r_3$, for which $f(x) = 0$: $-3 < r_1 < 0, 0 < r_2 < 1$, and $1 < r_3 < 2$. We will investigate this further in Example 3.10. ❖

We have seen that a simple substitution of the form $x = y - t$ can be used to convert the quadratic equation, $x^2 + bx + c = 0$, into the form $y^2 + C = 0$. So to solve the general cubic equation, $x^3 + ax^2 + bx + c = 0$, we use a substitution, $x = y - t$, to reduce it to an equivalent equation of the form $y^3 + py + q = 0$, which we can then solve by using the method described in Example 1. To determine t, we calculate

$(y - t)^3$:	y^3	$- 3y^2 t$	$+ 3yt^2$	$- t^3$
$a(y - t)^2$:		ay^2	$- 2aty$	$+ at^2$
$b(y - t)$:			by	$- bt$
c:				c

$$y^3 + (-3t + a)y^2 + (3t^2 - 2at + b)y + (-t^3 + at^2 - bt + c)$$

The coefficient y^2 will be 0 if we let $t = \frac{a}{3}$. So we substitute $x = y - \frac{a}{3}$ into the original equation and obtain the equation

$$y^3 + \left(b - \frac{a^2}{3}\right)y + \left(\frac{2a^3}{27} - \frac{ab}{3} + c\right) = 0.$$

We can then use the method presented above to find a solution $y = \gamma$ of this equation. Then $x = \gamma - \frac{a}{3}$ will solve the original equation, $x^3 + ax^2 + bx + c = 0$.

Example 2.2   To use this method to solve $x^3 + 3x^2 - 9x - 27 = 0$, first eliminate the second degree term by using the substitution $x = y - \frac{3}{3} = y - 1$:

$(y - 1)^3$:	y^3	$- 3y^2$	$+ 3y$	$- 1$
$3(y - 1)^2$:		$3y^2$	$- 6y$	$+ 3$
$- 9(y - 1)$:			$- 9y$	$+ 9$
$- 27$:				$- 27$

$$y^3 \qquad\qquad - 12y \quad - 16$$

Into the reduced equation $y^3 - 12y - 16 = 0$, substitute $y = w - \frac{(-12)}{3w} = w + \frac{4}{w}$:

$$\left(w + \frac{4}{w}\right)^3 - 12\left(w + \frac{4}{w}\right) - 16 = 0,$$

$$w^3 + 12w + \frac{48}{w} + \frac{64}{w^3} - 12w - \frac{48}{w} - 16 = 0,$$

$$w^6 - 16w^3 + 64 = 0,$$

$$(w^3 - 8)^2 = 0.$$

Then $w^3 = 8$, hence $y = 2 + \frac{4}{2} = 4$ and $x = 4 - 1 = 3$. Thus we have found one solution, $x = 3$, of $x^3 + 3x^2 - 9x - 27 = 0$.

To find the other solutions, we could use Cardano's formula to conclude that the solutions of $y^3 - 12y - 16 = 0$ are 4 and $2\left(-\frac{1}{2} - \frac{\sqrt{-3}}{2}\right) + 2\left(-\frac{1}{2} + \frac{\sqrt{-3}}{2}\right) = -2$ so that the solutions of $x^3 + 3x^2 - 9x - 27 = 0$ are $x = \pm 3$. Alternatively, we could observe that the polynomial $x - 3$ divides the polynomial $x^3 + 3x^2 - 9x - 27$, and by polynomial long division,

$$x^3 + 3x^2 - 9x - 27 = (x - 3)(x^2 + 6x + 9).$$

And thus since $x^2 + 6x + 9 = (x + 3)^2$, the solutions of $x^3 + 3x^2 - 9x - 27 = 0$ are $x = \pm 3$. In either case, we have

$$x^3 + 3x^2 - 9x - 27 = (x - 3)(x + 3)^2. \ \diamond$$

As previously noted, the algorithm does not always work this nicely; it can sometimes express simple solutions in complicated form.

Example 2.3   Let $p(x) = x^3 - 3x^2 + x - 3$. To use Cardano's method to solve $p(x) = 0$, substitute $x = y - \frac{(-3)}{3} = y + 1$:

$(y + 1)^3$:	y^3	$+ 3y^2$	$+ 3y$	$+ 1$
$- 3(y + 1)^2$:		$- 3y^2$	$- 6y$	$- 3$
$+ (y + 1)$:			$+ y$	$+ 1$
$- 3$:				$- 3$
	y^3		$- 2y$	$- 4$

In turn, to solve $y^3 - 2y - 4 = 0$, substitute $y = w - \frac{(-2)}{3w} = w + \frac{2}{3w}$:

$$\left(w + \frac{2}{3w}\right)^3 - 2\left(w + \frac{2}{3w}\right) - 4 = w^3 + \frac{8}{27w^3} - 4.$$

Using the quadratic formula to solve $(w^3)^2 - 4w^3 + \frac{8}{27} = 0$, we have

$$w^3 = \frac{4}{2} \pm \frac{\sqrt{16 - 4\frac{8}{27}}}{2} = 2 \pm \frac{10}{9}\sqrt{3},$$

and hence the solutions of $p(x) = 0$ are

$$\sqrt[3]{2 + \frac{10}{9}\sqrt{3}} + \sqrt[3]{2 - \frac{10}{9}\sqrt{3}} + 1,$$

$$\left(-\frac{1}{2} + \frac{\sqrt{-3}}{2}\right)\sqrt[3]{2 + \frac{10}{9}\sqrt{3}} + \left(-\frac{1}{2} - \frac{\sqrt{-3}}{2}\right)\sqrt[3]{2 - \frac{10}{9}\sqrt{3}} + 1,$$

and

$$\left(-\frac{1}{2} - \frac{\sqrt{-3}}{2}\right)\sqrt[3]{2 + \frac{10}{9}\sqrt{3}} + \left(-\frac{1}{2} + \frac{\sqrt{-3}}{2}\right)\sqrt[3]{2 - \frac{10}{9}\sqrt{3}} + 1.$$

However, if we notice that $p(3) = 3^3 - 3 \cdot 3^2 + 3 - 3 = 0$, then we can use polynomial long division to find that

$$p(x) = (x - 3)(x^2 + 1) = (x - 3)\ (x - \sqrt{-1})\ (x + \sqrt{-1})$$

and hence that the solutions of $p(x) = 0$ can also be written $x = 3, \sqrt{-1}, -\sqrt{-1}$. ❖

Quartic Polynomial Equations

To solve the general quartic equation,

$$x^4 + ax^3 + bx^2 + cx + d = 0,$$

we begin, as in the case of quadratic and cubic equations, by reducing it to an equivalent equation of the form

$$y^4 + py^2 + qy + r = 0.$$

In the case of the quadratic, $x^2 + ax + b = 0$, we used the substitution $x = y - \frac{a}{2}$, and in the case of the cubic, $x^3 + ax^2 + bx + c = 0$, we used $x = y - \frac{a}{3}$. So in the case of the quartic, we try the substitution $x = y - \frac{a}{4}$:

$(y - \frac{a}{4})^4:$	$y^4 - ay^3$	$+ \frac{3}{8}a^2y^2$	$- \frac{1}{16}a^3y$	$+ \frac{1}{256}a^4$
$a(y - \frac{a}{4})^3:$	ay^3	$- \frac{3}{4}a^2y^2$	$+ \frac{3}{16}a^3y$	$- \frac{1}{64}a^4$
$b(y - \frac{a}{4})^2:$		by^2	$- \frac{1}{2}aby$	$+ \frac{1}{16}a^2b$
$c(y - \frac{a}{4}):$			cy	$- \frac{1}{4}ac$
$d:$				d

$$y^4 \qquad + (b - \tfrac{3}{8}a^2)y^2 + (c - \tfrac{1}{2}ab + \tfrac{1}{8}a^3)y + (d - \tfrac{1}{4}ac + \tfrac{1}{16}a^2b - \tfrac{3}{256}a^4)$$

Thus it suffices to solve the equation $y^4 + py^2 + qy + r = 0$, where

$$p = b - \frac{3}{8}a^2, \quad q = c - \frac{1}{2}ab + \frac{1}{8}a^3, \quad r = d - \frac{1}{4}ac + \frac{1}{16}a^2b - \frac{3}{256}a^4.$$

The solution of the reduced equation

$$y^4 + py^2 + qy + r = 0$$

is based on the observation that it is easy to factor a difference of squares: $A^2 - B^2 = (A - B)(A + B)$. So if we can write the quartic $y^4 + py^2 + qy + r$ in the form $(y^2 + K)^2 - (Ly + M)^2$, then we can factor the resulting expression and write the reduced equation as

$$(y^2 - Ly + (K - M))(y^2 + Ly + (K + M)) = 0,$$

and hence we can solve the reduced equation by solving the quadratic equations

$$y^2 - Ly + (K - M) = 0 \quad \text{and} \quad y^2 + Ly + (K + M) = 0.$$

How can we find K, L, and M such that $y^4 + py^2 + qy + r = (y^2 + K)^2 - (Ly + M)^2$? We begin by observing that for any u,

$$y^4 + py^2 + qy + r = y^4 + uy^2 + \frac{u^2}{4} - uy^2 - \frac{u^2}{4} + py^2 + qy + r$$

$$= \left(y^2 + \frac{u}{2}\right)^2 - \left[(u - p)y^2 - qy + \left(\frac{u^2}{4} - r\right)\right].$$

So we need to find u for which there exist L and M such that

$$\left[(u - p)y^2 - qy + \left(\frac{u^2}{4} - r\right)\right] = (Ly + M)^2.$$

Since $(Ly + M)^2 = L^2y^2 + 2LMy + M^2$, L and M must satisfy

$$L^2 = u - p, \quad 2LM = -q, \quad \text{and} \quad M^2 = \frac{u^2}{4} - r,$$

and thus u must satisfy

$$q^2 = 4L^2M^2 = 4(u - p)\left(\frac{u^2}{4} - r\right),$$

i.e., u must solve the equation

$$u^3 - pu^2 - 4ru + (4pr - q^2) = 0.$$

This equation is a cubic and hence may be solved by the method described previously. Let v denote a solution of this equation. The reduced polynomial may then be written as follows, where the sign of $\sqrt{\frac{v^2}{4} - r}$ is chosen so that $2\sqrt{v - p}\sqrt{\frac{v^2}{4} - r} = -q$:

$$y^4 + py^2 + qy + r = \left(y^2 + \frac{v}{2}\right)^2 - \left(\sqrt{v-p}\,y + \sqrt{\frac{v^2}{4} - r}\right)^2$$

$$= \left(y^2 + \frac{v}{2} - \sqrt{v-p}\,y - \sqrt{\frac{v^2}{4} - r}\right)\left(y^2 + \frac{v}{2} + \sqrt{v-p}\,y + \sqrt{\frac{v^2}{4} - r}\right).$$

And hence solving the reduced quartic equation is equivalent to solving the quadratic equations

$$y^2 - \sqrt{v-p}\,y + \left(\frac{v}{2} - \sqrt{\frac{v^2}{4} - r}\right) = 0,$$

$$y^2 + \sqrt{v-p}\,y + \left(\frac{v}{2} + \sqrt{\frac{v^2}{4} - r}\right) = 0.$$

These equations can be solved by inspection (if possible) or by using the quadratic formula (if necessary). The solutions $y = \gamma$ obtained in this way will be a complete set of roots of the reduced quartic equation $y^4 + py^2 + qy + r = 0$. A complete set of roots of the original quartic equation may then be found by calculating $x = \gamma - \frac{a}{4}$ for each of the values $y = \gamma$.

Example 2.4   To solve

$$3x^4 + 12x^3 + 18x^2 + 24x = -24,$$

first rewrite it as

$$x^4 + 4x^3 + 6x^2 + 8x + 8 = 0.$$

Then reduce this equation by substituting $x = y - \frac{4}{4} = y - 1$:

$(y-1)^4$: y^4	$-4y^3$	$+6y^2$	$-4y$	$+1$
$4(y-1)^3$:	$4y^3$	$-12y^2$	$+12y$	-4
$6(y-1)^2$:		$6y^2$	$-12y$	$+6$
$8(y-1)$:			$8y$	-8
8:				8
y^4			$+4y$	$+3$

and note that for any u,

$$y^4 + 4y + 3 = y^4 + y^2u + \frac{u^2}{4} - \frac{u^2}{4} - y^2u + 4y + 3,$$

$$= \left(y^2 + \frac{u}{2}\right)^2 - \left[uy^2 - 4y + \left(\frac{u^2}{4} - 3\right)\right]. \tag{1}$$

For the square brackets to be a perfect square, there must exist L, M such that u satisfies the equation

$$(Ly + M)^2 = u\,y^2 - 4y + \left(\frac{u^2}{4} - 3\right).$$

Since $(Ly + M)^2 = L^2y^2 + 2LMy + M^2$, L and M must satisfy

$$L^2 = u, \quad 2LM = -4, \quad \text{and} \quad M^2 = \frac{u^2}{4} - 3,$$

and thus u must solve the equation

$$(-4)^2 = 4L^2M^2 = 4u\left(\frac{u^2}{4} - 3\right),$$

i.e., u must solve the equation

$$u^3 - 12u - 16 = 0.$$

In Example 2.2, we found that $u = 4$ solves this equation. Substituting $u = 4$ in equation (1), we have

$$
\begin{aligned}
y^4 + 4y + 3 &= (y^2 + 2)^2 - (4y^2 - 4y + 1) \\
&= (y^2 + 2)^2 - (2y - 1)^2 \\
&= (y^2 + 2y + 1)(y^2 - 2y + 3).
\end{aligned}
$$

Thus to solve the reduced equation $y^4 + 4y + 3 = 0$, it suffices to solve each of the quadratic equations $y^2 + 2y + 1 = 0$ and $y^2 - 2y + 3 = 0$ for y. This yields $y = -1, 1 \pm \sqrt{-2}$ as the solutions of the reduced equation, and thus, since $x = y - 1$, then $x = -2, \pm \sqrt{-2}$ are the solutions of the original equation $3x^4 + 12x^3 + 18x^2 + 24x = -24$. Notice that since

$$y^4 + 4y + 3 = (y + 1)^2\,(y - (1 + \sqrt{-2}))(y - (1 - \sqrt{-2})),$$

we have

$$3x^4 + 12x^3 + 18x^2 + 24x + 24 = 3(x + 2)^2(x + \sqrt{-2})(x - \sqrt{-2}).\ \clubsuit$$

Sometimes an appropriate value for u may be found by judicious observation, as in the next example.

Example 2.5   To solve

$$3x^4 + 12x^3 + 15x^2 = 9,$$

first rewrite it as $x^4 + 4x^3 + 5x^2 - 3 = 0$, and then reduce this equation by substituting $x = y - \frac{4}{4} = y - 1$:

$(y - 1)^4$:	y^4	$- 4y^3$	$+ 6y^2$	$- 4y$	$+ 1$
$4(y - 1)^3$:		$4y^3$	$- 12y^2$	$+ 12y$	$- 4$
$5(y - 1)^2$:			$5y^2$	$- 10y$	$+ 5$
$- 3$:					$- 3$
	y^4		$- y^2$	$- 2y$	$- 1$

So we must solve $y^4 - y^2 - 2y - 1 = 0$. For any u, this equation is equivalent to

$$y^4 + y^2u + \frac{u^2}{4} - \frac{u^2}{4} - y^2u - y^2 - 2y - 1 = 0,$$

which in turn is equivalent to

$$\left(y^2 + \frac{u}{2}\right)^2 - \left[(u + 1)y^2 + 2y + \left(\frac{u^2}{4} + 1\right)\right] = 0.$$

Observe that if $u = 0$, then the expression in the square brackets becomes $y^2 + 2y + 1$. This is easily seen to be a perfect square, and hence the above equation is easy to rewrite as

$$(y^2)^2 - (y + 1)^2 = 0, \quad \text{or} \quad (y^2 - y - 1)(y^2 + y + 1) = 0.$$

Setting each of the quadratic factors equal to zero, we have

$$y = \frac{1}{2} \pm \frac{\sqrt{5}}{2}, -\frac{1}{2} \pm \frac{\sqrt{-3}}{2},$$

and thus the solutions to the original equation are

$$x = -\frac{1}{2} \pm \frac{\sqrt{5}}{2}, -\frac{3}{2} \pm \frac{\sqrt{-3}}{2}. \quad \diamond$$

Higher Order Polynomial Equations

The next case is the quintic equation, $x^5 + ax^4 + bx^3 + cx^2 + dx + e = 0$. All attempts at solving this by similar methods failed and eventually Niels Henrik Abel (1802–1829) and Paolo Ruffini (1765–1822) independently proved that there was no method that could solve all such polynomials in a similar fashion. Evariste Galois (1811–1832) then attacked the more general question of determining which polynomial equations are solvable by these methods. We describe the modern version of Galois' approach in Parts 2–4. Their destination is the production of a non-solvable polynomial equation. Along the way, we will meet the major algebraic structures of modern mathematics. These structures, especially groups, have important implications far removed from the solving of polynomials. They are used in physics, chemistry, and economics, as well as in other parts of mathematics. This fundamental part of modern science started, however, with the problem described previously, viz., determining what polynomials may be solved by radicals.

Exercises

In Exercises 1–11, find all solutions of the given equation and determine the number of distinct real solutions.

1. $2x^3 + 6x + 2 = 0$

2. $2y^3 + 6y^2 + 18y = -10$

3. $4y^4 + 16y^3 + 20y^2 + 8y + 1 = 0$

4. $2x^4 - 8x^3 + 4x^2 + 16x = 16$

5. $x^4 - 4x^3 + x^2 - 4x + 1 = 0$ 6. $x^3 + x^2 + x + 1 = 0$

7. $x^4 - 4x^3 + x^2 + x + 1 = 0$ 8. $u^3 + 6u^2 + 12u + 8 = 0$

9. $2x^3 + 12x^2 + 18x + 8 = 0$ 10. $y^4 + 4y^3 - 16y - 16 = 0$

11. $x^4 + x^3 + x^2 + x + 1 = 0$

$$\left(\text{Hint: } \left(\frac{5}{6} + \frac{\sqrt{-15}}{6} \right)^3 = -\frac{25}{54} + \frac{\sqrt{-15}}{6}. \right)$$

12. Consider the fifth degree polynomial $y^5 + ay^4 + by^3 + cy^2 + dy + e$. Use a substitution $y = x - \alpha$ to reduce this polynomial to one of the form

$$x^5 + px^3 + qx^2 + rx + s.$$

Specify α and find p, q, r, s explicitly in terms of a, b, c, d, e.

13. Solve the sixth degree polynomial equation $x^6 + px^4 + qx^2 + r = 0$.

Exercises 14, 15, and 16 outline Lagrange's method of solving the quadratic equation. They refer to the following. Consider the quadratic polynomial equation $x^2 + px + q = 0$; denote its solutions by d_1, d_2. If the *resolvent* of a and b is $a - b$, then the only resolvents of the roots of $x^2 + px + q = 0$ are: $\delta_1 = d_1 - d_2$ and $\delta_2 = d_2 - d_1$. Construct the polynomial $t(x) = (x - \delta_1)(x - \delta_2)$.

14. Explain why the coefficient of the first power of x in $t(x)$ must be zero.

15. Show that $t(x)$ can be determined without knowing d_1 and d_2 by expressing the coefficients of $t(x)$ in terms of p and q.
    (Hint: $(x - d_1)(x - d_2) = x^2 + px + q$.)

16. By Exercise 14, the substitution $x^2 = w$ transforms $t(x)$ into a first degree polynomial $s(w)$. Let v denote the solution of $s(w) = 0$. Express the solutions of the original quadratic equation in terms of the solutions of the first degree equation $(w) = 0$. That is, express d_1 and d_2 in terms of $\sqrt{v}$.

Exercises 17, 18, and 19 outline Lagrange's method of solving the quartic equation. They refer to the following. Consider the quartic polynomial equation $x^4 + px^3 + qx^2 + rx + s = 0$. Denote its solutions by d_1, d_2, d_3, d_4. If the *resolvent* of $a, b, c,$ and d is $a - b + c - d$, then the resolvents of the roots of $x^4 + px^3 + qx^2 + rx + s = 0$ are:

$\delta_1 = d_1 - d_2 + d_3 - d_4$, $\delta_2 = d_1 - d_3 + d_2 - d_4$, $\delta_3 = d_1 - d_2 + d_4 - d_3$,

$\delta_4 = d_2 - d_1 + d_3 - d_4$, $\delta_5 = d_2 - d_1 + d_4 - d_3$, $\delta_6 = d_3 - d_1 + d_4 - d_2$.

Construct the polynomial $t(x) = (x - \delta_1)(x - \delta_2)(x - \delta_3)(x - \delta_4)(x - \delta_5)(x - \delta_6)$.

17. Explain why only even powers of x in $t(x)$ can have nonzero coefficients.

18. Show that $t(x)$ can be determined without knowing d_1, d_2, d_3, d_4 by expressing the

coefficients of $t(x)$ in terms of p, q, r, s. (*Hint:* $(x - d_1)(x - d_2)(x - d_3)(x - d_4) = x^4 + px^3 + qx^2 + rx + s$.)

19. By Exercise 17, the substitution $x^2 = w$ transforms $t(x)$ into a cubic polynomial $s(w)$. Let ν_1, ν_2, ν_3 denote the solutions of $s(w) = 0$. Express the solutions of the original quartic equation in terms of the solutions of the cubic equation $s(w) = 0$. That is, express d_1, d_2, d_3, d_4 in terms of $\sqrt{\nu_1}, \sqrt{\nu_2}, \sqrt{\nu_3}$.
    (*Hint:* Begin by showing that $d_1 + d_2 + d_3 + d_4 = -p$ and then that $\delta_1 = \sqrt{\nu_1}, \delta_2 = \sqrt{\nu_2}, \delta_3 = \sqrt{\nu_3}$.)

20. Explain why Lagrange's method described in the previous exercises cannot be used to solve the general quintic (degree 5) equation.

HISTORICAL NOTE

How the Cubic and Quartic Equations Were Solved

The story of the solving of the cubic and quartic equations involves several people, all of them from Italy. Fra Luca Pacioli (ca. 1445–1514) pessimistically summarized the conventional wisdom of 1494 in his book *Summa de arithmetica* as follows: "the means [for solving cubic equations] are not yet given, just as the means for squaring the circle are not given."[1]

Undeterred, Scipione del Ferro (1465–1526) of the University of Bologna found a way to solve the equation $x^3 + ax = b$, but he communicated it only to his son-in-law Annibale della Nave and his student Antonio Maria Fiore. At the time, reputations were made and lost in problem-solving duels; anyone could issue a challenge at any time and those who were challenged had to be prepared to defend themselves. Fiore decided to use his secret to advance himself in a duel with Niccolò Tartaglia.

Niccolò Fontana was born into a poor family in Brescia around 1500. In 1512, Brescia was captured by the French and Fontana was wounded in the larynx. Because he spoke with difficulty thereafter, he was nicknamed Tartaglia ("stutterer"). His family's poverty forced him to leave school before he had even learned the alphabet but he continued to study on his own and eventually taught at a private business college. By 1534, he was an experienced combatant and unafraid of Fiore. Even when he received a list of thirty equations of the form given above, he was not worried because he did not believe that Fiore could solve his own problems. However, he heard a rumor that Fiore had a method for solving the equations, and in a frantic search, as the deadline approached, he found such a method himself. Since Fiore was unable to solve even one of the equations that Tartaglia had given him, the duel ended with Tartaglia victorious. This duel was well publicized and many people wanted Tartaglia's secret. But he guarded it jealously and told it to no one . . . until he met Gerolamo Cardano.

Gerolamo Cardano was born in Pavia on September 24, 1501. His father was a well-educated lawyer who made sure his son was equally well educated. Gerolamo graduated from the University of Padua in 1526 and decided to become a physician. However, because of doubts about the legitimacy of his birth, he was denied admission to the College of Physicians in Milan and practiced in the provinces until 1539 when the college changed the rules so that he could be admitted. He wrote on a variety of subjects, including mathematics and astrology, and gambled at chess for forty years and at dice for twenty-five. The results were at times unpleasant, but eventually, in 1526, his experience allowed him to write *Liber de ludo aleae* (*The Book on Games of Chance*), which contained the beginnings of probability theory, including a preliminary statement of the law of large numbers. Toward the end of 1538, Cardano was completing a book that he hoped would supplant Pacioli's and desperately wanted Tartaglia's secret. On March 25, 1539, enticed by Cardano's promise of an introduction to the Spanish governor of Lombardy, Tartaglia met with Cardano.

According to Tartaglia's notes (which may of course be biased), the following exchange took place:

Tartaglia: "I plan to publish the work for practical application together with a new algebra. . . . If I give it some theorist (such as your Excellency), then he could easily find other chapters with the help of this explanation . . . and publish the fruit of my discovery under another name."

Cardano: "I swear to you by the Sacred Gospel, and on my faith as a gentleman, not only never to publish your discoveries, if you tell them to me, but I also promise and pledge my faith as a true Christian to put them down in cipher so that after my death no one shall be able to understand them."

Tartaglia: "If I did not believe an oath such as yours, then of course I myself would deserve to be considered an atheist."[2]

And after speaking these words, Tartaglia imparted his secret to Cardano.

Tartaglia left the meeting apprehensive, but on May 12, he received a copy of the *Practica arithmeticae generalis* without the secret method and an accompanying note from Cardano which said "I have verified the formula and believe that it has broad significance."[3] Over the next few years, Cardano worked hard to extend the method given him by Tartaglia. He managed to generalize it to equations of the form $x^3 = ax + b$ and $x^3 + b = ax$. (Since negative numbers were not in use at the time, these equations, along with the earlier one, were all considered different!) But he was bound by his oath not to publish any of these methods.

In 1536, Cardano took on **Lodovico Ferrari** (1522–1565) as a servant and student. Ferrari had a temper to match his phenomenal mathematical ability; when he was seventeen, he returned from a brawl with no fingers left on his right hand! Ferrari was devoted to Cardano and in 1543, the two of them traveled to Bologna, where del Ferro's son-in-law della Nave allowed them to peruse del Ferro's papers. They convinced themselves that del Ferro had had the solution before Tartaglia and that this absolved Cardano of the oath he had given Tartaglia. The result was *Ars magna* (*The Great Art*), which appeared in 1545 and included the solution of the general cubic equation, as well as Ferrari's contribution, the solution of the quartic equation. It presented the algorithms by example. For instance, the general solution of the equation $x^3 + ax = b$ was given by solving $x^3 + 6x = 20$. Cardano did not neglect his predecessors. He mentions del Ferro, Tartaglia, and Fiore and continues: "Then, however, having received Tartaglia's solution and seeking for the proof of it, I came to understand that there were a great many other things which could also be had. Pursuing this thought and with increased confidence, I discovered these others, partly by myself and partly through Lodovico Ferrari, formerly my pupil."[4]

Not surprisingly, Cardano's publication of *Ars magna* incensed Tartaglia. In 1546, he published his correspondence with Cardano and heaped abuse on him. Cardano did not react, but Ferrari challenged Tartaglia to a public debate. Hoping to draw out Cardano, Tartaglia demurred. But finally, in 1548, possibly to strengthen his position at Brescia, he agreed to a mathematical duel with Ferrari. The details of the duel are unknown, but it is certain that Tartaglia suffered a humiliating defeat. He returned to Venice a year later and died there in 1557. Ferrari became very famous, gave public lectures in Rome, headed the taxation department in Milan, and helped to bring up the emperor's son. However, he did no further scientific research. He died in 1565, possibly poisoned by his sister.

Cardano lived until 1576, but it was a difficult life. One of his sons poisoned his wife and was executed; another turned criminal and robbed his own father. In 1570, Cardano himself was sent to prison and his property confiscated, possibly on the initiative of the Inquisition. He finished his life in Rome, with a modest pension from the Pope, and spent his last years writing his autobiography. In its last pages, he finally replied to the now deceased Tartaglia when he wrote, "I confess that in mathematics I received a few suggestions, but very few, from brother Niccolò."[5]

References

1. Gindikin, Semyon Grigorevich. *Tales of Physicists and Mathematicians.* 2nd ed. Boston: Birkhäuser, 1985, p. 3.
2. *Ibid.,* pp. 9–10.
3. *Ibid.,* p. 10.
4. *Ibid.,* p. 12.
5. *Ibid.,* p. 18.

3

Complex Numbers

The solutions to the equations in Examples 2.1, 2.3, 2.4, and 2.5 all involve square roots of negative numbers. In general, the methods for solving polynomial equations given in Chapter 1 yield solutions of this kind, i.e., solutions that may not live in the real numbers. As described in the historical note at the end of this chapter, this observation spurred the development of the numbers that can encompass all such solutions, i.e., the complex numbers.

Formally adjoining the nonreal number $\sqrt{-1}$ to the real numbers struck some mathematicians as both mysterious and suspicious. But these doubts, along with the specific adjunction of $\sqrt{-1}$, disappeared with the discovery that the set of formal sums $a + b\sqrt{-1}$, for real numbers a and b, can be viewed, with no loss of algebraic content, as the set of ordered pairs (a,b). From this point of view, $\sqrt{-1}$ is just the ordered pair $(0,1)$ and thus presents neither mystery nor difficulty. In this chapter, we will adopt the standard notation $a + bi$, where i represents $\sqrt{-1}$. But it is important to note that this is merely convention; we are really looking at the set of ordered pairs (a,b) with particular algebraic operations of addition and multiplication. In fact, we will not abandon the notation (a,b) entirely. On the contrary, we will see that valuable geometric information can be gained by associating the complex number $a + bi$ with the point (a,b) on the Euclidean plane.

Notation. We let

$\mathbb{Z}$ denote the set of integers,

$\mathbb{Q}$ denote the set of rational numbers,

$\mathbb{R}$ denote the set of real numbers.

DEFINITION

The **complex numbers** are the elements of the set

$$\mathbb{C} = \{a + bi \mid a,b \in \mathbb{R}\}.$$

If $a + bi, c + di \in \mathbb{C}$, we add and multiply $a + bi$ and $c + di$ as follows:

$$(a + bi) + (c + di) = (a + c) + (b + d)i,$$
$$(a + bi)(c + di) = (ac - bd) + (ad + bc)i.$$

For notational convenience, we denote

$$a + 0i \quad \text{by} \quad a, \qquad 0 + bi \quad \text{by} \quad bi, \qquad a + (-b)i \quad \text{by} \quad a - bi.$$

This makes algebraic as well as notational sense because, as we show next, addition and multiplication in $\mathbb{C}$ obey many of the same algebraic laws that govern them in $\mathbb{Q}$ and $\mathbb{R}$.

Proposition 3.1   *The following laws hold in* $\mathbb{C}$.

A1. ADDITIVE CLOSURE: *For all* $a + bi, c + di \in \mathbb{C}, (a + bi) + (c + di) \in \mathbb{C}$.

A2. ADDITIVE ASSOCIATIVITY: *For all* $a + bi, c + di, e + fi \in \mathbb{C}$,
$(a + bi) + [(c + di) + (e + fi)] = [(a + bi) + (c + di)] + (e + fi)$.

A3. EXISTENCE OF AN ADDITIVE IDENTITY: *There is a complex number* 0, *such that for all* $a + bi \in \mathbb{C}, (a + bi) + 0 = a + bi = 0 + (a + bi)$.

A4. EXISTENCE OF ADDITIVE INVERSES: *For all* $a + bi \in \mathbb{C}$, *there is a complex number,* $-a - bi$, *such that* $(a + bi) + (-a - bi) = 0 = (-a - bi) + (a + bi)$.

A5. ADDITIVE COMMUTATIVITY: *For all* $a + bi, c + di \in \mathbb{C}, (a + bi) + (c + di) = (c + di) + (a + bi)$.

M1. MULTIPLICATIVE CLOSURE: *For all* $a + bi, c + di \in \mathbb{C}, (a + bi)(c + di) \in \mathbb{C}$.

M2. MULTIPLICATIVE ASSOCIATIVITY: *For all* $a + bi, c + di, e + fi \in \mathbb{C}$,
$(a + bi)[(c + di)(e + fi)] = [(a + bi)(c + di)](e + fi)$.

M3. EXISTENCE OF A MULTIPLICATIVE IDENTITY OR UNIT ELEMENT: *For all* $a + bi \in \mathbb{C}$, *there is a complex number,* 1, *such that* $(a + bi)(1) = a + bi = (1)(a + bi)$.

M4. EXISTENCE OF MULTIPLICATIVE INVERSES: *For all* $0 \neq a + bi \in \mathbb{C}$, *there is a complex number,* $x + yi$, *such that* $(x + yi)(a + bi) = 1 = (a + bi)(x + yi)$.

M5. MULTIPLICATIVE COMMUTATIVITY: *For all* $a + bi, c + di \in \mathbb{C}$,
$(a + bi)(c + di) = (c + di)(a + bi)$.

D1. LEFT DISTRIBUTIVITY: *For all* $a + bi, c + di, e + fi \in \mathbb{C}$,
$(a + bi)[(c + di) + (e + fi)] = (a + bi)(c + di) + (a + bi)(e + fi)$.

D2. RIGHT DISTRIBUTIVITY: *For all* $a + bi, c + di, e + fi \in \mathbb{C}$,
$[(a + bi) + (c + di)](e + fi) = (a + bi)(e + fi) + (c + di)(e + fi)$.

Proof By definition, $\mathbb{C}$ is additively and multiplicatively closed. With the exception of the existence of a multiplicative inverse, the remaining properties all involve similar calculations. As examples of such calculations, we prove additive commutativity and left distributivity. We conclude with a proof of the existence of multiplicative inverses.

A5. Addition of complex numbers is commutative because addition of real numbers is commutative:

$$(a + bi) + (c + di) = (a + c) + (b + d)i$$
$$= (c + a) + (d + b)i = (c + di) + (a + bi).$$

D1. Multiplication of complex numbers distributes over addition from the left because multiplication of real numbers distributes over addition from the left and because addition of real numbers is commutative:

$$(a + bi)((c + di) + (e + fi)) = (a + bi)((c + e) + (d + f)i)$$
$$= (a(c + e) - b(d + f)) + (b(c + e) + a(d + f))i$$
$$= ((ac - bd) + (ae - bf)) + ((bc + ad) + (be + af))i$$
$$= ((ac - bd) + (bc + ad)i) + ((ae - bf) + (be + af)i)$$
$$= (a + bi)(c + di) + (a + bi)(e + fi).$$

M4. Each nonzero complex number has a multiplicative inverse. Note that since $a + bi \neq 0$, either $a \neq 0$ or $b \neq 0$, and hence $a^2 + b^2 \neq 0$. Then

$$\frac{a}{a^2 + b^2} + \frac{-b}{a^2 + b^2} i$$

is a well-defined complex number, and

$$(a + bi)\left(\frac{a}{a^2 + b^2} + \frac{-b}{a^2 + b^2} i\right) = \left(\frac{a^2 + b^2}{a^2 + b^2}\right) + \left(\frac{-ab + ba}{a^2 + b^2}\right)i = 1,$$

$$\left(\frac{a}{a^2 + b^2} + \frac{-b}{a^2 + b^2} i\right)(a + bi) = \left(\frac{a^2 + b^2}{a^2 + b^2}\right) + \left(\frac{-ab + ba}{a^2 + b^2}\right)i = 1. \ \blacksquare$$

Note that associating the real number a and the complex number $a + 0i$ embeds the real numbers in the complex numbers in such a way that addition and multiplication of real numbers is the same as addition and multiplication of the corresponding complex numbers. This justifies our use of the notation a for $a + 0i$, bi for $0 + bi$, and $a - bi$ for $a + (-b)i$. And for this reason, in the sequel, we will always treat $\mathbb{R}$ as a subset of $\mathbb{C}$.

Note as well that, as we indicated before,

$$i^2 = (0 + i)(0 + i) = (0 - 1) + (0 + 0)i = -1.$$

In practice, it is usually easiest to multiply complex numbers by combining this observation with the distributive laws, rather than by using the definition. For example,

$$(2 + 3i)(1 + 5i) = 2(1 + 5i) + 3i(1 + 5i) = 2 \cdot 1 + 2 \cdot 5i + 3 \cdot 1i + 3 \cdot 5i^2$$
$$= (2 \cdot 1 - 3 \cdot 5) + (2 \cdot 5 + 3 \cdot 1)i = -13 + 13i.$$

We may divide complex numbers in several ways. The preceding proof supplies a formula for the multiplicative inverse of the nonzero complex number $a + bi$:

$$(a + bi)^{-1} = \frac{a}{a^2 + b^2} - \frac{b}{a^2 + b^2}i,$$

and we may use this formula to divide $c + di$ by $a + bi$:

$$\frac{c + di}{a + bi} = (c + di)(a + bi)^{-1} = \frac{(c + di)(a - bi)}{a^2 + b^2}.$$

An alternative method is to use an associated complex number called the complex conjugate.

DEFINITION

Let $z = a + bi \in \mathbb{C}$. The **complex conjugate** of z is the complex number

$$\bar{z} = \overline{a + bi} = a - bi.$$

We can use the complex conjugate to divide two complex numbers in the following way: multiply the quotient of two numbers written in standard form by 1 written in the form of the complex conjugate of the denominator over itself.

Example 3.2   We have

$$\frac{2 - 3i}{4 + i} = \frac{2 - 3i}{4 + i} \cdot \frac{4 - i}{4 - i} = \frac{(8 - 3) + (-2 - 12)i}{16 + 1} = \frac{5}{17} - \frac{14}{17}i. \quad ❖$$

We observe in passing that the operation of conjugation behaves very well with respect to multiplication and addition.

Proposition 3.3   *If $z, w \in \mathbb{C}$, then $\bar{z}\,\bar{w} = \overline{zw}$,   and $\bar{z} + \bar{w} = \overline{z + w}$.*

Proof   The proof is left to the reader (see Exercise 46). ∎

FIGURE 3.1

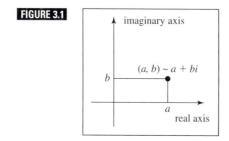

As previously noted, we can construct a picture of $\mathbb{C}$ by associating each $a + bi \in \mathbb{C}$ with the point (a,b) in the Euclidean plane $\mathbb{R}^2$ (as shown in Figure 3.1). When we think of complex numbers in this way, as points in the plane, we refer to $\mathbb{C}$ as the complex plane This identification allows us to view complex numbers geometrically.

If we further associate each point in the complex plane with its corresponding vector (as shown in Figure 3.2), then we can see that addition of complex numbers corresponds to addition of vectors:

$$(a + bi) + (c + di) = (a + c) + (b + d)i;$$
$$(a,b) + (c,d) = (a + c, b + d).$$

FIGURE 3.2

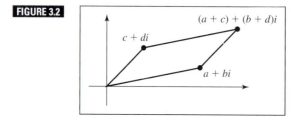

The complex conjugate of $a + bi$ is very easy to find on the complex plane; it is merely the reflection of $a + bi$ about the x-axis (as shown in Figure 3.3).

FIGURE 3.3

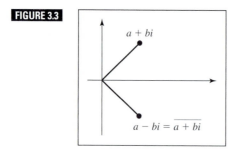

We can interpret multiplication, as well as addition, of complex numbers geometrically. To do this, recall that a nonzero vector (a, b) may be expressed in terms of its polar coordinates r (length) and Θ (angle). The analogous coordinates for complex numbers are the modulus and the argument. (See Figure 3.4.)

DEFINITION

Let $0 \neq a + bi \in \mathbb{C}$.

    (a)   The **modulus** of $a + bi$ is $|a + bi| = \sqrt{a^2 + b^2}$.

    (b)   If $a + bi \neq 0$, then the **argument** of $a + bi$, denoted $\arg(a + bi)$, is the angle in the appropriate quadrant whose tangent is $\frac{b}{a}$.

FIGURE 3.4

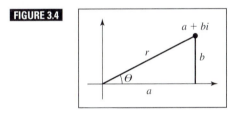

Whenever possible, the argument should be expressed as an angle between 0 and 2π radians. Note that the square of the modulus of a complex number is just the product of the number and its conjugate (see Exercise 33): $|z|^2 = z\bar{z}$. Note also that care must be taken to identify the proper quadrant for $\arg(a + bi)$. For example, if $a < 0$ and $b > 0$, then $\frac{\pi}{2} < \arg(a + bi) < \pi$, and if $a > 0$ and $b < 0$, then $\frac{3\pi}{2} < \arg(a + bi) < 2\pi$.

Example 3.4   We have

$$|1 + i| = \sqrt{1^2 + 1^2} = \sqrt{2}, \quad \frac{1}{1} = 1,$$

and $\arg(1 + i) = \frac{\pi}{4}$, while

$$|-1 - i| = \sqrt{(-1)^2 + (-1)^2} = \sqrt{2}, \quad \frac{-1}{-1} = 1,$$

but $\arg(-1 - i) = \frac{5\pi}{4}$ (see Figure 3.5). ❖

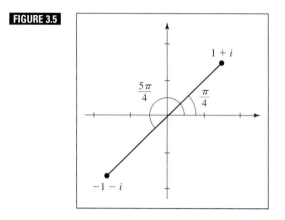

FIGURE 3.5

As indicated, the modulus and argument may be used to interpret multiplication of complex numbers geometrically. To describe this interpretation, let $z = a + bi$, $r = |z|$, and $\Theta = \arg(z)$. Observe that, as with polar coordinates, $a = r\cos\Theta$ and $b = r\sin\Theta$ (see Figure 3.4), and hence we may write

$$z = a + bi = r\cos\Theta + r\sin\Theta\, i = r(\cos\Theta + i\sin\Theta).$$

So if $z = r(\cos\Theta + i\sin\Theta)$ and $w = s(\cos\Psi + i\sin\Psi)$, then we may use the familiar formulas for the cosine and sine of the sum of two angles to see that

$$\begin{aligned}
zw &= r(\cos\Theta + i\sin\Theta)\, s(\cos\Psi + i\sin\Psi) \\
&= rs((\cos\Theta\cos\Psi - \sin\Theta\sin\Psi) + i(\cos\Theta\sin\Psi + \sin\Theta\cos\Psi)) \\
&= rs(\cos(\Theta + \Psi) + i\sin(\Theta + \Psi))
\end{aligned}$$

and therefore that $|zw| = |z|\,|w|$ and $\arg(zw) = \arg(z) + \arg(w)$. We may summarize these observations as follows.

Proposition 3.5   *The modulus of a product of complex numbers is the product of its moduli; and the argument of a product is the sum of its arguments.*

Example 3.6   Consider the complex numbers $1 + \sqrt{3}i$ and $-2\sqrt{3} + 2i$. Then

$$(1 + \sqrt{3}i)(-2\sqrt{3} + 2i) = (-2\sqrt{3} - 2\sqrt{3}) + (2 - 6)i = -4\sqrt{3} - 4i,$$

and we have:

$$|1 + \sqrt{3}i| = 2, \qquad\qquad \Theta = \arg(1 + \sqrt{3}i) = \frac{\pi}{3}$$

$$|2\sqrt{3} + 2i| = 4, \qquad\qquad \Psi = \arg(-2\sqrt{3} + 2i) = \frac{5\pi}{6}$$

$$|-4\sqrt{3} - 4i| = 8 = 2\cdot 4, \qquad Y = \arg(-4\sqrt{3} - 4i) = \frac{7\pi}{6} = \frac{\pi}{3} + \frac{5\pi}{6}.$$

Figure 3.6 shows these numbers plotted on the complex plane. ❖

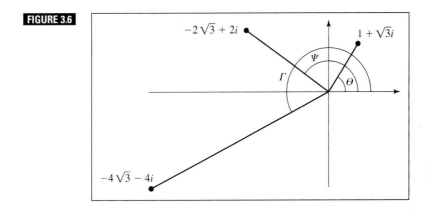

FIGURE 3.6

We have observed that multiplying complex numbers means adding arguments. Since powers are a special kind of product, we may apply this observation in the following special situation.

Proposition 3.7 **De Moivre's Theorem**

For any positive integer n,

$$(\cos\Theta + i\sin\Theta)^n = \cos(n\Theta) + i\sin(n\Theta).$$

Proof We proceed by induction on n, where $P(n)$ is the equality $(\cos\Theta + i\sin\Theta)^n = \cos(n\Theta) + i\sin(n\Theta)$.

(*i*) For $n = 1$, clearly the equality holds.

(*ii*) Suppose that k is a positive integer and that $P(k)$ holds, i.e., that $(\cos\Theta + i\sin\Theta)^k = \cos(k\Theta) + i\sin(k\Theta)$. Then, by the induction hypothesis and Proposition 3.5,

$$\begin{aligned}
(\cos\Theta + i\sin\Theta)^{k+1} &= (\cos\Theta + i\sin\Theta)^k (\cos\Theta + i\sin\Theta) \\
&= (\cos(k\Theta) + i\sin(k\Theta))(\cos\Theta + i\sin\Theta) \\
&= \cos((k+1)\Theta) + i\sin((k+1)\Theta),
\end{aligned}$$

and therefore De Moivre's theorem holds by induction. ■

De Moivre's theorem is obviously useful for computation; it is also useful for finding roots.

DEFINITION

Let $\alpha \in \mathbb{C}$. An ***nth root*** of α in $\mathbb{C}$ is a complex number z such that $z^n = \alpha$.

Note that z is an nth root of α if and only if z solves the equation $x^n - \alpha = 0$. Thus if we have found all nth roots of α, we have simultaneously found all solutions of the equation $x^n - \alpha = 0$. We will list all such nth roots, and hence all such solutions, in Proposition 3.8.

To phrase Propositon 3.8, we single out some special complex numbers. For any positive integer n, let

$$\zeta_n = \cos\frac{2\pi}{n} + i\sin\frac{2\pi}{n}.$$

These complex numbers all lie on the unit circle, $\frac{1}{n}$th of the way around it (see Figure 3.7). In particular, $\zeta_2 = -1$ is half way around the circle and $\zeta_4 = i$ is a quarter of the way around the circle.

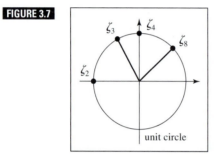

FIGURE 3.7

unit circle

The numbers ζ_n will be <u>extremely important</u> in much of what follows. Note in particular that by De Moivre's theorem, $(\zeta_n)^n = \cos 2\pi + i\sin 2\pi$, i.e., ζ_n is an nth root of 1, or in symbols

$$(\zeta_n)^n = 1.$$

As well, De Moivre's theorem allows us to express all the nth roots of a complex number in terms of any particular nth root and the numbers ζ_n as follows.

Propositon 3.8   *Let n be a positive integer and let α, β be nonzero complex members such that $\beta^n = \alpha$. Then the set*

$$\{\beta\zeta_n, \beta(\zeta_n)^2, \beta(\zeta_n)^3, \ldots, \beta(\zeta_n)^{n-1}, \beta\}$$

is the set of all nth roots of α in $\mathbb{C}$.

Proof   As we have observed, $(\zeta_n)^n = 1$, and hence for any positive integer k,

$$(\beta(\zeta_n)^k)^n = \beta^n((\zeta_n)^n)^k = \beta^n = \alpha.$$

Thus, every element of the given set is an nth root of α. Conversely, suppose that $z^n = \alpha$, and let $z = r(\cos \Psi + i \sin \Psi)$ and $\beta = s(\cos \Phi + i \sin \Phi)$. Then by De Moivre's theorem,

$$s^n(\cos(n\Phi) + i \sin(n\Phi)) = \beta^n = \alpha = z^n = r^n(\cos(n\Psi) + i \sin(n\Psi)).$$

This equation implies that $r^n = s^n$, and hence, since s and r are both positive real numbers, that $r = s$. This equation also implies that there exists $m \in \mathbb{Z}$ such that $n\Psi = n\Phi + 2m\pi$. Then $\Psi = \Phi + m\frac{2\pi}{n}$ and hence, by Proposition 3.5 and De Moivre's theorem,

$$
\begin{aligned}
z &= r(\cos \Psi + i \sin \Psi) = s\left(\cos\left(\Phi + m\frac{2\pi}{n}\right) + i \sin\left(\Phi + m\frac{2\pi}{n}\right)\right) \\
&= s(\cos \Phi + i \sin \Phi)\left(\cos\left(m\frac{2\pi}{n}\right) + i \sin\left(m\frac{2\pi}{n}\right)\right) \\
&= \beta\left(\cos\frac{2\pi}{n} + i \sin\frac{2\pi}{n}\right)^m = \beta(\zeta_n)^m.
\end{aligned}
$$

But by the division algorithm (Theorem 1.3), there exist $q, r \in \mathbb{Z}$ such that $m = nq + r$ and $0 \le r < n$, and hence

$$z = \beta(\zeta_n)^m = \beta(\zeta_n)^{nq+r} = \beta(\zeta_n)^{nq}(\zeta_n)^r = \beta(\zeta_n)^r.$$

Since $0 \le r < n$, we conclude that z must be a member of the given set. ∎

As we observed earlier, in listing all the nth roots of α, Proposition 3.8 simultaneously lists all solutions of the equation $x^n - \alpha = 0$.

A secondary consequence of Proposition 3.8 is that the notation $\sqrt[8]{137}$ or $\sqrt[5]{1 + i}$ is ambiguous in $\mathbb{C}$, because according to Proposition 3.8, there are eight possible meanings for $\sqrt[8]{137}$ and five possible meanings for $\sqrt[5]{1 + i}$. For real numbers r and positive integers n, we remove this ambiguity by adopting the usual conventions for $\sqrt[n]{r}$, conventions exemplified by the following:

$\sqrt[3]{-5}$ denotes the unique real number α such that $\alpha^3 = -5$;
$\sqrt{5}$ denotes the positive real number α such that $\alpha^2 = 5$;
$\sqrt{-5}$ denotes the complex number $i\sqrt{5}$.

We avoid using the notation $\sqrt[n]{z}$ in all other cases.

Note that Proposition 3.8 tells how to find all nth roots once one is known; it does not tell us how to find an initial nth root. However, the same sort of reasoning as that used in the proof of the proposition may be used to find this root as well.

Example 3.9   Suppose we want to find all the cube roots of the complex number $\alpha = -1 + i$. Writing α in terms of its modulus and argument, we have

$$\alpha = \sqrt{2}\left(-\frac{1}{\sqrt{2}} + i\frac{1}{\sqrt{2}}\right) = \sqrt{2}\left(\cos\frac{3\pi}{4} + i\sin\frac{3\pi}{4}\right).$$

Since we are interested in the cube roots, we extract the (real) cube root of the modulus, $\sqrt{2}$, and divide the argument, $\frac{3\pi}{4}$, by three:

$$s = \sqrt[3]{\sqrt{2}} = \sqrt[6]{2}, \quad\text{and}\quad \Phi = \frac{1}{3}\frac{3\pi}{4} = \frac{\pi}{4}.$$

We then take our initial root β to be

$$\beta = s(\cos\Phi + i\sin\Phi) = \sqrt[6]{2}\left(\frac{1}{\sqrt{2}} + i\frac{1}{\sqrt{2}}\right).$$

By direct calculation or De Moivre's theorem, $\beta^3 = \alpha$, i.e., β is a cube root of α. By Proposition 3.7, the complex cube roots of α are then

$$\beta = \sqrt[6]{2}\left(\cos\frac{\pi}{4} + i\sin\frac{\pi}{4}\right) = \sqrt[6]{2}\left(\frac{1}{\sqrt{2}} + i\frac{1}{\sqrt{2}}\right),$$

$$\beta\zeta_3 = \sqrt[6]{2}\left(\cos\frac{11\pi}{12} + i\sin\frac{11\pi}{12}\right),$$

$$\beta(\zeta_3)^2 = \sqrt[6]{2}\left(\cos\frac{19\pi}{12} + i\sin\frac{19\pi}{12}\right). \quad \diamondsuit$$

Example 3.10   According to De Moivre's theorem (Proposition 3.7),

$$(\zeta_9)^3 = \left(\cos\frac{2\pi}{9} + i\sin\frac{2\pi}{9}\right)^3 = \cos\frac{2\pi}{3} + i\sin\frac{2\pi}{3} = \zeta_3$$

and hence ζ_9 is a cube root of ζ_3. Thus by Proposition 3.8, the complex cube roots of ζ_3 are

$$\zeta_9, \quad \zeta_9\zeta_3 = \zeta_9(\zeta_9)^3 = (\zeta_9)^4 = (\zeta_9)^4, \quad\text{and}\quad \zeta_9(\zeta_3)^2 = \zeta_9(\zeta_9)^6 = (\zeta_9)^7.$$

We can use these roots to simplify the three real solutions of $x^3 - 3x + 1 = 0$, which we found in Example 2.1. For in that example, we showed by using Cardano's method that if $w^3 = -\frac{1}{2} + \frac{\sqrt{-3}}{2}$ then $x = w + \frac{1}{w}$ solves this equation. But if we write ζ_3 in the form $a + bi$, we find that $\zeta_3 = -\frac{1}{2} + i\frac{\sqrt{3}}{2} = -\frac{1}{2} + \frac{\sqrt{-3}}{2}$, and therefore if w is any cube root of ζ_3, then $x = w + \frac{1}{w}$ is a solution of $x^3 - 3x + 1 = 0$. So the solutions of this equation are

$$x = \zeta_9 + \frac{1}{\zeta_9}, \quad x = (\zeta_9)^4 + \frac{1}{(\zeta_9)^4}, \quad\text{and}\quad x = (\zeta_9)^7 + \frac{1}{(\zeta_9)^7}.$$

If $\alpha = \zeta_9$, then, since $\zeta_9(\zeta_9)^8 = (\zeta_9)^9 = 1$, we have $\frac{1}{\alpha} = \frac{1}{\zeta_9} = (\zeta_9)^8$, and thus one solution is $x = \zeta_9 + (\zeta_9)^8$. We can simplify this solution further by noting that $\zeta_9 = \cos\frac{2\pi}{9} + i\sin\frac{2\pi}{9}$ and hence, by De Moivre's theorem,

$$(\zeta_9)^8 = \cos\frac{16\pi}{9} + i\sin\frac{16\pi}{9} = \cos\left(-\frac{2\pi}{9}\right) + i\sin\left(-\frac{2\pi}{9}\right) = \cos\frac{2\pi}{9} - i\sin\frac{2\pi}{9}.$$

It then follows that the solution $x = \zeta_9 + (\zeta_9)^8 = 2\cos\frac{2\pi}{9}$ is in $\mathbb{R}$.

Note that according to these calculations, $\frac{1}{\zeta_9} = (\zeta_9)^8 = \cos\left(-\frac{2\pi}{9}\right) + i\sin\left(-\frac{2\pi}{9}\right) = \overline{\zeta_9}$ (see Figure 3.8).

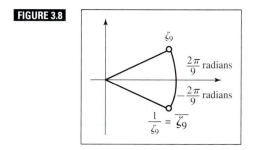

Geometrically this says that the multiplicative inverse of ζ_9 may be found by going backward around the unit circle $\frac{2\pi}{9}$ radians to the complex number $\overline{\zeta_9}$. As well this says that, in agreement with Example 2.1, the first solution may be written $x = \zeta_9 + \overline{\zeta_9}$ (cf. Exercise 51).

For $\alpha = (\zeta_9)^4$, a similar argument shows that the solution

$$x = (\zeta_9)^4 + \frac{1}{(\zeta_9)^4} = (\zeta_9)^4 + (\zeta_9)^5 = 2\cos\frac{8\pi}{9}$$

is in $\mathbb{R}$, and for $\alpha = (\zeta_9)^7$, that the corresponding solution is

$$x = (\zeta_9)^7 + \frac{1}{(\zeta_9)^7} = (\zeta_9)^7 + (\zeta_9)^2 = 2\cos\frac{14\pi}{9},$$

which is also in $\mathbb{R}$ (cf. Exercise 52).

Note as well that in agreement with Example 2.1, these solutions may also be written

$$x = (\zeta_9)^4 + \frac{1}{(\zeta_9)^4} = (\zeta_9)^4 + (\overline{\zeta_9})^4,$$

and

$$x = (\zeta_9)^7 + \frac{1}{(\zeta_9)^7} = (\zeta_9)^7 + (\overline{\zeta_9})^7. \; \diamondsuit$$

If our goal were merely to determine whether polynomial equations have solutions in the complex numbers, our next step would be to consider arbitrary polynomial equations with coefficients in $\mathbb{C}$, and eventually we would arrive at the following theorem.

Theorem 3.11   **Fundamental Theorem of Algebra (Gauss et al.)**

If $p(z)$ is a nonconstant polynomial with complex coefficients, then the equation $p(z) = 0$ has a complex solution.

Theorem 3.11 establishes the existence of a solution for every polynomial equation with complex coefficients but says nothing about the method by which a solution is to be found. And it is the method that concerns us here. Furthermore, during the nineteenth and twentieth centuries, this theorem was superseded by much more general results; it is called "the fundamental theorem of algebra" only as an historical artifact. Considering that this theorem is secondary to our major goal and that its proof requires sophisticated techniques, be they algebraic or analytic, we omit it (cf. Chapter IV, §3, pages 286–292 of *Mathematics: Its Content, Methods, and Meaning,* Volume 1). However, Theorem 3.11 does simplify the study of polynomials with rational coefficients because it ensures that such a polynomial always has a complete set of roots in $\mathbb{C}$ (Proposition 14.1). So for this reason, and because the complex numbers are familiar and form a very concrete setting for the construction of examples, we will sometimes restrict ourselves to the complex numbers and use Theorem 3.11 whenever it is convenient. Note that this means that some of our results will not be stated as generally as possible.

As previously noted, one consequence of this result (cf. Proposition 14.1) is that the polynomial $p(z)$ may be written

$$p(z) = a(z - c_1)(z - c_2) \cdots (z - c_n),$$

where $c_1, c_2, \ldots, c_n$ are the (not necessarily distinct) complex solutions of $p(z) = 0$. By Proposition 3.3, the coefficients of the polynomial

$$\bar{p}(z) = \bar{a}(z - \bar{c_1})(z - \bar{c_2}) \cdots (z - \bar{c_n})$$

are the complex conjugates of the corresponding coefficients of $p(z)$, and hence if $p(z)$ has real coefficients, then $\bar{p}(z) = p(z)$. In this case, for any solution c_i of the equation $p(z) = 0$, $\bar{c_i}$ is a solution of the equation $\bar{p}(z) = 0$ and hence, since $\bar{p}(z) = p(z)$, of the equation $p(z) = 0$ as well. These observations may be summarized as follows.

Proposition 3.12   *If $p(x)$ is a polynomial with real coefficients and c is a complex solution of the equation $p(x) = 0$, then $\bar{c}$ is also a solution of the equation $p(x) = 0$.*

Exercises

In Exercises 1–8, write the given complex number in the form $a + bi$.

1. $\dfrac{2i}{7 + 8i}$

2. $\dfrac{(2 - i)(1 + 2i)}{2 + 3i}$

3. $\dfrac{4i(2 + 5i)}{(1 + i)(3 - 2i)}$ **4.** ζ_4

5. $2\zeta_8$ **6.** $4(\zeta_4)^2$

7. $(\zeta_6)^5$ **8.** $\dfrac{(1 + i)(2 - 3i)}{(3 - i)(2 + i)}(2\zeta_9)^3$

In Exercises 9–22, find the modulus and argument of the given complex number and plot it on the complex plane.

9. $-i$ **10.** $-\sqrt{3} + i$

11. $-1 - i$ **12.** $2 - 2i\sqrt{3}$

13. $-2 + 2i$ **14.** ζ_1

15. $-\zeta_4$ **16.** $(\zeta_5)^4$

17. $\zeta_3(\zeta_5)^4$ **18.** $(1 + i)(5\zeta_3)^2$

19. $(\zeta_3)(-1 + i\sqrt{3})(4\zeta_6)^3$ **20.** $(3\zeta_7)^2(1 - i)(\zeta_5)$

21. $(-\sqrt{3} - i)(2\zeta_5)^3(1 + i)(3\zeta_{10})$ **22.** $(-1 - i)(\zeta_9)^3(2 + 2i)(2\zeta_6)$

In Exercises 23–30, write the given complex number in the form $r(\cos\Theta + i\sin\Theta)$.

23. $-3i$ **24.** $1 - i\sqrt{3}$

25. $-3 + 3i$ **26.** $5\zeta_7$

27. $7(\zeta_{11})^9$ **28.** $\zeta_7(\zeta_3)^5$

29. $(\sqrt{3} - i)(2\zeta_6)^4$ **30.** $(-6 + 6i)(\zeta_7)^4(2\zeta_4)$

31. Calculate $\left[-\dfrac{1}{2} + \dfrac{\sqrt{3}}{2}i\right]^2$ and $\left[-\dfrac{1}{2} + \dfrac{\sqrt{3}}{2}i\right]^3$. Interpret your results geometrically.

32. Show by direct calculation that ζ_3 and $(\zeta_3)^2$ are solutions of the equation $x^2 + x + 1 = 0$.

33. Show that for any $z \in \mathbb{C}$, $|z|^2 = z\bar{z}$.

34. Find all fourth roots of $-1 + i\sqrt{3}$.   **35.** Find all fifth roots of $-1 - i$.

36. Find all fourth roots of ζ_3.   **37.** Find all the cube roots of $\sqrt{3} - i$.

38. Find all sixth roots of $\sqrt{3} + i\sqrt{3}$.   **39.** Find all cube roots of ζ_5.

40. Show that $\zeta_2 = -1$.   **41.** Show that $\zeta_4 = i$.

42. Show that $\zeta_6^2 = \zeta_3$.   **43.** Show that $\zeta_{nk}{}^n = \zeta_k$.

44. Suppose that, for a polynomial $p(z)$ with rational coefficients, $p(-1 + 2i) = 0$. Find another complex number w such that $p(w) = 0$.

45. Suppose that, for a polynomial $f(z)$ with rational coefficients, $f(-7 + i\sqrt{3}) = 0$. Find another complex number w such that $f(w) = 0$.

46. Prove Proposition 3.3: If $z, w \in \mathbb{C}$, then $\bar{z}\,\bar{w} = \overline{zw}$, and $\bar{z} + \bar{w} = \overline{z + w}$.

47. Show by direct computation that if $z = a + bi \in \mathbb{C}$, then $z + \bar{z} = 2a$.

48. Use a geometric argument to show that for any $z = a + bi \in \mathbb{C}$, $z + \bar{z} = 2a$.

49. Show that multiplication of complex numbers is commutative.

50. Show that multiplication of complex numbers is associative.

51. Show that if z is a complex number of modulus 1, then $z^{-1} = \bar{z}$.

52. Show that if z is a complex number of modulus 1, then $z + z^{-1} \in \mathbb{R}$. (*Hint:* Use Exercise 51.)

53. (This exercise will be <u>very</u> useful for constructing examples.) Let $k, n \in \mathbb{Z}$. Show that if $1 \le k < n$, then $(\zeta_n)^k$ solves

$$x^{n-1} + x^{n-2} + \cdots + x + 1 = 0.$$

(*Hint:* First show that $x^n - 1 = (x^{n-1} + x^{n-2} + \cdots + x + 1)(x - 1)$ and that $(\zeta_n)^k$ solves $x^n - 1 = 0$.)

A complex number u such that $u^n = 1$ but $u^r \ne 1$ for any $0 < r < n$ is called a <u>primitive nth root of unity</u>. Exercises 54–56 concern these complex numbers.

54. Show that ζ_n is a primitive nth root of unity.

55. Show that if u is a primitive nth root of unity, then $u = \zeta_n^k$ for some positive integer k.

56. Show that the primitive nth roots of unity are precisely the numbers ζ_n^k, where $0 < k < n$ and the greatest common divisor of n and k is 1.

HISTORICAL NOTE

Highlights in the Development of the Complex Numbers

"For every equation, one can imagine as many roots as its degree indicates, but in many cases no quantity exists which corresponds to what one imagines."[1] So wrote the philosopher/mathematician René Descartes in 1637, simultaneously explaining why the complex numbers are both necessary (to state the fundamental theorem of algebra) and suspicious (because the nonreal complex numbers can exist only in the imagination).

These observations were not original to Descartes. From the time of the Babylonians (2000 to 600 B.C.) until the breakthroughs of Cardano and his contemporaries (ca. 1500 A.D.), quadratic equations had been classified in such a way that real solutions always existed: $x^2 + px = q$, $x^2 = px + q$, and $x^2 + q = px$, where p and q are both positive. Mathematicians had been content to label other quadratic equations as "unsolvable" and to ignore any nonreal solutions. However, in his *Algebra* of 1572, Bombelli pointed out that Cardano's method gives $x = \sqrt[3]{2 + \sqrt{-121}} + \sqrt[3]{2 - \sqrt{-121}}$ as a solution of $x^3 = 15x + 4$, whereas $x = 4$ was known to be the only positive solution. Expanding $(2 + \sqrt{-1})^3$, he showed that $\sqrt[3]{2 + \sqrt{-121}} = 2 + \sqrt{-1}$ and showed similarly that $\sqrt[3]{2 - \sqrt{-121}} = 2 - \sqrt{-1}$; he then concluded that $\sqrt[3]{2 + \sqrt{-121}} + \sqrt[3]{2 - \sqrt{-121}} = 4$, a "satisfying," if "sophisticated," result. (Cardano referred to such calculations as "mental torture."[2]) By the time of Descartes, examples such as this had led mathematicians to realize that it was possible at least to "imagine" solutions to all quadratic equations and perhaps even to all polynomial equations.

In this way, the work of Cardano and his contemporaries forced mathematicians to come to terms with square roots of negative numbers. Many mathematicians contributed to the subsequent clarification of the complex numbers and the proofs of the key theorems, and the biographies of a few of these people follow. In the seventeenth century, Albert Girard enunciated without proof the fundamental theorem of algebra, and then, in the middle of the eighteenth century, Jean Le Rond d'Alembert and Leonhard Euler separately published proofs of the theorem, both of which, however, had similar lacunae. Early in the eighteenth century, Abraham De Moivre proved his eponymous theorem by building on work of Roger Cotes. And at the beginning of the nineteenth century, Caspar Wessel and Jean-Robert Argand independently discovered the complex plane. It was left for Carl Friedrich Gauss in the nineteenth century to bring all these strands together by presenting the complex numbers as points on the complex plane and by giving four different proofs of the fundamental theorem of algebra. It should be emphasized that this list is only a sample: as the theory evolved, many other mathematicians not mentioned here also made important contributions.

Albert Girard was born in St. Mihiel, France, in the duchy of Lorraine, in 1595, and died on December 8, 1632, most likely in Leiden, the Netherlands. He was a member of the Reformed church and for that reason moved to the Netherlands, a country more hospitable to Protestants than France. It appears that he studied at Leiden and became an engineer in the army of Frederick Henry of Nassau, Prince of Orange, although in his publications he referred to himself only as a mathematician. In a posthumously published edition of his works, he complained of living in a foreign country, of lacking a patron, and of the hardships engendered by his large family. His widow explained in her preface to the book that the only inheritance to which she and her eleven children could lay claim was her late husband's faithful service and his dedication to discovering the "noble secrets of mathematics."[3]

Among Girard's mathematical innovations were the notation $\sqrt[3]{}$ and the abbreviations sin for sine, tan for tangent, and sec for secant. In 1629, he gave one of the first statements of the fundamental theorem of algebra and then observed, "One could say: What are these impossible solutions? I answer: For three things, for the certitude of the general rules, that there are no other solutions, and for their utility."[4]

Jean Le Rond d'Alembert was born in Paris, France, on November 17, 1717, and died in Paris on October 29, 1783. His mother, Madame de Tencin, a famous and vivacious Salon hostess, had previously abandoned her nun's vows and feared that a child might force her to return to a convent. So, soon after her son was born, she abandoned him on the steps of the church of St. Jean Baptiste le Rond, an establishment from which Jean Le Rond subsequently took his name. His father, a cavalry officer, the Chevalier Destouches-Canon, eventually located his son and arranged for a glazier and his wife to serve as foster parents, a couple with whom Jean Le Rond lived until he was forty-seven. When his mother attempted a reconciliation after he had become famous, he rejected her overtures because he preferred to be recognized as the son of his humble foster parents.

Unlike d'Alembert's mother, his father did not abandon him; on the contrary, he ensured that his son was educated at the Collège de Quatre-Nations, a Jansenist school which, together with classics and rhetoric, also offered an exceptionally thorough course in mathematics. His teachers wanted him to enter the church but he decided to take a secular path, first by studying law and medicine and later turning to mathematics.

In 1739, he announced his entry into Parisian intellectual life with the publication of several memoirs dealing with differential equations and the motions of bodies in resisting media. He was invited to join the Académie des Sciences in 1741, and in 1743 he published his most famous scientific work, the *Traité de dynamique*. While Newton had stated his three laws of motion verbally, d'Alembert gave algebraic formulations for his three laws and later in the book introduced his eponymous principle concerning change of inertial motion. In 1746, he published a geometric proof of the fundamental theorem of algebra which gave the form taken by the solutions of a polynomial equation but neglected to show that such solutions always exist. Indeed, d'Alembert never completely escaped from the thrall of geometry, and although he anticipated Cauchy's later advances by realizing that the differential was a limit, he never phrased this observation in algebraic terms.

In 1751, Diderot picked d'Alembert to be the science editor of the *Encyclopédie*, and d'Alembert eventually wrote the "Discours preliminaire" and many of the scientific articles as well. His interests were not limited to mathematics and physics. In 1752, he published *Éléments de musique théorique et pratique suivant les principes de M. Rameau* in which he popularized Rameau's secular, decidedly nonmystical ideas concerning the structure of music.

D'Alembert traveled little. He left France only once, in 1764, when he spent three months at the court of Frederick the Great. Frederick wanted him to head the Prussian Academy, but d'Alembert refused on the grounds that such an appointment would put him in a position superior to that of Euler, a situation he thought would be unfair. Catherine the Great of Russia, Frederick's daughter, invited d'Alembert to become a tutor to her son but again he refused, despite the enormous salary she offered.

In general, d'Alembert spent his days not unlike the other philosophes of the time. He worked in the morning and in the afternoon. And although his frame was slight and he had a high-pitched voice, his wit and gifts for mimicry and conversation ensured that in the evening he could be found in the salons, particularly those of Mme. de Deffand and Mlle. de Lespinasse. In fact, although he never married, he spent most of his life with Mlle. de Lespinasse. When he became ill in 1765, he moved into her house and she nursed him back to health; he remained there until her death in 1776. Earlier, in 1772, he had been elected perpetual secretary of the Académie Francaise. But during his tenure, the Académie produced nothing exceptional and d'Alembert spent much of his time writing obituaries as, one by one, his contemporaries passed away. The death of Mlle. de Lespinasse, his one true love, devastated him, and his last years were full of frustration and despair.

Leonhard Euler was born in Basel, Switzerland, on April 15, 1707, and died in St. Petersburg, Russia, on September 18, 1783. His father, Paul Euler, was a Calvinist minister who was fond of mathematics; his mother, born Margarete Brucker, was the daughter of another minister. Leonhard grew up in the village of Riehen, near Basel.

When he was growing up, Euler studied mathematics privately with a local amateur mathematician, and when he entered the University of Basel in 1720 he approached Johann Bernoulli to continue private tutoring. However, as Euler later remembered, Bernoulli "was very busy and so refused flatly to give me private lessons; but he gave me much more valuable advice to start reading more difficult mathematical books on my own and to study them as diligently as I could . . . and this, undoubtedly, is the best method to succeed in mathematical subjects."[5] Bowing to his father's wishes, Euler entered the department of theology in 1723 but with little success. He was spending too much of his time on mathematics and indeed in 1726 published his first paper.

Simultaneously, in Russia, Catherine I was fulfilling one of the wishes of her late husband, Peter the Great, by creating the St. Petersburg Academy of Sciences. Johann Bernoulli's sons, Nikolaus and Daniel, had recently gone there and they persuaded Catherine to offer a position to Euler as well. Faced with few opportunities in Switzerland, Euler accepted and left for St. Petersburg on April 5, 1727. His appointment was in physiology but he soon became an adjunct member of the Academy in mathematics. Although Catherine had founded the Academy, she was loath to fund it properly and as a result, by 1730, Euler was close to bankrupt. However, she died that year, and her successor, Anna Romanova, increased the Academy's funding.

A few years later, in 1733, when Daniel Bernoulli returned to Basel to become professor of mathematics, Euler was chosen to replace him in St. Petersburg, and soon after, at the end of the year, he married Katharina Gsell. Like Euler, Katharina was Swiss, her father having come to St. Petersburg to teach painting at the Gymnasium attached to the Academy. In 1738, Euler lost sight of his right eye due to disease, but otherwise his life continued pleasantly enough until the death of Anna

Romanova and the ascension in 1740 of Anna Leopoldovna, mother of Tzar Ivan VI, as Regent. The new Anna reversed the funding policies of the deceased Anna, but luckily as the purse strings were tightening in St. Petersburg, Euler received an invitation from Frederick the Great to join the newly rejuvenated Berlin Society of Sciences. He accepted the invitation and sailed for Berlin on June 19, 1741. He left Russia with his wife and two sons, and during his tenure in Berlin, Katharina had eleven more children, eight of whom died in infancy.

Euler had continued producing mathematics at an astounding rate during his time in St. Petersburg; by 1741, he had published fifty-five works and prepared many more. His work continued unabated in Berlin, and in 1749 he gave an entirely algebraic proof of the fundamental theorem of algebra. His proof, however, had the same fault as d'Alembert's and was therefore equally questionable. While he was in Berlin, he published another 275 works.

Euler's duties in Berlin were not merely scientific. He supervised the observatory and the botanical gardens and he managed various publications. He was asked to correct the level of a canal and even to improve the plumbing at the royal summer residence. He acquitted all his duties with aplomb but nonetheless, as time wore on, his relations with Frederick deteriorated. Euler was religious whereas Frederick was a freethinker; Frederick preferred practical mathematics and had little time for abstraction, while Euler's inclinations were just the opposite. By 1763, Euler was manager of the Academy but Frederick made it quite clear that he did not want him as president by offering that position to d'Alembert. Although d'Alembert declined the offer, relations between Euler and Frederick worsened as the king interfered more and more with the financial affairs of the Academy. Finally, in 1766, Euler left Berlin to return to St. Petersburg.

Soon after arriving in St. Petersburg, he developed a cataract in his left eye and, except for a few days after an operation in 1771, was completely blind for the rest of his life. When he had first returned, Catherine the Great had given him an elaborate house on the Neva. Just before the operation on his eye, this house burned to the ground and he was lucky to save himself and his manuscripts. And two years later, his wife, Katharina, died. However, he was not alone for long: three years after the death of Katharina, he married her half-sister, Salome Gsell. Throughout this turmoil, Euler's scientific work went on unabated; in fact, almost half his works were published after 1765.

Euler began September 18, 1783, in typical fashion: he gave a mathematics lesson, did some calculations, and discussed the recent discovery of the planet Uranus. At five o'clock, while having tea with a grandchild, he suffered a stroke. "I am dying,"[6] he said, and indeed six hours later he was dead.

Roger Cotes was born in Burbage, Leicestershire, England, on July 10, 1682, and died in Cambridge, England, on June 5, 1716. He was the second son of the Reverend Robert Cotes, the rector of Burbage, and Robert's second wife, the former Mary Chambers. The boy's mathematical ability was apparent from an early age, and he took private lessons from his uncle, the Reverend John Smith, rector of Gate-Burton on Lincolnshire.

Cotes graduated from Trinity College, Cambridge, in 1702, and in 1706, over the objections of Flamsteed, he became the college's first Plumian Professor. Plume's will had begun with the words "All this money I give and bequeath to erect an Observatory and to maintain a studious and learned Professor of Astronomy and Experimental Philosophy. . . ."[7] The observatory was duly built and Cotes lived there for the rest of his short life.

From 1709 to 1713, he edited the second edition of Newton's *Principia*. Relations between Cotes and Newton had begun cordially enough. However, as the work neared completion, Jean Bernoulli discovered (and communicated to Newton's rival Leibniz) an error in the first edition which had escaped both Cotes and Newton. Perhaps this is the reason that Newton omitted the following tribute, which appeared in the draft preface, from the preface of the printed edition: "In publishing all this, the very learned Mr. Roger Cotes, Professor of Astronomy at Cambridge, has been my collaborator: he corrected errors in the former edition and advised me to reconsider many points. Whence it came about that this edition is more correct than the former one."[8] In any case, on learning of Cotes's early death from fever, Newton had the courtesy to observe: "Had Cotes lived we might have known something."[9]

Cotes published only one paper, "Logometria," while he was alive. Its mention of a general method of factorization led Robert Smith, Cotes's cousin and successor as Plumian Professor, to search Cotes's papers which his untimely death had left in disarray. Smith found the method and published it posthumously with no proof. It anticipates De Moivre's theorem by observing that the factors of $x^\lambda \pm a^\lambda$ may be found by subdividing the circumference of the circle of radius a into λ parts.

Abraham De Moivre was born in Vitry-le-François, France, on May 26, 1667, and died in London, England, on November 27, 1754. He was a Protestant, the son of a provincial surgeon of modest means. He had an undistinguished education, first at the tolerant village Catholic school and later at the Protestant Academy at Sedan. When that school was closed because of its religious affiliation, he went to study at Saumur, and in 1684, he traveled to Paris to study mathematics under the supervision of Jacques Ozanam. However, in 1685, Louis XIV revoked the Edict of Nantes, thereby making Protestantism illegal, and according to some sources, De Moivre was imprisoned. What is certain is that he eventually emigrated to England and spent the rest of his life there. He was never able to obtain a position at a university and made his living as a private mathematics tutor. The astronomer Edmund Halley communicated De Moivre's first paper to the Royal Society in 1695 and gained him election in 1697. In his last year, 1754, he succumbed to lethargy and, according to a story of the time, declared that he had to sleep fifteen minutes more each day and that he would die when he slept the whole day through. He began that long sleep on November 27.

His premier mathematical work was a book on probability entitled *The Doctrine of Chances*. His premier mathematical discovery was the normal approximation to the binomial distribution, one of the most useful tools in all of probability theory. He wrote on other topics as well, and in 1730, he published the complex identity that bears his name. Curiously enough, the first published application of De Moivre's theorem was to prove the theorem of Cotes.

Caspar Wessel was born in Vestby, Norway, on June 8, 1745, and died in Copen-hagen, Denmark, on March 25, 1818. His grandfather was the pastor of the local parish and his father, Jonas Wessel, was a vicar in the parish; his mother was Maria Schumacher. Wessel attended the Christiana Cathedral School in Oslo and then the University of Copenhagen. He graduated in 1765 and began work as a cartographer. In 1778, he became survey superintendent and continued working as a cartographer and surveyor even after his retirement in 1805. Although he was frequently in finan-cial difficulty, his work was good enough to earn him a silver medal from the Dan-ish Academy and a knightship of Daneborg (in 1815).

In 1798, the Royal Danish Academy of Sciences published Wessel's only mathe-matical work, a work that he wrote in Danish and which he modestly labeled as "an attempt"[10] to define addition and multiplication of complex numbers represented as directed line segments. Because Scandinavia was a mathematical backwater at the time, his work went unnoticed until it was republished in a French translation in 1897.

Jean-Robert Argand was born in Geneva, Switzerland, on July 18, 1768, and died in Paris, France, on August 13, 1822. Little is known of his family: his parents were Jacques Argand and Èves Canac; he had a son who lived in Paris and a daughter who lived in Stuttgart. He made his living as a bookkeeper and was not a member of any mathematical organization.

Argand's mathematical work consists of several elaborations of his single mathe-matical discovery, the complex plane. He had a limited number of copies of his ini-tial results privately printed in 1806 but neglected to put his name in the book. So the matter might have ended there, with Argand in the same position as Wessel, if not for the honesty of J. F. Francais. Argand had shown his work to Legendre who in turn had mentioned it in a letter to Francais' brother. While going through his late brother's effects, Francais happened upon the letter, was intrigued by what he read there, developed the ideas further, and finally published his findings in 1813. He concluded that article with a plea to the unknown author to come forward and pub-lish any other results he had obtained. Argand responded by publishing eight more articles between 1813 and 1816, all in the journal in which Francais's article had appeared. He gave a fallacious proof of the fundamental theorem of algebra and asserted (correctly) that it held for complex as well as real coefficients.

Johann Friedrich Carl Gauss (known as Carl Friedrich) was born in Brunswick, Germany, on April 30, 1777, and died in Göttingen, Germany, on February 23, 1855. Gauss's grandparents had moved from the countryside into town and his parents had not climbed far on the social ladder; at best, they were minimally literate and knew basic arithmetic. Before her marriage, Gauss's mother had been a maid, and his father worked at a sequence of menial jobs, from a gardener to the treasurer of a small insurance fund. Gauss's teachers were consistently impressed with his abilities, and encouraged and helped him throughout his years as a student. In 1792, at the suggestion of E. A. W. Zimmermann, a professor at the local college and privy councilor to the Duke of Brunswick, the duke awarded Gauss a stipend that made him independent and continued with periodic increases until Napoleon's victory and the death of the duke in 1806.

From his earliest years, Gauss could calculate quickly and efficiently and his work typically began with extensive empirical investigation. He always worked independently and as time went on, he developed an intimate acquaintance with numbers and arithmetic that served him well throughout his life. When he entered the University of Göttingen in 1795, he finally had access to a good library and found that some of his discoveries were not new. However, in 1796, he found the result that gave him his first spectacular success: he proved that the regular 17-gon could be constructed with straightedge and compass alone. For the next four years, mathematical ideas came to him so quickly that he hardly had time to write them down.

In his doctoral dissertation of 1799, Gauss published the first of four proofs of the fundamental theorem of algebra. In this case, he gave an analytic proof that every polynomial with real coefficients can be factored into linear and quadratic factors; by including quadratic factors, he was able to dispense with the complex numbers. Ironically, while he rightly criticized the proofs of Euler and d'Alembert for including unproved assumptions about the existence of roots, his own proof included unproved assumptions about the branches of algebraic curves.

In 1798, Gauss returned to Brunswick, and in 1801, he published his famous treatise *Disquisitiones Arithmeticae*. It summarized previous work in a systematic way, solved many outstanding and difficult problems, and set the foundations of number theory for the future, introducing among other things modular arithmetic (and thereby equivalence relations). In the same year, 1801, a new planet, Ceres, had been observed but had disappeared behind the sun. Using the known observations and his phenomenal calculating ability, Gauss was able to predict correctly where it would emerge. One result of all this activity was an offer from St. Petersburg to become director of the Academy of Sciences. The Duke of Brunswick responded by raising Gauss's stipend and building him an observatory. So Gauss stayed in Brunswick.

In 1803, he met Johanna Osthof, the daughter of a prosperous tanner, whom he described as "exactly the kind of woman I have always wanted as a life companion."[11] He eventually proposed to her in a letter, but she had heard that he was engaged to someone else and neglected to answer. However, she finally accepted his entreaties and he became betrothed to "an angel who is almost too saintly for this world."[12] They were married on October 9, 1805, but four years later Johanna died soon after bearing their third child. Gauss was heartbroken; he "closed the angel eyes in which for five years I have found a heaven."[13] Less than a year later, he married Minna Waldeck, Johanna's best friend, but he was unable to regain a peaceful home life until 1831 when Minna died and his daughter Therese took over the household.

Then in 1806, the Duke of Brunswick lost both a battle and his life to Napoleon's Grande Armée. With the loss of his patron and friend, Gauss became worried about his livelihood, and in 1807, having carefully groomed himself for the position and with the support of several colleagues, he was appointed director of the observatory in Göttingen, a position he held until his death.

In 1816, he published his second and third proofs of the fundamental theorem of algebra. The second proof was a purely algebraic one which proceeded from two assumptions: (1) every polynomial of odd degree with real coefficients has a real root, and (2) every quadratic polynomial with complex coefficients has two complex roots. This proof is the one twentieth-century mathematicians adapted to much more general structures. The third proof was analytic but simpler than the second proof. Gauss published his fourth proof, which was a reworking of the first proof, in 1849. The geometric representation of complex numbers was implicit in his first proof of the fundamental theorem. However, it wasn't until 1831 that he published in detail his own description of the representation.

The poverty of Gauss's youth drove him to strive mightily for the security he desperately needed. He worked assiduously at mathematics, astronomy, geodesy, and physics, and seldom left Göttingen. Politically he was very conservative, his friendship with the duke making him a staunch nationalist and royalist. He believed in the priority of empiricism in science and did not adhere to the views of the idealist philosophers like Kant. Many found him cold and uncommunicative, his easiest dealings being with empirical scientists and technicians. He was a brilliant scientist who combined in one person the capabilities of many mathematicians.

References

1. van der Waerden, B. L. *A History of Algebra.* New York: Springer-Verlag, 1985, p. 177.
2. *Ibid.,* p. 177.
3. Itard, Jean. "Girard, Albert," *Dictionary of Scientific Biography,* 16 vols., Gillispie, Charles Coulston, ed. in chief. New York: Charles Scribner's Sons, 1970, vol. V, p. 408.

4. Bourbaki, Nicolas. *Eléments d'Histoire des Mathématiques.* Paris: Hermann, 1969, p. 31; see also Struik, D. J. *A Source Book in Mathematics 1200–1800.* Princeton, NJ: Princeton University Press, 1969 and 1986, p. 86.

5. Youschkevitsch, A. P. "Euler, Leonhard," *Dictionary of Scientific Biography.* 16 vols. Gillispie, Charles Coulston, ed. in chief. New York: Charles Scribner's Sons, 1970, vol. IV, p. 468.

6. *Ibid.,* p. 473.

7. Gowing, Ronald. *Roger Cotes: Natural Philosopher.* Cambridge: Cambridge University Press, 1983, p. 8.

8. *Ibid.,* p. 16.

9. Dubbey, J. M. "Cotes, Roger," *Dictionary of Scientific Biography,* 16 vols. Gillispie, Charles Coulston, ed. in chief. New York: Charles Scribner's Sons, 1970, vol. III, p. 430.

10. Jones, Phillip S. "Wessel, Caspar," *Dictionary of Scientific Biography,* 16 vols., Gillispie, Charles Coulston, ed. in chief. New York: Charles Scribner's Sons, 1970, vol. XIV, p. 280.

11. Muir, Jane. *Of Men and Numbers: The Story of the Great Mathematicians.* New York: Dover Publications, 1996 (originally published by Dodd Mead & Company, New York, 1961), p. 170.

12. *Ibid.,* p. 170.

13. May, Kenneth O. "Gauss, Carl Friedrich," *Dictionary of Scientific Biography,* 16 vols. Gillispie, Charles Coulston, ed. in chief. New York: Charles Scribner's Sons, 1970, vol. V, pp. 301–02.

4

Some Other Examples

The algebraic structures we have thus far considered—the integers, the rational numbers, the real numbers, and the complex numbers—should all be more or less familiar. In this chapter, we will present two other algebraic structures which will be useful for gaining insight into the abstract, axiomatically defined, structures on which we will base our future work.

The first of these structures is really a class of structures derived from the integers. Its elements are subsets of the integers that can be added and multiplied in a very natural way.

Example 4.1 Let n be a positive integer and let i be any integer. We let $[i]_n$ ("i mod n") denote the set of integers

$$[i]_n = \{i + kn \mid k \in \mathbb{Z}\}.$$

For $n = 3$, we have

$$
\begin{aligned}
[0]_3 &= \{\cdots, \quad -6, \quad -3, \quad 0, \quad 3, \quad 6, \quad 9, \quad \cdots\} \\
[1]_3 &= \{\cdots, \quad -5, \quad -2, \quad 1, \quad 4, \quad 7, \quad 10, \quad \cdots\} \\
[2]_3 &= \{\cdots, \quad -4, \quad -1, \quad 2, \quad 5, \quad 8, \quad 11, \quad \cdots\} \\
[3]_3 &= \{\cdots, \quad -3, \quad 0, \quad 3, \quad 6, \quad 9, \quad 12, \quad \cdots\} \\
[4]_3 &= \{\cdots, \quad -2, \quad 1, \quad 4, \quad 7, \quad 10, \quad 13, \quad \cdots\} \\
[5]_3 &= \{\cdots, \quad -1, \quad 2, \quad 5, \quad 8, \quad 11, \quad 14, \quad \cdots\} \\
[6]_3 &= \{\cdots, \quad 0, \quad 3, \quad 6, \quad 9, \quad 12, \quad 15, \quad \cdots\} \\
[7]_3 &= \{\cdots, \quad 1, \quad 4, \quad 7, \quad 10, \quad 13, \quad 16, \quad \cdots\}
\end{aligned}
$$

Note that the first three sets are all distinct, that they then repeat: $[0]_3 = [3]_3 = [6]_3$, $[1]_3 = [4]_3 = [7]_3$, and $[2]_3 = [5]_3$, and that the repetitions occur at multiples of 3:

$$[0]_3 = [0 + 1 \cdot 3]_3 = [0 + 2 \cdot 3]_3,$$
$$[1]_3 = [1 + 1 \cdot 3]_3 = [1 + 2 \cdot 3]_3,$$
$$[2]_3 = [2 + 1 \cdot 3]_3.$$

These observations are true in general. That is, for any positive integer n,

(1) $[i]_n = [i + kn]_n$ for all $k \in \mathbb{Z}$, and
(2)   if $0 \le i < j < n$, then $[i]_n \ne [j]_n$.

To prove these statements, we first show that the following test can be used to determine whether any two of these sets are equal:

(3) $[i]_n = [j]_n$ if and only if n divides $j - i$.

To prove the implication $\Rightarrow$, observe that if $[i]_n = [j]_n$, then since $j = j + 0n \in [j]_n$, $j \in [i]_n$ and hence $j = i + kn$ for some integer k. But then $j - i = kn$ and hence n divides $j - i$. To prove the other implication $\Leftarrow$, suppose that $j - i = dn$ for some integer d. We must show that the two sets $[i]_n$ and $[j]_n$ are equal; that is, we must show that each is contained in the other. So suppose that $x \in [j]_n$. Then $x = j + kn$ for some k and hence

$$x = j - dn + dn + kn = i + (d + k)n \in [i]_n;$$

so $[i]_n \supseteq [j]_n$. A similar argument shows that $[i]_n \subseteq [j]_n$, and thus that $[i]_n = [j]_n$.

Since n divides $(i + kn) - i$, statement (1) follows immediately from statement (3). For statement (2), observe that $0 \le i < j < n$, then $0 < j - i < n$, and hence n cannot divide $j - i$; so by statement (3), $[i]_n \ne [j]_n$.

Combining statements (1) and (2), we see that <u>there are exactly n distinct sets</u> $[i]_n$: $[0]_n, [1]_n, \ldots, [n - 1]_n$. We let $\mathbb{Z}_n$ denote this set of subsets of $\mathbb{Z}$:

$$\mathbb{Z}_n = \{\, [i]_n \mid i \in \mathbb{Z} \,\} = \{\, [0]_n, [1]_n, [2]_n, \ldots, [n - 1]_n \,\}.$$

What makes the set $\mathbb{Z}_n$ interesting to us is that it has a very natural addition and a very natural multiplication

$$[r]_n + [s]_n = [r + s]_n \quad \text{and} \quad [r]_n[s]_n = [rs]_n.$$

The apparent simplicity of these definitions hides a very subtle but important point. To see what might happen, consider $[2]_7$ and $[4]_7$ in $\mathbb{Z}_7$. It is easy to use test (3) to see that $[2]_7 = [58]_7$ and $[4]_7 = [46]_7$. According to our suggested definition of addition, $[2]_7 + [4]_7 = [6]_7$ and $[58]_7 + [46]_7 = [104]_7$. But if $[6]_7$ were <u>not</u> equal to $[104]_7$, our definition of $+$ wouldn't make sense, for then the sum of two elements in $\mathbb{Z}_7$ would depend on how those elements were represented. In general, when such an ambiguity occurs, the alleged operation may not be a properly defined function, and for this reason, a rule for which it has been proved that no such ambiguity exists is said to be "well defined." Of course the same difficulty can arise in trying to define any function. So, for later use (cf. Chapter 24), we phrase our definition in this general setting.

DEFINITION

Let $\lozenge$ be a rule that assigns elements of a set T to elements of a set S. Then $\lozenge$ is said to be **well defined** if it can be proved that whenever $X = Y$ in T, then $\lozenge$ assigns the same element of S to both X and Y.

An example of a rule that is <u>not</u> well defined is the following. Let $T = \mathbb{Q}$ and $S = \mathbb{Z}$. Let $\lozenge$ assign, to the rational number $\frac{a}{b}$, the integer $a + b$. Then $\frac{1}{2} = \frac{2}{4}$ but $1 + 2 \neq 2 + 4$. So the result of applying $\lozenge$ to the rational number $\frac{1}{2}$ varies with the representation of the number. That is, $\lozenge$ is not well defined.

In the case of the operations we are trying to define on $\mathbb{Z}_n$, the set T is $\{([r]_n, [s]_n) \mid [r]_n, [s]_n \in \mathbb{Z}_n\}$ and the set S is $\mathbb{Z}_n$; the rule for $+$ assigns to $([r]_n, [s]_n)$ the element $[r + s]_n$, and the rule for $\cdot$ assigns to $([r]_n, [s]_n)$ the element $[rs]_n$. So to show that $+$ is well defined, we must show that if $[r]_n = [u]_n$ and $[s]_n = [v]_n$, then $[r + s]_n = [u + v]_n$. We can use statement (3) to show these sets are equal. For statement (3) implies that since $[r]_n = [u]_n$, n divides $u - r$ and since $[s]_n = [v]_n$, n also divides $v - s$. But then n divides $(u - r) + (v - s) = (u + v) - (r + s)$, and hence $[r + s]_n = [u + v]_n$. So we conclude that $+$ is a well-defined operation. Once this has been established, it is not difficult to show that addition satisfies properties **A1–A5** of Proposition 3.1, i.e., that addition is closed, associative, and commutative, that $[0]_n$ is the additive identity in $\mathbb{Z}_n$, and that each element in $\mathbb{Z}_n$ has an additive inverse (Exercise 32).

A similar argument shows that $\cdot$ is also well defined (Exercise 31), and then it is not difficult to show that this multiplication satisfies properties **M1, M2, M3,** and **M5** of Proposition 3.1, i.e., that multiplication is closed, associative, and commutative, and that $[1]_n$ is the multiplicative identity in $\mathbb{Z}_n$ (Exercise 33). It is also not difficult to show that multiplication in $\mathbb{Z}_n$ distributes over addition from both right and the left (Exercise 34).

Then, for example,

$$[3]_8 + [6]_8 = [9]_8,$$
$$[3]_8[6]_8 = [18]_8,$$
$$[7]_8[5]_8 + ([4]_8 + [6]_8)[3]_8 = [35]_8 + [30]_8 = [65]_8,$$
$$([5]_7 + [4]_7)[6]_7 + [-4]_7([3]_7[6]_7 + [5]_7) = [54]_7 + [-92]_7 = [-38]_7.$$

While a calculation such as $[54]_7 + [-92]_7 = [-38]_7$ is correct as it stands, it is easier to understand if every $[k]_7$ is written in the form $[r]_7$ for $0 \le r < 7$. Rewriting

$[k]_n$ in this form is called "reducing k mod n" and amounts to applying the division algorithm (Theorem 1.3). For since $k \in [k]_7$ and $[k]_7 = [r]_7, k \in [r]_7$ so that $k = r + 7m$ for some $m \in \mathbb{Z}$ and $0 \le r < 7$. That is, r is the remainder determined by the division algorithm when k is divided by 7. For example, $54 = 7 \cdot 7 + 5$ so that $0 \le 5 < 7$ and $[54]_7 = [5]_7$. And $-92 = 7(-14) + 6$ so that $0 \le 6 < 7$ and $[-92]_7 = [6]_7$. Note that the minus sign cannot be ignored; the division algorithm applied to -92 and to 92 yields a different remainder in each case: $-92 = 7(-14) + 6$ (so that $[-92]_7 = [6]_7$) while $92 = 7(13) + 1$ (so that $[92]_7 = [1]_7$). For small numbers k, a quick way of reducing k mod n is to subtract or to add multiples of n to k until the result is between 0 and n. For instance, $38 - 7 \cdot 5 = 3$ so that $[38]_7 = [3]_7$, and $-38 + 7 \cdot 6 = 4$ so that $[-38]_7 = [4]_7$.

Successive calculations can sometimes be done more quickly by reducing each intermediate calculation. For instance, our previous calculation could be done as follows:

$$([5]_7 + [4]_7)[6]_7 + [-4]_7([3]_7[6]_7 + [5]_7)$$
$$= [2]_7[6]_7 + [-4]_7([4]_7 + [5]_7) = [5]_7 + [-4]_7[2]_7 = [5]_7 + [6]_7 = [4]_7$$

As $[4]_7 = [-38]_7$, this agrees with our earlier answer. Some other examples are the following:

$$([10]_{13} + [3]_{13})[9]_{13} + [12]_{13}[12]_{13}[8]_{13} = [0]_{13}[9]_{13} + [1]_{13}[8]_{13} = [8]_{13},$$
$$([27]_7 + [-12]_7 + [45]_7)[-65]_7 = ([6]_7 + [2]_7 + [3]_7)[5]_7 = [4]_7[5]_7 = [6]_7.$$

Subtraction is very simple because for any k and n, $[k]_n + [-k]_n = [0]_n$ and hence $-[k]_n = [-k]_n$. For instance,

$$[4]_{13} - [7]_{13} = [4 - 7]_{13} = [-3]_{13} = [10]_{13}.$$
$$[3]_9([1]_9 - [7]_9) - [8]_9[6]_9 = [3]_9[3]_9 - [3]_9 = [0]_9 + [-3]_9 = [6]_9,$$
$$([13]_7 + [-12]_7)([1]_7 - [25]_7) = ([6]_7 + [2]_7)([1]_7 - [4]_7) = [1]_7[4]_7 = [4]_7.$$

According to Proposition 4.3 below, if $n > 1$ is not prime, then not every element of $\mathbb{Z}_n$ will have a multiplicative inverse. So we postpone consideration of division to Example 4.4. ❖

The structures $\mathbb{Z}_n$ are extremely important and illustrate many of the abstract properties we will be considering. For instance, the nonzero elements $[2]_6$ and $[3]_6$ of $\mathbb{Z}_6$ have a zero product: $[2]_6[3]_6 = [6]_6 = [0]_6$. And a similar argument applies to all composite integers (Proposition 4.2).

Now if $[a]_n[b]_n = [0]_n$ and $[a]_n$ has a multiplicative inverse $[i]_n$ in $\mathbb{Z}_n$, then $[b]_n = [1]_n[b]_n = [i]_n[a]_n[b]_n = [i]_n[0]_n = [0]_n$. So if $\mathbb{Z}_n$ satisfies property **M4** of Proposition 3.1, i.e., if all its nonzero elements have multiplicative inverses, then it cannot contain any nonzero elements with zero product. We will use Theorem 1.4 on greatest common divisors to show that if p is prime, then $\mathbb{Z}_p$ always has this property, i.e., that nonzero elements have nonzero products (Proposition 4.3).

Proposition 4.2   *If m is a positive integer which is not a prime, then there exist nonzero elements $[a]_m$ and $[b]_m$ in $\mathbb{Z}_m$ whose product is zero.*

Proof   If m is not prime, then there exist integers $0 < k < m$ and $0 < l < m$ such that $m = kl$. Then $[k]_m \neq [0]_m$ and $[l]_m \neq [0]_m$, but $[k]_m[l]_m = [kl]_m = [0]_m$. ∎

Proposition 4.3   *Let p be a positive integer. Then every nonzero element of $\mathbb{Z}_p$ has a multiplicative inverse if and only if p is prime.*

Proof   If p is not prime, then by Proposition 4.2, $\mathbb{Z}_p$ contains nonzero elements whose product is zero, and we previously showed that this is impossible if every nonzero element of $\mathbb{Z}_p$ has a multiplicative inverse. Conversely, suppose that p is prime. We need to show that for all $0 < m < p$, there exists an integer a such that $[a]_p[m]_p = [1]_p$. Since p is prime, the greatest common divisor of m and p is 1, and therefore by the theorem on greatest common divisors (Theorem 1.4), there exist integers a and b such that $1 = am + bp$. Then $[1]_p = [am + bp]_p = [am]_p + [bp]_p = [a]_p[m]_p + [0]_p = [a]_p[m]_p$. ∎

Proposition 4.3 says that for p prime, any nonzero element $[k]_p$ in $\mathbb{Z}_p$ must have a multiplicative inverse. The next example gives a method for finding this inverse.

Example 4.4   In $\mathbb{Z}_7$, $[5]_7$ must have a multiplicative inverse, say, $[5]_7^{-1} = [k]_7$. Then $[1]_7 = [5]_7[k]_7 = [5k]_7$. Since $[1]_7 = [5k]_7$, $5k - 1$ is divisible by 7. Checking the possible values of k, we have $5 \cdot 1 - 1 = 4$, $5 \cdot 2 - 1 = 9$, $5 \cdot 3 - 1 = 14 = 2 \cdot 7$, and hence $k = 3$, i.e., $[5]_7^{-1} = [3]_7$. Thus, for example,

$$([1]_7 + [4]_7)^{-1}[6]_7 = [5]_7^{-1}[6]_7 = [3]_7[6]_7 = [18]_7 = [4]_7.$$

Similarly, since $4 \cdot 1 - 1 = 3$ and $4 \cdot 2 - 1 = 7 = 1 \cdot 7$, $[4]_7^{-1} = [2]_7$, and since $6 \cdot 1 - 1 = 5$, $6 \cdot 2 - 1 = 11$, $6 \cdot 3 - 1 = 17$, $6 \cdot 4 - 1 = 23$, $6 \cdot 5 - 1 = 29$, and $6 \cdot 6 - 1 = 35 = 5 \cdot 7$, $[6]_7^{-1} = [6]_7$. (Note that by Exercise 25, it is always the case that $[p - 1]_p^{-1} = [p - 1]_p$.) And thus

$$([9]_7 - [19]_7)^{-1}([10]_7[9]_7)^{-1} = ([2]_7 - [5]_7)^{-1}([3]_7[2]_7)^{-1}$$
$$= [4]_7^{-1}[6]_7^{-1} = [2]_7[6]_7 = [5]_7. ❖$$

Our second example differs substantially from the preceding ones. It is a generalization of the complex numbers which satisfies all the properties of Proposition 3.1 except **M5,** commutativity.

Example 4.5   **The Quaternions**   The quaternions are a noncommutative generalization of the complex numbers which Hamilton discovered in 1843. His crucial insight, which he scratched into the stone of an Irish bridge, was to consider not two, but three extra coordinates i, j, and k, such that $i^2 = j^2 = k^2 = -1 = ijk$. For then $ij = (-1)ij(kk) = (-1)(-k) = k$ while $ji = (-1)ji(ijk) = (-1)(j(ii)j)k = -k$ so that $ij \neq ji$.

Thus the set of quaternions, denoted by $\mathbb{H}$ (in honor of Hamilton), consists of expressions of the form

$$\alpha_0 + \alpha_1 i + \alpha_2 j + \alpha_3 k,$$

where all the α_i's come from $\mathbb{R}$. The same notational conventions that are used with the complex numbers are used with the quaternions: terms with coefficient 0 are ignored, as is the coefficient 1: e.g., $0 + 1i + 0j + 2k = i + 2k$. And, as with the complex numbers, quaternions are added componentwise:

$$(\alpha_0 + \alpha_1 i + \alpha_2 j + \alpha_3 k) + (\beta_0 + \beta_1 i + \beta_2 j + \beta_3 k)$$
$$= (\alpha_0 + \beta_0) + (\alpha_1 + \beta_1)i + (\alpha_2 + \beta_2)j(\alpha_3 + \beta_3)k.$$

They are multiplied in accordance with the left distributive law and the following rules for i, j, and k, which in turn derive from the relations given previously:

$$i^2 = j^2 = k^2 = -1, \quad ij = k, \quad jk = i, \quad ki = j, \quad ji = -k, \quad kj = -i, \quad ik = -j.$$

Using the distributive law and these equalities, we have

$$(\alpha_0 + \alpha_1 i + \alpha_2 j + \alpha_3 k)(\beta_0 + \beta_1 i + \beta_2 j + \beta_3 k)$$
$$= \alpha_0 \beta_0 + \alpha_0 \beta_1 i + \alpha_0 \beta_2 j + \alpha_0 \beta_3 k + \alpha_1 \beta_0 i + \alpha_1 \beta_1 ii + \alpha_1 \beta_2 ij + \alpha_1 \beta_3 ik$$
$$+ \alpha_2 \beta_0 j + \alpha_2 \beta_1 ji + \alpha_2 \beta_2 jj + \alpha_2 \beta_3 jk + \alpha_3 \beta_0 k + \alpha_3 \beta_1 ki + \alpha_3 \beta_2 kj + \alpha_3 \beta_3 kk$$
$$= \alpha_0 \beta_0 + \alpha_1 \beta_1 ii + \alpha_2 \beta_2 jj + \alpha_3 \beta_3 kk + \alpha_0 \beta_1 i + \alpha_1 \beta_0 i + \alpha_2 \beta_3 jk + \alpha_3 \beta_2 kj$$
$$+ \alpha_0 \beta_2 j + \alpha_2 \beta_0 j + \alpha_3 \beta_1 ki + \alpha_1 \beta_3 ik + \alpha_0 \beta_3 k + \alpha_3 \beta_0 k + \alpha_1 \beta_2 ij + \alpha_2 \beta_1 ji$$
$$= (\alpha_0 \beta_0 - \alpha_1 \beta_1 - \alpha_2 \beta_2 - \alpha_3 \beta_3) + (\alpha_0 \beta_1 + \alpha_1 \beta_0 + \alpha_2 \beta_3 - \alpha_3 \beta_2)i$$
$$+ (\alpha_0 \beta_2 + \alpha_2 \beta_0 + \alpha_3 \beta_1 - \alpha_1 \beta_3)j + (\alpha_0 \beta_3 + \alpha_3 \beta_0 + \alpha_1 \beta_2 - \alpha_2 \beta_1)k.$$

So we use the last expression to <u>define</u> quaternion multiplication:

$$(\alpha_0 + \alpha_1 i + \alpha_2 j + \alpha_3 k)(\beta_0 + \beta_1 i + \beta_2 j + \beta_3 k)$$
$$= (\alpha_0 \beta_0 - \alpha_1 \beta_1 - \alpha_2 \beta_2 - \alpha_3 \beta_3) + (\alpha_0 \beta_1 + \alpha_1 \beta_0 + \alpha_2 \beta_3 - \alpha_3 \beta_2)i$$
$$+ (\alpha_0 \beta_2 + \alpha_2 \beta_0 + \alpha_3 \beta_1 - \alpha_1 \beta_3)j + (\alpha_0 \beta_3 + \alpha_3 \beta_0 + \alpha_1 \beta_2 - \alpha_2 \beta_1)k.$$

With respect to these two operations, the quaternions have all the properties listed in Proposition 3.1 except **M5**, commutativity. The proof of this (Exercise 35) involves the obvious calculations, all of which are straightforward, except perhaps those for multiplicative associativity, which are merely tedious. (Note that while the definition for the multiplication was found by assuming the left distributive law, to show that the quaternions have this property, this law must be shown to hold for the general definition of the multiplication.) The quaternions have a multiplicative identity, namely $1 + 0i + 0j + 0k = 1$, and as we noted, they are not commutative because $ij = k \neq -k = ji$. However, nonzero quaternions do have multiplicative inverses.

Specifically, if $0 \neq \alpha_0 + \alpha_1 i + \alpha_2 j + \alpha_3 k \in \mathbb{H}$, then

$$\frac{1}{(\alpha_0{}^2 + \alpha_1{}^2 + \alpha_2{}^2 + \alpha_3{}^2)}(\alpha_0 - \alpha_1 i - \alpha_2 j - \alpha_3 k)$$

is its multiplicative inverse.

We noted that multiplication of quaternions can be viewed as the result of the distributive law applied in conjunction with the identities

$$i^2 = j^2 = k^2 = -1, \quad ij = k, \quad jk = i, \quad ki = j, \quad ik = -j, \quad kj = -i, \quad ji = -k.$$

These identities can themselves be generated by using the picture in Figure 4.1.

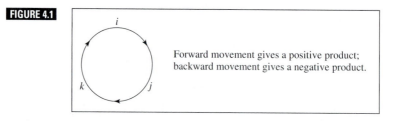

FIGURE 4.1

Forward movement gives a positive product;
backward movement gives a negative product.

Then the easiest way to multiply quaternions is as in the following example:

$$(1 + 2i + 3k)(2 - j + k) = (2 - j + k) +$$
$$(4i - 2ij + 2ik) + (6k - 3kj + 3k^2)$$
$$= (2 - j + k) + (4i - 2k + 2(-j)) +$$
$$(6k - 3(-i) + 3(-1))$$
$$= -1 + 7i - 3j + 5k. \; \text{❖}$$

Exercises

In Exercises 1–5, write the given expression in the form $[i]_n$, where $0 \leq i \leq n - 1$.

1. $([5]_{12} + [9]_{12})[4]_{12}$ **2.** $[7]_6 - [13]_6 + [4]_6$

3. $[15]_4 - [105]_4 + [-25]_4$ **4.** $[44]_{13}([45]_{13} + [123]_{13} - [2]_{13})$

5. $([54]_{22} - [-16]_{22})([7]_{22} + [-111]_{22})$

In Exercises 6–11, find the multiplicative inverse of the given element of $\mathbb{Z}_7$. Write the inverse in the form $[i]_7$, where $0 \leq i \leq 6$.

6. $[2]_7$ **7.** $[6]_7$ **8.** $[24]_7$

9. $[-9]_7$ **10.** $[-24]_7$ **11.** $[-102]_7$

In Exercises 12–18, write the given expression in the form $[i]_n$, where $0 \le i \le n - 1$.

12. $([2]_{13} - [7]_{13})([11]_{13})^{-1}[4]_{13}$ **13.** $[3]_5 + [-17]_5([-8]_5 - [34]_5)^{-1}$

14. $([5]_7[2]_7[3]_7)([11]_7 + [23]_7 + [13]_7)^{-1}$

15. $[14]_{13} + ([5]_{13} + [-19]_{13})([6]_{13} - [9]_{13})^{-1}$

16. $[-37]_{17}([12]_{17} - [7]_{17}[3]_{17})^{-1} - [100]_{17}$

17. $[24]_7 + [-43]_7([-16]_7 - [4]_7)^{-1} + [195]_7$

18. $[1]_{17} + [-17]_{17}(-[16541]_{17} + [2453]_{17})^{-1}$

In Exercises 19–24, write the given quaternion in the form $\alpha_0 + \alpha_1 i + \alpha_2 j + \alpha_3 k$ for real numbers $\alpha_0, \alpha_1, \alpha_2, \alpha_3 \in \mathbb{R}$.

19. $(2i - 3j + 7k) + (2 + 8j + 11k)$ **20.** $(2i + 4k)(7 - 3j + k)$

21. $(2 + 2i - k)(i - j + k)^{-1}$ **22.** $(i - j)(3i - 7j)(2 + k)$

23. $(i - 2j + k)(2 + i + 5j)(1 - k)^{-1}$

24. $(5 - i + 2j - k)(1 + i + j + k)^{-1}(1 - j + k)$

25. Let $1 < n \in \mathbb{Z}$. Show that $[n - 1]_n$ is its own multiplicative inverse in $\mathbb{Z}_n$.

In Exercises 26–28, the notation $n[k]_m$ denotes $[k]_m$ added to itself n times:

$$n[k]_m = [k]_m + \cdots + [k]_m \quad (n \text{ times}).$$

26. Show that for any integer n greater than 1 and any $[k]_n \in \mathbb{Z}_n$, $n[k]_n = [0]_n$.

27. Let p be a prime, let n be a positive integer, and let $[0]_p \ne [k]_p \in \mathbb{Z}_p$. Show that $n[k]_p = [0]_p$ if and only if p divides n.

28. Let c be a composite integer greater than 1 (i.e., c is <u>not</u> prime). Find a nonzero element $[k]_c$ of $\mathbb{Z}_c$ and a positive integer n such that $n[k]_c = [0]_c$ but c does not divide n.

29. Suppose that m and n are integers that are greater than 1 and whose greatest common divisor is 1. Show that if $[a]_m = [b]_m$ and $[a]_n = [b]_n$, then $[a]_{mn} = [b]_{mn}$.

30. Let d and m be integers greater than 1 and suppose that d divides m. Show that if $[a]_m = [b]_m$ in $\mathbb{Z}_m$, then $[a]_d = [b]_d$ in $\mathbb{Z}_d$.

31. For any positive integer n, show that multiplication in $\mathbb{Z}_n$ is well defined.

32. For any positive integer n, show that addition in $\mathbb{Z}_n$ satisfies properties **A1–A5** of Proposition 3.1.

33. For any positive integer n, show that multiplication in $\mathbb{Z}_n$ satisfies properties **M1, M2, M3,** and **M5** of Proposition 3.1.

34. For any positive integer n, show that multiplication and addition in $\mathbb{Z}_n$ satisfy properties **D1** and **D2** of Proposition 3.1.

35. Show in detail that the quaternions satisfy all the properties of Proposition 3.1 except **M5**.

36. Show that there are infinitely many quaternions that solve the equation $x^2 = -1$.

37. Show that the following rule $\lozenge$ is <u>not</u> well-defined. Let T be the set $\{(r,s) \mid r, s \in \mathbb{Q}\}$ and let $S = \mathbb{Q}$. Write elements of $\mathbb{Q}$ as fractions $\frac{a}{b}$, $b > 0$, and let $\lozenge$ be the rule: $\frac{a}{b} \lozenge \frac{c}{d} = \frac{a+b}{c+d}$.

38. For $x \in \mathbb{R}$, let $[x] = \{x + 2n\pi \mid n \in \mathbb{Z}\}$, and let $\mathbb{S} = \{[x] \mid x \in \mathbb{R}\}$. Multiply two elements $[x]$ and $[y]$ of $\mathbb{S}$ by letting

$$[x][y] = [xy].$$

Show that this multiplication is <u>not</u> well defined.

Exercises 39–42 prove Fermat's theorem: if a is an integer and p is a prime which does not divide a, then $[a^p]_p = [a]_p$. In each exercise, assume that p is a prime which does not divide a.

39. Show that for all $0 < r < p$, there exists a unique $0 < s < p$ such that $[sa]_p = [r]_p$.

40. Show that $[1 \cdots (p-1)]_p = [a(2a) \cdots ((p-1)a)]_p$.

41. Show that $[1]_p = [a^{p-1}]_p$.

42. Show that $[a]_p = [a^p]_p$.

HISTORICAL NOTE

William Rowan Hamilton

William Rowan Hamilton was born in Dublin, Ireland, on August 4, 1805, and died at Dunsink Observatory (near Dublin), on September 2, 1865. His father, Archibald Rowan Hamilton (ca.1779–1819) was a solicitor, and his mother, born Sarah Hutton (1780–1817), came from a prosperous coach-building family. Archibald's most important client was an Irish patriot of almost the same name, Archibald Hamilton Rowan (1751–1834), who bore such a striking resemblance to William that there is a suggestion that he was Archibald Rowan Hamilton's father and therefore William Rowan Hamilton's grandfather. It appears that the family had recognized William's precocity by 1808, for in that year, he was sent to Trim to be raised by his uncle who was curate there. William's father Archibald had fallen into financial difficulties, and in 1809, these became so severe that he was forced to declare bankruptcy. By the time he was five, William was familiar with Latin, Greek, and Hebrew, and eight years later, he could read all three languages quite easily. The story that by the time he was ten he had mastered Persian, Arabic, Chaldee, Hindu, Malay, and others, arose from an overenthusiastic letter from his uncle; his linguistic abilities, while substantial, have probably been overestimated. Both his parents died while he was in Trim, his mother in 1817 and his father in 1819.

As a boy, Hamilton was a rapid calculator, but he did not become engrossed in mathematics until 1822, the year before he entered Trinity College, Dublin. The mathematics curriculum at Trinity was excellent at the time, and Hamilton was such an exceptional student that in England Augustus De Morgan reported that he had "heard of the extraordinary attainments of a very young student of Trinity College, which were noised about at Cambridge."[1]

As Hamilton's scientific career was burgeoning, so was his personal life. His romantic tendencies had led him always to search for the Ideal and on August 17, 1824, he found its personification in Catherine Disney, daughter of a local landlord's agent. He was smitten, as apparently so was she, to the point where he sent her the following Valentine's Ode in 1825: "Forgive me, that on bliss so high/ Lingers thrilling phantasie:/ That the one Image, dear and bright,/ Feeds thought by day, and dreams by night:/ That hope presumes to mingle thee,/ With visions of my destiny!"[2] However, the following May, at her father's insistence and against her will, she married a man fifteen years her senior, a clergyman of a Dublin legal family. Hamilton sank into a depression so severe that it affected his performance on the June examinations.

Hamilton the scientist was pursuing the study of optics at the time and submitted "On caustics" to the Royal Irish Academy in 1824. He received the report rejecting the paper soon after Catherine's marriage. While the referees had found the contents acceptable, they complained that the paper was difficult to read, the first inkling of a problem that was to plague Hamilton all his life. He resubmitted his results in 1827 under the title "Theory of systems of rays," which treated optics algebraically and, when it was published, brought accolades from Britain. His reputation was such that in 1827, even without his degree, he was appointed Astronomer Royal at Dunsink Observatory and Professor of Astronomy at Trinity College. He seems to have been a concerned and enthusiastic lecturer but one who had difficulty expressing his arguments clearly, a tendency that, as we have seen, occurred in his writing as well.

In 1831, he fell in love again, this time with Ellen de Vere, the daughter of a local country gentleman from Curragh Chase. Hamilton's salary had just been doubled and he was contemplating marriage until she casually remarked that she "could not live happily anywhere but at Curragh."[3] He construed the remark literally, dropped his suit, and entered into what proved to be a long-lived and fruitful but ultimately unhappy marriage with Helen Bayley in 1833.

Triads were of great philosophical interest at the time, and Hamilton had pondered their significance off and on over the years. Mathematically, it was easy to add and subtract them but not to multiply them. Then, as he was walking over the Royal Canal on the Brougham Bridge near the observatory, as a plaque on the bridge now reads, "on the 16th of October 1843 Sir William Rowan Hamilton in a flash of genius discovered the fundamental formula for quaternion multiplication $i^2 = j^2 = k^2 = ijk = -1$ & cut it on a stone of this bridge."[4] The key was of course that triplets were inadequate; four, not three, coordinates were needed, and he promptly named his discovery the quaternions. Simultaneously, Hermann Günther Grassmann was initiating the study of vector spaces, but since both authors had difficulty expressing themselves, it took Peter Guthrie Tait's *Elementary Treatise on Quaternions* (1867) and Herman Hankel's *Vorlesungen über die complexen Zahlen und ihre Funktionen* (also 1867) to bring their discoveries widespread recognition.

In 1845, Hamilton became reacquainted with Catherine Disney. Their romance flowered anew and the stress on Catherine was overwhelming. She confessed her love for William to her husband and then took an overdose of laudanum. She did not die, but she left her husband, as well as her religious principles, and went to live with her family in Dublin. William's strong sense of religion prevented him from seeing her, until October 1853, when she sent him a pencil case bearing the inscription "From one whom you must never forget, nor think unkindly of, and who would have died more contented if we had once more met."[5] He rushed to her side, knelt, and in his own words, "offered to her the Book (*Lectures on Quaternions*) which represented the scientific labours of my life. Rising, I received, or took as my reward, all that she could lawfully give—a kiss, nay many kisses—for the *known* and *near* approach of *death* made such communion holy. It could not be indeed, without *agitation* on both sides, that for the first time in our *lives*, our lips then met. . . ."[6] She died two weeks later, and for the rest of his life, Hamilton never tired of discreetly retelling the story of his long romance.

In 1865, a combination of worry, overwork, and alcohol brought about his final fatal illness. And on September 2, in accordance with his romantic nature, in the words of a friend who was present at his death, "he breathed his last, having first, as I learned the following day, solemnly stretched himself at full length upon his bed, and symmetrically disposed his arms and hands, thus calmly to await his death."[7]

References

1. O'Donnell, Sèan. *William Rowan Hamilton,* Dublin: Boole Press, 1983, p. 48.
2. *Ibid.,* p. 53.
3. *Ibid.,* p. 90.
4. *Ibid.,* p. 137.
5. *Ibid.,* p. 159.
6. *Ibid.,* p. 160.
7. *Ibid.,* p. 185.

PART 2

Algebraic Extension Fields

5

Fields

The fundamental theorem of algebra allows us to be more specific about our goal of finding a polynomial equation that is not solvable by the methods of Chapter 2. We wish to produce a polynomial equation with integral coefficients whose solutions in the complex numbers may not all be obtained "by radicals." In the next chapter, we will define precisely what we mean by the expression "solvable by radicals." In this chapter, we will develop the structures that provide the framework for that definition.

The solutions to polynomial equations produced by the methods of Chapter 2 all may be expressed—albeit in a very complicated fashion—by using rational numbers and the coefficients of the original equation. These coefficients are combined by using addition, subtraction, multiplication, division, and the taking of nth roots. The trouble arises from the taking of the roots. Combining rational numbers using addition, subtraction, multiplication, and division produces another rational number. Extracting roots may produce a nonrational, or even nonreal, number.

For instance, according to the methods of Chapter 2, the number

$$a = \sqrt[3]{-1 + \sqrt{2}} - \frac{1}{\sqrt[3]{-1 + \sqrt{2}}}$$

solves the equation $x^3 + 3x + 2 = 0$. Although in this case, there is nothing ambiguous in our definition of a, we noted at the end of Chapter 3 that, for complex numbers in general, the notation $\sqrt[n]{}$ is ambiguous. We can avoid this type of ambiguity by picking a number x such that $x^n = y$ rather than "defining" $x = \sqrt[n]{y}$. In this case, we could build a from the rational numbers in stages: first pick a real number s such that $s^2 = 2$, then pick a real number r such that $r^3 = -1 + s$, and finally let $a = r - \frac{1}{r}$.

Note that although both the coefficients of the original equation and the square of s are rational numbers, s itself is not. And similarly, although the cube of r is an algebraic combination of s and a rational number, r itself is not. However, the solution a is an algebraic combination of r and the rational number 1. The solution a may thus be considered as living in a subset of the real numbers which is built from the rational numbers, where the coefficients live, by using addition, subtraction, multiplication, and division to adjoin successively first a square root s of the rational number 2, and then a cube root r of $-1 + s$.

That is, the rational numbers $\mathbb{Q}$ may be viewed as a somewhat sparse subset of the real line, and we have $s^2 \in \mathbb{Q}$ but $s \notin \mathbb{Q}$:

So form a subset F_1 of $\mathbb{R}$ by using addition, subtraction, multiplication, and division to combine s with the elements of $\mathbb{Q}$, and note that $r^3 \in F_1$ but $r \notin F_1$:

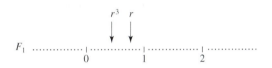

So form a bigger subset F_2 of $\mathbb{R}$ by using addition, subtraction, multiplication, and division to combine r with the elements of F_1. Since 1 and r are both in F_1, $\frac{1}{r}$ is in F_2, and thus $a = r - \frac{1}{r}$ is also in F_2:

Phrased in such a vague fashion, this construction does us little good. We need to make the construction of F_1 and F_2 precise. We will do this in Chapter 6, but first we need to specify exactly what we mean by using addition, subtraction, multiplication, and division to create a subset of the complex numbers. The approach we will take is to isolate those subsets that are closed with respect to these operations and then to demand that the subsets we create fall into this restricted class. The subsets contained in the resulting class will be those that satisfy the algebraic laws which, according to Proposition 3.1, govern the complex numbers.

One of the most powerful methods of abstract algebra is the procedure that gives it its name, viz., abstraction, that is, isolating as axioms those properties that are fundamental to the structure under consideration. For this reason, we want to phrase the algebraic laws in which we are interested in a general axiomatic setting.

To accomplish this, it is best to view the operations of addition and multiplication as functions. For instance, addition on the real numbers is technically a function from the plane $\mathbb{R}^2$ to the real numbers $\mathbb{R}$, $(x,y) \mapsto x + y$; and similarly multiplication is the function $(x,y) \mapsto xy$. Functions such as these, which combine two elements of a given set into a single element, are called binary operations.

DEFINITION

Let F be a set and let $F^2 = \{(x,y) \,|\, x,y \in F\}$. A **binary operation** on F is a function from F^2 to F.

In general, we will have a set F and two binary operations defined on F, one of which we will want to represent addition and hence we will write as $+$, and the other of which we will want to represent multiplication and hence we will write as $\cdot$. Following the conventions for $\mathbb{Q}$, $\mathbb{R}$, and $\mathbb{C}$, we will write $x + y$ instead of $+(x,y)$ and either $x \cdot y$, or more simply xy, instead of $\cdot(x,y)$. We want a list of axioms which will characterize these operations as addition and multiplication.

Note that since $+$ and $\cdot$ are binary operations and hence functions into F, it follows from the definition that both $x + y$ and xy are in F. However, in practice, to check whether a given pair, $+$ and $\cdot$, does indeed satisfy the axioms, it is important to remember to check first that they are both binary operations. Therefore, although it is technically redundant, we include closure for each operation in our list of axioms.

The complete list is then the following.

A1. *ADDITIVE CLOSURE: For all $x, y \in F$, $x + y \in F$.*

A2. *ADDITIVE ASSOCIATIVITY: For all $x, y, z \in F$, $(x + y) + z = x + (y + z)$.*

A3. *EXISTENCE OF AN ADDITIVE IDENTITY: There exists $0 \in F$ such that for all $x \in F$, $x + 0 = x = 0 + x$.*

A4. *EXISTENCE OF ADDITIVE INVERSES: For all $x \in F$, there exists $y \in F$ such that $x + y = 0 = y + x$.*

A5. *ADDITIVE COMMUTATIVITY: For all $x, y \in F$, $x + y = y + x$.*

M1. *MULTIPLICATIVE CLOSURE: For all $x, y \in F$, $xy \in F$.*

M2. *MULTIPLICATIVE ASSOCIATIVITY: For all $x, y, z \in F$, $x(yz) = (xy)z$.*

M3. *EXISTENCE OF A MULTIPLICATIVE IDENTITY (OR UNIT ELEMENT): There exists $1 \in F$ such that for all $x \in F$, $x1 = x = 1x$.*

M4. *EXISTENCE OF MULTIPLICATIVE INVERSES: For all $0 \neq x \in F$, there exists $z \in F$ such that $xz = 1 = zx$.*

M5. *MULTIPLICATIVE COMMUTATIVITY: For all $x, y \in F$, $xy = yx$.*

D1. *LEFT DISTRIBUTIVITY: For all $x, y, z \in F$, $x(y + z) = xy + xz$.*

D2. *RIGHT DISTRIBUTIVITY: For all $x, y, z \in F$, $(x + y)z = xz + yz$.*

> **DEFINITION**
>
> A **field** is a set F which contains at least two elements and which has two operations $+$ and $\cdot$ satisfying all the axioms in our list.

The requirement that a field have at least two elements is included to ensure that $0 \neq 1$ (see Proposition 5.1).

We observed previously that $\mathbb{Q}$, $\mathbb{R}$, and $\mathbb{C}$ are all fields, and according to Example 4.1 and Proposition 4.3, for any prime p, $\mathbb{Z}_p$ is a field. Other examples are given in Examples 5.3, 5.4, 5.5, and 5.6 and in the exercises.

On the other hand, the integers, $\mathbb{Z}$, form a familiar structure which has almost no multiplicative inverses and hence is not a field. And as well, by Proposition 4.3, for composite integers m, $\mathbb{Z}_m$ also lacks multiplicative inverses and hence is not a field. These are very important algebraic structures and we will not neglect them. The general algebraic structure of which these are the primary examples will be considered in Chapter 7.

Of course, $\mathbb{Q}$, $\mathbb{R}$, and $\mathbb{C}$ all have more properties than the twelve familiar ones in our list. So it is quite possible that these twelve axioms may encompass all sorts of pathological examples. Certainly, there are some that may appear at first to be unusual (e.g., the fields $\mathbb{Z}_p$ for primes p). However, most of the familiar properties of elements in $\mathbb{Q}$, $\mathbb{R}$, and $\mathbb{C}$ also hold for elements of fields in general. For instance, there is only one multiplicative identity and each element has only one additive inverse. These properties are summarized in the next result. They allow much of the intuition gained from knowledge of $\mathbb{Q}$, $\mathbb{R}$, and $\mathbb{C}$ to be transferred to any field.

Proposition 5.1   *For any field F,*

    (i)   the additive identity is unique: if $a = a + z$ for all $a \in F$, then $z = 0$;

    (ii)   for all $a \in F$, $a0 = 0 = 0a$;

    (iii)   additive inverses are unique:
        if $a, c, d \in F$ are such that $a + c = 0 = d + a$, then $c = d$;
        we use $-a$ to denote the unique additive inverse of a;

    (iv)   for all $a, b \in F$, $(-a)b = -(ab) = a(-b)$;

    (v)   the multiplicative identity is unique: if $a = ae$ for all $a \in F$, then $e = 1$;

    (vi)   multiplicative inverses are unique:
        if $a, c, d \in F$ are such that $a \neq 0$ and $ac = 1 = da$, then $c = d$;
        we use a^{-1} to denote the unique multiplicative inverse of a;

    (vii) $0 \neq 1$.

Proof   (i)   If $a = a + z$ for all $a \in F$, then $0 = 0 + z = z$.

(ii)   Let $a \in F$. Since $a0$ has an additive inverse, there exists $w \in F$ such that $a0 + w = 0$. Furthermore, $a0 = a(0 + 0) = a0 + a0$ by the distributive law and thus (by the associative law for addition)

$$0 = a0 + w = (a0 + a0) + w = a0 + (a0 + w) = a0 + 0 = a0.$$
Similarly, $0 = 0a$.

(*iii*)   Let $a, c, d \in F$. If $a + c = 0$ and $a + d = 0$, then by the commutative and associative laws for addition $c = 0 + c = (d + a) + c = d + (a + c) = d + 0 = d$.

(*iv*)   Let $a, b \in F$. By the distributive law and part (*ii*), $ab + a(-b) = a(b - b) = a0 = 0$. Thus by part (*iii*), $a(-b) = -(ab)$. Similarly, $(-a)b = -(ab)$.

(*v*)   Similarly to part (*i*), if $ae = a$ for all $a \in F$, then $e = 1e = 1$.

(*vi*)   Similarly to part (*iii*), let $a, c, d \in F$. If $ac = 1 = ad$, then $c = 1c = (da)c = d(ac) = d1 = d$.

(*vii*)   Since F has at least two elements, there exists $0 \neq a \in F$. If $0 = 1$, then $a0 = a1$, and hence by (*ii*) and the definition of 1, $0 = a0 = a1 = a$. This contradicts our choice of a, and thus $0 \neq 1$. ∎

The sets F_1 and F_2 mentioned at the beginning of the chapter are fields (cf. Example 5.4). They are special fields in the sense that they arise inside another field, viz., the complex numbers. Fields arising in such a way occur frequently and are given their own special names.

DEFINITION

Let U be a field. A **subfield** of U is a subset F of U which is a field with respect to the operations which it inherits from U. If F is a subfield of a field U, we also say that U is an **extension field** of F.

For instance, $\mathbb{R}$ is an extension field of $\mathbb{Q}$ and a subfield of $\mathbb{C}$. Note that a given field is always both a subfield and an extension field of itself. So in general, if we have fields $F \subseteq K \subseteq L$, all with the same operations, then F, K, and L are all subfields of L and extension fields of F. And K is a subfield of L and an extension field of F. (See Figure 5.1.)

FIGURE 5.1

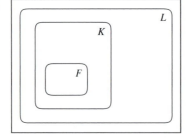

To determine whether a given subset of a field is a subfield by checking each of the twelve field axioms can be tedious in the extreme. Luckily, much of this tedium can be avoided. For instance, the associative law holds in any subset. For if S is a subset of a field U and $x, y, z \in S$, then $x, y, z \in U$ as well and hence $(x + y) + z = x + (y + z)$. This sort of argument will work for the associative laws, the distributive laws, and the commutative laws. Judicious use of other laws can reduce the amount of work still further so that we are left with only five things to check.

Proposition 5.2   *A subset F of a field U is a subfield of U if and only if it satisfies the following conditions:*

    (i)   *F contains at least two elements;*
    (ii)   *if $x, y \in F$, then $x + y \in F$;*
    (iii)   *if $x \in F$ and $-x$ is the additive inverse of x in U, then $-x \in F$;*
    (iv)   *if $x, y \in F$, then $xy \in F$;*
    (v)   *if $0 \neq x \in F$ and x^{-1} is the multiplicative inverse of x in U, then $x^{-1} \in F$.*

Proof   If F is a subfield, then it is in fact a field, and hence it satisfies (i)–(v). Conversely, suppose that F has at least two elements and satisfies (ii), (iii), (iv), and (v). That F is additively and multiplicatively closed and has additive and multiplicative inverses follows from these assumptions about F. Furthermore, as noted above, any subset of U is additively and multiplicatively associative, additively and multiplicatively commutative, and satisfies both distributive laws. It thus remains to show that F has both an additive and a multiplicative identity. Since F contains at least two elements, we may pick an element $0 \neq x \in F$. By (iii), the additive inverse, $-x$, of x is in F. By (ii), $0 = x + (-x) \in F$. Since 0 is an additive identity for U, it is also an additive identify for F and thus F has an additive identity. Similarly, since $0 \neq x$, (v) implies that the multiplicative inverse, x^{-1}, of x is in F, and thus by (iv), $1 = xx^{-1} \in F$. But since 1 is a multiplicative identity for U, it is also a multiplicative identity for F and thus F has a multiplicative identity. We conclude that F is a subfield of U. ∎

The next few examples illustrate the use of Proposition 5.2. Note that the elegant way in which the elements of the subfield are expressed creates some difficulty in proving (v).

Example 5.3   We will show that the set

$$F = \{a + b\sqrt{-2} \mid a, b \in \mathbb{Q}\}$$

is a subfield of $\mathbb{C}$. Clearly $F \subseteq \mathbb{C}$, and since $0 = 0 + 0\sqrt{-2} \in F$ and $1 = 1 + 0\sqrt{-2} \in F$ and $0 \neq 1$, F contains at least two elements. Let $a + b\sqrt{-2}$, $c + d\sqrt{-2} \in F$. Then

$$(a + b\sqrt{-2}) + (c + d\sqrt{-2}) = (a + c) + (b + d)\sqrt{-2} \in F$$

because $a + c, b + d \in \mathbb{Q}$;

$$-(a + b\sqrt{-2}) = (-a) + (-b)\sqrt{-2} \in F$$

because $-a, -b \in \mathbb{Q}$; and

$$(a + b\sqrt{-2})(c + d\sqrt{-2}) = (ac - 2bd) + (ad + bc)\sqrt{-2} \in F$$

because $ac - 2bd, ad + bc \in \mathbb{Q}$.

Thus F satisfies (ii), (iii), and (iv) of Proposition 5.2. To see that F satisfies (v), suppose that $0 \neq a + b\sqrt{-2} \in F$.

Some preliminary work can show us what the inverse of $a + b\sqrt{-2}$ should look like. If $x + y\sqrt{-2}$ is an inverse of $a + b\sqrt{-2}$, then

$$1 = (a + b\sqrt{-2})(x + y\sqrt{-2}) = (ax - 2by) + (ay + bx)\sqrt{-2},$$

and this equation will certainly hold if $ax - 2by = 1$ and $ay + bx = 0$. But since $a + b\sqrt{-2} \neq 0$ and a and b are rational, a and b are not both zero, and hence these two equations have solutions

$$x = \frac{a}{a^2 + 2b^2} \qquad \text{and} \qquad y = \frac{-b}{a^2 + 2b^2}.$$

Since $\dfrac{a}{a^2 + 2b^2}$ and $\dfrac{-b}{a^2 + 2b^2}$ are both in $\mathbb{Q}$,

$$\frac{a}{a^2 + 2b^2} + \frac{-b}{a^2 + 2b^2}\sqrt{-2} \in F.$$

So since

$$(a + b\sqrt{-2})\left(\frac{a}{a^2 + 2b^2} + \frac{-b}{a^2 + 2b^2}\sqrt{-2}\right) = 1,$$

$$(a + b\sqrt{-2})^{-1} = \frac{a}{a^2 + 2b^2} + \frac{-b}{a^2 + 2b^2}\sqrt{-2} \in F,$$

and thus F satisfies (v) of Proposition 5.2. We conclude that F is a subfield of $\mathbb{C}$ by Proposition 5.2. ❖

Example 5.4   We will show that the set

$$S = \{a + b\sqrt{2} \,|\, a, b \in \mathbb{Q}\}$$

is a subfield of $\mathbb{R}$. The first part of the argument parallels that of Example 5.3. Clearly, $S \subseteq \mathbb{R}$, and since $0 = 0 + 0\sqrt{2} \in S$ and $1 = 1 + 0\sqrt{2} \in S$ and $0 \neq 1$, S contains at least two elements. If $a + b\sqrt{2}, c + d\sqrt{2} \in S$, then

$$(a + b\sqrt{2}) + (c + d\sqrt{2}) = (a + c) + (b + d)\sqrt{2} \in S$$

because $a + c, b + d \in \mathbb{Q}$;

$$-(a + b\sqrt{2}) = (-a) + (-b)\sqrt{2} \in S$$

because $-a, -b \in \mathbb{Q}$; and

$$(a + b\sqrt{2})(c + d\sqrt{2}) = (ac + 2bd) + (ad + bc)\sqrt{2} \in S$$

because $ac + 2bd, ad + bc \in \mathbb{Q}$.

So S satisfies (*ii*), (*iii*), and (*iv*) of Proposition 5.2. There is a subtle difficulty which complicates the proof of property (*v*) of Proposition 5.2.

Suppose that $0 \neq a + b\sqrt{2} \in S$. To find the inverse of $a + b\sqrt{2}$, we proceed as in Example 5.3. If $x + y\sqrt{2}$ is an inverse of $a + b\sqrt{2}$, then

$$1 = (a + b\sqrt{2})(x + y\sqrt{2}) = (ax + 2by) + (ay + bx)\sqrt{2},$$

and this equation will certainly hold if $ax + 2by = 1$ and $ay + bx = 0$. In this case, however, the solutions take the form

$$x = \frac{a}{a^2 - 2b^2} \quad \text{and} \quad y = \frac{-b}{a^2 - 2b^2},$$

and thus will exist in $\mathbb{R}$ only if $a^2 - 2b^2 \neq 0$, a possibility we cannot ignore. So we must provide a proof that if $a + b\sqrt{2} \neq 0$, then $a^2 - 2b^2 \neq 0$.

To see this, suppose the contrary, that $a^2 - 2b^2 = 0$. If $b = 0$, then $a^2 = 2 \cdot 0^2 = 0$ and hence $a + b\sqrt{2} = 0$, a contradiction of our choice of $a + b\sqrt{2}$. But if $b \neq 0$, then $\frac{a^2}{b^2} = 2$ and hence $\sqrt{2} = \pm\frac{a}{b} \in \mathbb{Q}$, a contradiction of Proposition 1.18. We thus conclude that $a^2 - 2b^2 \neq 0$, hence that $\frac{a}{a^2 - 2b^2}$ and $\frac{-b}{a^2 - 2b^2}$ are both in $\mathbb{Q}$.

Then, similarly to Example 5.3, it follows that

$$\frac{a}{a^2 - 2b^2} + \frac{-b}{a^2 - 2b^2}\sqrt{2} \in S,$$

and since

$$(a + b\sqrt{2})\left(\frac{a}{a^2 - 2b^2} - \frac{-b}{a^2 - 2b^2} + \sqrt{2}\right) = 1,$$

$$(a + b\sqrt{2})^{-1} = \frac{a}{a^2 - 2b^2} + \frac{-b}{a^2 - 2b^2}\sqrt{2} \in S.$$

So S satisfies condition (*v*) of Proposition 5.2 and hence is a subfield of $\mathbb{R}$. ❖

Note that the subfield S of Example 5.4 is the set F_1, the first of the two sets constructed at the beginning of the chapter that we used to capture Cardano's solution of the equation $x^3 + 3x + 2 = 0$. That is, if s is the positive real number such that $s^2 = 2$, then S is the set of all real numbers that can be created from s and elements of $\mathbb{Q}$ by using addition, subtraction, multiplication, and division. For certainly each element of S is created in this way from s and elements of $\mathbb{Q}$, and since $s \in S$ and $\mathbb{Q} \subseteq S$ and S is closed with respect to addition, subtraction, multiplication, and division, the only real numbers that can be created from s and elements of $\mathbb{Q}$ in this way are those in S.

Example 5.5   We will show that the set

$$H = \{(a + b\sqrt{2}) + (c + d\sqrt{2})i \,|\, a, b, c, d \in \mathbb{Q}\}$$

is a subfield of $\mathbb{C}$. Clearly $H \subseteq \mathbb{C}$. A method similar to that of Examples 5.3 and 5.4 could be used to show that H has at least two elements, is additively and multiplicatively closed, and has additive inverses. However, it would be quite complicated to show that H has multiplicative inverses by using the form of the elements of H given above. A way to avoid this difficulty is to observe that $H = \{A + Bi \,|\, A, B \in S\}$, where S is the set which we defined in Example 5.4 and which we have already shown to be a subfield of $\mathbb{R}$, and hence of $\mathbb{C}$.

Then since $0 = 0 + 0i \in H$ and $1 = 1 + 0i \in H$ and $0 \neq 1$, H contains at least two elements. Let $A + Bi, C + Di \in H$. Then

$$(A + Bi) + (C + Di) = (A + C) + (B + D)i \in H$$

because $A + C, B + D \in S$;

$$-(A + Bi) = (-A) + (-B)i \in H$$

because $-A, -B \in S$; and

$$(A + Bi)(C + Di) = (AC - BD) + (AD + BC)i \in H$$

because $AC - BD, AD + BC \in S$.

So H satisfies (ii), (iii), and (iv) of Proposition 5.2. To see that H satisfies (v), suppose that $0 \neq A + Bi \in H$. Then since $A, B \in \mathbb{R}$, we know from the proof of Proposition 3.1 that the multiplicative inverse of $A + Bi$ in $\mathbb{C}$ is

$$\frac{A}{A^2 + B^2} + \frac{-B}{A^2 + B^2}i,$$

and that, since S is a subfield of $\mathbb{R}$,

$$\frac{A}{A^2 + B^2}, \frac{-B}{A^2 + B^2} \in S.$$

It follows that

$$(A + Bi)^{-1} = \frac{A}{A^2 + B^2} + \frac{-B}{A^2 + B^2}i \in H,$$

and hence that H is a subfield of $\mathbb{C}$ by Proposition 5.2. ❖

Example 5.6 We will show that the set

$$K = \{(a + bi) + (c + di)\sqrt{2} \mid a, b, c, d \in \mathbb{Q}\}$$

is a subfield of $\mathbb{C}$. Clearly, $K \subseteq \mathbb{C}$. A method similar to that of the previous examples could be used to show that K satisfies (i)–(v) of Proposition 5.2. However, it is much easier to observe that

$$\begin{aligned} K &= \{(a + bi) + (c + di)\sqrt{2} \mid a, b, c, d \in \mathbb{Q}\} \\ &= \{(a + c\sqrt{2}) + (b + d\sqrt{2})i \mid a, b, c, d \in \mathbb{Q}\} = H, \end{aligned}$$

and we have already shown in Example 5.5 that H is a subfield of $\mathbb{C}$. ❖

Exercises

In Exercises 1–10, show that the given set is a subfield of $\mathbb{C}$.

1. $\{a + b\sqrt{7} \mid a, b \in \mathbb{Q}\}$ 2. $\{a + bi \mid a, b \in \mathbb{Q}\}$
3. $\{a + b\sqrt{5} \mid a, b \in \mathbb{Q}\}$ 4. $\{a + b\sqrt{-7} \mid a, b \in \mathbb{Q}\}$
5. $\{a + b\sqrt{-5} \mid a, b \in \mathbb{Q}\}$ 6. $\{2a + 3b\sqrt{13} \mid a, b \in \mathbb{Q}\}$
7. $\{(a + b\sqrt{7}) + (c + d\sqrt{7})i \mid a, b, c, d \in \mathbb{Q}\}$
8. $\{(a + bi) + (c + di)\sqrt{7} \mid a, b, c, d \in \mathbb{Q}\}$
9. $\{(a + b\sqrt{5}) + (c + d\sqrt{5})i \mid a, b, c, d \in \mathbb{Q}\}$
10. $\{(a + bi) + (c + di)\sqrt{5} \mid a, b, c, d \in \mathbb{Q}\}$

11. Show that the field axioms are redundant in the sense that if F is a set with binary operations $+$ and $\cdot$ which satisfy axioms **A1–A5** and **M1–M5,** then $+$ and $\cdot$ satisfy axiom **D1** if and only if $+$ and $\cdot$ satisfy axiom **D2.**

12. Prove or give a counterexample: If F and K are subfields of a field E, then $F \cap K$ is also a subfield of E.

13. Prove or give a counterexample: If F and K are subfields of a field E, then $F \cup K$ is also a subfield of E.

In Exercises 14–18, suppose that F is a field and prove the given statement.

14. For all $a \in F$, $-(-a) = a$.

15. For all $0 \neq a \in F$, $-a \neq 0$ and $(-a)^{-1} = -(a^{-1})$.

16. For all $0 \neq a \in F$ and $0 \neq b \in F$, $ab \neq 0$.

17. For all nonzero $a, b \in F$, $ab \neq 0$ and $(ab)^{-1} = a^{-1}b^{-1}$.

18. For all nonzero $a, b \in F$, $ab \neq 0$ and $a^{-1} + b^{-1} + (a + b)(ab)^{-1}$.
    $\left(\text{Note that for the field } \mathbb{R}, \text{ this equation is usually written: } \frac{1}{a} + \frac{1}{b} = \frac{a + b}{ab}.\right)$

19. Suppose that F is a field and that S is a subset of F which satisfies conditions (*ii*), (*iii*), (*iv*), and (*v*) of Proposition 4.2. Show that S has at least two elements if and only if S contains both 0 and 1.

20. Show that the only subfield of $\mathbb{Q}$ is $\mathbb{Q}$ itself.

21. According to Example 5.4, $S = \{a + b\sqrt{2} \mid a, b \in \mathbb{Q}\}$ is a subfield of $\mathbb{R}$. Show that the only subfields of S are $\mathbb{Q}$ and S itself.

22. According to Example 5.5, $H = \{(a + b\sqrt{2}) + (c + d\sqrt{2})i \mid a, b, c, d \in \mathbb{Q}\}$ is a subfield of $\mathbb{C}$. Find all subfields of H.

23. Show that the field axioms are redundant in the sense that if F is a set with binary operations $+$ and $\cdot$ that satisfy all the axioms except **A5,** then $+$ also satisfies axiom **A5.** (*Hint:* Consider $(1 + 1)(x + y)$.)

24. Determine whether $\{a + b\sqrt[3]{3} \mid a, b \in \mathbb{Q}\}$ is a subfield of $\mathbb{C}$. (Assume that for all $a, b \in \mathbb{Q}$, $\sqrt[3]{9} \neq a + b\sqrt[3]{3}$.)

25. Show that a subset S of a <u>finite</u> field is a subfield if and only if S contains more than one element and is both additively and multiplicatively closed.

6

Solvability by Radicals

According to Chapter 5, the notion of solvability by radicals hinges on extension fields generated by nth roots. The objective of this chapter is to phrase this notion rigorously. To do this, we need to know first that the appropriate extension fields exist. Then we can investigate how to use these fields to create the precise definition we require.

So we begin by showing that the necessary extension fields exist; that is, we show that there will always be a smallest subfield containing a given subset of a field. The following result provides this smallest subfield.

Proposition 6.1   *If S is a subset of a field U, then the intersection of all the subfields of U that contain S is the smallest subfield of U which contains S.*

Proof   Let $\mathbb{S}$ denote the set of all the subfields of U that contain S, and let F denote the intersection of all the subfields in $\mathbb{S}$. Note that certainly $U \in \mathbb{S}$, and thus $\mathbb{S}$ is non-empty. Furthermore, since 0 and 1 are in each of the subfields of $\mathbb{S}$, 0 and 1 must also be in F, and thus since $0 \neq 1$ by Proposition 5.1, F contains at least two elements. If $x, y \in F$ and $0 \neq z \in F$, then x, y, and z are all members of each subfield in $\mathbb{S}$, and hence so are $x + y$, xy, $-x$, and z^{-1}. Then $x + y$, xy, $-x$, and z^{-1} are all in F, and therefore by Proposition 5.2, F is a subfield of U. Since by definition of F, every subfield of U which contains S also contains F, F must be the smallest subfield of U which contains S. ∎

The subsets S in which we are primarily interested are sets similar to $\mathbb{Q} \cup \{\sqrt{2}\}$ in $\mathbb{C}$; that is, subsets which consist of a subfield together with a finite number of other elements. We therefore adopt the following notation.

> **DEFINITION**
> Let F be a subfield of a field U and let $r_1, r_2, \ldots, r_n$ be elements of U. The smallest subfield of U containing $F \cup \{r_1, r_2, \ldots, r_n\}$ is called the subfield of U **generated** by F and $r_1, r_2, \ldots, r_n$ and is denoted by $F(r_1, r_2, \ldots, r_n)$.

Example 6.2 We will show that $\mathbb{Q}(\sqrt{2}) = \{\alpha + \beta\sqrt{2} \mid \alpha, \beta \in \mathbb{Q}\}$ in $\mathbb{C}$ by showing (1) that $\mathbb{Q}(\sqrt{2}) \supseteq \{\alpha + \beta\sqrt{2} \mid \alpha, \beta \in \mathbb{Q}\}$ and (2) that $\mathbb{Q}(\sqrt{2}) \subseteq \{\alpha + \beta\sqrt{2} \mid \alpha, \beta \in \mathbb{Q}\}$.

(1) We know that $\mathbb{Q} \subseteq \mathbb{Q}(\sqrt{2})$ and that $\sqrt{2} \in \mathbb{Q}(\sqrt{2})$ by definition of $\mathbb{Q}(\sqrt{2})$. Hence for any $\alpha, \beta \in \mathbb{Q}$, the elements α, β, and $\sqrt{2}$ are all in $\mathbb{Q}(\sqrt{2})$. So since $\mathbb{Q}(\sqrt{2})$ is a field, $\alpha + \beta\sqrt{2}$ must also be in $\mathbb{Q}(\sqrt{2})$. It follows that $\mathbb{Q}(\sqrt{2}) \supseteq \{\alpha + \beta\sqrt{2} \mid \alpha, \beta \in \mathbb{Q}\}$.

(2) Since we do not know the form of the elements of $\mathbb{Q}(\sqrt{2})$, we cannot use a straightforward proof of the sort we used for case (1). Instead, we rely on the definition of $\mathbb{Q}(\sqrt{2})$ as the <u>smallest</u> subfield of $\mathbb{C}$ which contains both $\mathbb{Q}$ and $\sqrt{2}$. For by this property of $\mathbb{Q}(\sqrt{2})$, any subfield of $\mathbb{C}$ which contains both $\mathbb{Q}$ and $\sqrt{2}$ must also contain $\mathbb{Q}(\sqrt{2})$.

So we must show that $\{\alpha + \beta\sqrt{2} \mid \alpha, \beta \in \mathbb{Q}\}$ is a subfield of $\mathbb{C}$ which contains both $\mathbb{Q}$ and $\sqrt{2}$. Luckily, we have already shown in Example 5.4 that $\{\alpha + \beta\sqrt{2} \mid \alpha, \beta \in \mathbb{Q}\}$ is a subfield of $\mathbb{C}$; otherwise we would be faced with a rather lengthy argument. As well, $\sqrt{2} = 0 + 1\sqrt{2} \in \{\alpha + \beta\sqrt{2} \mid \alpha, \beta \in \mathbb{Q}\}$ and for all $q \in \mathbb{Q}$, $q = q + 0\sqrt{2} \in \{\alpha + \beta\sqrt{2} \mid \alpha, \beta \in \mathbb{Q}\}$. Therefore, since $\mathbb{Q}(\sqrt{2})$ is the smallest field that contains both $\mathbb{Q}$ and $\sqrt{2}$, we conclude that $\mathbb{Q}(\sqrt{2}) \subseteq \{\alpha + \beta\sqrt{2} \mid \alpha, \beta \in \mathbb{Q}\}$. ❖

Example 6.3 Similarly to Example 6.2, we will show that $\mathbb{Q}(-3 + i\sqrt{2}, 2 - \sqrt{8}) = \mathbb{Q}(i, \sqrt{2})$ by showing (1) that $\mathbb{Q}(-3 + i\sqrt{2}, 2 - \sqrt{8}) \subseteq \mathbb{Q}(i, \sqrt{2})$ and (2) that $\mathbb{Q}(-3 + i\sqrt{2}, 2 - \sqrt{8}) \supseteq \mathbb{Q}(i, \sqrt{2})$. In this case, both proofs require the indirect method used to prove the second containment of Example 6.2.

(1) Since $\mathbb{Q}(-3 + i\sqrt{2}, 2 - \sqrt{8})$ is the <u>smallest</u> subfield of $\mathbb{C}$ containing the subfield $\mathbb{Q}$, and the elements $-3 + i\sqrt{2}$ and $2 - \sqrt{8}$, it suffices to show that $\mathbb{Q}(i, \sqrt{2})$ is a subfield of $\mathbb{C}$ which contains $\mathbb{Q}, -3 + i\sqrt{2}$, and $2 - \sqrt{8}$. The first assertions follow from the definition of $\mathbb{Q}(i, \sqrt{2})$: certainly $\mathbb{Q}(i, \sqrt{2})$ is a subfield of $\mathbb{C}$ and certainly $\mathbb{Q} \subseteq \mathbb{Q}(i, \sqrt{2})$. Also by definition, $-3, 2, i$, and $\sqrt{2}$ are all in $\mathbb{Q}(i, \sqrt{2})$. So since $\mathbb{Q}(i, \sqrt{2})$ is a subfield, both $-3 + i\sqrt{2}$ and $2 - \sqrt{8} = 2 - (\sqrt{2})^3$ are in $\mathbb{Q}(i, \sqrt{2})$ as well. We conclude that since $\mathbb{Q}(-3 + i\sqrt{2}, 2 - \sqrt{8})$ is the smallest subfield of $\mathbb{C}$ containing $\mathbb{Q}, -3 + i\sqrt{2}$, and $2 - \sqrt{8}$, $\mathbb{Q}(-3 + i\sqrt{2}, 2 - \sqrt{8}) \subseteq \mathbb{Q}(i, \sqrt{2})$.

(2) Conversely, since $\mathbb{Q}(i, \sqrt{2})$ is the <u>smallest</u> subfield of $\mathbb{C}$ containing the subfield $\mathbb{Q}$, and the elements i and $\sqrt{2}$, it suffices to show that

$\mathbb{Q}(-3 + i\sqrt{2}, 2 - \sqrt{8})$ is a subfield of $\mathbb{C}$ containing $\mathbb{Q}$, i, and $\sqrt{2}$. As in step (1), by definition, $\mathbb{Q}(-3 + i\sqrt{2}, 2 - \sqrt{8})$ is a subfield of $\mathbb{C}$ which contains $\mathbb{Q}$. Furthermore, since 2 and $2 - \sqrt{8}$ are in $\mathbb{Q}(-3 + i\sqrt{2}, 2 - \sqrt{8})$,

$$\sqrt{2} = (2 - (2 - \sqrt{8}))(2)^{-1} \in \mathbb{Q}(-3 + i\sqrt{2}, 2 - \sqrt{8}),$$

and since 3, $-3 + i\sqrt{2}$, and $\sqrt{2}$ are all in $\mathbb{Q}(-3 + i\sqrt{2}, 2 - \sqrt{8})$,

$$i = ((-3 + i\sqrt{2}) + 3)(\sqrt{2})^{-1} \in \mathbb{Q}(-3 + i\sqrt{2}, 2 - \sqrt{8}).$$

It follows that since $\mathbb{Q}(i, \sqrt{2})$ is the smallest subfield of $\mathbb{C}$ containing $\mathbb{Q}$, i, and $\sqrt{2}$, $\mathbb{Q}(-3 + i\sqrt{2}, 2 - \sqrt{8}) \supseteq \mathbb{Q}(i, \sqrt{2})$. ❖

The field that served as the foundation for the construction described at the beginning of Chapter 5 was the field from which the equation's coefficients were drawn. In the general situation, we will consider a field and ask whether polynomial equations with coefficients from that field are solvable by radicals.

DEFINITION

Let F be a field. The set of all polynomials.

$$p(x) = a_0 + a_1 x + a_2 x^2 + \cdots + a_n x^n,$$

where $a_i \in F$, is denoted by $F[x]$. If $p(x) \in F[x]$, U is an extension field of F, and $r \in U$, then $p(r)$ is the element.

$$p(r) = a_0 + a_1 r + a_2 r^2 + \cdots + a_n r^n$$

of U. If $p(r) = 0$, then r is said to be a **root** of $p(x)$ or a **solution** of the equation $p(x) = 0$.

Note that a polynomial $p(x) = a_0 + a_1 x + \cdots a_n x^n$ may be viewed either as a formal sum or as a function from U to itself. In the former case, x is considered as a symbol unrelated to the elements of U, whereas in the latter case, it is considered to be a variable which represents an arbitrary element of U. Both points of view are useful. Thinking of polynomials as formal sums allows us to see immediately that two polynomials are equal if and only if they have the same coefficients; it also allows us to add polynomials easily by adding coefficients. On the other hand, when we ask for a root of $p(x)$, we are thinking of $p(x)$ as a function from U to itself and asking for an element $r \in U$ such that $p(r) = 0$.

Note also that the word "root" was used in Chapter 3 in a different sense. If $c \in \mathbb{C}$ is given and $r^n = c$, then r is an "nth root" of c. However, to say that $r^n = c$

is to say that r solves the equation $z^n - c = 0$, and thus to say that r is an nth root of c is to say that r is a root of the polynomial $z^n - c$. From this point of view, the use of the word root in reference to polynomials is a generalization of its use in reference to complex numbers.

We now have the machinery we need to define precisely what it means to solve an equation by means of the methods of Chapter 2. Those methods find the solutions of a polynomial equation by using addition, subtraction, multiplication, division, and the extraction of nth roots to combine the coefficients of the polynomial. Since the coefficients are all rational and since the rational numbers are a subfield of $\mathbb{C}$, we can phrase this in the language of field extensions by saying that the solutions live in an extension field of $\mathbb{Q}$ generated by nth roots.

For instance, the solutions of the equation $x^2 - 2x - 1 = 0$ are of course $x = 1 \pm \sqrt{2}$ which are built from elements of $\mathbb{Q} \cup \{\sqrt{2}\}$ and which thus, in the language of field extensions, live in $\mathbb{Q}(\sqrt{2})$.

However, we want our phrasing to be precise, and more complicated examples point out some subtleties we will have to take into consideration.

Example 6.4 Consider the equation $x^4 - 8x^3 + 22x^2 - 24x + 7 = 0$. Eliminating the x^3 by substituting $x = y + 2$ yields the equation $y^4 - 2y^2 - 1 = 0$ which is a quadratic in y^2 and which thus can be solved by using the quadratic formula: $y^2 = 1 \pm \sqrt{2}$. Then $y = \pm\sqrt{1 + \sqrt{2}}, \pm\sqrt{1 - \sqrt{2}}$, and thus the solutions to the original equation are $x = 2 \pm \sqrt{1 + \sqrt{2}}, 2 \pm \sqrt{1 - \sqrt{2}}$. We can place these solutions in an appropriate extension field by using successive extensions to correspond to the nested square roots. Specifically, we can first adjoin the radical $\sqrt{2}$ and then adjoin the radicals $\sqrt{1 + \sqrt{2}}$ and $\sqrt{1 - \sqrt{2}}$, finally forming $\mathbb{Q}(\sqrt{2}, \sqrt{1 + \sqrt{2}}, \sqrt{1 - \sqrt{2}})$, as shown in Figure 6.1. Note that the numbers $\sqrt{1 + \sqrt{2}}$ and $\sqrt{1 - \sqrt{2}}$ are not roots of elements of $\mathbb{Q}$ and hence if we wish to limit our adjoined elements to roots, we cannot adjoin them without first adjoining $\sqrt{2}$. ❖

FIGURE 6.1

$$\mathbb{Q}$$
$$\cap$$
$$\mathbb{Q}(\sqrt{2})$$
$$\cap$$
$$\mathbb{Q}(\sqrt{2}, \sqrt{1 + \sqrt{2}})$$
$$\cap$$
$$\mathbb{Q}(\sqrt{2}, \sqrt{1 + \sqrt{2}}, \sqrt{1 - \sqrt{2}})$$

Example 6.5 The solutions of the equation $x^4 + 4x^3 + 4x^2 + 1 = 0$ illustrate other difficulties. Proceeding as before, we substitute $x = y - 1$ and obtain $y^4 - 2y^2 + 2 = 0$. Solving this for y^2, we have $y^2 = 1 \pm i$. We would like to write $y = \pm\sqrt{1 + i}$, $\pm\sqrt{1 - i}$ and then $x = -1 \pm \sqrt{1 + i}, -1 \pm \sqrt{1 - i}$. However, as noted at the end of Chapter 2, the notation $\sqrt{1 + i}$ is ambiguous. Which complex square root of $1 + i$ is meant? We can avoid this difficulty by using powers instead of roots. Specifically, we can let r_1 be a complex number such that $r_1{}^2 = 1 + i$ (so that r_1 is one of the two complex square roots of $1 + i$ determined by the methods of Chapter 2) and r_2 be a complex number such that $r_2{}^2 = 1 - i$ (so that r_2 is one of the two complex square roots of $1 - i$). Then, whichever square roots we have chosen for r_1 and r_2, the other square roots are $-r_1$ and $-r_2$ respectively and the previous argument shows that the solutions of $x^4 + 4x^3 + 4x^2 + 1 = 0$ are $x = -1 \pm r_1, -1 \pm r_2$. It follows that $\mathbb{Q}(i, r_1, r_2)$ is an extension of $\mathbb{Q}$ that contains these solutions and which is built from $\mathbb{Q}$ by adjoining a series of nth roots, as shown in Figure 6.2. Note that, as in the previous example, r_1 and r_2 are not roots of elements of $\mathbb{Q}$, and hence if we wish to limit our adjoined elements to roots, we cannot adjoin them without first adjoining i. ❖

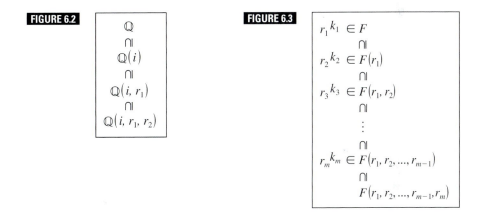

In view of this discussion, we adopt the following definition.

DEFINITION

(See Figure 6.3.) Suppose that F is a field, that $p(x) \in F[x]$, and that U is an extension field of F which contains all the roots of $p(x)$. Let $\alpha_1, \alpha_2, \ldots, \alpha_n$ be the roots of $p(x)$ in U, and let

$$F^{p(x)} = F(\alpha_1, \alpha_2, \ldots, \alpha_n).$$

Then the equation $p(x) = 0$ is **solvable by radicals** if there exist nonzero $r_1, r_2, \ldots, r_m \in U$ and positive integers $k_1, \ldots, k_m$ such that

(a) $F^{p(x)} \subseteq F(r_1, r_2, \ldots, r_m)$

(b) $(r_1)^{k_1} \in F$, and $(r_i)^{k_i} \in F(r_1, r_2, \ldots, r_{i-1})$ for $1 < i \leq m$.

The elements $r_1, r_2, \ldots, r_m \in U$ are not the solutions of $p(x) = 0$; rather they are the radicals from which the solutions are built. They are best imagined as the parts of the solutions that require the extraction of nth roots. This is precisely what part (b) says; for to require that

$$(r_i)^{k_i} \in F(r_1, r_2, \ldots, r_{i-1})$$

is to require (imprecisely speaking) that r_i be the k_ith root of the element $(r_i)^{k_i}$ in $F(r_1, r_2, \ldots, r_{i-1})$, i.e., that

$$r_i = \sqrt[k_i]{\rho} \text{ for some } \rho \in F(r_1, r_2, \ldots, r_{i-1}).$$

Example 6.6   Consider $p(x) = x^2 + 4x + 5 \in \mathbb{Q}[x]$. According to the quadratic formula, the two complex solutions of the equation $p(x) = 0$ are $-2 + i$, and $-2 - i$, and hence $\mathbb{Q}^{p(x)} = \mathbb{Q}(-2 + i, -2 - i)$. Both these solutions are algebraic combinations of the rational number -2 and the square root of the rational number -1. Thus, $i = \sqrt{-1}$ is the obvious choice for r_1, along with the power $k_1 = 2$. It is clear from the definition that $\mathbb{Q}(i)$ is a subfield of $\mathbb{C}$ containing $\mathbb{Q}$ and certainly -2 and i are elements of $\mathbb{Q}(i)$. But then both $-2 + i$ and $-2 - i$ are in $\mathbb{Q}(i)$ and hence, since $\mathbb{Q}^{p(x)}$ is the smallest subfield of $\mathbb{C}$ containing $\mathbb{Q}$, $-2 + i$ and $-2 - i$, $\mathbb{Q}^{p(x)}$ must be contained in $\mathbb{Q}(i)$. Since $r_1{}^{k_1} = i^2 = -1 \in \mathbb{Q}$, we conclude that the equation $x^2 + 4x + 5 = 0$ is solvable by radicals according to the definition.

Note that to satisfy the definition, it is only necessary to show that $\mathbb{Q}^{p(x)} \subseteq \mathbb{Q}(i)$. However, in this case, we can say more. For by definition, $\mathbb{Q}^{p(x)}$ is a subfield of $\mathbb{C}$ which contains $\mathbb{Q}$, and as well, since $\mathbb{Q}^{p(x)}$ is a field containing $\frac{1}{2}, -1 + i$, and $-2 - i$, we have that $i = \frac{1}{2}((-2 + i) - (-2 - i)) \in \mathbb{Q}^{p(x)}$. We can therefore conclude that, since $\mathbb{Q}(i)$ is the smallest subfield of $\mathbb{C}$ containing $\mathbb{Q}$ and i, $\mathbb{Q}(i) \subseteq \mathbb{Q}^{p(x)}$ and hence that in fact $\mathbb{Q}^{p(x)} = \mathbb{Q}(i)$. ❖

Example 6.7   Let $p(x) = 3x^4 + 12x^3 + 15x^2 - 9 \in \mathbb{Q}[x]$. According to Example 2.5, the numbers

$$\frac{-1 + \sqrt{5}}{2}, \quad \frac{-1 - \sqrt{5}}{2}, \quad \frac{-3 + \sqrt{-3}}{2}, \quad \frac{-3 - \sqrt{-3}}{2}$$

are the four complex solutions of the equation $p(x) = 0$, and hence

$$\mathbb{Q}^{p(x)} = \mathbb{Q}\left(\frac{-1 + \sqrt{5}}{2}, \frac{-1 - \sqrt{5}}{2}, \frac{-3 + \sqrt{-3}}{2}, \frac{-3 - \sqrt{-3}}{2}\right).$$

The radicals required to form these solutions are $\sqrt{5}$ and $\sqrt{-3}$, and thus we choose $r_1 = \sqrt{5}, k_1 = 2$, and $r_2 = \sqrt{-3}, k_2 = 2$.

We need to show that $\mathbb{Q}^{p(x)} \subseteq \mathbb{Q}(r_1, r_2)$ and that $r_1{}^{k_1} \in \mathbb{Q}$ and $r_2{}^{k_2} \in \mathbb{Q}(r_1)$. By definition, $\mathbb{Q}(r_1, r_2)$ is a subfield of $\mathbb{C}$ which contains $\mathbb{Q}$. Furthermore, since $\mathbb{Q}(r_1, r_2)$ is a field containing $-1, 2, -3, \sqrt{5}$, and $\sqrt{-3}$, it must contain all four of the solutions of $p(x) = 0$. Thus, since $\mathbb{Q}^{p(x)}$ is the smallest subfield of $\mathbb{C}$ containing $\mathbb{Q}$ and the four solutions of $p(x) = 0$, $\mathbb{Q}^{p(x)} \subseteq \mathbb{Q}(r_1, r_2)$. Clearly, $r_1{}^{k_1} = (\sqrt{5})^2 = 5 \in \mathbb{Q}$, and since $r_2{}^{k_2} = (\sqrt{-3})^2 = -3 \in \mathbb{Q}$ and $\mathbb{Q}$ is a subset of $\mathbb{Q}(\sqrt{5})$, certainly $r_2{}^{k_2} \in \mathbb{Q}(r_1)$. It follows that the equation $3x^4 + 12x^3 + 15x^2 - 9 = 0$ is solvable by radicals according to the definition.

As in Example 6.6, to satisfy the definition, it is only necessary to show, as we have done, that $\mathbb{Q}^{p(x)} \subseteq \mathbb{Q}(r_1, r_2)$. However, in this case as well, equality holds. For $\mathbb{Q}^{p(x)}$ is a subfield of $\mathbb{C}$ which contains $\mathbb{Q}$, and as well, $\dfrac{-1 + \sqrt{5}}{2}, \dfrac{-1 - \sqrt{5}}{2}, \dfrac{-3 + \sqrt{3}}{2}$, and $\dfrac{-3 - \sqrt{-3}}{2}$ are all in $\mathbb{Q}^{p(x)}$. Then

$$\sqrt{5} = \frac{-1 + \sqrt{5}}{2} - \frac{-1 - \sqrt{5}}{2} \in \mathbb{Q}^{p(x)},$$

and

$$\sqrt{-3} = \frac{-3 + \sqrt{-3}}{2} - \frac{-3 - \sqrt{-3}}{2} \in \mathbb{Q}^{p(x)},$$

and thus, since $\mathbb{Q}(r_1, r_2)$ is the smallest subfield of $\mathbb{C}$ which contains $\mathbb{Q}$, $\sqrt{5}$, and $\sqrt{-3}$, we also have that $\mathbb{Q}(r_1, r_2) \subseteq \mathbb{Q}^{p(x)}$ and hence in fact that $\mathbb{Q}^{p(x)} = \mathbb{Q}(r_1, r_2)$. ❖

Example 6.8   We noted in Example 6.4 that the solutions of $x^4 - 8x^3 + 22x^2 - 24x + 7 = 0$ are $x = 2 \pm \sqrt{1 + \sqrt{2}}, 2 \pm \sqrt{1 - \sqrt{2}}$, so that $\mathbb{Q}^{p(x)} = \mathbb{Q}(2 \pm \sqrt{1 + \sqrt{2}}, 2 \pm \sqrt{1 - \sqrt{2}})$. To see that this equation is solvable by radicals, let

$$\begin{aligned} r_1 &= \sqrt{2}, k_1 = 2, \\ r_2 &= \sqrt{1 + \sqrt{2}}, k_2 = 2, \text{ and} \\ r_3 &= \sqrt{1 - \sqrt{2}}, k_3 = 2. \end{aligned}$$

We must show that $\mathbb{Q}^{p(x)} \subseteq \mathbb{Q}(r_1, r_2, r_3)$ and that $r_1{}^{k_1} \in \mathbb{Q}$, $r_2{}^{k_2} \in \mathbb{Q}(r_1)$, and $r_3{}^{k_3} \in \mathbb{Q}(r_1, r_2)$. By definition, $\mathbb{Q}(r_1, r_2, r_3)$ is a subfield of $\mathbb{C}$ which contains $\mathbb{Q}$, and as well $2, \sqrt{1 + \sqrt{2}}$, and $\sqrt{1 - \sqrt{2}}$ are all in $\mathbb{Q}(r_1, r_2, r_3)$. So the solutions $2 \pm \sqrt{1 + \sqrt{2}}$ and $2 \pm \sqrt{1 - \sqrt{2}}$ are all in $\mathbb{Q}(r_1, r_2, r_3)$, and thus since $\mathbb{Q}^{p(x)}$ is the smallest subfield of $\mathbb{C}$ which contains $\mathbb{Q}$ and the four solutions of $p(x) = 0$, $\mathbb{Q}^{p(x)} \subseteq \mathbb{Q}(r_1, r_2, r_3)$. Furthermore,

$$\begin{aligned} r_1{}^{k_1} &= (\sqrt{2})^2 = 2 \in \mathbb{Q}, \\ r_2{}^{k_2} &= (\sqrt{1 + \sqrt{2}})^2 = 1 + \sqrt{2} \in \mathbb{Q}(r_1), \text{ and} \\ r_3{}^{k_3} &= (\sqrt{1 - \sqrt{2}})^2 = 1 - \sqrt{2} \in \mathbb{Q}(r_1, r_2). \end{aligned}$$

So we conclude that the equation $x^4 - 8x^3 + 22x^2 - 24x + 7 = 0$ is solvable by radicals according to the definition.

The argument of the previous paragraph shows that the equation is solvable by radicals. However, as in the previous two examples, we can say more in this case as well. Specifically, we can show that $\mathbb{Q}^{p(x)} = \mathbb{Q}(r_1, r_2, r_3)$. For by definition, $\mathbb{Q}^{p(x)}$ is a subfield of $\mathbb{C}$ which contains $\mathbb{Q}$ and $2 + \sqrt{1 + \sqrt{2}}$ and $2 + \sqrt{1 - \sqrt{2}}$, and hence

$$r_1 = \sqrt{2} = ((2 + \sqrt{1 + \sqrt{2}}) - 2)^2 - 1 \in \mathbb{Q}^{p(x)},$$
$$r_2 = \sqrt{1 + \sqrt{2}} = (2 + \sqrt{1 + \sqrt{2}}) - 2 \in \mathbb{Q}^{p(x)}, \text{ and}$$
$$r_3 = \sqrt{1 - \sqrt{2}} = (2 + \sqrt{1 - \sqrt{2}}) - 2 \in \mathbb{Q}^{p(x)}.$$

So since $\mathbb{Q}(r_1, r_2, r_3)$ is the smallest subfield of $\mathbb{C}$ which contains $\mathbb{Q}$, $\sqrt{2}$, $\sqrt{1 + \sqrt{2}}$, and $\sqrt{1 - \sqrt{2}}$, we have that $\mathbb{Q}^{p(x)} \supseteq \mathbb{Q}(r_1, r_2, r_3)$ and, thus, since we have shown that $\mathbb{Q}^{p(x)} \subseteq \mathbb{Q}(r_1, r_2, r_3)$, in fact we have that $\mathbb{Q}^{p(x)} = \mathbb{Q}(r_1, r_2, r_3)$. ❖

Example 6.9   Let $p(x) = x^3 - 5 \in \mathbb{Q}[x]$. Since $\sqrt[3]{5}$ solves $p(x) = 0$, Proposition 3.8 implies that the other solutions of the equation are $\sqrt[3]{5}\zeta_3$ and $\sqrt[3]{5}\zeta_3^2$, and thus $\mathbb{Q}^{p(x)} = \mathbb{Q}(\sqrt[3]{5}, \sqrt[3]{5}\zeta_3, \sqrt[3]{5}\zeta_3^2)$. Let $r_1 = \sqrt[3]{5}$, $k_1 = 3$, and $r_2 = \zeta_3$, $k_2 = 3$. Then by definition, $\mathbb{Q}(r_1, r_2)$ is a subfield of $\mathbb{C}$ containing $\mathbb{Q}$. As well, since $\mathbb{Q}(r_1, r_2)$ is a field containing $\sqrt[3]{5}$ and ζ_3, then $\sqrt[3]{5}, \sqrt[3]{5}\zeta_3$, and $\sqrt[3]{5}\zeta_3^2$ are all in $\mathbb{Q}(r_1, r_2)$ so that $\mathbb{Q}^{p(x)} \subseteq \mathbb{Q}(r_1, r_2)$. Furthermore $r_1^{k_1} = (\sqrt[3]{5})^3 = 5 \in \mathbb{Q}$ and $r_2^{k_2} = (\zeta_3)^3 = 1 \in \mathbb{Q}(r_1)$. So it follows that the equation $p(x) = 0$ is solvable by radicals according to the definition.

As in Examples 6.6, 6.7, and 6.8, to satisfy the definition, it is only necessary to show, as we have done, that $\mathbb{Q}^{p(x)} \subseteq \mathbb{Q}(r_1, r_2)$. However, in this case as well, we can say more. For $\mathbb{Q}^{p(x)}$ is a subfield of $\mathbb{C}$ by definition, and as well $\sqrt[3]{5} \in \mathbb{Q}^{p(x)}$ and $\zeta_3 = (\sqrt[3]{5})^{-1}(\sqrt[3]{5}\zeta_3) \in \mathbb{Q}^{p(x)}$. Thus, since $\mathbb{Q}(r_1, r_2)$ is the smallest subfield of $\mathbb{C}$ which contains $\mathbb{Q}$, $\sqrt[3]{5}$, and ζ_3, we also have that $\mathbb{Q}(r_1, r_2) \subseteq \mathbb{Q}^{p(x)}$ and hence that $\mathbb{Q}^{p(x)} = \mathbb{Q}(r_1, r_2)$. ❖

In Examples 6.6, 6.7, 6.8, and 6.9, it turned out that the extension field $\mathbb{Q}^{p(x)}$ was not only contained in the extension field generated by the radicals but in fact equal to it. This need not happen. For instance, consider the equation $x^2 + 2 = 0$. Its solutions are $x = \pm i\sqrt{2}$, and if $r_1 = i$, $k_1 = 2$, and $r_2 = \sqrt{2}$, $k_2 = 2$, then $\mathbb{Q}^{p(x)} \subseteq \mathbb{Q}(r_1, r_2)$, $r_1^{k_1} \in \mathbb{Q}$ and $r_2^{k_2} \in \mathbb{Q}(r_1)$, but $\mathbb{Q}^{p(x)} \neq \mathbb{Q}(r_1, r_2)$ (see Exercise 33).

In Example 6.2, we saw that $\mathbb{Q}(\sqrt{2})$ may be expressed very concisely in terms of $\sqrt{2}$ and elements of $\mathbb{Q}$:

$$\mathbb{Q}(\sqrt{2}) = \{\alpha + \beta\sqrt{2} \mid \alpha, \beta \in \mathbb{Q}\}.$$

The hard part of the proof was showing that the set $\{\alpha + \beta\sqrt{2} \mid \alpha, \beta \in \mathbb{Q}\}$ was a field, the proof of which was given in Example 5.4. Even in such a simple case, the existence of multiplicative inverses was difficult to prove. We did not even attempt such a description for $\mathbb{Q}(\sqrt{5}, \sqrt{-3})$ in Example 6.7. In most cases, trying to describe multiple extensions or extensions by roots other than square roots is very difficult by the direct methods of Examples 5.3, 5.4, 5.5, and 6.2. Determining the structure of so simple an extension field as that generated by a single solution of an arbitrary polynomial equation may even appear to be impossible; after all, we are trying to show that the techniques developed in Chapter 2 for solving polynomial equations are unavailable in general. In Chapter 10 (Theorem 10.4), we will see that the elements of such an extension field do indeed have a simple form analogous to what we found above for elements of $\mathbb{Q}(\sqrt{2})$. The proof that multiplicative inverses of such elements have the same simple form will not be easy; in fact it will require us to know something of the algebraic structure of the set of polynomials $F[x]$. We begin the general study of such structures in the next chapter.

Exercises

In Exercises 1–7, for the polynomial $p(x)$, find $\mathbb{Q}^{p(x)}$.

1. $p(x) = 2x - 13$
2. $p(x) = x^2 - 2x + 5$
3. $p(x) = x^2 + x + 1$
4. $p(x) = x^3 - 1$
5. $p(x) = x^3 - 7$
6. $p(x) = 4x^4 + 16x^3 + 20x^2 + 8x + 1$
7. $p(x) = x^4 + 4x^3 - x^2 + 2x + 3$
(*Hint:* $y^4 - 7y^2 + 12y - 3 = y^4 + 2y^2 + 1 - 9y^2 + 12y - 4$)

In Exercises 8–23, prove the given equality in detail.

8. $\mathbb{Q}(\sqrt{7}) = \{a + b\sqrt{7} \mid a, b \in \mathbb{Q}\}$
9. $\mathbb{Q}(\sqrt{5}) = \{a + b\sqrt{5} \mid a, b \in \mathbb{Q}\}$
10. $\mathbb{Q}(\sqrt{-7}) = \{a + b\sqrt{-7} \mid a, b \in \mathbb{Q}\}$
11. $\mathbb{Q}(i) = \{a + bi \mid a, b \in \mathbb{Q}\}$
12. $\mathbb{Q}(i\sqrt{5}) = \{a + b\sqrt{-5} \mid a, b \in \mathbb{Q}\}$
13. $\mathbb{Q}(\sqrt{7}, i) = \{(a + b\sqrt{7}) + (c + d\sqrt{7})i \mid a, b, c, d \in \mathbb{Q}\}$
14. $\mathbb{Q}(\sqrt{7}, i) = \{(a + bi) + (c + di)\sqrt{7} \mid a, b, c, d \in \mathbb{Q}\}$
15. $\mathbb{Q}(i, \sqrt{5}) = \{(a + b\sqrt{5}) + (c + d\sqrt{5})i \mid a, b, c, d \in \mathbb{Q}\}$
16. $\mathbb{Q}(i, \sqrt{5}) = \{(a + bi) + (c + di)\sqrt{5} \mid a, b, c, d \in \mathbb{Q}\}$
17. $\mathbb{Q}(2 + i) = \mathbb{Q}(i)$
18. $\mathbb{Q}(3 - \sqrt{2}, 5 + i) = \mathbb{Q}(\sqrt{2}, i)$
19. $\mathbb{Q}(i, \sqrt{7}, \sqrt{-6}) = \mathbb{Q}(\sqrt{-6}, i, \sqrt{7})$
20. $\mathbb{Q}(i, \sqrt{7}, \sqrt{-6}) = \mathbb{Q}(\sqrt{6}, i, \sqrt{7})$
21. $\mathbb{Q}(1 + i, 3 - i\sqrt{8}, 8 + i\sqrt{6}) = \mathbb{Q}(i, i\sqrt{8}, i\sqrt{6})$
22. $\mathbb{Q}(2 + i, 1 + i\sqrt{12}, 12 - i\sqrt{8}) = \mathbb{Q}(\sqrt{8}, \sqrt{12}, i)$
23. $\mathbb{Q}(7 - i, -3 + i\sqrt{8}, 4 - i\sqrt{10}) = \mathbb{Q}(i, \sqrt{2}, \sqrt{5})$

In Exercises 24–32, show that the given polynomial equation is solvable by radicals over $\mathbb{Q}$ according to the definition.

24. $x^2 + 2x + 3 = 0$
25. $x^3 - 11 = 0$
26. $x^7 - 5 = 0$
27. $x^2 + 3x + 5 = 0$

28. $x^4 - 1 = 0$

29. $x^4 + x^3 + x^2 + x + 1 = 0$

30. $x^4 + 4x^3 + 4x^2 + 1 = 0$

31. $x^n - 1 = 0$, where $n > 0$

32. $x^4 + 4x^3 - x^2 + 2x + 3 = 0$

(*Hint:* $y^4 - 7y^2 + 12y - 3 = y^4 + 2y^2 + 1 - 9y^2 + 12y - 4$)

33. Show that $x^2 + 2 = 0$ is solvable by radicals by using $r_1 = i$ and $r_2 = \sqrt{2}$. Show also that if $p(x) = x^2 + 2$, then $\mathbb{Q}^{px} \neq \mathbb{Q}(r_1, r_2)$.

34. Prove or give a counterexample: If F and K are subfields of a field E, then $\{k + f \mid k \in K, f \in F\}$ is also a subfield of E.

35. Prove or give a counterexample: If F and K are subfields of a field E, then $\{kf \mid k \in K, f \in F\}$ is also a subfield of E.

36. Suppose that $p(x)$ is a quadratic polynomial with rational coefficients. Show that $p(x) = 0$ is solvable by radicals according to the definition.

37. Suppose that $p(x)$ is a cubic polynomial with rational coefficients. Show that $p(x) = 0$ is solvable by radicals according to the definition.

38. Suppose that $p(x)$ is a quartic polynomial with rational coefficients. Show that $p(x) = 0$ is solvable by radicals according to the definition.

39. Prove or give a counterexample: For all nonzero complex numbers z and all nonzero rational numbers a and b, $\mathbb{Q}(az + b) = \mathbb{Q}(z)$.

40. Prove or give a counterexample: For all nonzero complex numbers z and all nonzero rational numbers a and b, $\mathbb{Q}(az^2 + b) = \mathbb{Q}(z)$.

41. Prove or give a counterexample: For all nonzero complex numbers z and all nonzero rational numbers a and b, $\mathbb{Q}(a^2z - b) = \mathbb{Q}(z)$.

42. Suppose that F is a subfield of a field K and that $r, s, t \in K$. Show that $F(r, s, t) = F(s, t, r)$.

43. Let F be a field, let $p(x) \in F[x]$, and suppose that U is an extension field of F which contains all the roots of $p(x)$. Show that $p(x) = 0$ is solvable by radicals if and only if there exist $s_1, \ldots, s_k \in U$ and positive integers $t_1, \ldots, t_k$ such that if $F_0 = F$ and $F_i = F(s_1, \ldots, s_i)$ for $1 \leq i \leq k$, then (a) $F^{p(x)} \subseteq F_k$, (b) for all $1 \leq i \leq k$, $s_i{}^{t_i} \in F_{i-1}$, and (c) for all $1 \leq i \leq k$, either $s_i = \zeta_{t_i}$ or $\zeta_{t_i} \in F_{i-1}$.

44. Show that $\mathbb{Q}$ and $\mathbb{Q}(\zeta_4)$ are the only subfields of $\mathbb{Q}(\zeta_4)$.

45. Show that $\mathbb{Q}$ and $\mathbb{Q}(\sqrt[4]{7})$ are <u>not</u> the only subfields of $\mathbb{Q}(\sqrt[4]{7})$.

HISTORICAL NOTE

Niels Henrik Abel

Niels Henrik Abel was born in Finnöy, Norway (an island near Stavanger), on August 5, 1802, and died in Froland, Norway, on April 6, 1829. His father, Sören Georg Abel, was a Lutheran minister and the son of a Lutheran minister; his mother, Ane Marie Simonson, was the daughter of a wealthy merchant from Risör, on the southern coast. Niels was the second of seven children. Finnöy was Sören Georg's first parish and he found the work quite difficult. He left in 1804 when his father retired and he was appointed his successor in Gjerstad, near Risör. Niels Henrik received his education at home until 1815 when he and his older brother were sent to the Cathedral School in Kristiana (now Oslo).

His marks were satisfactory but nothing more, until 1817 when he entered the mathematics class of Bernt Holmboe, a new teacher only seven years older than Abel. Holmboe quickly realized Abel's potential and gave him special assignments and outside reading. Indeed soon it was Abel who was teaching Holmboe rather than the other way round, Holmboe becoming so enthusiastic that the school's rector had to force him to moderate his comments in the record book. Of course, this did not keep Holmboe's opinions from reaching the mathematicians at the university and they soon made Abel's acquaintance. It was during his last year at school that Abel attacked the problem of the solvability of the quintic equation. He found what he thought was a solution and because no one in Kristiana could understand his method, he sent his paper to Ferdinand Degen in Copenhagen. Degen could find nothing wrong with Abel's solution but wanted examples. And in constructing examples, Abel found that his method did not work. Degen also believed that trying to solve the quintic equation was not a fruitful way to begin a mathematical career and suggested that Abel take up the theory of elliptic integrals. Indeed he did just this, and within a few years he had probably already discovered his revolutionary contributions to the field.

Some years earlier, Abel's father had entered politics, and in 1818, as a member of the Norwegian parliament, he began hurling unfounded accusations at his opponents. By this time, he was an alcoholic (as was Abel's mother) and was forced to return to his parish in disgrace. He died in 1820, leaving his wife with a pension which was barely sufficient to support her and their many children. So when Niels Henrik entered the university in 1821, he did so with a destitute family and no means of support. The university had only begun operation in 1813 and there was no fellowship money available. But in recognition of his phenomenal ability and desperate financial circumstances, several professors used parts of their own salaries to support him.

In 1823, one of Abel's benefactors gave him enough money to travel to Copenhagen to meet with the Danish mathematicians. He found not only mathematical inspiration in Copenhagen, but also Christine Kemp (nicknamed Crelly), the woman who was to become his fiancée. Ironically, in view of Degen's opinions, his return to Norway coincided with his return to the problem of solving the quintic equation. This time, he succeeded in showing (correctly) that it is not always solvable. To reach a larger audience, he wrote his proof in French, and to save expenses, he compressed his argument to six pages. He sent copies to foreign mathematicians but there was no reaction. Indeed after the death of Gauss, the copy he received in Göttingen was found among his papers, unopened.

With all his financial worries and mathematical work, Abel had not forgotten Crelly Kemp. He found, through his own pupils, a situation for her as governess for the children of a family in Son, only a short day's travel from Kristiana. She came to Norway, and at Christmas 1824, the couple publicly announced their engagement.

Abel had long felt the desire to visit the scientific centers of Europe and in the spring of 1825, encouraged by the similar plans of several friends, he felt the time was ripe. He applied for, and eventually won, a grant for two years of foreign study. Before departing, he made every effort to ensure the welfare of his family while he was away, and just prior to his departure, he paid a last visit to Crelly Kemp in Son.

His original itinerary began with a stay in Paris. However, his friends were on their way to Berlin and, somewhat apprehensive of what lay before him, he decided to join them. This apparently unimportant decision was a turning point in Abel's mathematical career. For one of the Danish mathematicians had given him a letter of introduction to an influential Berlin engineer, one Leopold Crelle, and Abel visited him soon after his arrival in Berlin. Crelle had long intended to begin a German mathematical journal to rival those already available in France, and during their meeting, Abel gave him a copy of his paper on the unsolvability of the quintic. Finding the compressed proof very difficult to understand, Crelle asked for an expanded version and, after he received it, published it in the first volume of his journal. Crelle gave his new journal the name *Journal für die reine und angewandte Mathematik*, but it was, and still is, known by its nickname, *Crelle's Journal*. Many of Abel's most influential papers appeared in its initial volumes and helped to establish it as the leading German mathematical journal of the nineteenth century.

Abel was unaware that twenty-five years earlier, Ruffini had published the same result. However, in a posthumous paper, Abel credits Ruffini by writing: "If I am not mistaken, the first to attempt a proof of the impossibility of an algebraic solution of the general equation was the mathematician Ruffini; but his memoir is so complicated that it is difficult to judge the correctness of his argument."[1] Ruffini of course could have made a similar comment about Abel's initial paper.

Amidst the joy of finding both a friend and a publisher, Abel received very disappointing news from Norway. One of the Norwegian mathematicians had resigned his position and the faculty had appointed Abel's teacher Holmboe to fill the vacant post. Abel's friends found this most unfair and it is hard to believe that Abel did not agree. Although he wrote a warm letter of congratulations to his teacher and they remained friends, it is certain that he began to worry about the future again. It would be very difficult for him to marry Crelly Kemp if he did not have a position.

In retrospect, it seems a calamitous mistake that Abel was not appointed to the vacant position. However, from the point of view of the faculty at the time, the issue is hardly so clear-cut. They were faced with an undoubtedly brilliant but young and inexperienced candidate on the one hand, and an excellent and experienced teacher who would be able to take up his duties immediately on the other. They made the less risky choice and it cost mathematics dearly.

In the spring of 1826, Abel traveled to Austria and Italy and then finally to Paris. However, he found the French mathematicians aloof and life in Paris difficult. He had reserved his work on elliptic integrals for presentation to the French Academy and after several months more work, he submitted it for publication. The referees were Legendre, still active at seventy-four years of age, and Cauchy, a mathematician whose prodigious output was matched only by his ambition. He had the habit of extending the work presented in manuscripts given to him for review and then reporting on both the manuscripts and his own extensions simultaneously, and this may have had something to do with his treatment of Abel's paper. Whatever the reason, Cauchy put the manuscript aside while Abel waited in vain for a report, a report which was finally issued only when necessitated by his death. It is true that the paper was long and contained entirely new and extremely general ideas, and without doubt it presented its referees with a difficult task. However, this does not excuse the harm Cauchy's indifference caused.

Abel returned to Berlin as soon as the terms of his fellowship permitted but became ill soon after he arrived, probably with the tuberculosis that eventually killed him. He desperately wanted to return to Norway but the terms of his fellowship required that he live abroad until it expired. Crelle wanted him to stay in Berlin until he could obtain a post for him at a German university, but in May 1827, Abel returned to Norway. He found his prospects even more dismal than he had expected. He had no prospect of a position and no fellowship, and he was in debt. The university granted him a small stipend but specified that it be deducted from any future salary. For the next six months, he eked out a living as a tutor in Kristiana, until Christmas when he was given a two-year appointment as a substitute.

Abel used well the spare time which his poverty gave him. He was in the process of determining exactly which equations are solvable by radicals when he found that he had a competitor, Karl Jacobi, in the field of elliptic functions. Spurred on, he interrupted his work on solvability by radicals and quickly wrote a series of papers on elliptic functions. In the meantime, Crelle continued his efforts to obtain a position for him in Berlin. While Crelle's efforts had yet to bear fruit, they did ensure that the European mathematicians were aware of Abel's precarious situation. And in September 1828, four prominent members of the French Academy requested that the king of Sweden and Norway create a suitable position for him.

However, Abel's health was deteriorating. Crelly Kemp was now a governess in Froland, and he could visit her only during the holidays. Against his doctor's advice, he went to Froland for the Christmas holidays of 1828. He was waiting for the sled that was to take him back to Kristiana when he suffered a violent hemorrhage. Abel obeyed the doctor's orders for prolonged bed rest but nevertheless died from tuberculosis in April and was buried in Froland during a blizzard.

Shortly after and with great difficulty, Crelly Kemp wrote to inform their mutual friends in Copenhagen of Abel's death. "My Abel is dead!" she wrote, "he died on April 6 at four o'clock in the afternoon. I have lost all on earth. Nothing, nothing have I left."[2] While Crelly Kemp had nothing left, the world still had the mathematical legacy of Niels Henrik Abel.

References

1. Ore, Oystein. *Niels Henrik Abel*. New York: Chelsea. 1957, p. 74.
2. *Ibid.,* p. 225.

7 Rings

In Chapter 6, we stated precisely what we mean when we say that a polynomial equation is "solvable by radicals." Having a rigorous definition is a good first step, but it will do us no good unless we can put it to use. And that is the difficulty. It is not enough to be able to show that a given polynomial equation is solvable by radicals; we want to show that there are polynomial equations that are <u>not</u> solvable by radicals. That is, we want to find polynomials $p(x) \in \mathbb{Q}[x]$ such that for <u>any</u> radicals $r_1, \ldots, r_m$, the extension field $\mathbb{Q}^{p(x)}$ is <u>not</u> contained in $\mathbb{Q}(r_1, \ldots, r_m)$. This is of course a much more difficult problem, and, to solve it, we will need to know as much as we can about the structure of extension fields of $\mathbb{Q}$.

In fact, we do know a little about such extensions. In the cases that interest us, the elements to be adjoined are always roots of polynomials: the elements adjoined to $\mathbb{Q}$ to form $\mathbb{Q}^{p(x)}$ are all roots of the polynomial $p(x)$ in $\mathbb{Q}[x]$, and the elements adjoined to $\mathbb{Q}$ to form $\mathbb{Q}(r_1, \ldots, r_m)$ are all roots of the polynomials $x^{k_i} - r_i^{k_i}$ in $\mathbb{Q}(r_1, \ldots, r_{i-1})[x]$. For this reason, we now turn our attention to the sets $F[x]$ for fields F.

We will see in the next chapter that these sets are algebraically very much like the integers, an observation we will exploit in subsequent chapters. In this chapter, we introduce the general class of algebraic structures to which $\mathbb{Z}$, $\mathbb{Z}_m$, and $F[x]$ belong, that of rings.

If F is a field, then a typical element in $F[x]$ is the polynomial

$$p(x) = a_0 + a_1x + a_2x^2 + a_3x^3 + \cdots + a_{m-1}x^{m-1} + a_mx^m,$$

where $m \geq 0$ and each $a_i \in F$. Viewing these polynomials as formal sums, we see that

$$a_0 + a_1x + a_2x^2 + \cdots + a_{m-1}x^{m-1} + a_mx^m =$$
$$b_0 + b_1x + b_2x^2 + \cdots + b_{n-1}x^{n-1} + b_nx^n$$

if and only if $n = m$ and $a_i = b_i$ for all i.

Since F is a field, we can add and multiply the coefficients of the polynomials and thus in the usual way we can add and multiply the polynomials themselves. In particular, if

$$p(x) = a_0 + a_1 x + a_2 x^2 + a_3 x^3 + \cdots + a_{m-1} x^{m-1} + a_m x^m,$$
$$q(x) = b_0 + b_1 x + b_2 x^2 + b_3 x^3 + \cdots + b_{n-1} x^{n-1} + b_n x^n.$$

Then we define

$$p(x) + q(x) = (a_0 + b_0) + (a_1 + b_1)x + \cdots +$$
$$(a_{r-1} + b_{r-1})x^{r-1} + (a_r + b_r)x^r,$$

where r is the maximum of m and n and we take $a_i = 0$ or $b_j = 0$ as necessary.

Multiplication, of course, involves the distributive law and may appear to be a notational horror. However, there is a pattern to it. Consider

$$a_0 + a_1 x + a_2 x_2, \quad b_0 + b_1 x + b_2 x_2 \in \mathbb{Q}[x].$$

Then

$$(a_0 + a_1 x + a_2 x_2)(b_0 + b_1 x + b_2 x_2)$$
$$= a_0 b_0 + (a_0 b_1 + a_1 b_0)x + (a_0 b_2 + a_1 b_1 + a_2 b_0)x^2 +$$
$$(a_1 b_2 + a_2 b_1)x^3 + (a_2 b_2)x^4.$$

Notice that the coefficient of each x^k is a sum of products $a_i b_j$, where the sum of the indices $i + j$, is exactly k. So, in general, for the polynomials $p(x)$ and $q(x)$, we define multiplication in $F[x]$ by letting

$$p(x)q(x) = \sum_{k=0}^{m+n} \left(\sum_{i+j=k} a_i b_j \right) x^k.$$

Since $F[x]$ is built from a field, one might suspect that, with these operations, it too is a field. Indeed it satisfies most of the axioms for a field . . . but not all. Most polynomials do <u>not</u> have multiplicative inverses.

Proposition 7.1   *If F is a field, then F[x] is additively closed, associative, and commutative, has an additive identity and additive inverses, is multiplicatively closed, associative, and commutative, and has a multiplicative identity; as well, multiplication distributes over addition.*

Proof   By definition of the operations, $F[x]$ is additively and multiplicatively closed. The other properties follow from the corresponding axioms for the field; some samples are provided below, the rest being left to the reader (Exercises 11–16).

EXISTENCE OF AN ADDITIVE INVERSE: For $p(x) = a_0 + a_1 x + \cdots + a_m x^m$, let $q(x) = b_0 + b_1 x + \cdots + b_m x^m$, where $a_i + b_i = 0$ for all $i = 1, \ldots, m$. Then

$$p(x) + q(x) = (a_0 + b_0) + (a_1 + b_1)x + \cdots + (a_m + b_m)x^m$$
$$= 0 + 0x + \cdots + 0x^r = 0.$$

MULTIPLICATIVE COMMUTATIVITY: Let $p(x) = a_0 + a_1 x + \cdots + a_m x^m$, and $q(x) = b_0 + b_1 x + \cdots + b_n x^n$. Then, since multiplication in F is commutative,

$$p(x)q(x) = \sum_{k=0}^{m+n}\left(\sum_{i+j=k} a_i b_j\right)x^k = \sum_{k=0}^{n+m}\left(\sum_{i+j=k} b_j a_i\right)x^k = q(x)p(x). \ \blacksquare$$

Proposition 7.2   *For any field F, $a_0 + a_1 x + \cdots + a_m x^m$ has a multiplicative inverse in $F[x]$ if and only if $a_0 \neq 0$ and $a_i = 0$ for all $i \geq 1$.*

Proof   If $a_0 \neq 0$ and $a_i = 0$ for all $i \leq 1$, then $a_0 + a_1 x + \cdots + a_m x^m = a_0$. As an element of F, a_0 has an inverse a_0^{-1}. But a_0^{-1} is also a polynomial and as polynomials, $a_0 a_0^{-1} = 1$. Since the polynomial 1 is the multiplicative identity in $F[x]$, we have shown that a_0 has a multiplicative inverse in $F[x]$. Conversely, suppose that $p(x) = a_0 + a_1 x + \cdots + a_m x^m a_m \neq 0$, has a multiplicative inverse in $F[x]$, say, $q(x) = b_0 + b_1 x + \cdots + b_n x^n$. Certainly $q(x) \neq 0$ and hence we may assume that $b_n \neq 0$. Suppose by way of contradiction that $m \geq 1$. Then $p(x)q(x) = c_0 + c_1 x + \cdots + c_{m+n} x^{m+n}$, where $c_{m+n} = a_m b_n$. If $a_m b_n = 0$, then by Proposition 5.1, $b_n = (a_m^{-1} a_m)b_n = a_m^{-1} 0 = 0$, a contradiction; hence $c_{m+n} = a_m b_n \neq 0$. But then, since $m + n \geq 1$, $p(x)q(x) \neq 1$ and hence $q(x)$ cannot be a multiplicative inverse of $p(x)$. This contradicts our choice of $q(x)$, and hence $m = 0$, i.e., $a_i = 0$ for all $i \geq 1$. But since $p(x) \neq 0$, $a_0 \neq 0$. $\blacksquare$

By Proposition 7.2, only the constant polynomials in $F[x]$ have multiplicative inverses, and hence $F[x]$ with its two natural operations cannot be a field. However, it is nonetheless a very important structure which we want to be able to consider in general. The structures it exemplifies are called rings.

DEFINITION

A **ring** is a set R with two binary operations $+$ and $\cdot$ which satisfy the following axioms:

A1.   ADDITIVE CLOSURE: *For all $x, y \in R$, $x + y \in R$.*

A2.   ADDITIVE ASSOCIATIVITY: *For all $x, y, z \in R$, $(x + y) + z = x + (y + z)$.*

A3.   EXISTENCE OF AN ADDITIVE IDENTITY: *There exists $0 \in R$ such that for all $x \in R$, $x + 0 = x = 0 + x$.*

A4.   EXISTENCE OF ADDITIVE INVERSES: *For all $x \in R$, there exists $y \in R$ such that $x + y = 0 = y + x$.*

A5.   ADDITIVE COMMUTATIVITY: *For all $x, y \in R$, $x + y = y + x$.*

M1.   MULTIPLICATIVE CLOSURE: *For all $x, y \in R$, $xy \in R$.*

M2.   MULTIPLICATIVE ASSOCIATIVITY: *For all $x, y, z \in R$, $x(yz) = (xy)z$.*

D1.   LEFT DISTRIBUTIVITY: *For all $x, y, z \in R$, $x(y + z) = xy + xz$.*

D2.   RIGHT DISTRIBUTIVITY: *For all $x, y, z \in R$, $(x + y)z = xz + yz$.*

Like fields, rings have two operations. But rings have fewer restrictions on these operations. So by definition, since a field satisfies all the ring axioms and more, any field is certainly a ring. On the other hand, not every ring is a field. For $\mathbb{Z}$ is obviously a ring but not a field, and for any field F, Propositions 7.1 and 7.2 imply that $F[x]$ is also a ring but not a field. Furthermore, for any positive integer n, $\mathbb{Z}_n$ is a ring, and by Proposition 4.3, if n is not prime, $\mathbb{Z}_n$ is not a field.

DEFINITION

If F is a field, then the **polynomial ring** over F is the ring $F[x]$ of polynomials

$$p(x) = a_0 + a_1 x + \cdots + a_m x^m, \qquad q(x) = b_0 + b_1 x + \cdots + b_n x^n,$$

where $a_i, b_i \in F$. These polynomials are added and multiplied as follows:

$$p(x) + q(x) = (a_0 + b_0) + (a_1 + b_1)x + \cdots +$$
$$(a_{r-1} + b_{r-1})x^{r-1} + (a_r + b_r)x^r,$$

$$p(x)q(x) = \sum_{k=0}^{m+n} \left(\sum_{i+j=k} a_i b_j \right) x^k,$$

where r is the maximum of m and n.

We've seen that for any field F, $F[x]$ satisfies more of the field axioms than just those which define rings: for instance, multiplication in $F[x]$ is commutative and $F[x]$ has a multiplicative identity (or unit element). Rings like $F[x]$ which satisfy additional axioms are important enough that they are given special names. (For rings satisfying axiom **M4,** see Exercises 22 and 23.)

DEFINITION

A **commutative ring** is a ring whose multiplication is commutative (i.e., which satisfies field axiom **M5**). A **ring with unit element** is a ring which has a multiplicative identity (i.e., which satisfies field axiom **M3**).

Note that a commutative ring need <u>not</u> have a unit element (Exercise 24) and that a ring with unit element need <u>not</u> be commutative (Example 7.3).

Clearly $\mathbb{Z}$ is a commutative ring with unit element, and as we saw in Example 4.1, for any positive integer n, $\mathbb{Z}_n$ is also a commutative ring with unit element. Furthermore, by Proposition 7.1, for any field F, the polynomial ring $F[x]$ is a commutative ring with unit element as well.

There are also many noncommutative rings. For instance, matrix rings similar to the one described in the following example are rings with unit element which are not commutative.

Example 7.3 Let $M_2(\mathbb{R})$ be the set of 2×2 matrices with real entries. Add and multiply matrices in the usual way:

$$\begin{bmatrix} a_{11} & a_{12} \\ a_{21} & a_{22} \end{bmatrix} + \begin{bmatrix} b_{11} & b_{12} \\ b_{21} & b_{22} \end{bmatrix} = \begin{bmatrix} a_{11} + b_{11} & a_{12} + b_{12} \\ a_{21} + b_{21} & a_{22} + b_{22} \end{bmatrix}$$

$$\begin{bmatrix} a_{11} & a_{12} \\ a_{21} & a_{22} \end{bmatrix} \begin{bmatrix} b_{11} & b_{12} \\ b_{21} & b_{22} \end{bmatrix} = \begin{bmatrix} a_{11}b_{11} + a_{12}b_{21} & a_{11}b_{12} + a_{12}b_{22} \\ a_{21}b_{11} + a_{22}b_{21} & a_{21}b_{12} + a_{22}b_{22} \end{bmatrix}.$$

It is straightforward to show that $M_2(\mathbb{R})$ is a ring in which the additive identity is $\begin{bmatrix} 0 & 0 \\ 0 & 0 \end{bmatrix}$, the unit element (multiplicative identity) is $\begin{bmatrix} 1 & 0 \\ 0 & 1 \end{bmatrix}$, and the additive inverse of $\begin{bmatrix} a_{11} & a_{12} \\ a_{21} & a_{22} \end{bmatrix}$ is $\begin{bmatrix} -a_{11} & -a_{12} \\ -a_{21} & -a_{22} \end{bmatrix}$. However, $M_2(\mathbb{R})$ is not a commutative ring; for example,

$$\begin{bmatrix} 1 & 2 \\ 3 & 4 \end{bmatrix}\begin{bmatrix} 1 & 1 \\ 0 & 1 \end{bmatrix} = \begin{bmatrix} 1 & 3 \\ 3 & 7 \end{bmatrix} \text{ and } \begin{bmatrix} 1 & 1 \\ 0 & 1 \end{bmatrix}\begin{bmatrix} 1 & 2 \\ 3 & 4 \end{bmatrix} = \begin{bmatrix} 4 & 6 \\ 3 & 4 \end{bmatrix},$$

and hence

$$\begin{bmatrix} 1 & 2 \\ 3 & 4 \end{bmatrix}\begin{bmatrix} 1 & 1 \\ 0 & 1 \end{bmatrix} \neq \begin{bmatrix} 1 & 1 \\ 0 & 1 \end{bmatrix}\begin{bmatrix} 1 & 2 \\ 3 & 4 \end{bmatrix}.$$

Furthermore, $M_2(\mathbb{R})$ possesses nonzero elements whose product is zero:

$$\begin{bmatrix} 0 & 1 \\ 0 & 0 \end{bmatrix} \neq \begin{bmatrix} 0 & 0 \\ 0 & 0 \end{bmatrix} \text{ and } \begin{bmatrix} 1 & 0 \\ 0 & 0 \end{bmatrix} \neq \begin{bmatrix} 0 & 0 \\ 0 & 0 \end{bmatrix}$$

$$\text{but } \begin{bmatrix} 0 & 1 \\ 0 & 0 \end{bmatrix}\begin{bmatrix} 1 & 0 \\ 0 & 0 \end{bmatrix} = \begin{bmatrix} 0 & 0 \\ 0 & 0 \end{bmatrix}. \quad ❖$$

By Proposition 4.2, if n is a positive integer which is not prime, then $\mathbb{Z}_n$ also possesses nonzero elements whose product of zero. On the other hand, for the familiar rings $\mathbb{Z}$, $\mathbb{R}$, $\mathbb{Q}$, and $\mathbb{C}$, the product of nonzero elements cannot be zero. This property is important enough that rings which satisfy it are given their own name. Since we are interested primarily in commutative rings with unit element, we restrict ourselves to that case.

DEFINITION

A nonzero element x of a ring R is a **zero divisor** if there exists a nonzero element y of R such that $xy = 0$. A commutative ring R with unit element which has at least two elements but no zero divisors is called an **integral domain**. That is, a commutative ring R with unit element and at least two elements is an integral domain if it satisfies the additional axiom.

DM. *If $x, y \in R$ satisfy $xy = 0$, then $x = 0$ or $y = 0$.*

We have already observed that $\mathbb{Z}$ is an integral domain; in fact, integral domains are so named because they share many properties with $\mathbb{Z}$. As well, we have noted that $\mathbb{Q}$, $\mathbb{R}$, and $\mathbb{C}$ are also integral domains. We will show below that, in general, any field F is an integral domain (Proposition 7.5), as is its associated polynomial ring $F[x]$ (Proposition 7.6). So in particular, for any prime p, both $\mathbb{Z}_p$ and $\mathbb{Z}_p[x]$ are integral domains.

To prove these results, we will need the following analogue of Proposition 5.1.

Proposition 7.4   *For any ring R,*

     (*i*)   *the additive identity is unique: if $a = a + z$ for all $a \in R$, then $z = 0$;*
     (*ii*) *for all $a \in R$, $a0 = 0 = 0a$;*
     (*iii*) *additive inverses are unique:*
              if $a, c, d \in R$ are such that $a + c = 0 = d + a$, then $c = d$;
              we use $-a$ to denote the unique additive inverse of a;
     (*iv*) *for all $a, b \in R$, $(-a)b = -(ab) = a(-b)$;*
     (*v*) *if R has a multiplicative identity, then it is unique:*
              if $a = ae$ for all $a \in R$ or if $a = ea$ for all $a \in R$, then $e = 1$.

Proof   Appropriately modify the proof of Proposition 5.1 (cf. Exercises 17–21). ■

Proposition 7.5   *Every field is an integral domain.*

Proof   By definition, every field F is a commutative ring with unit element which has at least two elements. Note that the axiom **DM** is of the form "if A, then B or C," and that to prove such a statement, it suffices to show that if both A and not B hold, then C holds. So assume that $a, b \in F$ are such that $ab = 0$ and $a \neq 0$. Then $a^{-1} \in F$ and by Proposition 5.1, $b = (a^{-1}a)b = a^{-1}(ab) = a^{-1}0 = 0$. We conclude that axiom **DM** holds and hence that F is an integral domain. ■

Proposition 7.6   *For any field, F, F[x] is an integral domain.*

Proof   By Proposition 7.1, $F[x]$ is a commutative ring with unit element, and since F has at least two elements, so does $F[x]$. We will show that axiom **DM** holds by proving its contrapositive. That is, we will show that if $p(x)$ and $q(x)$ are nonzero polynomials in $F[x]$, say $p(x) = a_0 + a_1x + \cdots + a_mx^m$, $a_m \neq 0$, and $q(x) = b_0 + b_1x + \cdots + b_nx^n$, $b_n \neq 0$, then $p(x)q(x) \neq 0$. For by the definition of polynomial multiplication, $p(x)q(x) = c_0 + c_1x + \cdots + c_{m+n}x^{m+n}$, where $c_{m+n} = a_mb_n$. Since F is an integral domain by Proposition 7.5 and $a_m \neq 0$ and $b_n \neq 0$, then $a_mb_n \neq 0$, and hence $p(x)q(x) \neq 0$. It follows that $F[x]$ is an integral domain. ■

Exercises

1. Show in detail that $M_2(\mathbb{R})$ is a ring with unit element.

2. Do nonzero elements of the ring $M_2(\mathbb{R})$ have multiplicative inverses? Justify your answer.

3. Show that, with respect to pointwise addition and multiplication of functions, the set of all differentiable functions $f: \mathbb{R} \to \mathbb{R}$ is a commutative ring with unit element.

4. Show that if R is a ring and $r \in R$, then $-(-r) = r$.

5. Show that if R is a ring with unit element and $1 = 0$, then $R = \{0\}$.

6. Show that in any integral domain, $0 \neq 1$.

7. Show that if R is a ring with unit element and $a, b, d \in R$ are such that $ba = 1$ and $ad = 1$, then $b = d$.

8. Show that if R is a ring with no zero divisors and $a, b, c, d \in R$ are such that $ab = c \neq 0$ and $ad = c \neq 0$, then $b = d$.

9. Let R be an integral domain, let $a, x, y \in R$, and suppose that $a \neq 0$. Show that if $ax = ay$ or if $xa = ya$, then $x = y$. (This is the <u>cancellation law.</u>)

10. Give an example of a ring in which the cancellation law does not hold. That is, find a ring R with nonzero elements a, x, y such that $ax = ay$ but $x \neq y$.

Exercises 11–16 complete the proof of Proposition 7.1. In each exercise, F denotes an arbitrary field with polynomial ring $F[x]$.

11. Show that $F[x]$ has an additive identity.

12. Show that $F[x]$ has a multiplicative identity.

13. Show that multiplication in $F[x]$ distributes over addition from the left.

14. Show that multiplication in $F[x]$ distributes over addition form the right.

15. Show that addition in $F[x]$ is associative.

16. Show that multiplication in $F[x]$ is associative.

Exercises 17–21 provide a detailed proof of Proposition 7.4. In each exercise, R is any ring.

17. If $x \in R$ is such that $a = a + z$ for all $a \in R$, then $z = 0$.

18. For all $a \in R$, $a0 = 0 = 0a$.

19. If $a, c, d \in R$ are such that $a + c = 0 = d + a$, then $c = d$.

20. For all $a, b \in R$, $(-a)b = -(ab) = a(-b)$.

21. If R has a multiplicative identity 1, and if either $a = ae$ for all $a \in R$ or if $a = ea$ for all $a \in R$, then $e = 1$.

A <u>division ring</u> is a ring with unit element which has at least two elements and all of whose nonzero elements have multiplicative inverses (i.e., all the field axioms hold except **M5**, multiplicative commutativity). For Exercises 22 and 23, let D be an arbitrary division ring.

22. Show that multiplicative inverses in D are unique, i.e., show that if $a, c, d \in D$ are such that $ac = 1 = da$, then $c = d$.

23. Show that D has no zero divisors, i.e., show that if $x, y \in D$ satisfy $xy = 0$, then $x = 0$ or $y = 0$.

24. Construct an example of a commutative ring that does not have a unit element. Remember to justify your answer.

25. Suppose that R is a ring and that $a^2 = 0$ for all $a \in R$. Show that for all $x \in R$, $ax + xa$ commutes with a (i.e., show that $a(ax + xa) = (ax + xa)a$).

26. Suppose that R is a set with two operations $+$ (addition) and $\cdot$ (multiplication) with respect to which it is additively closed and associative, is multiplicatively closed and associative, has an additive identity and a multiplicative identity, and has additive inverses. Suppose as well that multiplication distributes over addition from both the left and the right. Show that addition is commutative.

27. Let S be a set and let R be the set of subsets of S. For subsets, $A, B \in R$, the <u>symmetric difference</u> of A and B is

$$A \nabla B = (\bar{A} \cap B) \cup (A \cap \bar{B}),$$

where $\bar{A}$ is the complement of A in S. Show that $(R, \nabla, \cap)$ is a commutative ring with unit element. (That is, show that R is a commutative ring with unit element when ∇ is taken as $+$ and $\cap$ is taken as $\cdot$.)
(*Hint:* Recall De Morgan's laws for intersection, union, and complement:

$$A \cap (B \cup C) = (A \cap B) \cup (A \cap C),$$
$$A \cup (B \cap C) = (A \cup B) \cap (A \cup C),$$
$$\overline{A \cap B} = \bar{A} \cup \bar{B}, \text{ and}$$
$$\overline{A \cup B} = \bar{A} \cap \bar{B}).)$$

A <u>Boolean ring</u> is a ring R with unit element in which $x^2 = x$ for all $x \in R$. Exercises 28 and 29 concern these rings.

28. Show that the ring $(R, \nabla, \cap)$ of Exercise 27 is a Boolean ring.

29. Show that a Boolean ring is commutative. (*Hint:* $(x + y)^2 = x + y$.)

Ways in Which Polynomial Rings Are Like the Integers

At the end of Chapter 6, we set a goal of finding a concise description of extension fields of the form $F(r)$. Our hope was to accomplish this by investigating the structure of the polynomial rings $F[x]$. We have the advantage of knowing precisely what the elements in these rings look like: they are just polynomials with coefficients in F. So it is not surprising that we can find out a lot about how these elements behave. What is surprising is how many properties polynomial rings share with the integers.

Specifically, we will see that there is a variation on the division algorithm for the integers which is true for polynomial rings (Theorem 8.3), that polynomials as well as integers have greatest common divisors (Theorem 8.7), that the Euclidean algorithm can be translated into the language of polynomial rings (Example 8.10), and that prime numbers have polynomial analogues (the irreducible polynomials—Propositions 8.11 and 8.12).

We begin by considering division.

The Division Algorithm

We can transfer the notion of division directly, without change, from the integers to polynomial rings.

> **DEFINITION**
>
> If F is a field and if $f(x)$ and $g(x)$ are polynomials in $F[x]$, then $g(x)$ **divides** (or is a **divisor** of) $f(x)$ if there exists a polynomial $d(x) \in F[x]$ such that $f(x) = d(x)g(x)$.

Example 8.1   So the polynomial $x + 1$ in $\mathbb{Q}[x]$ divides $x^2 - 1$ because $x^2 - 1 = (x - 1)(x + 1)$ and $x - 1$ is also in $\mathbb{Q}[x]$. On the other hand, $x^2 - 1$ cannot divide $x + 1$ because any nonzero product $d(x)(x^2 + 1)$ must have a nonzero term $a_k x^k$ with $k \geq 2$. ❖

In the case of the integers, while a positive integer n may not divide another integer m, the division algorithm ensures that m can be written $m = nq + r$ for $0 \leq r < n$. This of course merely asserts that long division is possible for the integers, and we know that long division works for polynomials with rational coefficients as well. Thus finding a division algorithm for arbitrary polynomial rings amounts to showing that long division also works for polynomials with coefficients drawn from any field.

Specifically, long division in $\mathbb{Z}$ can be expressed in the following ways:

$$
\begin{array}{r}
15 \\
13 \overline{)197} \\
13 \\
\hline
67 \\
65 \\
\hline
2
\end{array}
\qquad \text{or} \qquad \frac{197}{13} = 15 + \frac{2}{13} \qquad \text{or} \qquad 197 = 15 \times 13 + 2
$$

The last equation, which is of course the division algorithm, has the advantage that it is an equation not involving division but rather an equation phrased entirely inside the ring of integers. The same point of view can be taken for polynomials with rational coefficients:

$$
\begin{array}{r}
x^2 \quad + x + 1 \\
x^2 + 2 \overline{)x^4 + x^3 + 3x^2 + 2x - 1} \\
x^4 \qquad + 2x^2 \\
\hline
x^3 + x^2 + 2x \\
x^3 \qquad + 2x \\
\hline
x^2 \qquad - 1 \\
x^2 \qquad + 2 \\
\hline
- 3
\end{array}
$$

or

$$
\frac{x^4 + x^3 + 3x^2 + 2x - 1}{x^2 + 2} = (x^2 + x + 1) + \frac{-3}{x^2 + 2}
$$

or

$$
x^4 + x^3 + 3x^2 + 2x - 1 = (x^2 + x + 1)(x^2 + 2) + (-3)
$$

Here too, the last expression has the advantage of being an equation phrased entirely inside the polynomial ring $\mathbb{Q}[x]$. We want to generalize this process to arbitrary polynomial rings.

The only difficulty arises in trying to describe the remainder. In the case of the integers, the remainder was required to be less than the divisor. However, we do not know what it means for one polynomial to be less than another and hence we cannot directly translate this restriction on the remainder from integers to polynomials. However, we do know that the process of polynomial long division stops when it is impossible for the divisor to divide what remains, i.e., when the highest power of x in the divisor is greater than the highest power of x in the remainder. That is, the process depends on the <u>degrees</u> of the remainder and divisor.

DEFINITION

If $0 \neq f(x) \in F[x]$, then the **degree** of $f(x)$, written deg $f(x)$, is the highest power of x in $f(x)$ with a nonzero coefficient.

Note that (1) the zero polynomial is assigned <u>no</u> degree, and (2) the degree of any polynomial is always a <u>nonnegative</u> integer.

We will need the following important property of the degree.

Proposition 8.2   *For any field F and nonzero polynomials $f(x)$ and $g(x)$ in $F[x]$, deg $f(x)g(x) =$ deg $f(x) +$ deg $g(x)$.*

Proof   Let $f(x) = a_0 + a_1 x + \cdots + a_m x^m$, $a_m \neq 0$, and $g(x) = b_0 + b_1 x + \cdots + b_n x^n$, $b_n \neq 0$. Then deg $f(x) = m$, deg $g(x) = n$, and, according to the definition of polynomial multiplication given in Chapter 7, $f(x)g(x) = c_0 + c_1 x + \cdots + c_{m+n} x^{m+n}$, where $c_{m+n} = \sum_{i+j=m+n} a_i b_j = a_m b_n$. By Proposition 7.5, F is an integral domain and hence, since both a_m and b_n are nonzero, $c_{m+n} = a_m b_n$ is also nonzero. It follows that deg $f(x)f(x) = m + n =$ deg $f(x) +$ deg $g(x)$. ∎

The notion of degree allows us to say concisely when the process of long division stops. Either the remainder is the zero polynomial (and therefore has no degree) or it is not the zero polynomial (and therefore has a degree) and its degree is less than the degree of the divisor. What we have done is to translate the condition on the integer remainder to a condition on the <u>degree</u> of the polynomial remainder. Note (1) that we must treat the zero polynomial separately because it has no degree and (2) that since polynomials do not have negative degrees, we can ignore the lower bound on the degree of the remainder. Specifically, we have the following.

Theorem 8.3   **Division Algorithm**

Let F be a field and let $f(x)$, $g(x) \in F[x]$, $g(x) \neq 0$. Then there exist $q(x)$, $r(x) \in F[x]$ such that

$$f(x) = g(x)q(x) + r(x) \quad \text{and} \quad r(x) = 0 \quad \text{or} \quad \deg r(x) < \deg g(x).$$

Proof If $g(x)$ divides $f(x)$, then the conclusion holds with $r(x) = 0$. So suppose that $g(x)$ does not divide $f(x)$. The rest of the proof closely follows that of the division algorithm for the integers.

Let S be the following subset of the integers

$$S + \{\deg(f(x) - g(x)p(x)) \,|\, p(x) \in F[x]\}.$$

(Note that by our assumption, $f(x) - g(x)p(x)$ is never 0 and hence its degree is always defined.) Since degrees are never negative, S is bounded below by 0, and thus, since S is not empty, the well-ordering principle (Chapter 1) implies that S contains a minimal element m. Let $q(x) \in F[x]$ be such that $\deg(f(x) - g(x)q(x)) = m$ and let $r(x) = f(x) - g(x)q(x)$. Then $f(x) = g(x)q(x) + r(x)$, and hence, since $r(x)$ cannot be 0, it suffices to show that $\deg r(x) < \deg g(x)$. Suppose by way of contradiction that $\deg r(x) \geq \deg g(x)$, and let $r(x) = \rho_0 + \cdots + \rho_k x^k$, $\rho_k \neq 0$, and $g(x) = \gamma_0 + \cdots + \gamma_t x^t$, $\gamma_t \neq 0$. Then $k \geq t$ and $h(x) = r(x) - g(x)(\rho_k \gamma_t^{-1} x^{k-t})$ is a polynomial in $F[x]$. If $h(x) = 0$, then $r(x) = g(x)(\rho_k \gamma_t^{-1} x^{k-t})$ and hence $f(x) = g(x)(q(x) + \rho_k \gamma_t^{-1} x^{k-t})$, a contradiction of our assumption that $g(x)$ does not divide $f(x)$. So $h(x)$ has a degree and since

$$h(x) = (\rho_0 + \cdots + \rho_k x^k) - (\rho_k \gamma_t^{-1} \gamma_0 x^{k-t} + \cdots + \rho_k \gamma_t^{-1} \gamma_{t-1} x^{k-1} + \rho_k x^k),$$

$\deg h(x) < k \leq m$, and since

$$h(x) = f(x) - g(x)(q(x) - \rho_k \gamma_t^{-1} x^{k-t}),$$

and $q(x) - \rho_k \gamma_t^{-1} x^{k-t} \in F[x]$, $\deg h(x) \in S$. But this contradicts our choice of m as minimal in S. We thus conclude that $\deg r(x) < \deg g(x)$. ∎

Example 8.4   If $f(x) = 2x^5 + 2x^4 + 3x^3 + x^2 + x + 3$ and $g(x) = x^3 - x^2 + 3x + 1$ in $\mathbb{Q}[x]$, then polynomial long division of $f(x)$ by $g(x)$ gives a quotient $q(x) = 2x^2 + 4x + 1$ and a remainder $r(x) = -12x^2 - 6x + 2$. Then $f(x) = q(x)g(x) + r(x)$ and $r(x) \neq 0$ and $\deg r(x) = 2 < 3 = \deg g(x)$. According to Theorem 8.3, this method works equally well in any field (e.g., Exercises 19–24). ❖

The division algorithm is an essential tool for the study of polynomials. The proof of the next result shows how it is typically used.

Proposition 8.5   *Let F be a field and let $0 \neq f(x) \in F[x]$. If $f(a) = 0$ for some $a \in F$, then the polynomial $x - a$ divides $f(x)$ in $F[x]$.*

Proof By the division algorithm, there exist polynomials $q(x), r(x) \in F[x]$ such that $f(x) = q(x)(x - a) + r(x)$ and $r(x) = 0$ or $\deg r(x) < 1$. If $r(x) \neq 0$, then $\deg r(x) = 0$ and hence $r(x) = r_0$ for some $0 \neq r_0 \in F$. But then

$$0 = f(a) = q(a)(a - a) + r(a) = 0 + r(a) = r_0,$$

a contradiction. Thus, $r(x) = 0$ and hence $f(x) = q(x)(x - a)$, i.e., $x - a$ divides $f(x)$. ∎

Greatest Common Divisors

One of the most important consequences of the division algorithm for the integers is that two nonzero integers have a unique greatest common divisor (Theorem 1.4). An analogous proof shows that the same result holds for arbitrary polynomial rings, a result that will be essential for our characterization of extension fields $F(r)$.

To phrase the desired result, we need a precise definition of a greatest common divisor, $d(x)$, of two polynomials $f(x)$ and $g(x)$. We will certainly need to require that $d(x)$ be a common divisor of $f(x)$ and $g(x)$ and that any common divisor of $f(x)$ and $g(x)$ be a divisor of $d(x)$. To ensure uniqueness of the greatest common divisor of two integers, we required it to be positive. In the case of polynomials, ensuring uniqueness is more complicated.

Example 8.6   In $\mathbb{Q}[x]$, since $x + 1$ divides itself and $x^2 + x$, $x + 1$ is a common divisor of itself and $x^2 + x$. And certainly any common divisor of $x + 1$ and $x^2 + x$ divides $x + 1$. But the same can be said of $2x + 2$. For $x + 1 = \left(\frac{1}{2}\right)(2x + 2)$ and $x^2 + x = \left(\frac{1}{2}x\right)(2x + 2)$. And if $x + 1 = p(x)q(x)$, then $2x + 2 = p(x)(2q(x))$ so that any common divisor of $x + 1$ and $x^2 + x$, being in particular a divisor of $x + 1$, must divide $2x + 2$. So $x + 1$ and $x^2 + x$ have at least two common divisors ($x + 1$ itself and $2x + 2$) which are divisible by every common divisor. ❖

We will ensure that there is only <u>one</u> greatest common divisor by restricting the coefficient of the highest power of x.

DEFINITION

Let F be a field, let $f(x), g(x), d(x)$ be nonzero polynomials in $F[x]$, and let $k = \deg d(x)$. If the coefficient of x^k in $d(x)$ is 1, then $d(x)$ is **monic.** The polynomial $d(x)$ is a **greatest common divisor** of $f(x)$ and $g(x)$ if

   (i) $d(x)$ is monic;

   (ii) $d(x)$ divides both $f(x)$ and $g(x)$;

   (iii) if $h(x)$ divides both $f(x)$ and $g(x)$, then $h(x)$ divides $d(x)$.

The proof that any two nonzero polynomials have a unique greatest common divisor is similar to the proof of the analogous statement (Theorem 1.4) for two nonzero integers.

Theorem 8.7   *Let F be a field and let $f(x)$ and $g(x)$ be two nonzero polynomials in $F[x]$. Then there exists a unique greatest common divisor $d(x)$ of $f(x)$ and $g(x)$, and for this polynomial $d(x)$ there exist polynomials $\alpha(x)$, $\beta(x) \in F[x]$ such that*

$$d(x) = \alpha(x)f(x) + \beta(x)g(x).$$

Proof   Let I denote the set

$$I = \{a(x)f(x) + b(x)g(x) \,|\, a(x), b(x) \in F[x]\},$$

and let S be the set of degrees of nonzero elements of I:

$$S = \{\deg p(x) \,|\, 0 \neq p(x) \in I\}.$$

Since polynomials do not have negative degrees, S is bounded below by 0, and hence by the well-ordering principle (Chapter 1), S contains a minimal element m. Let $p(x) \in I$ have degree m, $p(x) = \pi_0 + \cdots + \pi_m x^m$, $\pi_m \neq 0$, and let $d(x) = \pi_m^{-1} p(x)$. We will show that $d(x)$ is the unique greatest common divisor of $f(x)$ and $g(x)$ and that $d(x) = \alpha(x)f(x) + \beta(x)g(x)$ for $\alpha(x), \beta(x) \in F[x]$.

By definition, $d(x)$ is monic and has degree m. Since $p(x) \in I$, $p(x) = a(x)f(x) + b(x)g(x)$ for some $a(x), b(x) \in F[x]$, and thus, for $\alpha(x) = \pi_m^{-1} a(x)$ and $\beta(x) = \pi_m^{-1} b(x)$, $d(x) = \alpha(x)f(x) + \beta(x)g(x) \in I$. By the division algorithm (Theorem 8.2), there exist $q(x), r(x) \in F[x]$ such that $f(x) = d(x)q(x) + r(x)$ and either $r(x) = 0$ or $\deg r(x) < \deg d(x) = m$. But

$$\begin{aligned}
r(x) &= f(x) - d(x)q(x) = f(x) - (\alpha(x)f(x) + \beta(x)g(x))q(x) \\
&= f(x) - \alpha(x)q(x)f(x) + \beta(x)q(x)g(x) \\
&= (1 - \alpha(x)q(x))f(x) + (-\beta(x)q(x))g(x) \in I,
\end{aligned}$$

and hence if $r(x) \neq 0$, I would contain a polynomial of degree less than m, a contradiction of our choice of m. We conclude that $r(x) = 0$, i.e., that $d(x)$ divides $f(x)$. Similarly $d(x)$ divides $g(x)$. Suppose next that $h(x)$ divides both $f(x)$ and $g(x)$. Then $f(x) = w(x)h(x)$ and $g(x) = v(x)g(x)$, and hence

$$d(x) = \alpha(x)w(x)h(x) + \beta(x)v(x)h(x) = (\alpha(x)w(x) + \beta(x)v(x))h(x)$$

so that $h(x)$ divides $d(x)$. Thus $d(x)$ is a greatest common divisor of $f(x)$ and $g(x)$.

Suppose finally that $k(x)$ is also a greatest common divisor of $f(x)$ and $g(x)$. Since $d(x)$ divides $f(x)$ and $g(x)$, $d(x)$ divides $k(x)$, i.e., $k(x) = s(x)d(x)$ for some $s(x) \in F[x]$. Thus by Proposition 8.1, $\deg k(x) = \deg s(x) + \deg d(x) \geq \deg d(x)$. Similarly, $\deg d(x) \geq \deg k(x)$ so that in fact $\deg k(x) = \deg d(x)$. But then $\deg s(x) = 0$, i.e., $s(x) = s_0$ for some $0 \neq s_0 \in F$. Then $k(x) = s_0 d(x)$ and in particular, if $\deg d(x) = \deg k(x) = e$, then the coefficients of x^e in $k(x)$ and $s_0 d(x)$ must be equal, i.e., $1 = s_0 1$. It follows that $s_0 = 1$ and hence that $k(x) = d(x)$. That is, $d(x)$ is the unique greatest common divisor of $f(x)$ and $g(x)$. ∎

The proof of Theorem 8.7 shows that the greatest common divisor of polynomials $f(x)$ and $g(x)$ is the monic polynomial of smallest degree in the set $I = \{a(x)f(x) + b(x)g(x) \mid a(x), b(x) \in F[x]\}$. We next give two other characterizations of greatest common divisors which involve this set I. (The analogous result concerning the integers is Proposition 1.6.) We will use these characterizations to show that greatest common divisors do not change when the field is extended (Proposition 8.9) and to derive the polynomial version of the Euclidean algorithm (Example 8.10).

Proposition 8.8   *Let F be a field, let $f(x)$ and $g(x)$ be two nonzero polynomials in $F[x]$, and let $I = \{a(x)f(x) + b(x)g(x) \mid a(x), b(x) \in F[x]\}$. Then for any polynomial $d(x) \in F[x]$, the following statements are equivalent:*

    (i) $d(x)$ is the greatest common divisor of $f(x)$ and $g(x)$;

    (ii) $d(x)$ is a monic divisor of $f(x)$ and $g(x)$ and $d(x) \in I$;

    (iii) $d(x)$ is monic and $I = \{p(x)d(x) \mid p(x) \in F[x]\}$.

Proof   The proof is similar enough to that of Proposition 1.6 that we relegate it to the exercises (Exercise 42). ∎

If K is an extension field of a field that is not F itself, then the polynomial ring $K[x]$ will contain more polynomials than $F[x]$. That means that if $f(x)$ and $g(x)$ are polynomials in $F[x]$, then there will be more polynomials in $K[x]$ which divide both $f(x)$ and $g(x)$ than there are in $F[x]$ itself, and hence that the greatest common divisor of $f(x)$ and $g(x)$ in $K[x]$ may be different from the greatest common divisor of $f(x)$ and $g(x)$ in $F[x]$. The next result shows that this doesn't happen, i.e., that the greatest common divisor of two polynomials does not depend on what field generates the polynomial ring.

Proposition 8.9   *Let F be a subfield of a field K and suppose that $f(x)$ and $g(x)$ are nonzero polynomials in $F[x]$. If $d(x)$ is the greatest common divisor of $f(x)$ and $g(x)$ in $F[x]$, then $d(x)$ is also the greatest common divisor of $f(x)$ and $g(x)$ in $K[x]$.*

Proof   Since $d(x)$ is a monic divisor of both $f(x)$ and $g(x)$ in $F[x]$, then $d(x)$ is also a monic divisor of both $g(x)$ and $f(x)$ in $K[x]$. Furthermore, by Theorem 8.7, there exist $\alpha(x)$, $\beta(x) \in F[x]$ such that $d(x) = \alpha(x)f(x) + \beta(x)g(x)$. Since $\alpha(x), \beta(x) \in K[x]$, $d(x)$ must therefore be in the set $\{a(x)f(x) + b(x)f(x) \mid a(x), b(x) \in K[x]\}$. Hence by Proposition 8.8, $d(x)$ is the greatest common divisor of $f(x)$ and $g(x)$ in $Kx]$. ∎

As mentioned before, we can also use Proposition 8.8 to find the polynomial ring version of the Euclidean algorithm. That is, Proposition 8.8 can help us to find a process that can be used to compute not only the greatest common divisor $d(x)$ of two polynomials $f(x)$ and $g(x)$ but also to find polynomials $\alpha(x)$ and $\beta(x)$ such that $d(x) = \alpha(x)f(x) + \beta(x)g(x)$.

Example 8.10   **(Euclidean Algorithm for Polynomials)** Consider the polynomials

$$f(x) = 4x^4 - 2x^3 - 4x^2 - 2,$$
$$g(x) = 2x^3 + 2x^2$$

in $\mathbb{Q}[x]$, and, as in Proposition 8.8, let

$$I = \{a(x)f(x) + b(x)g(x) \,|\, a(x), b(x) \in \mathbb{Q}[x].$$

As in the case of the integers, apply the division algorithm first to $f(x)$ and $g(x)$ and then to succeeding divisors and remainders until obtaining a remainder of 0:

$$f(x) = g(x)(2x - 3) + (2x^2 - 2), \tag{1}$$
$$g(x) = (2x^2 - 2)(x + 1) + (2x + 2), \tag{2}$$
$$2x^2 - 2 = (2x + 2)(x - 1) + 0. \tag{3}$$

Then divide the last nonzero remainder by the coefficient of its highest power of x to make it monic:

$$d(x) = \frac{1}{2}(2x + 2) = x + 1.$$

Since $2x + 2$ divides $2x^2 - 2$ by equation (3), $d(x)$ divides $2x^2 - 2$ as well. But then in equation (2), $d(x)$ divides both $2x + 2$ and $2x^2 - 2$ and hence must also divide $g(x)$. Applying the same argument to equation (1) shows that $d(x)$ also divides $f(x)$. As well, we may write $d(x)$ in terms of $f(x)$ and $g(x)$ by solving each equation for its remainder and then substituting:

$$
\begin{aligned}
d(x) &= \frac{1}{2}(2x + 2) \\
&= \frac{1}{2}(g(x) - (2x^2 - 2)(x + 1)) = \frac{1}{2}g(x) - \frac{1}{2}(x + 1)(2x^2 - 2) \\
&= \frac{1}{2}g(x) - \frac{1}{2}(x + 1)(f(x) - g(x)(2x - 3)) \\
&= \frac{1}{2}(1 + (x + 1)(2x - 3))g(x) - \frac{1}{2}(x + 1)f(x) \\
&= \left(-\frac{1}{2}x - \frac{1}{2}\right)f(x) + \left(x^2 - \frac{1}{2}x - 1\right)g(x) \in I.
\end{aligned}
$$

Then by Proposition 8.8, $d(x)$ is the greatest common divisor of $f(x)$ and $g(x)$. ❖

Irreducible Polynomials

We conclude this chapter by considering what will be for our purposes the most important property of polynomial rings, viz., the existence of polynomials that are analogous to the prime numbers. These polynomials, the irreducible polynomials, are as important to the study of polynomials as the prime numbers are to the study of the integers; they will be crucial to our eventual description of extension fields $F(r)$ in Chapter 10.

We first note that 1 is not considered a prime number because it divides every integer, and since the prime numbers are meant to be the building blocks of the integers, 1 is redundant from this point of view. This situation is more complicated in polynomial rings because there are more polynomials with this property than just the constant polynomial 1. In particular, every nonzero constant polynomial a_0 divides every polynomial $p(x)$: $p(x) = a_0(a_0^{-1}p(x))$. So every polynomial has many polynomial divisors of degree zero. For this reason, we look for the analogues of prime numbers among polynomials of positive degree and, when looking for factors, we exclude polynomials of degree zero.

DEFINITION

Let F be a field and let $p(x) \in F[x]$ be of positive degree. Then $p(x)$ is **reducible** if there exist $d(x)$, $e(x) \in F[x]$ such that $p(x) = d(x)e(x)$ and $\deg d(x) < \deg p(x)$ and $\deg e(x) < \deg p(x)$. A polynomial of positive degree which is not reducible is **irreducible.**

The prime numbers can be characterized in terms of greatest common divisors (Proposition 1.8), and it is frequently very useful to have a similar characterization of irreducible polynomials.

Proposition 8.11   *Let F be a field and let $p(x) \in F[x]$ be of positive degree. Then $p(x)$ is irreducible if and only if for any nonzero polynomial $f(x)$, either $p(x)$ divides $f(x)$ or the greatest common divisor of $p(x)$ and $f(x)$ is 1.*

Proof   The proof is analogous to that of Proposition 1.8 and is therefore relegated to the exercises (Exercise 43). ∎

Because irreducible polynomials will be essential for our description of extension fields $F(r)$, we want to be able to determine when a given polynomial is irreducible. In general, this is a very difficult problem. However, for polynomials of sufficiently low degree and for certain polynomials in $\mathbb{Q}[x]$, there are tests that are relatively easy to apply. The tests for polynomials of low degree are given next; the test for polynomials in $\mathbb{Q}[x]$ (called Eisenstein's criterion) is given in Chapter 11.

Proposition 8.12   *Let F be a field and let $p(x) \in F[x]$.*

   (i)   If $\deg p(x) = 1$, then $p(x)$ is irreducible in $F[x]$.

(ii)   If $1 < \deg p(x) \le 3$, then $p(x)$ is irreducible in $F[x]$ if and only if $p(x) = 0$ has no solutions in F.

(iii)   If F is a subfield of $\mathbb{R}$ and $p(x) = ax^2 + bx + c, a \le 0$, then $p(x)$ is irreducible if and only if $\sqrt{b^2 - 4ac} \notin F$.

Proof   (i)   Suppose that $p(x)$ is reducible. Then there exist $d(x), e(x) \in F[x]$ such that $p(x) = d(x)e(x)$ and $\deg d(x) < \deg p(x)$ and $\deg e(x) < \deg p(x)$. Then $\deg d(x) = 0 = \deg e(x)$ and by Proposition 8.1, $\deg p(x) = \deg d(x) + \deg e(x) = 0$, a contradiction. So $p(x)$ is irreducible.

(ii)   We will prove the contrapositive of (ii): $p(x)$ is reducible in $F[x]$ if and only if $p(x) = 0$ has a solution in F. If $p(x)$ is reducible in $F[x]$, then $p(x) = d(x)e(x)$, where $\deg d(x) < \deg p(x)$ and $\deg e(x) < \deg p(x)$. By Proposition 8.1, $\deg p(x) = \deg d(x) + \deg e(x)$, and hence, since $\deg p(x) = 2$ or $\deg p(x) = 3$, either $\deg d(x) = 1$ or $\deg e(x) = 1$, say $\deg d(x) = 1$. Then $p(x) = (\alpha x + \beta)e(x)$, where $\alpha, \beta \in F, \alpha \ne 0$, and thus $-\alpha^{-1}\beta \in F$ and $p(-\alpha^{-1}\beta) = 0$ so that $p(x) = 0$ has a solution in F. Conversely, suppose that $p(a) = 0$ for $a \in F$. Then by Proposition 8.5, $p(x) = (x - a)q(x)$, and by hypothesis, $\deg (x - a) = 1 < \deg p(x)$. Furthermore, by Proposition 8.1, $\deg p(x) = \deg (x - a) + \deg q(x) = 1 + \deg q(x)$ so that as well $\deg q(x) < \deg p(x)$. It follows that $p(x)$ is reducible in $F[x]$.

(iii)   Since $\deg p(x) = 2$, (ii) implies that $p(x)$ is irreducible in F if and only if F contains no solutions of $p(x) = 0$. But we have already observed in Chapter 2 that the solutions of $p(x) = 0$ are $\dfrac{-b + \sqrt{b^2 - 4ac}}{2a}$ and $\dfrac{-b - \sqrt{b^2 - 4ac}}{2a}$, and hence $p(x) = 0$ has no solutions in F if and only if $\sqrt{b^2 - 4ac} \notin F$. ∎

Applying Proposition 8.12 is straightforward.

Example 8.13   We have that $\pi x - e \in \mathbb{R}[x]$ is irreducible because it has degree one, and $x^2 + x + 1 \in \mathbb{Q}[x]$ is irreducible because $\sqrt{1^2 - 4 \cdot 1 \cdot 1} = \sqrt{-3} \notin \mathbb{Q}$. ❖

Note that to apply test (ii), it is essential that <u>no</u> roots be in F.

Example 8.14   We have that $x^3 - 2 \in \mathbb{Q}[x]$ is irreducible because <u>none</u> of its roots, $\sqrt[3]{2}, \sqrt[3]{2}\,\zeta_3$, and $\sqrt[3]{2}\,\zeta_3{}^2$, are in $\mathbb{Q}$. However, $x^3 - 8 \in \mathbb{Q}[x]$ is <u>reducible</u> because, although two of its roots, $2\zeta_3$, and $2\zeta_3{}^2$, are not in $\mathbb{Q}$, the remaining root, 2, is in $\mathbb{Q}$. ❖

Note also that test (ii) does <u>not</u> apply to polynomials of degree greater than three.

Example 8.15   We have that $x^4 - 4 \in \mathbb{Q}[x]$ is reducible because $x^4 - 4 = (x^2 - 2)(x^2 + 2)$ but the roots of $x^4 - 4$ are $\pm\sqrt{2}$ and $\pm\sqrt{-2}$ and hence $x^4 - 4 = 0$ has no solutions in $\mathbb{Q}$. ❖

The fundamental theorem of arithmetic (Theorem 1.17) says that every integer greater than one can be uniquely written as a product of primes. An analogous result is true for polynomials, namely that every polynomial $p(x)$ of positive degree can be uniquely written in the form $p(x) = a_0 m_1(x) \cdots m_n(x)$, where a_0 is a nonzero constant and $m_1(x), \ldots, m_n(x)$ are irreducible and monic. Since we will have no need of the uniqueness of the decomposition, we will not prove it. However, we will need to know that every polynomial of positive degree can be written as a product of irreducible polynomials. So we conclude the chapter with this result.

Proposition 8.16   *If F is a field, then any polynomial of positive degree in $F[x]$ may be written as a product of polynomials irreducible in $F[x]$.*

Proof   Let $p(x) \in F[x]$. The proof by induction on $\deg p(x)$.

   (i)   If $\deg p(x) = 1$, then by Proposition 8.12 (i), $p(x)$ is already irreducible.
   (ii)   Suppose that $p(x)$ is a polynomial in $F[x]$ of degree $n > 0$, and suppose further that any polynomial in $F[x]$ of positive degree less than n can be written as a product of irreducible polynomials. If $p(x)$ is not already irreducible, it is reducible, and hence $p(x) = a(x)b(x)$ for polynomials $a(x), b(x) \in F[x]$, each of degree less than n. By the induction hypothesis, both $a(x)$ and $b(x)$ may each be written as a product of irreducible polynomials. Multiplying these products together expresses $p(x)$ as a product of irreducible polynomials. So in all cases, $p(x)$ can be written as a product of irreducible polynomials. ■

Since irreducible polynomials will turn out to be the key to our description of extension fields of the form $F(r)$ (cf. Theorem 10.4), it is important to know all we can about them. As noted before, Chapter 11 will be devoted to developing a test for irreducibility of polynomials in $\mathbb{Q}[x]$; in Chapter 14, we will show, among other things, that polynomials irreducible over a subfield of $\mathbb{C}$ have no multiple roots (Proposition 14.2).

Exercises

In Exercises 1–12, determine whether the given polynomial is irreducible in $\mathbb{Q}[x]$. Remember to justify your answer.

1. $x - 4$ **2.** $3x - 12$ **3.** $2x^2 - 4x - 6$

4. $x^2 - 3x + 5$ **5.** $x^2 + 3x - 5$ **6.** $x^2 + x + 2$

7. $x^2 + 9x + 23$ **8.** $x^3 - 1$ **9.** $x^3 - 13$

10. $x^3 - 6$ **11.** $x^3 - 3x^2 + 3x - 1$ **12.** $x^3 + x^2 + x + 1$

In Exercises 13–18, for the given polynomial $g(x)$ and $f(x) = x^5 - x^3 + 3x - 5$, find the polynomials $q(x)$ and $r(x)$ in $\mathbb{Q}[x]$ whose existence is guaranteed by Theorem 8.3.

13. $g(x) = x + 2$ **14.** $g(x) = x - 2$

15. $g(x) = x^2 + 7$ **16.** $g(x) = 4x^2 + 13x - 19$

17. $g(x) = x^3 + x - 1$ **18.** $g(x) = 3x^3 + 2x^2 - 7$

In Exercises 19–24, for the given polynomial $g(x)$ and $f(x) = x^5 - x^3 + [3]_{11}x - [5]_{11}$, find the polynomials $q(x)$ and $r(x)$ in $\mathbb{Z}_{11}[x]$ whose existence is guaranteed by Theorem 8.3. Express the coefficients in the form $[i]_{11}$, where $0 \le i \le 10$.

19. $g(x) = x + [2]_{11}$ **20.** $g(x) = x - [2]_{11}$

21. $g(x) = x^2 + [7]_{11}$ **22.** $g(x) = x^3 + x - [1]_{11}$

23. $g(x) = [4]_{11}x^2 + [13]_{11}x - [19]_{11}$ **24.** $g(x) = [14]_{11}x^3 + [13]_{11}x^2 - [7]_{11}$

In Exercises 25–31, find the greatest common divisor in $\mathbb{Q}[x]$ of the given polynomials.

25. $3x - 12, x + 6$ **26.** $x - 3, x^2 - 3x + 5$

27. $x - 11, x^2 + 3x - 5$ **28.** $x^3 - 6, x^2 + 3x - 1$

29. $2x^3 + x^2 - 2x - 2, 2x^2 - x - 3$

30. $2x^5 - x^4 - 5x^3 + 8x^2 + 4, x^4 - 3x^2 + 3x + 2$

31. $x^5 + 7x^4 - x^3 - 13x^2 - 2x - 2, x^4 + 6x^3 - 8x^2 - 12x + 12$

32. If F is a field, if $p(x)$ is irreducible in $F[x]$, and if $p(x)$ does not divide $f(x) \ne 0$ in $F[x]$, prove that there exist $\alpha(x), \beta(x) \in F[x]$ such that $1 = \alpha(x)f(x) + \beta(x)p(x)$.

33. Prove or give a counterexample: If F is a field and $f(x), g(x)$ are nonzero polynomials in $F[x]$ such that $g(x)$ does not divide $f(x)$, then there exists $a(x) \in F[x]$ such that $\deg(f(x) + a(x)g(x)) < \deg g(x)$.

34. Let F be a field. Find all divisors of the polynomial x in $F[x]$. Remember to justify your answer.

In Exercises 35 and 36, let F be a field and let $f(x)$ and $g(x)$ be nonzero polynomials in $F[x]$ such that $f(x) + g(x)$ is nonzero.

35. Show that $\deg(f(x) + g(x)) \le \max\{\deg f(x), \deg g(x)\}$.

36. For what polynomials is $\deg(f(x) + g(x)) < \max\{\deg f(x), \deg g(x)\}$? (Remember to justify your answer.)

37. Let F be a field and let $f(x) \in F[x]$. Suppose that $f(a) = 0 = f(b)$ for $a \ne b$ in F. Show that $f(x)$ is divisible by $(x - a)(x - b)$.

38. Let F be a field and let $a \in F$. Suppose that $f(x), g(x), q(x), r(x)$ are all as in the statement of Theorem 8.3. If $g(x) = x - a$, show that $r(x) = f(a)$.

39. Let F be a field and let $f(x)$ and $g(x)$ be nonzero polynomials in $F[x]$ such that deg $g(x) = 1$. If $f(a) = 0 = g(a)$ for some $a \in F$, show that $g(x)$ divides $f(x)$.

40. Suppose that F is a field and that $f(x)$ and $g(x)$ are irreducible polynomials of different degrees in $F[x]$. Show that for all $p(x) \in F[x]$, there exist $\alpha(x), \beta(x) \in F[x]$ such that $p(x) = \alpha(x)f(x) + \beta(x)g(x)$.

41. Let F be a field and let $p(x) \in F[x]$ be ireducible. If $a(x)$ and $b(x)$ are nonzero polynomials in $F[x]$ and $p(x)$ divides their product $a(x)b(x)$, show that either $p(x)$ divides $a(x)$ or $p(x)$ divides $b(x)$.

42. Prove Proposition 8.8: If F is a field, if $f(x)$ and $g(x)$ are nonzero polynomials in $F[x]$, if $I = \{a(x)f(x) + b(x)g(x) \,|\, a(x), b(x) \in F[x]\}$, and if $d(x) \in F[x]$, then the following statements are equivalent:
  (i) $d(x)$ is the greatest common divisor of $f(x)$ and $g(x)$;
  (ii) $d(x)$ is a monic divisor of $f(x)$ and $g(x)$ and $d(x) \in I$;
  (iii) $d(x)$ is monic and $I = \{p(x)d(x) \,|\, p(x) \in F[x]\}$.

43. Prove Proposition 8.11: If F is a field and $p(x) \in F[x]$ is of positive degree, then $p(x)$ is irreducible if and only if for any nonzero polynomial $f(x)$, either $p(x)$ divides $f(x)$ or the greatest common divisor of $p(x)$ and $f(x)$ is 1.

Let F be a field and let $f(x)$ and $g(x)$ be two zero polynomials of $f[x]$. A polynomial $t(x)$ is a <u>least common multiple</u> of $f(x)$ and $g(x)$ if
  (1) $t(x)$ is monic;
  (2) both $f(x)$ and $g(x)$ divide $t(x)$;
  (3) if both $f(x)$ and $g(x)$ divide $h(x)$, then $t(x)$ divides $h(x)$.

Exercises 44 and 45 prove the analogue of Theorem 8.7 for least common multiples.

44. Show that nonzero polynomials $f(x)$ and $g(x)$ in $F[x]$ have a least common multiple. (*Hint:* Let $I = \{p(x) \in F[x] \,|\, \text{both } f(x) \text{ and } g(x) \text{ divide } p(x)\}$, and then appropriately modify the proof of Theorem 8.7.)

45. Show that nonzero polynomials $f(x)$ and $g(x)$ in $F[x]$ have only one least common multiple.

46. Give a general statement of the Euclidean algorithm for polynomials used in Example 8.10. Prove that your general algorithm does indeed find the greatest common divisor.

47. Describe the differences and the similarities between the proof of Theorem 8.3 and that of Theorem 1.3.

48. Describe the differences and the similarities between the proof of Theorem 8.7 and that of Theorem 1.4.

49. The proof of Proposition 8.16 is by induction.

   (*a*)   Which principle of mathematical induction, the first or the second, is used?

   (*b*)   What statement $P(n)$ is being proved?

   (*c*)   What is the induction hypothesis in step (*ii*)?

HISTORICAL NOTE

Julia Robinson

Julia Bowman Robinson was born Julia Bowman on December 8, 1919, in St. Louis, Missouri, USA; she died on July 30, 1985. She was the second of two daughters born to Ralph and Helen Hall Bowman, and her sister, Constance Reid, is a writer whose works include biographies of the mathematicians David Hilbert, Richard Courant, and Jerzy Neyman. Helen died two years after giving birth to Julia, and Ralph sent Julia and Constance to a village outside Phoenix to stay with their grandmother. He remarried, and having as he thought saved enough money to retire, he and his new wife moved first to Arizona and then, with Julia and Constance, to San Diego. When Julia was nine years old, she contracted scarlet fever and then rheumatic fever. Her illness forced her to miss over two years of school but nonetheless, she graduated in 1936 at the age of sixteen with honors in all the sciences she had elected to take.

In the fall, she entered San Diego State College, intent on studying mathematics. The Depression had wiped out most of her father's savings, and the next year he committed suicide. In spite of this, both Julia and Constance were able to continue their education and, when Constance found a job as a teacher at the beginning of Julia's senior year, Julia was able to transfer to the University of California at Berkeley. This was a turning point in her life. "I was very happy, really blissfully happy, at Berkeley," she later recalled to Constance, "In San Diego, there had been no one at all like me. If, as Bruno Bettelheim has said, everyone has his or her own fairy story, mine is the story of the ugly duckling. Suddenly, at Berkeley, I found that I was really a swan. There were lots of people, students as well as faculty members, just as excited as I was about mathematics. I was elected to the honorary mathematics fraternity, and there was quite a bit of departmental social activity in which I was included. Then there was Raphael."[1]

"Raphael" was Assistant Professor Raphael M. Robinson who taught the number theory course she was taking. Two years later, a few weeks before the Japanese attack on Pearl Harbor, they were married. At the time, there was a nepotism rule at Berkeley, that members of the same family could not teach in the same department. Since Julia was already a teaching assistant, the rule did not immediately apply, and she was happy to audit mathematics courses and plan her family life. Soon she became pregnant, much to her delight, only to lose her child a few months later. And then, on a visit to her stepmother in San Diego, she came down with pneumonia and was told by the attending physician that the rheumatic fever she had had as a child had scarred her heart and that she should not become pregnant again.

She was quite depressed and in 1946, Raphael suggested that she take up mathematics again. With his support, she became a student of Alfred Tarski's and received her Ph.D. in 1948. It was at this time, during a discussion with Tarski, that she became interested in one of the famous problems Hilbert had proposed at the International Congress of Mathematicians in 1900. The problem that attracted her interest was the tenth one: "to devise a process according to which it can be determined by a finite number of operations whether the equation is solvable in rational integers."[2] This problem occupied the largest portion of her professional career. She presented her initial results at a meeting of the American Mathematical Society in 1950 but then became involved in politics for much of the 1950s. In the summer of 1959, Martin Davis and Hilary Putnam sent her a copy of a paper of theirs which utilized some of her previous work. According to Davis, "almost by return mail," she "greatly simplified the proof, which had become quite intricate. In the published version, the proof was elementary and elegant."[3] The resulting joint paper showed that there was a nonsolvable polynomial equation provided that a certain hypothesis, dubbed the Robinson hypothesis, was true.

Her heart was giving her trouble by this time, but after a successful operation in 1961, her health improved dramatically, although she could still only manage to teach one graduate course a quarter. Then, on February 15, 1970, Davis phoned to tell her that a twenty-two year old mathematician (Yuri Matijasevič) from Leningrad had proved the Robinson hypothesis and hence solved Hilbert's Tenth Problem. For her work, in 1975, Julia became the first woman mathematician to be elected to the National Academy of Sciences, and as a consequence of that honor was finally given a full professorship at Berkeley. And then in 1982 she became the first woman to be elected president of the American Mathematical Society.

In 1984, when she was presiding over the summer meeting of the American Mathematical Society in Eugene, Oregon, she learned that she had leukemia. Although the disease went into remission for a short time, she died a year later.

References

1. Reid, Constance. "Julia Robinson," *More Mathematical People*. Albers, Donald J., Alexanderson, Gerald L., and Reid, Constance, eds. Boston: Harcourt Brace Jovanovich, 1990, p. 271.

2. Hilbert. David. "Mathematical Problems," trans. by Mary Winston Newson. *Mathematical developments arising from Hilbert Problems,* Browder, Felix, ed. Proceedings of Symposia **28**, American Mathematical Society, Providence, 1976, p. 18 (originally published in the *Bull. Amer. Math. Soc.* **8** (1902), pp. 437–79).

3. Reid, Constance, with Robinson, Raphael M. "Julia Bowman Robinson (1919-1985)". *Women of Mathematics*. Grinstein, Louise S., and Campbell, Paul J., ed. New York: Greenwood Press, 1987, p. 184.

9

Principal Ideals

The proof of Theorem 8.7 shows that the greatest common divisor $d(x)$ of two polynomials $f(x)$ and $g(x)$ is the monic polynomial of minimal degree in the set $I = \{a(x)f(x) + b(x)g(x) \mid a(x), b(x) \in F[x]\}$. According to Proposition 8.8, the set I characterizes $d(x)$ as the monic polynomial whose set of multiples is exactly I, i.e., such that $I = \{p(x)d(x) \mid p(x) \in F[x]\}$. That is, one way of finding $d(x)$ is to form the set I and then to find the monic polynomial which generates I in the sense that I consists of all its products. Processes like this are in general very useful; in the next chapter we will use a similar one as the key step in determining the elements in the extension fields $F(r)$.

In this chapter, we look at such sets in the general context of a commutative ring R with unit element. In this general setting, such sets, i.e., sets of the form $\{rd \mid r \in R\}$ for a fixed element $d \in R$, are rings in their own right (i.e., they are subrings of R) which have the special property that any product of an element from the set by an element of R remains in the set again. Subrings with this property are called ideals, and ideals that take the form of all multiples of a single fixed element are called principal ideals. What makes polynomial rings special is that every ideal in a polynomial ring must be a principal ideal, a result we will prove at the end of the chapter (Theorem 9.10). Then to show that a subset of a polynomial ring is the set of all multiples of a single polynomial, it suffices to show merely that the subset is an ideal, and as we will see, this amounts to checking that the subset is closed with respect to addition, subtraction, and multiplication by arbitrary polynomials (Proposition 9.2).

We begin with the general notion of subrings of rings, i.e., subsets of rings which are to rings as subfields are to fields.

DEFINITION

A **subring** of a ring R is a subset S of R such that S is a ring with respect to the operations which it inherits from R.

It is not difficult to check that for all $k \in \mathbb{Z}$, the subsets $\{nk \mid n \in \mathbb{Z}\}$ form subrings of $\mathbb{Z}$ (cf. Proposition 9.6). It is also obvious that $\mathbb{Q}$ is a subring of $\mathbb{R}$ and that $\mathbb{R}$ is a subring of $\mathbb{C}$, and we will show in Example 9.5 that for any field F, the set of polynomials with constant term zero is a subring of $F[x]$.

Notice that when an integer of the form nk is multiplied by another integer, say w, then the result is an integer of the same form, viz., $(wn)k$. Similarly, when a polynomial with $a_0 = 0$ is multiplied by another polynomial, the result is a polynomial of the same form, viz., one with constant term 0. However, when a rational number is multiplied by a real number, the result is usually <u>not</u> another rational number, and when a real number is multiplied by a complex number, the result is usually <u>not</u> another real number. This property is what distinguishes ideals from other subrings.

DEFINITION

An **ideal** of a ring R is a subring I of R such that for all $x \in R$ and $y \in I$, both $xy \in I$ and $yx \in I$.

As previously indicated, we will show below that the subrings $\{nk \mid n \in \mathbb{Z}\}$ are ideals of $\mathbb{Z}$ and that the subring $\{a_0 + \cdots + a_n x^n \mid a_0 = 0\}$ is an ideal of $F[x]$. On the other hand, we have the following.

Example 9.1 While $\mathbb{Q}$ is a subring of $\mathbb{R}$, $\mathbb{Q}$ is not an ideal of $\mathbb{R}$ because $1 \in \mathbb{Q}$ and $\pi \in \mathbb{R}$ but $1 \cdot \pi \notin \mathbb{Q}$. Similarly, $\mathbb{R}$ is a subring of $\mathbb{C}$ which is not an ideal of $\mathbb{C}$ because $1 \in \mathbb{R}$ and $i \in \mathbb{C}$ but $1 \cdot i \notin \mathbb{R}$. ❖

As previously indicated, we will need to determine in general whether a given subset is an ideal. On the surface, this could involve checking nine axioms. However, as in the case of subfields, many of these axioms automatically hold because the operation is inherited from the ring. Specifically we have the following analogue of Proposition 5.2.

Proposition 9.2 *A subset I of a ring R is an ideal of R if and only if I satisfies*

   (i) $I \neq \varnothing$;
   (ii)   if $x, y \in I$, then $x + y \in I$;
   (iii) if $x \in I$ and $-x$ is the additive inverse of x in R, then $-x \in I$;
   (iv)   for all $x \in R$ and $y \in I$, both $xy \in I$ and $yx \in I$.

Proof Appropriately modify the proof of Proposition 5.2. ∎

Obviously for any ring R, both $\{0\}$ and R itself satisfy (i)–(iv) of Proposition 9.2 and hence both are always ideals of R. (And in the case of a field, these are the only ideals—see Exercise 10.) That is, we have the following result (Exercise 5).

Proposition 9.3   *For any ring R, both {0} and R itself are ideals of R.*

It is usually easy to determine when an ideal is the zero ideal. When the ring has a unit element, it is also easy to determine whether an ideal is the ring itself by applying the following test.

Proposition 9.4   *Let R be a ring with unit element 1 and let I be an ideal of R. Then $I = R$ if and only if $1 \in I$.*

Proof   If $I = R$, then certainly $1 \in I$. Conversely, suppose that $1 \in I$. Since I is an ideal of R, $I \subseteq R$. On the other hand, since I is an ideal, if $x \in R$, then $x = x1 \in I$, and thus $R \subseteq I$. It follows that $R = I$. ∎

According to Exercise 10, the only ideals of a field F are $\{0\}$ and F itself. On the other hand, the ring $\mathbb{Z}$ of integers has many ideals which are nontrivial and proper in the sense that they are neither $\{0\}$ nor $\mathbb{Z}$; for, as we will show, $\{nk \mid n \in \mathbb{Z}\}$ is an ideal of $\mathbb{Z}$ for any $k \in \mathbb{Z}$, and certainly $\{0\} \neq \{nk \mid n \in \mathbb{Z}\} \neq \mathbb{Z}$ whenever $0 \neq k \neq \pm 1$. Polynomial rings also have many proper ideals, some of which may be constructed in a manner similar to the following (cf. Exercise 16).

Example 9.5   Consider $I = \{a_0 + a_1 x + \cdots + a_n x^n \in \mathbb{Q}[x] \mid a_0 = 0\}$. We will use Proposition 9.2 to show that I is an ideal of $\mathbb{Q}[x]$. Since $0 \in I, I \neq \varnothing$. If $p(x) = a_0 + \cdots + a_n x^n \in I$ and $q(x) = b_0 + \cdots b_k x^k \in I$, then $a_0 = 0 = b_0$ and hence $a_0 + b_0 = 0$ and $-a_0 = 0$. Since $a_0 + b_0$ is the constant term of $p(x) + q(x)$, it follows that $p(x) + q(x) \in I$, and since $-a_0$ is the constant term of $-p(x)$, it also follows that $-p(x) \in I$. Finally suppose that $r(x) = \rho_0 + \cdots \rho_m x^m \in \mathbb{Q}[x]$. Since the constant term of $p(x)r(x)$ is $a_0 \rho_0$ and $a_0 \rho_0 = 0\rho_0 = 0$, $p(x)r(x) \in I$. Since $\mathbb{Q}[x]$ is commutative, this implies that $r(x)p(x) \in I$ as well, and therefore by Proposition 9.2, I is an ideal of $\mathbb{Q}[x]$. ❖

We asserted above that for any commutative ring R with unit element and for any $d \in R$, the set $\{rd \mid r \in R\}$ is an ideal of R. (So in particular, for each $k \in \mathbb{Z}$, the subsets $\{nk \mid n \in \mathbb{Z}\}$ are ideals of $\mathbb{Z}$.) In fact, we will prove more than this below; we will show that $\{rd \mid r \in R\}$ is an ideal of R which is contained in all ideals of R which contain d. Similar ideals exist in an arbitrary ring but there they are more complicated to define (see Exercises 29 and 30; see also Exercises 34–36).

Proposition 9.6   *Let R be a commutative ring with a unit element. If $d \in R$, then*

$$(d) = \{rd \mid r \in R\}$$

is the smallest ideal of R which contains d.

Proof   Since $0 = 0d \in (d)$, $(d) \neq \varnothing$. Let $rd, sd \in (d)$ and $t \in R$. Then

$$rd + sd = (r + s)d \in (d);$$
$$-(rd) = (-r)d \in (d);$$
$$(rd)t = t(rd) = (tr)d \in (d).$$

Thus by Proposition 9.2, (d) is an ideal of R, and since $1 \in R, d = 1d \in (d)$. Finally, observe that if I is an ideal of R which contains d, then $rd \in I$ for all $r \in R$, and hence $(d) \subseteq I$. Thus (d) is the smallest ideal of R containing d. ∎

As we indicated above, ideals that take the form given in Proposition 9.6 are very important and are therefore given their own name.

DEFINITION

Let R be a commutative ring with a unit element. An ideal I of R is **principal** if there exists $d \in R$ such that $I = (d) = \{rd \mid r \in R\}$. In this case, d is said to **generate** I.

As we observed, multiplying every integer by 3 produces a principal ideal strictly contained in the integers (see Figure 9.1). Although this is what we would usually expect to happen, a principal ideal need not always be <u>strictly</u> contained in its ring.

FIGURE 9.1

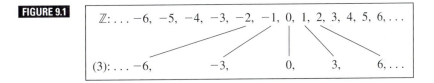

Example 9.7   In the ring $\mathbb{Q}$, $1 = \frac{1}{3} \cdot 3 \in (3)$ and hence $(3) = \mathbb{Q}$ by Proposition 9.4. (Another way to see this is to observe that (3) is a nonzero ideal of the field $\mathbb{Q}$ and then to use Exercise 10.) In general, since in any commutative ring R with unit element 1, $1 = 1 \cdot 1 \in (1)$, Proposition 9.4 also implies that $(1) = R$. ❖

Note that an ideal can be principal without being expressed in terms of a generator. For instance, as we show next, the ideal of Example 9.5 is principal.

Example 9.8   We showed in Example 9.5 that $I = \{a_0 + a_1x + \cdots + a_nx^n \in \mathbb{Q}[x] \mid a_0 = 0\}$ is an ideal of $\mathbb{Q}[x]$. To prove that I is principal, we need to exhibit a generator. Since every polynomial in I is a multiple of x: $a_1x + \cdots + a_nx^n = (a_1 + \cdots + a_nx^{n-1})x$, we choose x as a prospective generator. We have observed that the polynomials in I are multiples of x and hence that $I \subseteq (x)$. On the other hand, any multiple of x has the form $(a_0 + \cdots + a_n x^n)x = a_0 x + \cdots + a_n x^{n+1}$ and hence is in I. So $(x) \subseteq I$. We conclude that $I = (x)$ and hence that I is principal.

Note that a similar proof could be used to show that in fact any polynomial qx for $0 \neq q \in \mathbb{Q}$ is also a generator of I (cf. Exercises 11 and 12). This also follows easily from Corollary 9.11.

Note also that showing that $I = (x)$ shows simultaneously both that I is an ideal and that I is principal. For once we know that $I = (x)$, then Proposition 9.6 implies that I is an ideal. ❖

Some other subsets that are principal ideals but may be concisely defined without reference to a generator are those that arose in our study of greatest common divisors and which we mentioned earlier.

Example 9.9   Let F be a field, let $f(x), g(x) \in F[x]$, and let $I = \{a(x)f(x) + b(x)g(x) \,|\, a(x),\ b(x) \in F[x]\}$. If $g(x) = 0$, then $I = \{a(x)f(x) \,|\, a(x) \in F[x]\} = (f(x))$ and hence I is a principal ideal of $F[x]$. Similarly if $f(x) = 0$, then $I = (g(x))$ and hence is also a principal ideal of $F[x]$. If both $f(x)$ and $g(x)$ are nonzero, then by Theorem 8.7 they have a greatest common divisor $d(x)$ and by Proposition 8.8 $I = \{p(x)d(x) \,|\, p(x) \in F[x]\} = (d(x))$. Thus, in this case as well, I is a principal ideal of $F[x]$. ❖

Not every ideal is principal; a nonprincipal ideal is constructed in Exercises 34–36. However, as mentioned before, in the rings in which we are most interested, the polynomial rings, every ideal is principal. Integral domains with this property are very important and are given their own name.

DEFINITION

An integral domain R is a **principal ideal domain** if every ideal of R is principal.

We are now ready to prove the result mentioned at the beginning of the chapter, namely that the polynomial rings $F[x]$ are all principal ideal domains. A similar proof, which we leave to the reader, shows that $\mathbb{Z}$ is also a principal ideal domain (Exercise 25).

Theorem 9.10   *For any field F, $F[x]$ is a principal ideal domain.*

Proof   By Proposition 7.6, $F[x]$ is an integral domain. To see that every ideal is principal, let I be an ideal of $F[x]$. If $I = (0)$, then I is principal. Suppose, therefore, that $I \neq (0)$. We need to find a generator of I. Since a generator has to divide all the polynomials in I, its degree must be less than or equal to that of all the nonzero polynomials in I. As a candidate for a generator, we therefore choose a nonzero polynomial $f(x)$ of minimal degree in I. (Note that such a polynomial exists. For the set $\{\deg p(x) \mid 0 \neq p(x) \in I\}$ is a nonempty subset of $\mathbb{Z}$ which is bounded below by 0 and which therefore contains a minimal element m by the well-ordering principle; for $f(x)$, choose one of the polynomials in I whose degree is m.)

We will show that $I = (f(x))$. Since $f(x) \in I$ and I is an ideal, $f(x)h(x) \in I$ for any $h(x) \in F[x]$; thus $I \supseteq (f(x))$. Conversely, suppose that $g(x) \in I$. By the division algorithm, there exist $q(x), r(x) \in F[x]$ such that $g(x) = q(x)f(x) + r(x)$ and $r(x) = 0$ or $\deg r(x) < \deg f(x)$. Since $f(x) \in I$ and I is an ideal, $q(x)f(x) \in I$; thus, since $g(x) \in I$,

$$r(x) = g(x) - q(x)f(x) \in I.$$

If $r(x) \neq 0$, then $\deg r(x) < \deg f(x)$, a contradiction of our choice of $f(x)$ as a nonzero element of I whose degree is less than or equal to the degree of all other nonzero elements of I. We conclude that $r(x) = 0$ and hence that $g(x) = q(x)f(x) \in (f(x))$. It follows that $I \subseteq (f(x))$ and therefore that $I = (f(x))$. ■

The proof of Theorem 9.10 gives important information about the generators of the ideals of $F[x]$; viz., any nonzero polynomial in an ideal whose degree is minimal among nonzero polynomials in the ideal must generate the ideal. In fact, this property characterizes the generators of the ideal.

Corollary 9.11 *Suppose that F is a field and that I is a nonzero ideal of $F[x]$. Then a nonzero polynomial in I is a generator of I if and only if it is of minimal degree in I.*

Proof Let $0 \neq f(x) \in I$. If $\deg f(x)$ is minimal among nonzero elements of I, then the proof of Theorem 9.10 shows that $f(x)$ generates I. Conversely, suppose that $f(x)$ generates I and let $0 \neq g(x) \in I$. Since $f(x)$ generates $I, I = (f(x))$, i.e., $g(x) = a(x)f(x)$ for some $a(x) \in F[x]$ and hence by Proposition 8.1, $\deg g(x) = \deg a(x) + \deg f(x) \geq \deg f(x)$. Therefore, $f(x)$ must be of minimal degree in I. ■

One immediate consequence of Corollary 9.11 is that we can always choose a generator to be monic.

Corollary 9.12 *If F is a field, then any nonzero ideal I of $F[x]$ has a monic generator.*

Proof By Theorem 9.10, I has a generator

$$p(x) = \pi_0 + \pi_1 x + \cdots + \pi_r x^r, \quad \pi_r \neq 0.$$

Let $m(x) = \pi_r^{-1} p(x)$. Clearly $m(x)$ is nonzero and monic, and since $p(x) \in I$, since $\pi_r^{-1} \in F[x]$, and since I is an ideal, $m(x) \in I$. Since $\deg \pi_r^{-1} = 0$, $\deg m(x) = \deg p(x)$ by Proposition 8.1, and hence by Corollary 9.11, $m(x)$ is a generator of I. ■

We will show in the next chapter that the elements of an extension field $F(r)$ are of the form $a_0 + \cdots + a_{n-1}r^{n-1}$, where n is fixed. The number n of terms in the sum will be the degree of the generator of a particular ideal. Corollary 9.11 says that it will therefore be the minimal degree of all nonzero polynomials in the ideal.

Note that Corollary 9.11 also provides yet another characterization of greatest common divisors (see Exercise 24): $d(x)$ is the greatest common divisor of $f(x)$ and $g(x)$ if and only if $d(x)$ is a monic polynomial of minimal degree in $I = \{a(x)f(x) + b(x)g(x) \mid a(x), b(x) \in F[x]\}$.

Exercises

1. Recall (Example 7.3, and Exercise 1 in Chapter 7) that $M_2(\mathbb{R})$ is a ring with multiplicative identity $\begin{bmatrix} 1 & 0 \\ 0 & 1 \end{bmatrix}$. Show that $T = \left\{ \begin{bmatrix} r & r \\ r & r \end{bmatrix} \middle| r \in \mathbb{R} \right\}$ is a subring of $M_2(\mathbb{R})$ which has multiplicative identity $\begin{bmatrix} \frac{1}{2} & \frac{1}{2} \\ \frac{1}{2} & \frac{1}{2} \end{bmatrix}$.

2. Prove or give a counterexample: If R is a ring, S is a subring of R, and I is an ideal of R, then $I \cap S$ is an ideal of S.

3. Prove or give a counterexample: If R is a ring, S is a subring of R, and I is an ideal of R, then $I \cup S$ is an ideal of R.

4. For $0 \neq k \in \mathbb{Z}$, find two distinct generators of the ideal (k).

5. Prove Proposition 9.3 in detail: For any ring R, both $\{0\}$ and R itself are ideals of R.

6. Suppose that I is an ideal of $\mathbb{Q}[x]$ which contains both $x^2 + 2x + 4$ and $x^3 - 3$. Show that $I = \mathbb{Q}[x]$.
   (*Hint:* Begin by showing that $x^3 - 3$ is irreducible in $\mathbb{Q}[x]$.)

7. Let F be a field and suppose that I is an ideal of $F[x]$ which contains an irreducible polynomial of degree n and a nonzero polynomial of degree less than n. Show that $I = F[x]$.

For Exercises 8 and 9, suppose that F is a field and that $f(x)$ and $g(x)$ are nonzero elements of $F[x]$.

8. Prove or give a counterexample: If $(f(x)) = (g(x))$, then $\deg f(x) = \deg g(x)$.

9. Prove or give a counterexample: If $\deg f(x) = \deg g(x)$, then $(f(x)) = (g(x))$.

10. Let R be a commutative ring with a unit element. Show that R is a field if and only if R has exactly two ideals.

11. Show directly (without using Corollary 9.11) that $2x$ generates (x).

12. Show directly (without using Corollary 9.11) that $-3x - 3$ generates $(x + 1)$.

For Exercises 13–15, let F be a field and let $I \subseteq F[x]$ be the set

$$I = \{a_0 + a_1x + \cdots + a_nx^n \,|\, a_0 = 0 = a_1\}.$$

13. Show that I is an ideal of $F[x]$.

14. Show that I is a principal ideal of $F[x]$.

15. If F is infinite, find three distinct generators of the ideal I.

16. Find an ideal J of $\mathbb{Q}[x]$ such that $\{0\} \neq J \neq \mathbb{Q}[x]$ and for any $0 \leq k \in \mathbb{Z}$ $J \neq \{a_0 + a_1x + \cdots + a_nx^n \,|\, a_0 = 0 = a_1 = \cdots = a_n\}$.

17. Prove Proposition 9.2: A subset I of a ring R is an ideal of R if and only if I satisfies
   (i) $I \neq \varnothing$;
   (ii) if $x, y \in I$, then $x + y \in I$;
   (iii) if $x \in I$ and $-x$ is the additive inverse of x in R, then $-x \in I$;
   (iv) for all $x \in R$ and $y \in I$, both $xy \in I$ and $yx \in I$.

18. Prove the following variant of Proposition 9.2 (cf. Exercise 35 in Chapter 18):
   A subset I of a ring R is an ideal of R if and only if I satisfies
   (i) $I \neq \varnothing$;
   (ii) if $x, y \in I$, then $x - y \in I$;
   (iii) for all $x \in R$ and $y \in I$, both $xy \in I$ and $yx \in I$.

19. Show that a subset S of a ring R is a subring of R if and only if S satisfies
   (i) $S \neq \varnothing$;
   (ii) if $x, y \in S$, then $x - y \in S$;
   (iii) if $x, y \in S$, then $xy \in S$.

20. Suppose that R is a ring with a unit element. Show that a subset I of R is an ideal of R if and only if I satisfies.
   (i) $I \neq \varnothing$;
   (ii) if $x, y \in I$, then $x + y \in I$;
   (iii) for all $x \in R$ and $y \in I$, both $xy \in I$ and $yx \in I$.

21. Show that if R is a commutative ring and $a \in R$, then $\{ra \,|\, r \in R\}$ is an ideal of R.

22. Explain why, in spite of Exercise 21, it was assumed in Proposition 9.6 that R had a unit element.

23. Let R be a commutative ring with a unit element and let a and b be nonzero elements of R. Then a <u>divides</u> b if there exists $c \in R$ such that $ac = b$. If a and b are nonzero elements of a commutative ring R with a unit element, show that a divides b if and only if $(a) \supseteq (b)$.

24. Let F be a field, let $f(x)$ and $g(x)$ be two nonzero polynomials in $F[x]$, and let $I = \{a(x)f(x) + b(x)g(x) \,|\, a(x), b(x) \in F[x]\}$. Show that $d(x) \in F[x]$ is the greatest common divisor of $f(x)$ and $g(x)$ if and only if $d(x)$ is a monic polynomial of minimal degree in I.

25. Prove that $\mathbb{Z}$ is a principal ideal domain.

26. Find all the ideals of $\mathbb{Z}$ which contain 7.

27. Let R be a commutative ring with unit element and fix $s_1, s_2, \ldots, s_n \in R$. Let I be the set of elements of R of the form $s_1x_1 + s_2x_2 + \cdots + s_nx_n$, where $x_1, x_2, \ldots, x_n \in R$. Show that I is an ideal of R. (So I is the ideal generated by $s_1, s_2, \ldots, s_n$.)

28. Find a ring R with unit element in which there exists elements s_1, s_2 such that

$$\{s_1x_1 + s_2x_2 \,|\, x_1, x_2 \in R\}$$

is not an ideal of R.

Let R be a ring and let $a \in R$. For $0 < n \in \mathbb{Z}$, let $na = a + \cdots + a$ (n times); for $0 > n \in \mathbb{Z}$, let $na = (-a) + \cdots + (-a)$ ($-n$ times); and let $0a = 0$. Exercises 29 and 30 concern the following subset of R:

$$I_a = \left\{ na + a\alpha + \beta a + \sum_{i=1}^{m} \gamma_i a \delta_i \,\middle|\, n \in \mathbb{Z}, \alpha, \beta, \gamma_i, \delta_i \in R \right\}$$

29. Show that I_a is the smallest ideal of R which contains a.

30. If R is commutative and has a unit element, show that $I_a = (a)$.

Exercises 31–33 concern greatest common divisors in principal ideal domains. Specifically, for a principal ideal domain R, an element $a \in R$ <u>divides</u> an element $b \in R$ if there exists $c \in R$ such that $ac = b$. A <u>greatest common divisor</u> of two nonzero elements a and b in R is then an element $d \in R$ such that d divides both a and b, and, whenever $k \in R$ is such that k divides both a and b, k also divides d.

31. Show that any nonzero elements a and b in a principal ideal domain R have a greatest common divisor d.

32. If a and b are nonzero elements in a principal ideal domain R, and if d is a greatest common divisor of a and b, show that there exist elements x and y in R such that $d = xa + yb$.

33. According to Exercise 31, nonzero elements a and b in a principal ideal domain R have a greatest common divisor. Is this greatest common divisor unique? (Remember to justify your answer.)

The objective of Exercises 34–36 is to find an ideal which is not principal. The ideal is constructed as follows. Let $\mathbb{Z}[x]$ be the set of all polynomials in one variable with integral coefficients:

$$\mathbb{Z}[x] = \left\{ \sum_{i=0}^{n} a_i x^i \,\middle|\, a_i \in \mathbb{Z} \right\}$$

Define addition and multiplication on $\mathbb{Z}[x]$ in the usual way:

$$\sum_{i=0}^{m} a_i x^i + \sum_{j=0}^{n} b_j x^j = \sum_{k=0}^{\max\{m,n\}} (a_k + b_k) x^k$$

$$\left(\sum_{i=0}^{m} a_i x^i \right)\left(\sum_{j=0}^{n} b_j x^j \right) = \sum_{k=0}^{m+n} \left(\sum_{i+j=k} a_i b_j \right) x^k$$

Let I be the subset of $\mathbb{Z}[x]$:

$$I = \{ 2\,a(x) + x\,b(x) \,|\, a(x), b(x) \in \mathbb{Z}[x] \}.$$

34. Show that $\mathbb{Z}[x]$ is an integral domain.

35. Show that I is an ideal of $\mathbb{Z}[x]$.

36. Show that I is <u>not</u> a principal ideal of $\mathbb{Z}[x]$.

HISTORICAL NOTE

Emmy Noether

Amalie Emmy Noether was born in Erlangen, Germany, on March 23, 1882, and died in Bryn Mawr, Pennsylvania, on April 14, 1935. Her father, Max Noether, had been born in 1844 in Mannheim into a family of Jewish merchants, and, after a difficult bout with polio at the age of fourteen, he had become a mathematician at the University of Erlangen. Her mother, Ida Amalia Kaufmann, had been born in 1852 into a wealthy Jewish family in Cologne. Emmy Noether had three brothers, one of whom, Alfred, was a chemist and one of whom, Fritz, was an applied mathematician; Fritz's son, Gottfried, also became a mathematician. From 1900 to 1902, Emmy studied mathematics and foreign languages at the University of Erlangen and then in 1903, she transferred to the University of Göttingen to specialize in mathematics. Because she was a woman, she was not allowed to enroll officially at either university and thus only audited the lectures. However, in 1904, she was permitted to matriculate at the University of Erlangen and received her Ph.D., summa cum laude, in 1907. Following her graduation, she worked without compensation at the Mathematical Institute in Erlangen.

Her doctoral dissertation on algebraic invariants had been very computational, but under the influence of Ernst Fischer, a new arrival at Erlangen, her approach to algebra became much more abstract and therefore much more in accord with the approach of one of the premier mathematicians of the time, David Hilbert of Göttingen. Her work brought her to the attention of Hilbert and he invited her to Göttingen in 1915. In the face of strong prejudice against women, Hilbert tried repeatedly to obtain a position for her at the university, at one point arguing: "I do not see that the sex of the candidate is an argument against her admission as Privatdozent. After all, we are a university, not a bathing establishment."[1] However, she was unable to obtain a position until the sweeping changes brought about by the end of World War I.

In 1919, she presented her *Habilitation* lecture at Göttingen, and in 1922, she became an "unofficial associate professor." The position was "unofficial" because it came with neither responsibilities nor salary. However, after a year of effort on the part of Richard Courant, the new head of the mathematics department, she finally received, in 1923, a modest remuneration for teaching.

It is interesting that her *Habilitation* thesis, *"Invariante Variationsprobleme,"* is better known to physicists than to mathematicians. In it, she showed how the invariance of a system under a continuous symmetry forces the existence of a conservation law.

She spent most of the next ten years at Göttingen, with the exception of visiting professorships at Moscow (winter 1929) and Frankfurt (summer 1930). In 1924, B. L. van der Waerden, then 22 years old, arrived from Holland. He enthusiastically absorbed her approach to algebra and promulgated it in his famous and influential book *Moderne Algebra*, published in 1931. As van der Waerden wrote at the end of his moving obituary, "throughout the world today the triumphant progress of modern algebra which developed from her ideas seems to be unending."[2]

In 1933, she and the other Jewish faculty were summarily dismissed from Göttingen. Her brother Alfred had died in 1918. But Fritz worked at the Technical Institute in Breslau and in 1934 the Nazis forced him to resign as well. He secured a position in Tomsk, Siberia, and Emmy was offered a visiting professorship at Bryn Mawr College in the United States. She worked there and at the Institute for Advanced Study at Princeton until her untimely death in 1935 from complications following surgery for the removal of a pelvic tumor. Her ashes are buried under the brick walk in the cloisters of the Bryn Mawr library.

It is difficult to overstate the importance of Emmy Noether's work. Her influence pervades all of algebra and most other branches of mathematics. In a letter written in support of her reappointment at Bryn Mawr, Norbert Wiener wrote: "Miss Noether . . . is one of the ten or twelve leading mathematicians of the present generation in the entire world. . . . Even after she was deprived of her position in Germany on account of her sex, race and liberal attitude, numbers of students (men as well as women) continued to meet in her rooms for mathematical instruction. Of all the cases of German refugees, whether in this country or elsewhere, that of Miss Noether is without doubt the first to be considered."[3]

References

1. Kimberling, Clark. "Emmy Noether and Her Influence," *Emmy Noether, a Tribute to Her Life and Work.* Brewer, James K., and Smith, Martha K., eds. New York and Basel: Marcel Dekker, Inc., 1981, p. 14.
2. van der Waerden. "Obituary of Emmy Noether," trans. By Christina M. Mynhart. *Emmy Noether, a Tribute to Her Life and Work.* Brewer, James K., and Smith, Martha K., eds. New York and Basel: Marcel Dekker, Inc., 1981, p. 98 (originally published in *Math. Ann.* **111** (1935), 469-76)
3. Kimberling, *op. cit.,* p. 35.

10

Algebraic Elements

If F is a subfield of $\mathbb{C}$, and if r is a root of a polynomial $m(x)$ with coefficients in F, then $F(r)$ is defined to be the smallest subfield of $\mathbb{C}$ which contains both F and r. The problem with this definition is its indirectness: while we know that such a field exists, we can't work with it easily because we don't know what form its elements take. For this reason, at the end of Chapter 6, we set out to find a simple description of the elements of $F(r)$. The intervening chapters have provided us with a wealth of information about the polynomial ring $F[x]$, and we now want to use this information to investigate the structure of $F(r)$ in detail. We begin by doing this in the following special case.

Example 10.1   In Example 6.2, we determined that the elements of $\mathbb{Q}(\sqrt{2})$ are all of the form $\alpha + \sqrt{2}\,\beta$, for $\alpha, \beta \in \mathbb{Q}$, by showing that $\mathbb{Q}(\sqrt{2}) = \{\alpha + \beta\sqrt{2} \mid \alpha, \beta \in \mathbb{Q}\}$. Now $\sqrt{2}$ is a root of the polynomial $x^2 - 2$ in $\mathbb{Q}[x]$, and hence if $r \in \mathbb{C}$ is a root of the polynomial $m(x) = x^2 + 3x - 5$ in $\mathbb{Q}[x]$, then perhaps analogously $\mathbb{Q}(r) = \{\alpha + \beta r \mid \alpha, \beta \in \mathbb{Q}\}$.

How can we prove this? First we can simplify our notation by letting $V = \{\alpha + \beta r \mid \alpha, \beta \in \mathbb{Q}\}$; then we want to show that $\mathbb{Q}(r) = V$. Well certainly, for any α, β in $\mathbb{Q}$, α, β and r are all in $\mathbb{Q}(r)$, and hence since products and sums of elements of $\mathbb{Q}(r)$ are back in $\mathbb{Q}(r)$, $\alpha + \beta r \in \mathbb{Q}(r)$; so $\mathbb{Q}(r) \supseteq V$. On the other hand, if $\alpha \in \mathbb{Q}$, then $\alpha = \alpha + 0 \cdot r \in V$ so that $\mathbb{Q} \subseteq V$, and similarly $r = 0 + 1 \cdot r$ so that also $r \in V$. So if V is a field, then $\mathbb{Q}(r) \subseteq V$ because $\mathbb{Q}(r)$ is the smallest subfield of $\mathbb{C}$ which contains both $\mathbb{Q}$ and r.

Thus it remains to show that V is a subfield of $\mathbb{C}$, i.e., (by Proposition 5.2) that V has at least two elements and is closed with respect to addition and multiplication and the formation of additive and multiplicative inverses. Since $0 = 0 + 0 \cdot r$ and $1 = 1 + 0 \cdot r$ are both certainly in V, V contains at least two elements, and as well, for $\alpha, \beta, \gamma, \delta \in \mathbb{Q}$, $(\alpha + \beta r) + (\gamma + \delta r) = (\alpha + \gamma) + (\beta + \delta)r \in V$, and $-(\alpha + \beta r) = (-\alpha) + (-\beta)r \in V$ because $\alpha + \gamma, \beta + \delta, -\alpha$, and $-\beta$ are all in

$\mathbb{Q}$. So V is closed with respect to addition and the formation of additive inverses. Showing that a product of elements in V is back in V is more difficult; for $(\alpha + \beta r)(\gamma + \delta r) = \alpha\gamma + (\alpha\delta + \beta\gamma)r + \beta\delta r^2$ and it is far from obvious that this expression is in V. However, if we observe that r is a root of $m(x)$ and thus that $r^2 + 3r - 5 = 0$, we can see that $r^2 = 5 - 3r$ and hence that

$$\begin{aligned} (\alpha + \beta r)(\gamma + \delta r) &= \alpha\gamma + (\alpha\delta + \beta\gamma)r + \beta\delta r^2 \\ &= (\alpha\gamma + 5\beta\delta) + (\alpha\delta + \beta\gamma - 3\beta\delta)r \in V. \end{aligned}$$

Finally, we must show that V is closed with respect to the formation of multiplicative inverses, i.e., that if $0 \neq \alpha + \beta r$ is in V, then $(\alpha + \beta r)^{-1}$ can be written in the form $\delta_0 + \delta_1 r$ for rational numbers δ_0 and δ_1. This is the point in the argument where the work of the previous chapters comes into play. First form the polynomial $p(x) = \alpha + \beta x \in \mathbb{Q}[x]$ and note that if $p(x) = 0$, then $\alpha = 0 = \beta$ and hence $\alpha + \beta r = 0$, a contradiction. So $p(x)$ is nonzero, and since its degree is less than that of $m(x)$, $m(x)$ cannot divide it. But since 29 is prime, $\sqrt{3^2 + 4 \cdot 1 \cdot 5} = \sqrt{29} \notin \mathbb{Q}$, and thus by Proposition 8.12, $m(x) = x^2 + 3x - 5$ is irreducible in $\mathbb{Q}[x]$. So by Proposition 8.11, the greatest common divisor of $m(x)$ and $p(x)$ is 1, and hence by Theorem 8.7, there exist polynomials $f(x)$ and $g(x)$ in $\mathbb{Q}[x]$ such that $1 = f(x)m(x) + g(x)p(x)$. But then

$$1 = f(r)m(r) + g(r)p(r) = f(x) \cdot 0 + g(r)(\alpha + \beta r) = g(r)(\alpha + \beta r),$$

and hence $g(r)$ is the multiplicative inverse of $\alpha + \beta r$ in $\mathbb{C}$. Is $g(r)$ in V? Since $g(x)$ is a polynomial in $\mathbb{Q}[x]$, $g(x) = \gamma_0 + \gamma_1 x + \cdots + \gamma_k x^k$ for rational numbers $\gamma_0, \gamma_1, \ldots, \gamma_k$, and hence $g(r) = \gamma_0 + \gamma_1 r + \cdots + \gamma_k r^k$. But if $k \geq 2$, then $g(r)$ might not be in V. To show $g(r)$ is in V, we must find rational numbers δ_0 and δ_1 such that $g(r) = \delta_0 + \delta_1 r$. As an example of how to do this, suppose that $g(r) = \gamma_0 + \gamma_1 r + \gamma_2 r^2 + \gamma_3 r^3$. We observed before that $r^2 = 5 - 3r$. But then $r^3 = 5r - 3r^2 = 5r - 3(5 - 3r) = -15 + 14r$, and thus

$$\begin{aligned} g(r) &= \gamma_0 + \gamma_1 r + \gamma_2(5 - 3r) + \gamma_3(-15 + 14r) \\ &= (\gamma_0 + 5\gamma_2 - 15\gamma_3) + (\gamma_1 - 3\gamma_2 + 14\gamma_3)r. \end{aligned}$$

In the same fashion, we can rewrite any expression $\gamma_0 + \gamma_1 r + \cdots + \gamma_k r^k$ in the form $\delta_0 + \delta_1 r$. So we conclude that $g(r) \in V$ and hence that V is a subfield of $\mathbb{C}$. And thus, with some difficulty, we have shown that $\mathbb{Q}(r) = V = \{\alpha + \beta r \mid \alpha, \beta \in \mathbb{Q}\}$. ❖

What were the essential ingredients of the argument given in Example 10.1? First, in order to show that V was multiplicatively closed, we needed r to be the root of some monic polynomial $m(x)$ so that we could write powers of r that were greater than or equal to the degree of $m(x)$ in terms of lower powers of r. And second, in order to show that multiplicative inverses of elements in V were themselves in V, we needed $m(x)$ to be irreducible. The rest of the chapter is devoted to generalizing this argument and showing (Theorem 10.8) that if r is the root of a monic irreducible polynomial $m(x)$ of degree n in $F[x]$, then $F(r)$ has the simple form $F(r) = \{\alpha_0 + \alpha_1 r + \cdots + \alpha_{n-1} r^{n-1} \mid \alpha_i \in F\}$.

We begin by introducing the following terminology.

DEFINITION

Let F be a subfield of a field U. An element $a \in U$ is **algebraic** over F if there exists a nonzero polynomial $p(x) \in F[x]$ such that $p(a) = 0$.

Example 10.2   For instance, $\sqrt[3]{5} \in \mathbb{C}$ is algebraic over $\mathbb{Q}$ because $p(x) = x^3 - 5 \in \mathbb{Q}[x]$ and $p(\sqrt[3]{5}) = 0$. And similarly $i \in \mathbb{C}$ is algebraic over $\mathbb{R}$ because $p(x) = x^2 + 1 \in \mathbb{R}[x]$ and $p(i) = 0$.

Note that there will always be many polynomials which can be used to show that an element is algebraic. For instance, $\sqrt[3]{5} \in \mathbb{C}$ can also be shown to be algebraic over $\mathbb{Q}$ by observing that $p(x) = 4x^3 - 20 \in \mathbb{Q}[x]$ and $p(\sqrt[3]{5}) = 0$, and $i \in \mathbb{C}$ can also be shown to be algebraic over $\mathbb{Q}$ by observing that $p(x) = x^4 - 1 \in \mathbb{Q}[x]$ and $p(i) = 0$. ❖

Our objective is to show that if r is algebraic over F, then we can write $F(r)$ in the simple form given here. According to the heuristic reasoning employed above, that means we need a monic irreducible polynomial of which r is a root. We find this polynomial by showing that the set of polynomials which have r as a root is an ideal and that the monic generator of this ideal is the polynomial we seek.

Proposition 10.3   *Let F be a subfield of a field U, let $r \in U$ be algebraic over F, and let $I = \{g(x) \in F[x] \mid g(r) = 0\}$. Then I is an ideal of $F[x]$ and its monic generator $m(x)$ is the unique monic irreducible polynomial in $F[x]$ such that $m(r) = 0$.*

Proof   We show that I is an ideal by checking the conditions of Proposition 9.2. Note first that since $0 \in I, I \neq \varnothing$. Next let $g(x), h(x) \in I$ and $f(x) \in F[x]$. Then $g(x) + h(x) \in I$ because $g(r) + h(r) = 0$, $-g(x) \in I$ because $-g(r) = 0$, and $g(x)f(x), f(x)g(x) \in I$ because $g(r)f(r) = 0f(r) = 0$ and $f(r)g(r) = f(r)0 = 0$. Then I is an ideal of $F[x]$.

Now since r is algebraic over F, I is nonzero, and thus Corollary 9.12 implies that I has a monic generator $m(x)$ whose degree, according to Corollary 9.11, is minimal in I. Note that since $m(x) \in I, m(a) = 0$ and hence $\deg m(x) \geq 1$. Suppose by way of contradiction that $m(x)$ is reducible. Then there exist polynomials $d(x), e(x) \in F[x]$ such that $m(x) = d(x)e(x)$, $\deg d(x) < \deg m(x)$, and $\deg e(x) < \deg m(x)$. But $e(r)d(r) = m(r) = 0$ and since, by Proposition 7.5, U is an integral domain, either $e(r) = 0$ or $d(r) = 0$. In either case, we have found a nonzero polynomial in I of degree less than $\deg m(x)$. This contradicts our choice of $m(x)$ and thus $m(x)$ must be irreducible.

To see that $m(x)$ is unique, suppose that $f(x)$ is a monic irreducible polynomial in $F[x]$ such that $f(r) = 0$. Since $f(r) = 0, f(x) \in I$, and hence, since $m(x)$ generates $I, f(x) = m(x)h(x)$ for some $h(x) \in F[x]$. We previously noted that $\deg m(x) > 0$, and by Proposition 8.2, $\deg f(x) = \deg m(x) + \deg h(x)$. Then $\deg h(x) < \deg f(x)$, and thus since $f(x)$ is not reducible, $\deg m(x) = \deg f(x)$. Then $\deg h(x) = 0$, i.e., $h(x) = h_0$ for some $0 \neq h_0 \in F$. Equating coefficients of highest powers of x in $f(x)$ and $m(x)h(x)$, we have $1 = 1h_0$ and hence $h(x) = 1$, i.e., $f(x) = m(x)$. ∎

Example 10.1 indicates that the polynomial $m(x)$ of Proposition 10.3 is the key to the description of $F(r)$. With this in mind, we make the following definition. (The adjective "minimum" is used because the degree of the polynomial is minimum among degrees of polynomials in the ideal I.)

> **DEFINITION**
>
> Let F be a subfield of a field U and let $r \in U$ be algebraic over F. The unique monic irreducible polynomial $m(x)$ which has r as a root is called the **minimum polynomial** of r over F. The **degree** of r over F, written $[r:F]$, is the degree of the minimum polynomial $m(x)$, i.e., $[r:F] = \deg m(x)$.

Example 10.4   By Example 10.2, $\sqrt[3]{5}$ is algebraic over $\mathbb{Q}$. We will show that $m(x) = x^3 - 5$ is the minimum polynomial of $\sqrt[3]{5}$ over $\mathbb{Q}$. Clearly $m(x) \in \mathbb{Q}[x], m(x)$ is monic, and $m(\sqrt[3]{5}) = 0$. Furthermore, by Proposition 3.8, the roots of $x^3 - 5$ are $\sqrt[3]{5}, \sqrt[3]{5}\zeta_3$, and $\sqrt[3]{5}\zeta_3^2$, and none of these are in $\mathbb{Q}$. So $m(x)$ is irreducible by Proposition 8.12 and hence is the minimum polynomial of $\sqrt[3]{5}$ over $\mathbb{Q}$.

We also showed in Example 10.2 that i is algebraic over $\mathbb{Q}$. We will show that $p(x) = x^2 + 1$ is the minimum of i over $\mathbb{Q}$. Clearly $p(x) \in \mathbb{Q}[x], p(x)$ is monic, and $p(i) = 0$. Furthermore, neither of the roots, $\pm i$, of $p(x)$ are in $\mathbb{Q}$. So $p(x)$ is irreducible by Proposition 8.12 and hence is the minimum polynomial of i over $\mathbb{Q}$. ❖

Example 10.5   For ζ_3 and ζ_3^2, consider the polynomial $m(x) = x^2 + x + 1$. Since $\arg(\zeta_3) = \frac{2\pi}{3}, \zeta_3 - 1 \neq 0$. Thus, since $(\zeta_3 - 1)(\zeta_3^2 + \zeta_3 + 1) = \zeta_3^3 - 1 = 0$, and since $\mathbb{C}$ is an integral domain, $\zeta_3^2 + \zeta_3 + 1 = 0$, i.e., $m(\zeta_3) = 0$. A similar argument shows that $m(\zeta_3^2) = 0$. So since $m(x) \in \mathbb{Q}[x]$, both ζ_3 and ζ_3^2 are algebraic over $\mathbb{Q}$.

Now we have observed that $m(x) \in \mathbb{Q}[x]$ and that both ζ_3 and ζ_3^2 are roots of $m(x)$. Obviously $m(x)$ is monic, and furthermore, since $\sqrt{1^2 - 4 \cdot 1 \cdot 1} = \sqrt{-3} \notin \mathbb{Q}, m(x)$ is also irreducible in $\mathbb{Q}[x]$ by Proposition 8.12. So $m(x)$ is the minimum polynomial of both ζ_3 and ζ_3^2 in $\mathbb{Q}[x]$. ❖

Proposition 10.3 finds the irreducible polynomial whose degree will determine the length of the sums $\alpha_0 + \alpha_1 r + \cdots + \alpha_{n-1}r^{n-1}$. To show that the product of two such sums is another such sum and to show that the multiplicative inverse of such a sum is another such sum, we need to be able to write powers of r that are greater than $n - 1$ in terms of powers of r that are less than or equal to $n - 1$. The method we used in Example 10.1 applies generally.

Example 10.6   Let $p(x) = x^3 - 2x + 3 \in \mathbb{Q}[x]$ and   suppose   that $p(r) = 0$.   Then $r^3 - 2r + 3 = 0$ and hence $r^3 = -3 + 2r$. As well,

$$r^4 = r(-3 + 2r) = -3r + 2r^2,$$
$$r^5 = r(-3r + 2r^2) = -3r^2 + 2r^3 = -3r^2 + 2(-3 + 2r) = -6 + 4r - 3r^2,$$
$$r^6 = r(-6 + 4r - 3r^2) = -6r + 4r^2 - 3r^3$$
$$= -6r + 4r^2 - 3(-3 + 2r) = 9 - 12r + 4r^2,$$
$$r^7 = r(9 - 12r + 4r^2) = 9r - 12r^2 + 4r^3$$
$$= 9r - 12r^2 + 4(-3 + 2r) = -12 + 17r - 12r^2,$$

and so forth. By using these equations, sums of the form $\alpha_0 + \alpha_1 r + \cdots + \alpha_{n-1}r^{n-1}$ may always be reduced to sums of the form $\beta_0 + \beta_1 r + \beta_2 r^2$. For example,

$$1 + 2r + 3r^3 = 1 + 2r + 3(-3 + 2r) = -8 + 8r$$
$$-3 + 4r - 2r^4 + 7r^6 = -3 + 4r - 2(-3r + 2r^2) + 7(9 - 12r + 4r^2)$$
$$= 60 - 74r + 24r^2. \ \diamond$$

The division algorithm shows quickly that reductions such as those in Examples 10.1 and 10.6 can always be done, although it does not give the exact form the resulting sums will take. Happily, the only thing we need to know is that the process always results in a polynomial of appropriately low degree.

Lemma 10.7   Let F be a subfield of a field U, let $r \in U$ be algebraic over F, and let $m(x) = \mu_0 + \cdots + \mu_{n-1}x^{n-1} + x^n \in F[x]$ be such that $m(r) = 0$. For any $b_0, \ldots, b_k \in F$, there exist $\beta_0, \ldots, \beta_{n-1} \in F$ such that $b_0 + \cdots + b_k r^k = \beta_0 + \cdots + \beta_{n-1}r^{n-1}$.

Proof   Form the polynomial $p(x) = b_0 + \cdots + b_k x^k$ in $F[x]$. By the division algorithm (Theorem 8.3), there exist $q(x), R(x) \in F[x]$ such that $p(x) = m(x)q(x) + R(x)$ and $R(x) = 0$ or $\deg R(x) < \deg m(x) = n$. Then there exist $\beta_0, \ldots, \beta_{n-1} \in F$ (some or possibly all 0) such that $R(x) = \beta_0 + \cdots + \beta_{n-1}x^{n-1}$ and hence

$$b_0 + \cdots + b_k r^k = p(r) = m(r)q(r) + R(r) = R(r) = \beta_0 + \cdots + \beta_{n-1}r^{n-1}. \ \blacksquare$$

We are now ready to prove that if r is algebraic over F and if n is the degree of the minimum polynomial of r, then $F(r) = \{\alpha_0 + \cdots + \alpha_{n-1}r^{n-1} \mid \alpha_i \in F\}$. The proof generalizes the discussion in Example 10.1 for a root r of the polynomial

$x^2 + 3x - 5$. Note how Lemma 10.7 is used to reduce powers of r and how the irreducibility of the minimum polynomial is used to show that multiplicative inverses may be written in the desired form.

Theorem 10.8   *Suppose that F is a subfield of a field U and that $r \in U$ is algebraic over F, and let $n = [r:F]$. Then*

$$F(r) = \{\alpha_0 + \alpha_1 r + \alpha_2 r^2 + \cdots + \alpha_{n-1} r^{n-1} \mid \alpha_i \in F\}.$$

Proof   Let $V = \{\alpha_0 + \cdots + \alpha_{n-1} r^{n-1} \mid \alpha_i \in F\}$. If $\alpha_i \in F$ for $i = 0, \ldots, n-1$, then $\alpha_i \in F(r)$, and since also $r \in F(r)$ and $F(r)$ is a field, $\alpha_0 + \cdots + \alpha_{n-1} r^{n-1} \in F(r)$. It follows that $V \subseteq F(r)$. Conversely, it is clear that $F \subseteq V$ and $r \in V$, and thus to show that $F(r) \subseteq V$, it suffices merely to show that V is a field because $F(r)$ is the smallest subfield of U containing both F and r. We will use Proposition 5.2 to show that V is a subfield. Note first that V clearly contains both 0 and 1 and hence V has at least two elements by Proposition 5.1 (*vii*). Furthermore, if $\alpha_0 + \cdots + \alpha_{n-1} r^{n-1}$, $\beta_0 + \cdots + \beta_{n-1} r^{n-1} \in V$, then

$$(\alpha_0 + \cdots + \alpha_{n-1} r^{n-1}) + (\beta_0 + \cdots + \beta_{n-1} r^{n-1})$$
$$= (\alpha_0 + \beta_0) + \cdots + (\alpha_{n-1} + \beta_{n-1}) r^{n-1} \in V$$

because for all i, $\alpha_i + \beta_i \in F$, and

$$-(\alpha_0 + \cdots + \alpha_{n-1} r^{n-1}) = (-\alpha_0) + \cdots + (-\alpha_{n-1}) r^{n-1} \in V$$

because for all i, $-\alpha_i \in F$. To show that V is multiplicatively closed, note that since $\deg m(x) = [r:F] = n$ and $m(r) = 0$, Lemma 10.7 implies that there exist $\gamma_0, \ldots, \gamma_{n-1} \in F$ such that

$$(\alpha_0 + \cdots + \alpha_{n-1} r^{n-1})(\beta_0 + \cdots + \beta_{n-1} r^{n-1}) = \sum_{k=0}^{2n-2} \sum_{i+j=k} \alpha_i \beta_j r^k$$
$$= \gamma_0 + \cdots + \gamma_{n-1} r^{n-1} \in V.$$

To show that V has multiplicative inverses, suppose that $0 \neq \alpha_0 + \cdots + \alpha_{n-1} r^{n-1} \in V$, and let $p(x) \in F[x]$ be the polynomial $p(x) = \alpha_0 + \cdots + \alpha_{n-1} x^{n-1}$. Since $\deg p(x) \leq n - 1 < n = \deg m(x)$, $m(x)$ does not divide $p(x)$, and thus since $m(x)$ is irreducible, the greatest common divisor of $m(x)$ and $p(x)$ must be 1. Therefore, by Proposition 8.7, there exist $f(x), g(x) \in F[x]$ such that $1 = f(x)m(x) + g(x)p(x)$. Then, in particular, since $m(r) = 0$, $1 = f(r)m(r) + g(r)p(r) = g(r)p(r)$, and hence, since $p(r) = \alpha_0 + \cdots + \alpha_{n-1} r^{n-1}$, $g(r) = (\alpha_0 + \cdots + \alpha_{n-1} r^{n-1})^{-1}$ in U. But by Lemma 10.7, there exist $\lambda_0, \ldots, \lambda_{n-1} \in F$ such that $g(r) = \lambda_0 + \cdots + \lambda_{n-1} r^{n-1}$, and thus $(\alpha_0 + \cdots + \alpha_{n-1} r^{n-1})^{-1} = \lambda_0 + \cdots + \lambda_{n-1} r^{n-1} \in V$. We conclude from Proposition 5.2 that V is a field. ∎

In Example 10.1, we showed that if r is a root of $x^2 + 3x - 5$, then $\mathbb{Q}(r) = \{\alpha + \beta r \mid \alpha, \beta \in \mathbb{Q}\}$. Theorem 10.8 allows us to conclude this with hardly any work at all.

Example 10.9 Let $m(x) = x^2 + 3x - 5$ and let $r \in \mathbb{C}$ be a root of $m(x)$. Clearly $m(x) \in \mathbb{Q}[x]$, and thus, since $m(r) = 0$, r is algebraic over $\mathbb{Q}$. We have observed that $m(x) \in \mathbb{Q}[x]$ and that $m(r) = 0$, and obviously $m(x)$ is monic. Furthermore, since $\sqrt{3^2 + 4 \cdot 1 \cdot 5} = \sqrt{29} \notin \mathbb{Q}$, $m(x)$ is irreducible in $\mathbb{Q}[x]$ by Proposition 8.12, and thus $m(x)$ is the minimum polynomial of r over $\mathbb{Q}$. So, in agreement with Example 10.1, by Theorem 10.8,

$$\mathbb{Q}(r) = \{\alpha + \beta r \mid \alpha, \beta \in \mathbb{Q}\}.$$

The method described in Example 10.6 and Lemma 10.7 allows us to convert products of elements of $\mathbb{Q}(r)$ into the standard form $\alpha + \beta r$ as follows. Since r solves $x^2 + 3x - 5 = 0$, we have $r^2 = 5 - 3r$, and thus

$$(1 + 4r)(2 - r) = 2 + 7r - 4r^2 = 2 + 7r - 4(5 - 3r) = -18 + 19r. \; \diamond$$

Of course, solvability by radicals involves multiple extensions of the form $F(r_1, \ldots, r_k)$. The form of the elements in such fields can be found by applying Theorem 10.8 several times.

Example 10.10 As in Example 10.9, let $m(x) = x^2 + 3x - 5$ and let $r \in \mathbb{C}$ be a root of $m(x)$. We will show that

$$\mathbb{Q}(r, i) = \{(\alpha_0 + \alpha_1 r) + (\beta_0 + \beta_1 r)i \mid \alpha_i, \beta_j \in \mathbb{Q}\}.$$

We first note that $\mathbb{Q}(r, i) = \mathbb{Q}(r)(i)$, where $\mathbb{Q}(r)(i)$ is the smallest subfield of $\mathbb{C}$ which contains i and $\mathbb{Q}(r)$. For clearly $\mathbb{Q}(r, i)$ is a subfield of $\mathbb{C}$ which contains i and $\mathbb{Q}(r)$, and hence $\mathbb{Q}(r, i) \supseteq \mathbb{Q}(r)(i)$, and similarly $\mathbb{Q}(r)(i)$ is a subfield of $\mathbb{C}$ which contains r, i, and $\mathbb{Q}$ and hence $\mathbb{Q}(r, i) \subseteq \mathbb{Q}(r)(i)$.

We have seen in Example 10.9 that the elements of $\mathbb{Q}(r)$ have the form $\alpha + \beta r$ for α and β in $\mathbb{Q}$. Now $g(x) = x^2 + 1 \in \mathbb{Q}(r)[x]$ and $g(i) = 0$ so that i is algebraic over $\mathbb{Q}(r)$. So to find the elements in $\mathbb{Q}(r, i) = \mathbb{Q}(r)(i)$, we apply Theorem 10.8 with $\mathbb{Q}(r)$ as the field and i as the algebraic element. We have observed that $g(x) \in \mathbb{Q}(r)[x]$ and $g(i) = 0$, and clearly $g(x)$ is monic. Furthermore, the quadratic formula shows that both roots of $m(x)$ are real and hence in particular that r must be real. So every element of $\mathbb{Q}(r)$ must be real. But the complex roots of $g(x)$ are $x = \pm i$, and neither of these is real. So $g(x)$ has no roots in $\mathbb{Q}(r)$, and hence by Proposition 8.9, $g(x)$ is irreducible in $\mathbb{Q}(r)[x]$. It follows that $g(x)$ is the minimum polynomial of i over $\mathbb{Q}(r)$ and thus that $[i : \mathbb{Q}(r)] = 2$. By Theorem 10.8, the elements of $\mathbb{Q}(r, i) = \mathbb{Q}(r)(i)$ thus have the form $A_0 + A_1 i$, where $A_0, A_1 \in \mathbb{Q}(r)$. But the elements of $\mathbb{Q}(r)$ all have the form $\alpha_0 + \alpha_1 r$ for $\alpha_i \in \mathbb{Q}$, and thus

$$\mathbb{Q}(r, i) = \mathbb{Q}(r)(i) = \{A_0 + A_1 i \mid A_0, A_1 \in \mathbb{Q}(r)\}$$
$$= \{(\alpha_0 + \alpha_1 r) + (\beta_0 + \beta_1 r)i \mid \alpha_i, \beta_j \in \mathbb{Q}\}.$$

To multiply in $\mathbb{Q}(r, i)$, use the method described in Example 10.6 and Lemma 10.7 in the following way. Since r solves $x^2 + 3x - 5 = 0$, $r^2 = 5 - 3r$, and since i solves $x^2 + 1 = 0$, $i^2 = -1$. Thus

$$(r + (2 - 3r)i)((3 - r) - ri)$$
$$= 3r - r^2 - r^2 i + 6i - 2ri - 2ri^2 - 9ri + 3r^2 i + 3r^2 i^2$$
$$= 3r - (5 - 3r) - (5 - 3r)i + 6i - 2ri$$
$$\quad - 2r(-1) - 9ri + 3(5 - 3r)i + 3(5 - 3r)(-1)$$
$$= (-20 + 17r) + (16 - 17r)i. \; \diamondsuit$$

Examples 10.9 and 10.10 illustrate how to multiply in small extension fields. Dividing in these fields is somewhat more complicated because it involves solving polynomial equations of the form $1 = f(x)m(x) + g(x)p(x)$ for $f(x)$ and $g(x)$. The Euclidean algorithm for polynomials (Example 8.10) provides one way of solving such equations directly. Another method of calculating such inverses, based on considering $F(r)$ as a vector space over F, will be given in Chapter 12 (see Example 12.3).

Exercises

In Exercises 1–6, show that the given complex number is algebraic over $\mathbb{Q}$.

1. $\sqrt{7}$ **2.** $\sqrt[3]{2}$ **3.** $\sqrt[3]{5}$

4. $2 + i$ **5.** $1 + 2i$ **6.** $5 - 6i$

In Exercises 7–13, show that the complex number c is algebraic over $\mathbb{Q}$ and find its minimum polynomial.

7. $c = i$ **8.** $c = \sqrt[3]{7}$ **9.** $c = \zeta_3$

10. $c = 3 + 2i$ **11.** c is a solution of $x^2 + 2x + 3 = 0$

12. $c = 1 - 10i$ **13.** c is a solution of $3x^2 + 4x + 13 = 0$

In Exercises 14–20, show that the complex number r is algebraic over $\mathbb{Q}$ and express the elements of $\mathbb{Q}(r)$ in terms of elements of $\mathbb{Q}$ and powers of r.

14. $r = \zeta_3$ **15.** $r = \sqrt{13}$ **16.** $r = \sqrt[3]{11}$

17. $r = 4 + i$ **18.** r is a solution of $x^2 + 6x + 3 = 0$

19. $r = 2 - 3i$ **20.** r is a solution of $2x^2 + 9x + 6 = 0$

In Exercises 21–26, describe the elements of $\mathbb{Q}(r, s)$ in terms of r, s, and elements of $\mathbb{Q}$.

21. $\mathbb{Q}(\sqrt{2}, i)$ **22.** $\mathbb{Q}(i, \sqrt{5})$ **23.** $\mathbb{Q}(\sqrt{2}, \zeta_3)$

24. $\mathbb{Q}(\sqrt[3]{2}, i)$ **25.** $\mathbb{Q}(\sqrt[3]{11}, \zeta_3)$ **26.** $\mathbb{Q}(\zeta_3, \sqrt[3]{7})$

27. Prove or give a counterexample: If r is algebraic over a field F and if

$$I = \{g(x) \in F[x] \,|\, g(r) = 1\},$$

then I is an ideal of $F[x]$.

28. Let F be a field and let $r \in F$. Show that r is algebraic over F, find the minimum polynomial of r, and determine $[r{:}F]$. (Remember to justify your answers.)

29. Let $I = \{g(x) \in \mathbb{Q}[x] \,|\, g(\sqrt[3]{5}) = 0\}$. Show that I is an ideal of $\mathbb{Q}[x]$ and find a monic irreducible generator of I.

30. Let $I = \{g(x) \in \mathbb{Q}[x] \,|\, g(\zeta_3) = 0\}$. Show that I is an ideal of $\mathbb{Q}[x]$ and find a monic irreducible generator of I.

For Exercises 31–36, let $p(x) = x^4 + 5x^2 + 10 \in \mathbb{Q}[x]$ and suppose that $r \in \mathbb{C}$ solves $p(x) = 0$. Assume that $p(x)$ is irreducible in $\mathbb{Q}[x]$.

31. Show that r is algebraic over $\mathbb{Q}$ with $p(x)$ as its minimum polynomial.

32. Show that $\mathbb{Q}(r) = \{\alpha_0 + \alpha_1 r + \alpha_2 r^2 + \alpha_3 r^3 \,|\, \alpha_i \in \mathbb{Q}\}$.

In Exercises 33–36, write the given product of elements of $\mathbb{Q}(r)$ in the form of $\alpha_0 + \alpha_1 r + \alpha_2 r^2 + \alpha_3 r^3$ for $\alpha_i \in \mathbb{Q}$.

33. $(r^3 + 2)(r^3 + 3r)$ **34.** $(2 - 3r + r^2 - 2r^3)(5 + r^2)$

35. $(5r + 7r^2)(1 - 4r^3)$ **36.** $(2r^3)(3 - r + 2r^3)(3 - r)$

For Exercises 37–42, let $q(x) = x^5 + 2x + 2 \in \mathbb{Q}[x]$ and suppose that $s \in \mathbb{C}$ solves $q(x) = 0$. Assume that $q(x)$ is irreducible in $\mathbb{Q}[x]$.

37. Show that s is algebraic over $\mathbb{Q}$ with $q(x)$ as its minimum polynomial.

38. Show that $\mathbb{Q}(s) = \{\alpha_0 + \alpha_1 s + \alpha_2 s^2 + \alpha_3 s^3 + \alpha_4 s^4 \,|\, \alpha_i \in \mathbb{Q}\}$.

In Exercises 39–42, write the given product of elements of $\mathbb{Q}(s)$ in the form $\alpha_0 + \alpha_1 s + \alpha_2 s^2 + \alpha_3 s^3 + \alpha_4 s^4$ for $\alpha_i \in \mathbb{Q}$.

39. $(s^3 + 2)(s^3 + 3s)$ **40.** $(2 - s + s^2 - s^3 + 2s^4)(1 - s^3)$

41. $(7s + 3s^4)(2 - 5s^3)$ **42.** $(3s^4)(2 + s^3)(5 - s + s^2)$

43. Show directly (without using Theorem 10.8) that

$$\mathbb{Q}(\sqrt[3]{5}) = \{\alpha_0 + \alpha_1 \sqrt[3]{5} + \alpha_2 \sqrt[3]{25} \,|\, \alpha_i \in \mathbb{Q}\}.$$

44. Show directly (without using Theorem 10.8) that

$$\mathbb{Q}(\sqrt[3]{2}) = \{\alpha_0 + \alpha_1 \sqrt[3]{2} + \alpha_2 \sqrt[3]{4} \,|\, \alpha_i \in \mathbb{Q}\}.$$

For Exercises 45 and 46, let $f(x) = 2x^4 + 10x^2 + 20 \in \mathbb{Q}[x]$ and suppose that $u \in \mathbb{C}$ solves $f(x) = 0$. Assume that $f(x)$ is irreducible in $\mathbb{Q}[x]$.

45. Show that u is algebraic over $\mathbb{Q}$ and find the minimum polynomial of u over $\mathbb{Q}$.

46. Show directly (without using Theorem 10.8) that

$$\mathbb{Q}(u) = \{\alpha_0 + \alpha_1 u + \alpha_2 u^2 + \alpha_3 u^3 \mid \alpha_i \in \mathbb{Q}\}.$$

For Exercises 47 and 48, let $g(x) = 3x^5 + 6x + 6 \in \mathbb{Q}[x]$ and suppose that $v \in \mathbb{C}$ solves $g(x) = 0$. Assume that $g(x)$ is irreducible in $\mathbb{Q}[x]$.

47. Show that v is algebraic over $\mathbb{Q}$ and find the minimum polynomial of v over $\mathbb{Q}$.

48. Show directly (without using Theorem 10.8) that

$$\mathbb{Q}(v) = \{\alpha_0 + \alpha_1 v + \alpha_2 v^2 + \alpha_3 v^3 + \alpha_4 v^4 \mid \alpha_i \in \mathbb{Q}\}.$$

49. Let F be a subfield of a field U, let $r \in U$ be algebraic over F, and let $m(x) = \mu_0 + \mu_1 x + \cdots + \mu_{n-1} x^{n-1} + x^n \in F[x]$ be such that $m(r) = 0$. According to Lemma 10.7, for any $b_0, \ldots, b_k \in F$, there exist $\beta_0, \ldots, \beta_{n-1} \in F$ such that $b_0 + \cdots + b_k r^k = \beta_0 + \cdots + \beta_{n-1} r^{n-1}$. Find an explicit form for the coefficients $\beta_0, \ldots, \beta_{n-1}$. (*Hint:* Define the β_i inductively.)

11 Eisenstein's Irreducibility Criterion

Theorem 10.8 described the elements in the extension fields $F(r)$ for r algebraic over F. The key to this description is the ability to determine the minimum polynomial of r. In turn, the key to determining the minimum polynomial is the ability to determine irreducibility. This chapter is devoted to establishing a useful test, known as Eisenstein's criterion, for determining when certain polynomials in $\mathbb{Q}[x]$ are irreducible. The proof is more number-theoretic than algebraic and is in that sense, a digression.

Eisenstein's criterion applies to polynomials $a(x) = \alpha_0 + \cdots + \alpha_n x^n \in \mathbb{Q}[x]$ with integral coefficients. It says that $a(x)$ is irreducible if there is a prime p such that p does not divide α_n, p does divide α_i for all $i < n$, and p^2 does not divide α_0.

Now if $a(x)$ were reducible, then $a(x)$ could be written as a product of polynomials in $\mathbb{Q}[x]$ of lower degrees. But to use the hypotheses, the factors of $a(x)$ must have integral coefficients. So the first step in deriving Eisenstein's criterion is to show that if there are factors with rational coefficients, then there are also factors of the same degrees with integral coefficients.

Example 11.1   We have that

$$6x^4 + 20x^3 + 4x^2 - 4x - 90 = \left(\frac{12}{5}x^2 + \frac{24}{5}x - 12\right)\left(\frac{5}{2}x^2 + \frac{10}{3}x + \frac{15}{2}\right).$$

However,

$$\frac{12}{5}x^2 + \frac{24}{5}x - 12 = \frac{12}{5}(x^2 + 2x - 5) \qquad \text{and}$$

$$\frac{5}{2}x^2 + \frac{10}{3}x + \frac{15}{2} = \frac{5}{6}(3x^2 + 4x + 9)$$

so that

$$6x^4 + 20x^3 + 4x^2 - 4x - 90 = (2x^2 + 4x - 10)(3x^2 + 4x + 9). \; \diamondsuit$$

The proof of Proposition 11.3 shows that if the original polynomial has integral coefficients, then this process will always yield factors with integral coefficients. For this proof, we need the following lemma.

We will use reduction mod p to prove many of the results of this chapter. This is a convenient strategy because $\mathbb{Z}_p$ is an integral domain by Propositions 4.3 and 7.5, and, as shown in Example 4.1, p divides n if and only if $[n]_p = [0]_p$.

Lemma 11.2   **Gauss's Lemma**

Let $f(x)$, $g(x)$ be polynomials with integral coefficients and suppose that the coefficients of $f(x)$ have no common prime factor and that the coefficients of $g(x)$ have no common prime factor. Then the coefficients of $f(x)g(x)$ also have no common prime factor.

Proof   Let $f(x) = \alpha_0 + \cdots + \alpha_m x^m$, $g(x) = \beta_0 + \cdots + \beta_n x^n$, and $f(x)g(x) = \lambda_0 + \cdots + \lambda_{m+n} x^{m+n}$. And recall that the coefficients λ_k are all of the form $\lambda_k = \sum_{i+j=k} \alpha_i \beta_j$. Suppose by way of contradiction that a prime p divides all the coefficients of $f(x)g(x)$, and let α_r and β_s be the first coefficients of $f(x)$ and $g(x)$ respectively which are not divisible by p. Then $[\alpha_r]_p \neq [0]_p$, $[\beta_s]_p \neq [0]_p$, $[\alpha_i]_p = [0]_p$ for all $i < r$, and $[\beta_j]_p = [0]_p$ for all $j < s$. Thus, $[\alpha_i]_p[\beta_j]_p = [0]_p$ whenever $i < r$ or $j < s$, and since by Propositions 4.3 and 7.5 $\mathbb{Z}_p$ is an integral domain, $[\alpha_r]_p[\beta_s]_p \neq [0]_p$. It follows that

$$[\lambda_{r+s}]_p = \left[\sum_{i+j=k} \alpha_i \beta_j \right]_p = \sum_{i+j=k} [\alpha_i]_p[\beta_j]_p = [\alpha_r]_p[\beta_s]_p \neq [0]_p$$

and hence that p does not divide λ_{r+s}, a contradiction. ∎

Proposition 11.3   *A polynomial with integral coefficients which can be factored into polynomials with rational coefficients can also be factored into polynomials of the same degrees with integral coefficients.*

Proof   Let $f(x)$ be a polynomial with integral coefficients and suppose that $f(x) = a(x)b(x)$ for $a(x), b(x) \in \mathbb{Q}[x]$. Write the coefficients of $a(x)$ in lowest terms and let α_d be the product of the resulting denominators, and let α_n be the product of any common prime factors of the numerators. Then $a(x) = \alpha_n \alpha_d^{-1} \tilde{a}(x)$, where $\tilde{a}(x)$ has integral coefficients with no common prime factor. Similarly $b(x) = \beta_n \beta_d^{-1} \tilde{b}(x)$, where $\tilde{b}(x)$ has integral coefficients with no common prime factor. By Gauss's lemma, $\tilde{a}(x)\tilde{b}(x)$ is a polynomial with integral coefficients having no common prime factor. But $\alpha_d \beta_d f(x) = \alpha_n \beta_n \tilde{a}(x)\tilde{b}(x)$, and therefore, since the coefficients of $\tilde{a}(x)\tilde{b}(x)$

have no common prime factor, $\alpha_d \beta_d$ divides $\alpha_n \beta_n$. Then $f(x) = k\tilde{a}(x)\tilde{b}(x)$ for some $k \in \mathbb{Z}$, and hence $k\tilde{a}(x)$ and $\tilde{b}(x)$ are the desired factors of $f(x)$. ∎

We now have the necessary tools to prove Eisenstein's criterion. To see how the proof works, consider the following example.

Example 11.4   Let $a(x) = 252x^3 - 225x^2 + 180 \in \mathbb{Q}[x]$. Note that the prime 5 does not divide 252 but does divide the other coefficients ($-225, 0$, and 180) and that $5^2 = 25$ does not divide the constant coefficient 180. Suppose by way of contradiction that $a(x)$ is reducible. Then $a(x) = b(x)c(x)$ for polynomials $b(x)$ and $c(x)$, each of degree less than three, say $b(x) = \beta_0 + \beta_1 x$ and $c(x) = \gamma_0 + \gamma_1 x + \gamma_2 x^2$. Then $\beta_0 \gamma_0 = 180$ and $\beta_1 \gamma_2 = 252$ so that 5 divides exactly one of β_0 and γ_0 and divides neither β_1 nor γ_2. Suppose first that 5 does not divide β_0. Then $[\beta_0]_5 \neq [0]_5$, $[\gamma_0]_5 = [0]_5$, $[\beta_1]_5 \neq [0]_5$, and $[\gamma_2]_5 \neq [0]_5$. Equating the coefficients of x in $a(x)$ and $b(x)c(x)$ and reducing them modulo 5, we have that

$$[0]_5 = [\beta_0]_5[\gamma_1]_5 + [\beta_1]_5[\gamma_0]_5 = [\beta_0]_5[\gamma_1]_5$$

and $[\beta_0]_5 \neq [0]_5$. But $\mathbb{Z}_5$ has no zero divisors and hence $[\gamma_1]_5 = [0]_5$. So

$$[0]_5 = [-225]_5 = [\beta_0]_5[\gamma_2]_5 + [\beta_1]_5[\gamma_1]_5 = [\beta_0]_5[\gamma_2]_5$$

and thus, again since $[\beta_0]_5 \neq [0]_5$ and $\mathbb{Z}_5$ has no zero divisors, $[\gamma_2]_5 = [0]_5$, a contradiction. The other possibility is that 5 does not divide γ_0. Then $[\gamma_0]_5 \neq [0]_5$, $[\beta_0]_5 = [0]_5$, $[\beta_1]_5 \neq [0]_5$, and $[\gamma_2]_5 \neq [0]_5$. As before, equating the coefficients of x in $a(x)$ and $b(x)c(x)$ and reducing them mod 5, we have

$$[0]_5 = [\beta_0]_5[\gamma_1]_5 + [\beta_1]_5[\gamma_0]_5 = [\beta_1]_5[\gamma_0]_5$$

and $[\gamma_0]_5 \neq [0]_5$ so that $[\beta_1]_5 = [0]_5$. This leads us to the same contradiction as before:

$$[0]_5 = [-225]_5 = [\beta_0]_5[\gamma_2]_5 + [\beta_1]_5[\gamma_1]_5 = [\beta_0]_5[\gamma_2]_5$$

and hence $[\gamma_2]_5 = [0]_5$. We therefore conclude that $a(x)$ is irreducible. ❖

In the general case, the details of the proof are as follows.

Proposition 11.5   **Eisenstein's Irreducibility Criterion**

For a given prime p, let $a(x) = \alpha_0 + \cdots + \alpha_n x^n$ be a polynomial with integral coefficients. Suppose that

> *p does not divide α_n,*
> *p does divide all the other α_i, and*
> *p^2 does not divide α_0.*

Then $a(x)$ is irreducible in $\mathbb{Q}[x]$.

Proof. Suppose that $a(x)$ is in fact reducible. Then there exist $b(x), c(x) \in \mathbb{Q}[x]$ such that $a(x) = b(x)c(x)$, where $b(x) = \beta_0 + \cdots + \beta_k x^k, \beta_k \neq 0, k < n$, and $c(x) = \gamma_0 + \cdots + \gamma_m x^m, \gamma_m \neq 0, m < n$. Recall that the coefficients α_r are all of the form $\alpha_r = \sum_{i+j=r} \beta_i \gamma_j$, and note that by Proposition 11.3, we can assume that the coefficients β_i and γ_j are integers. Since p^2 does not divide α_0 and $\alpha_0 = \beta_0 \gamma_0$, either p does not divide β_0 or p does not divide γ_0. Without loss of generality, suppose that p does not divide β_0. Then since p divides α_0 and $\alpha_0 = \beta_0 \gamma_0$, p must divide γ_0. Also, since p does not divide α_n and $\alpha_n = \beta_k \gamma_m$, p divides neither β_k nor γ_m. That is, we have

$$[\beta_0]_p \neq [0]_p, \qquad [\gamma_0]_p = [0]_p, \qquad [\beta_k]_p \neq [0]_p, \qquad [\gamma_m]_p \neq [0]_p.$$

Let r be the least j such that $[\gamma_j]_p \neq [0]_p$, i.e., let r be such that $[\gamma_r]_p \neq [0]_p$ while $[\gamma_j]_p = [0]_p$ for all $j < r$. Then $r \leq m < n$ and hence by hypothesis $[\alpha_r]_p = [0]_p$. And thus, since $[\beta_i]_p [\gamma_j]_p = [0]_p$ whenever $j < r$,

$$[\beta_0]_p [\gamma_r]_p = \sum_{i+j=r} [\beta_i]_p [\gamma_j]_p = \left[\sum_{i+j=r} \beta_i \gamma_j \right]_p = [\alpha_r]_p = [0]_p.$$

However, $[\gamma_r]_p \neq [0]_p$ and $[\beta_0]_p \neq [0]_p$ and by Propositions 4.3 and 6.5, $\mathbb{Z}_p$ is an integral domain. So $[\beta_0]_p [\gamma_r]_p \neq [0]_p$. This is a contradiction and hence $a(x)$ is irreducible. ■

As previously noted, we are interested solely in applications of Eisenstein's criterion. In particular, we want to be able to use it in conjunction with Theorem 10.8, as in the following examples.

Example 11.6   Consider $\sqrt[5]{7} \in \mathbb{R}$. Certainly $\sqrt[5]{7}$ is a root of $m(x) = x^5 - 7$, and clearly $m(x) \in \mathbb{Q}[x]$ and $m(x)$ is monic. As well, by Eisenstein's criterion (with prime 7), $m(x)$ is irreducible, and hence $m(x)$ is the minimum polynomial of $\sqrt[5]{7}$ over $\mathbb{Q}$. Thus, $[\sqrt[5]{7} : \mathbb{Q}] = 5$, and by Theorem 10.8,

$$\mathbb{Q}(\sqrt[5]{7}) = \{\alpha_0 + \alpha_1 \sqrt[5]{7} + \alpha_2 \sqrt[5]{7}^{\,2} + \alpha_3 \sqrt[5]{7}^{\,3} + \alpha_4 \sqrt[5]{7}^{\,4} \mid \alpha_j \in \mathbb{Q}\}. ❖$$

Example 11.7   We may describe $\mathbb{Q}(\sqrt[5]{7}, i)$ by using the method of Example 10.10. We have seen in Example 11.6 how to write the elements of $\mathbb{Q}(\sqrt[5]{7})$ in terms of $\sqrt[5]{7}$ as elements of $\mathbb{Q}$, and as in Example 10.10, it is easy to see that $\mathbb{Q}(\sqrt[5]{7}, i) = \mathbb{Q}(\sqrt[5]{7})(i)$. So we need to apply Theorem 10.8 again, this time to the element i over the field $\mathbb{Q}(\sqrt[5]{7})$. That is, we need to find the minimum polynomial of i over $\mathbb{Q}(\sqrt[5]{7})$. Certainly i is a root of $p(x) = x^2 + 1$, and clearly $p(x) \in \mathbb{Q}(\sqrt[5]{7})[x]$ and $p(x)$ is monic. As well, the roots of $p(x)$ are $\pm i$, and since $\mathbb{Q}(\sqrt[5]{7}) \subseteq \mathbb{R}$, neither of these roots is in $\mathbb{Q}(\sqrt[5]{7})$. So $p(x)$ is irreducible in $\mathbb{Q}(\sqrt[5]{7})[x]$ by Proposition 8.12, and therefore $p(x)$ is the minimum polynomial of i over $\mathbb{Q}(\sqrt[5]{7})$. Then $[i : \mathbb{Q}(\sqrt[5]{7})] = 2$, and by Theorem 10.8,

$$\mathbb{Q}(\sqrt[5]{7})(i) = \{A_0 + A_1 i \,|\, A_k \in \mathbb{Q}(\sqrt[5]{7})\}.$$

But by Example 11.6, the elements of $\mathbb{Q}(\sqrt[5]{7})$ may all be expressed in the form $\alpha_0 + \alpha_1 \sqrt[5]{7} + \alpha_2 \sqrt[5]{7}^{\,2} + \alpha_3 \sqrt[5]{7}^{\,3} + \alpha_4 \sqrt[5]{7}^{\,4}$, where $\alpha_j \in \mathbb{Q}$, and therefore

$$
\begin{aligned}
\mathbb{Q}(\sqrt[5]{7}, i) &= \mathbb{Q}(\sqrt[5]{7})(i) \\
&= \{(\alpha_0 + \alpha_1 \sqrt[5]{7} + \alpha_2 \sqrt[5]{7}^{\,2} + \alpha_3 \sqrt[5]{7}^{\,3} + \alpha_4 \sqrt[5]{7}^{\,4}) \\
&\quad + (\beta_0 + \beta_1 \sqrt[5]{7} + \beta_2 \sqrt[5]{7}^{\,2} + \beta_3 \sqrt[5]{7}^{\,3} + \beta_4 \sqrt[5]{7}^{\,4})i \,|\, \alpha_j, \beta_k \in \mathbb{Q}\}. ❖
\end{aligned}
$$

Example 11.8   Consider the polynomial $f(x) = x^5 - 15x^4 + 25x^2 + 10x - 5$. Since $f(0) = -5 < 0 < 16 = f(1)$, the intermediate value theorem of single variable calculus implies that the equation $f(x) = 0$ has a solution $r \in \mathbb{R}$. We will write the elements of $\mathbb{Q}(r, \zeta_3)$ in terms of r, ζ_3, and elements of $\mathbb{Q}$.

We first observe that $f(x)$ is the minimum polynomial of r over $\mathbb{Q}$. For certainly $f(x) \in \mathbb{Q}[x]$ and $f(x)$ is monic, and by our choice of $r, f(r) = 0$. As well, by Eisenstein's criterion (with prime 5), $f(x)$ is irreducible over $\mathbb{Q}$, and hence $f(x)$ is the minimum polynomial of r over $\mathbb{Q}$. So $[r : \mathbb{Q}] = 5$, and thus by Theorem 10.8,

$$\mathbb{Q}(r) = \{\alpha_0 + \alpha_1 r + \alpha_2 r^2 + \alpha_3 r^3 + \alpha_4 r^4 \,|\, \alpha_j \in \mathbb{Q}\}.$$

We next observe that $g(x) = x^2 + x + 1$ is the minimum polynomial of ζ_3 over $\mathbb{Q}(r)$. For certainly $g(x) \in \mathbb{Q}(r)[x]$ and $g(x)$ is monic, and by Example 10.5, $g(\zeta_3) = 0$. As well, since $r \in \mathbb{R}, \mathbb{Q}(r) \subseteq \mathbb{R}$, and thus $\sqrt{1^2 - 4 \cdot 1 \cdot 1} = \sqrt{-3} \notin \mathbb{Q}(r)$. So by Proposition 8.12, $g(x)$ is irreducible over $\mathbb{Q}(r)$, and hence $g(x)$ is the minimum polynomial of ζ_3 over $\mathbb{Q}(r)$. So $[\zeta_3 : \mathbb{Q}(r)] = 2$, and thus by Theorem 10.8,

$$\mathbb{Q}(r)(\zeta_3) = \{A_0 + A_1 \zeta_3 \,|\, A_k \in \mathbb{Q}(r)\}.$$

But as in Example 10.10, $\mathbb{Q}(r, \zeta_3) = \mathbb{Q}(r)(\zeta_3)$, and thus

$$
\begin{aligned}
\mathbb{Q}(r, \zeta_3) = \{&(\alpha_0 + \alpha_1 r + \alpha_2 r^2 + \alpha_3 r^3 + \alpha_4 r^4) \\
&+ (\beta_0 + \beta_1 r + \beta_2 r^2 + \beta_3 r^3 + \beta_4 r^4)\zeta_3 \,|\, \alpha_j, \beta_k \in \mathbb{Q}\}.
\end{aligned}
$$

Note that since $\mathbb{Q}(r, \zeta_3) = \mathbb{Q}(r)(\zeta_3) = \mathbb{Q}(\zeta_3)(r)$, it is also possible to construct $\mathbb{Q}(r, \zeta_3)$ by first adjoining ζ_3 to $\mathbb{Q}$ and then adjoining r. However, since Eisenstein's criterion applies only to polynomials in $\mathbb{Q}[x]$ and not to polynomials in $\mathbb{Q}(\zeta_3)[x]$, this result could not be used to show that $f(x)$ is irreducible over $\mathbb{Q}(r)$. So we do not have the tools to find the minimum polynomial of r over $\mathbb{Q}(\zeta_3)$, and hence we do not have the tools to carry out this construction. We will see in the next chapter how to make such calculations indirectly. ❖

The construction given in Examples 10.10, 11.7, and 11.8 may be carried out in general. Proposition 12.5 in the next chapter phrases it in the language of vector spaces.

Exercises

In Exercises 1–4, find factors with integral coefficients of the same degree as the given factors.

1. $2x^4 - 7x^2 - 15 = \left(\frac{14}{10}x^2 - \frac{21}{3}\right)\left(\frac{20}{14}x^2 + \frac{15}{7}\right)$

2. $3x^4 - 6x^3 + 10x^2 - 14x + 7 = \left(\frac{1}{2}x^2 - x + \frac{1}{2}\right)\left(\frac{18}{3}x^2 + \frac{28}{2}\right)$

3. $15x^9 + 12x^6 - 45x^5 - 24x^2 + 30x = \left(\frac{20}{14}x^5 - \frac{20}{7}x\right)\left(\frac{42}{4}x^4 + \frac{42}{5}x - \frac{21}{2}\right)$

4. $2x^5 - 7x^4 + 15x^2 - 4x - 3 =$
$$\left(\frac{9}{3}x^3 - \frac{15}{2}x^2 + \frac{3}{2}x + \frac{6}{4}\right)\left(\frac{4}{6}x^2 - \frac{6}{9}x - \frac{10}{5}\right)$$

In Exercises 5–10, determine whether the given polynomial is irreducible according to Eisenstein's criterion. Explain the way in which the conditions are or are not fulfilled.

5. $x^5 + 15x^4 - 50x^2 + 25x - 45$ **6.** $x^5 + 15x^4 - 50x^2 + 45x - 25$

7. $3x^4 + 6x^3 + 12x^2 - 6x + 18$ **8.** $x^2 - 182x + 143$

9. $2x^7 + 6x^6 - 12x^4 + 24x^3 + 6x^2 - 18$ **10.** $x^4 + 6x^2 - 12x + 9$

In Exercises 11–16, use the method of Examples 11.7 and 11.8 to express the elements of $\mathbb{Q}(r, s)$ in terms of elements of $\mathbb{Q}$ and powers of r and s.

11. $r = \sqrt{5}, s = i$ **12.** $r = \sqrt[7]{12}, s = i$ **13.** $r = \sqrt[3]{12}, s = \zeta_3$

14. $r = -\sqrt[4]{6}, s = \zeta_3$ **15.** $r = \sqrt[5]{11}, s = i$ **16.** $r = \sqrt[3]{16}, s = \zeta_3$

17. Suppose that I is an ideal of $\mathbb{Q}[x]$ which contains both

$$x^4 + 2x^3 + 3x^2 + 2x + 2 \quad \text{and} \quad 3x^6 + 12x^4 - 18x^3 + 6.$$

Show that $I = \mathbb{Q}[x]$.

18. Suppose that I is an ideal of $\mathbb{Q}[x]$ which contains both $5x^7 + 12x^6 + 18x^3 + 6x^2 + 12$ and a polynomial $f(x)$ of degree less than seven. Show that $I = \mathbb{Q}[x]$.

Exercises 19 and 20 will be used in some of the exercises in Chapter 12.

19. Show that $x^4 + x^3 + x^2 + x + 1$ is irreducible over $\mathbb{Q}$. (*Hint:* First show that if $p(x) \in \mathbb{Q}[x]$ is reducible and $s(x) = p(x + 1)$, then $s(x)$ is also reducible over $\mathbb{Q}$. Then show that if $p(x) = x^4 + x^3 + x^2 + x + 1$ in $\mathbb{Q}[x]$ and $s(x) = p(x + 1)$, then $s(x)$ is irreducible over $\mathbb{Q}$.)

20. Show that $x^6 + x^5 + x^4 + x^3 + x^2 + x + 1$ is irreducible over $\mathbb{Q}$.
(*Hint:* Generalize the hint in Exercise 19.)

21. If p is prime, show that $x^{p-1} + x^{p-2} + \cdots + x + 1$ is irreducible over $\mathbb{Q}$.
(*Hint:* Use Exercises 43 and 44 in Chapter 1 and generalize the hint in Exercise 19.)

22. Suppose that p is an odd prime and that $[n]_p \neq [1]_p$. Show that $[n^2]_p = [1]_p$ if and only if $[n]_p = [-1]_p$.

23. Suppose that p is a prime and that $[n]_p \neq [1]_p$ but $[n^3]_p = [1]_p$. Show that $[1]_p + [n]_p + [n^2]_p = [0]_p$.

24. Suppose that $f(x)$ and $g(x)$ are polynomials with integral coefficients and let $h(x) = f(x)g(x)$. Show that if the coefficients of $h(x)$ have no common prime factor, then the coefficients of $f(x)$ have no common prime factor and the coefficients of $g(x)$ have no common prime factor.

HISTORICAL NOTE

Gotthold Eisenstein

Ferdinand Gotthold Max Eisenstein was born in Berlin, Germany, on April 16, 1823, and died in Berlin on October 11, 1852. His parents converted from Judaism to Protestantism before Eisenstein was born, and his father had served in the Prussian army and then gone into business, at which he became successful only late in his life. All of Gotthold's siblings died in childhood, most of meningitis, which Gotthold also contracted. His poor health led his parents to send him first, from 1833 to 1837, to the Cauer Academy near Berlin as a residential student and then, from 1837 to 1842, to the Friedrich Werder Gymnasium. In his final year, he attended mathematical lectures at the university as well. Eisenstein's mathematical talent surfaced when he was six years old and gave him little rest. He later wrote that the mathematical "way of deducing and discovering new truths from old ones, and the extraordinary clarity and self-evidence of the theorems, the ingeniousness of the ideas . . . had an irresistible fascination for me."[1]

In 1840, his father had traveled to England in the hope of financial reward, and in the summer of 1842, Eisenstein and his mother went there to join him. While financial success eluded his father, Eisenstein had the time to study Gauss's *Disquisitiones Arithmeticae* and early in 1843 he met William Rowan Hamilton in Dublin. Hamilton gave him a copy of his paper "On the argument of Abel, respecting the impossibility of expressing a root of any general equation above the fourth degree" to present to the Berlin Academy.

In 1843, he passed his final high school examination and enrolled at the University of Berlin, and in January 1844, he delivered Hamilton's paper to the Berlin

Academy, along with a treatise of his own. Leopold Crelle read Eisenstein's paper, and spotting mathematical genius at an early stage (as he had done with Abel), he accepted it for publication in his journal. Crelle also introduced Eisenstein to Alexander von Humboldt who became his life-long protector and mentor. Altogether in 1844, Eisenstein published a phenomenal twenty-five papers in *Crelle's Journal* and established himself as a rising star in the world of mathematics.

In June 1844, armed with a glowing letter of recommendation from Humboldt, he traveled to Göttingen to visit Gauss. It was a very successful trip but none of his mathematical successes could rid him of constant bouts of depression, not even the awarding in February 1845 of an honorary doctorate in philosophy from the University of Breslau. The suggestion for awarding this degree came from Karl Jacobi (possibly acting for Humboldt) who ironically accused Eisenstein of plagiarism the very next year. Eisenstein admitted in response that he had failed to acknowledge Jacobi's work but pleaded "naïve innocence."[2]

Over the next two years, he continued publishing his mathematical results, but early in 1848, he became involved in politics. On March 19, he was in a house from which shots had been fired and the police took him to the Citadel in Spandau. Because of the harsh treatment he received en route, his health immediately took a turn for the worse. In addition, he was branded as a "republican" and his financial support dwindled. Even though his health prevented him from delivering many lectures, he continued producing mathematical results and in 1851, on Gauss's recommendation, he was elected a corresponding member of the Göttingen Society, and in 1852, on Dirichlet's recommendation, he was elected to membership in the Berlin Academy. But none of these mathematical rewards could help his deteriorating physical condition. In July 1852, he suffered a severe hemorrhage and in October he died of pulmonary tuberculosis.

References   1. Bierman, Kurt-R. "Eisenstein, Ferdinand Gotthold Max," *Dictionary of Scientific Biography,* 16 vols. Gillispie, Charles Coulston, ed. in chief. New York: Charles Scribner's Sons, 1970, vol. IV, p. 340.
2. *Ibid.,* p. 341.

12

Extension Fields as Vector Spaces

As we will see in the following discussion, any extension field K of a field F may be viewed as a vector space over F. In this chapter, we will investigate some of the ramifications of this observation. In particular, we will be interested in interpreting Theorem 10.8 from this point of view.

DEFINITION

A set V is a **vector space** over a field F if V has defined on it a closed, associative, and commutative binary operation $+$ with respect to which it has an identity and all of its elements have inverses (cf. Chapter 5), and if, for every $\alpha \in F$ and $v \in V$, there exists an element $\alpha v \in V$ for which the following conditions hold for $\alpha, \beta \in F$ and $v, w \in V$:

    (i) $\alpha(v + w) = \alpha v + \alpha w$
    (ii) $(\alpha + \beta)v = \alpha v + \beta v$
    (iii) $\alpha(\beta v) = (\alpha\beta)v$
    (iv) $1v = v$.

A vector space is **nontrivial** if it has at least two elements. The elements of V are called **vectors** and the elements of F are called **scalars.** A subset of V that is a vector space with respect to the inherited operations is called a **subspace** of V over F.

The most familiar examples of vector spaces are of course the Euclidean spaces $\mathbb{R}^2 = \{(x,y) \mid x,y \in \mathbb{R}\}$ and $\mathbb{R}^3 = \{(x,y,z) \mid x,y,z \in R\}$ over the field $\mathbb{R}$, where addition and scalar multiplication are defined coordinate by coordinate. These spaces are built from the standard basis vectors: $(1,0)$ and $(0,1)$ in the case of $\mathbb{R}^2$, and $(1,0,0)$, $(0,1,0)$ and $(0,0,1)$ in the case of $\mathbb{R}^3$. That is, any vector $(x,y) \in \mathbb{R}^2$ may be uniquely written $(x,y) = x(1,0) + y(0,1)$, and any vector $(x,y,z) \in \mathbb{R}^3$

may be uniquely written $(x, y, z) = x(1,0,0) + y(0,1,0) + z(0,0,1)$. The subset $\{(x, y, 0) \mid x, y \in \mathbb{R}\}$ of $\mathbb{R}^3$ is a subspace of $\mathbb{R}^3$ which can be identified with $\mathbb{R}^2$.

Next recall the definition of the basis of a vector space.

DEFINITION

Let V be a vector space over a field F. The subset $\{v_1, \ldots, v_n\}$ of V is **linearly independent** if for all scalars $\alpha_1, \ldots, \alpha_n \in F$,

$$\alpha_1 v_1 + \cdots + \alpha_n v_n = 0 \text{ implies that } \alpha_1 = \alpha_2 = \cdots = \alpha_n = 0;$$

$\{v_1, \ldots, v_n\}$ **spans** V if for all $x \in V$, there exist $\beta_1, \ldots, \beta_n \in F$ such that

$$x = \beta_1 v_1 + \cdots + \beta_n v_n.$$

A **basis** of V is a linearly independent subset of V which spans V. An expression of the form $\beta_1 v_1 + \cdots + \beta_n v_n$, where $\beta_1, \ldots, \beta_n \in F$, is called a **linear combination** of the vectors $v_1, \ldots, v_n$.

To require that a basis $\{b_1, \ldots, b_n\}$ spans V is to require that every vector $v \in V$ can be written as a linear combination of members of the basis: $v = \alpha_1 b_1 + \cdots + \alpha_n b_n$ for $\alpha, \ldots, a_n \in F$. To require further that $\{b_1, \ldots, b_n\}$ be linearly independent is to require that there be only one such expression. For if $\alpha_1 b_1 + \cdots + \alpha_n b_n = \gamma_1 b_1 + \cdots + \gamma_n b_n$, then $(\alpha_1 - \gamma_1) b_1 + \cdots + (\alpha_n - \gamma_n) b_n = 0$ and hence by linear independence $\alpha_i = \gamma_i$ for all i.

We next collect together the results we will need from the theory of bases. Since we assume that the reader has already seen their proofs in another setting, we relegate them to Appendix C.

Proposition 12.1 *Suppose that V is a nontrivial vector space which is spanned by the finite set $\{w_1, \ldots, w_k\}$ and which has a basis with n elements. Then*

(i)   some subset $\{b_1, \ldots, b_n\}$ of $\{w_1, \ldots, w_k\}$ is a basis of V;

(ii)   every basis of V has n elements;

(iii)   every subspace of V has a basis with no more than n elements;

(iv)   the only subspace of V containing a basis with n elements is V itself.

Vector spaces that have finite bases are the ones that will be important for our purposes. We therefore recall the following definition.

DEFINITION

A nontrivial vector space V over a field F is **finite dimensional** if it has a basis with a finite number of elements. The number of elements in any such basis is called the **dimension** of V over F and is denoted $n = [V:F]$.

Now consider an extension field K of a field F. We can consider K as a vector space over F by forgetting multiplications ab for $a \notin F$. That is, take the operation $+$ on the field K to be the operation $+$ on the vector space K, and use the operation $\cdot$ on the field K to define multiplication of vectors in K by scalars in F. Then the field axioms (Chapter 5) show that K is a vector space over F.

For instance, the extension field $\mathbb{C}$ of the field $\mathbb{R}$ is a vector space over $\mathbb{R}$ with respect to the addition: $(a + bi) + (c + di) = (a + c) + (b + d)i$ and the multiplication: $r(a + bi) = (ra) + (rb)i$. We note in passing that as vector spaces over $\mathbb{R}$, $\mathbb{C}$ and $\mathbb{R}^2$ are indistinguishable.

In particular, suppose that r is algebraic over F and consider $F(r)$ as a vector space over F. By Theorem 10.8, $F(r) = \{\alpha_0 + \cdots + \alpha_{n-1}r^{n-1} \mid \alpha_i \in F\}$, where $n = [r:F]$. So certainly the set of vectors $\{1, r, \ldots, r^{n-1}\}$ spans $F(r)$ as a vector space over F. The following lemma shows that this set is linearly independent and hence that it is a basis (Proposition 12.4). The lemma can also be used to derive an alternative to the Euclidean algorithm for finding multiplicative inverses in $F(r)$ (Example 12.3).

Lemma 12.2   *Let r be algebraic over F and let $n = [r:F]$. If*

$$\alpha_0 + \alpha_1 r + \cdots + \alpha_{n-1}r^{n-1} = \beta_0 + \beta_1 r + \cdots + \beta_{n-1}r^{n-1}$$

for $\alpha_i, \beta_i \in F$, then $\alpha_i = \beta_i$ for all i.

Proof   Define a polynomial $f(x) \in F[x]$ by

$$f(x) = (\alpha_0 - \beta_0) + (\alpha_1 - \beta_1)x + \cdots + (\alpha_{n-1} - \beta_{n-1})x^{n-1}.$$

Then $f(r) = 0$, and hence by Proposition 10.3, $f(x) \in (m(x))$, where $m(x)$ is the minimum polynomial of r over F. However, if $f(x) \neq 0$, then $\deg f(x) \leq n - 1 < n = \deg m(x)$ and hence $m(x)$ cannot divide $f(x)$. This contradicts our observation that $f(x) \in (m(x))$, and thus $f(x) = 0$. It then follows that $\alpha_i = \beta_i$ for all i. ❖

Lemma 12.2 yields the following algorithm for calculating multiplicative inverses in $F(a)$.

Example 12.3   **Method of Undetermined Coefficients**
Let r be a solution of the equation $x^4 + 3x - 3 = 0$ in $\mathbb{C}$. Obviously $x^4 + 3x - 3$ is monic and has r as a root, and by Eisenstein's criterion with prime $p = 3$, $x^4 + 3x - 3$ is irreducible over $\mathbb{Q}$. So by Theorem 10.8, the elements of $\mathbb{Q}(r)$ may all be written in the form

$$\alpha_0 + \alpha_1 r + \alpha_2 r^2 + \alpha_3 r^3.$$

Consider $1 + 2r^2$ and $3 - r^3$ in $\mathbb{Q}(r)$. As in Examples 10.6, 10.9, and 10.10, we may multiply these elements by using the equation which r solves. For since

$r^4 + 3r - 3 = 0$, we have $r^4 = 3 - 3r$ and $r^5 = 3r - 3r^2$, and thus

$$(1 + 2r^2)(3 - r^3) = 3 + 6r^2 - r^3 - 2r^5 = 3 - 6r + 12r^2 - r^3.$$

This technique, when combined with Lemma 12.2, also provides a method for calculating inverses. For if $(1 + 2r^2)^{-1} = \alpha_0 + \alpha_1 r + \alpha_2 r^2 + \alpha_3 r^3$, then

$$\begin{aligned}
1 &= (1 + 2r^2)(\alpha_0 + \alpha_1 r + \alpha_2 r^2 + \alpha_3 r^3) \\
&= \alpha_0 + \alpha_1 r + (\alpha_2 + 2\alpha_0)r^2 + (\alpha_3 + 2\alpha_1)r^3 + 2\alpha_2 r^4 + 2\alpha_3 r^5 \\
&= (\alpha_0 + 6\alpha_2) + (\alpha_1 + 6\alpha_2 - 6\alpha_3)r + (\alpha_2 + 2\alpha_0 - 6\alpha_3)r^2 + (\alpha_3 + 2\alpha_1)r^3,
\end{aligned}$$

and by Lemma 12.2,

$$\begin{aligned}
\alpha_0 + 6\alpha_2 &= 1, \\
\alpha_1 - 6\alpha_2 + 6\alpha_3 &= 0, \\
\alpha_2 + 2\alpha_0 - 6\alpha_3 &= 0, \\
\alpha_3 + 2\alpha_1 &= 0.
\end{aligned}$$

Solving these four equations for the four unknowns α_0, α_1, α_2, and α_3, we find that

$$\alpha_0 = \frac{61}{193}, \qquad \alpha_1 = \frac{-12}{193}, \qquad \alpha_2 = \frac{22}{193}, \qquad \alpha_3 = \frac{24}{193},$$

and hence that

$$(1 + 2r^2)^{-1} = \frac{61}{193} - \frac{12}{193}r + \frac{22}{193}r^2 + \frac{24}{193}r^3. \quad \diamondsuit$$

We next confirm that $\{1, r, \ldots, r^{n-1}\}$ is indeed a basis of $F(r)$ over F.

Proposition 12.4   *If r is algebraic over F and $n = [r:F]$, then $\{1, r, \ldots, r^{n-1}\}$ is a basis of the vector space $F(r)$ over F and hence $[F(r):F] = [r:F]$.*

Proof   By Theorem 10.8, for all $\xi \in F(r)$, there exist $\alpha_0, \ldots, \alpha_{n-1} \in F$ such that $\xi = \alpha_0 + \cdots + \alpha_{n-1}r^{n-1}$, and hence $\{1, r, \ldots, r^{n-1}\}$ spans $F(r)$. Suppose that $c_0, \ldots, c_{n-1} \in F$ are such that $c_0 + \cdots + c_{n-1}r^{n-1} = 0$. Then since also $0 = 0 + \cdots + 0 \cdot r^{n-1}$, Lemma 12.2 implies that $c_i = 0$ for all i, and hence $\{1, r, \ldots, r^{n-1}\}$ is linearly independent. We conclude that $\{1, r, \ldots, r^{n-1}\}$ is a basis for $F(r)$ and that, since $\{1, r, \ldots, r^{n-1}\}$ has n elements, $[F(r):F] = n = [r:F]$. $\quad \diamondsuit$

Can we also determine $[F(r, s):F]$? To see what can happen, consider Examples 10.9 and 10.10. If r is a root of $x^2 + 3x - 5 \in \mathbb{Q}[x]$, then according to Example 10.9, $x^2 + 3x - 5$ is the minimum polynomial of r over $\mathbb{Q}$ and hence by Proposition 12.4, $\{1, r\}$ is a basis of $\mathbb{Q}(r)$ over $\mathbb{Q}$. Similarly, according to Example 10.10, $x^2 + 1 \in \mathbb{Q}(r)[x]$ is the minimum polynomial of i over $\mathbb{Q}(r)$ and hence by Proposition 12.4, $\{1, i\}$ is a basis of $\mathbb{Q}(r, i)$ over $\mathbb{Q}(r)$. Example 10.10 shows further that

$\{1, r, i, ir\}$ spans $\mathbb{Q}(r, i)$ as a vector space over $\mathbb{Q}$. Note that $\{1, r, i, ir\}$ is the set of <u>all</u> multiplications of elements of $\{1, r\}$ by elements of $\{1, i\}$. We show next that, in general, spanning sets formed in this way are in fact bases and therefore that for finite-dimensional extension fields $F \subseteq L \subseteq K$, $[K:F] = [K:L][L:F]$. In particular, for r algebraic over F and s algebraic over $F(r)$, $[F(r, s):F] = [F(r, s):F(r)][F(r):F]$.

Proposition 12.5 *Suppose that $\{a_1, \ldots, a_m\}$ is a basis of the extension field K over the field L and that $\{b_1, \ldots, b_n\}$ is a basis of the extension field L over the field F. Then $\{a_1b_1, \ldots, a_1b_n, a_2b_1, \ldots, a_mb_n\}$ is a basis of the extension field K over the field F and hence $[K:F] = [K:L][L:F]$.*

Proof Let $B = \{a_1b_1, \ldots, a_1b_n, a_2b_1, \ldots, a_mb_n\}$. To see that B spans K, let $v \in K$. Then $v = \sum_{j=1}^{n}\beta_j b_j$, where $\beta_j \in L$. But each $\beta_j = \sum_{i=1}^{m}\alpha_{ij}a_i$, where $\alpha_{ij} \in F$. Hence $v = \sum_{j=1}^{n}\sum_{i=1}^{m}\alpha_{ij}(a_ib_j)$, and thus B spans K. To see that B is linearly independent over F, suppose that $\alpha_{11}, \ldots, \alpha_{mn} \in F$ are such that

$$0 = \sum_{j=1}^{n}\sum_{i=1}^{m}\alpha_{ij}(a_ib_j) = \sum_{j=1}^{n}\left(\sum_{i=1}^{m}\alpha_{ij}a_i\right)b_j$$

Since $\sum_{i=1}^{m}\alpha_{ij}a_i \in L$ for all j, and since $\{b_1, \ldots, b_n\}$ is linearly independent over L, $\sum_{i=1}^{m}\alpha_{ij}a_i = 0$ for all j. But also $\alpha_{ij} \in F$ for all i and j and $\{a_1, \ldots, a_m\}$ is linearly independent over F. Therefore $\alpha_{ij} = 0$ for all i and j, and thus B is linearly independent over F. It follows that B is a basis for K over F and therefore that $[K:F] = [K:L][L:F]$. ❖

Example 12.6 Consider $\sqrt[5]{7} \in \mathbb{R}$. As in Example 11.6, we may think of $\mathbb{Q}(\sqrt[5]{7}, i)$ as being built in two stages

$$\mathbb{Q} \subseteq \mathbb{Q}(\sqrt[5]{7}) \subseteq \mathbb{Q}(\sqrt[5]{7}, i).$$

As in Example 11.6, $x^5 - 7 \in \mathbb{Q}[x]$ is the minimum polynomial of $\sqrt[5]{7}$ over $\mathbb{Q}$ and, as in Example 11.7, $x^2 + 1 \in \mathbb{Q}(\sqrt[5]{7})[x]$ is the minimum polynomial of i over $\mathbb{Q}(\sqrt[5]{7})$. So by Theorem 10.8,

$$\mathbb{Q}(\sqrt[5]{7}) = \{\alpha_0 + \alpha_1\sqrt[5]{7} + \alpha_2\sqrt[5]{7}^2 + \alpha_3\sqrt[5]{7}^3 + \alpha_4\sqrt[5]{7}^4 \mid \alpha_j \in \mathbb{Q}\},$$
$$\mathbb{Q}(\sqrt[5]{7}, i) = \{A_0 + A_1i \mid A_0, A_1 \in \mathbb{Q}(t)\}.$$

By Proposition 12.4, $\{1, \sqrt[5]{7}, \sqrt[5]{7}^2, \sqrt[5]{7}^3, \sqrt[5]{7}^4\}$ is a basis for $\mathbb{Q}(\sqrt[5]{7})$ over $\mathbb{Q}$, and $\{1, i\}$ is a basis for $\mathbb{Q}(\sqrt[5]{7}, i)$ over $\mathbb{Q}(\sqrt[5]{7})$. And by Proposition 12.5, $[\mathbb{Q}(\sqrt[5]{7}, i):\mathbb{Q}] = [\mathbb{Q}(\sqrt[5]{7}, i):\mathbb{Q}(\sqrt[5]{7})][\mathbb{Q}(\sqrt[5]{7}):\mathbb{Q}] = 2 \cdot 5 = 10$ and

$$\{1, \sqrt[5]{7}, \sqrt[5]{7}^2, \sqrt[5]{7}^3, \sqrt[5]{7}^4, i, \sqrt[5]{7}i, \sqrt[5]{7}^2i, \sqrt[5]{7}^3i, \sqrt[5]{7}^4i\}$$

is a basis for $\mathbb{Q}(\sqrt[5]{7}, i)$ over $\mathbb{Q}$. This basis is reflected in the final description of the elements of $\mathbb{Q}(\sqrt[5]{7}, i)$ that we found in Example 11.7:

$$(\alpha_0 + \alpha_1 \sqrt[5]{7} + \alpha_2 \sqrt[5]{7}^{\,2} + \alpha_3 \sqrt[5]{7}^{\,3} + \alpha_4 \sqrt[5]{7}^{\,4})$$
$$+ (\beta_0 + \beta_1 \sqrt[5]{7} + \beta_2 \sqrt[5]{7}^{\,2} + \beta_3 \sqrt[5]{7}^{\,3} + \beta_4 \sqrt[5]{7}^{\,4})i$$
$$= \alpha_0 + \alpha_1 \sqrt[5]{7} + \alpha_2 \sqrt[5]{7}^{\,2} + \alpha_3 \sqrt[5]{7}^{\,3} + \alpha_4 \sqrt[5]{7}^{\,4}$$
$$+ \beta_0 i + \beta_1 i \sqrt[5]{7} + \beta_2 i \sqrt[5]{7}^{\,2} + \beta_3 i \sqrt[5]{7}^{\,3} + \beta_4 i \sqrt[5]{7}^{\,4}. \; \diamond$$

In Example 12.6, we computed $[\mathbb{Q}(\sqrt[5]{7}, i):\mathbb{Q}]$ by first adjoining $\sqrt[5]{7}$ and then adjoining i. We could in principle construct $\mathbb{Q}(\sqrt[5]{7}, i)$ equally well by first adjoining i and then adjoining $\sqrt[5]{7}$. However, this would involve using the equation $[\mathbb{Q}(\sqrt[5]{7}, i):\mathbb{Q}] = [\mathbb{Q}(\sqrt[5]{7}, i):\mathbb{Q}(i)][\mathbb{Q}(i):\mathbb{Q}]$, and would hence require us to calculate $[\mathbb{Q}(\sqrt[5]{7}, i):\mathbb{Q}(i)]$. But we cannot use Proposition 12.4 to determine this dimension because neither Proposition 8.12 (because the degree is too large) nor Eisenstein's criterion (because the field is not $\mathbb{Q}$) can be used to show that $x^5 - 7$, the obvious candidate for the minimum polynomial of $\sqrt[5]{7}$ over $\mathbb{Q}(i)$, is irreducible. The following example shows how to calculate $[\mathbb{Q}(\sqrt[5]{7}, i):\mathbb{Q}(i)]$ indirectly by using $[\mathbb{Q}(\sqrt[5]{7}, i):\mathbb{Q}]$, and, as a byproduct, shows that $x^5 - 7$ is the minimum polynomial of $\sqrt[5]{7}$ over $\mathbb{Q}(i)$.

Example 12.7   Consider $\sqrt[5]{7} \in \mathbb{R}$. In Example 12.6, we showed that $[\mathbb{Q}(\sqrt[5]{7}, i):\mathbb{Q}] = 10$. Now certainly i is algebraic over $\mathbb{Q}$ because i is a root of $x^2 + 1 \in \mathbb{Q}[x]$, and $\sqrt[5]{7}$ is algebraic over $\mathbb{Q}(i)$ because $\sqrt[5]{7}$ is a root of $x^5 - 7 \in \mathbb{Q}(i)[x]$. So by Proposition 12.5, $[\mathbb{Q}(\sqrt[5]{7}, i):\mathbb{Q}] = [\mathbb{Q}(\sqrt[5]{7}, i):\mathbb{Q}(i)][\mathbb{Q}(i):\mathbb{Q}]$. Furthermore, $x^2 + 1 \in \mathbb{Q}[x]$, $x^2 + 1$ is monic, and since the roots $\pm i$ of $x^2 + 1$ are not in $\mathbb{Q}$, $x^2 + 1$ is irreducible over $\mathbb{Q}$ by Proposition 8.12. So $x^2 + 1$ is the minimum polynomial of i over $\mathbb{Q}$, and thus by Proposition 12.4, $[\mathbb{Q}(i):\mathbb{Q}] = 2$. But then

$$10 = [\mathbb{Q}(\sqrt[5]{7}, i):\mathbb{Q}] = [\mathbb{Q}(\sqrt[5]{7}, i):\mathbb{Q}(i)][\mathbb{Q}(i):\mathbb{Q}] = [\mathbb{Q}(\sqrt[5]{7}, i):\mathbb{Q}(i)] \cdot 2$$

and hence $[\mathbb{Q}(\sqrt[5]{7}, i):\mathbb{Q}(i)] = 5$.

We previously noted that we do not have the tools to find the minimum polynomial, $m(x)$, of $\sqrt[5]{7}$ over $\mathbb{Q}(i)$ directly. By extending our argument, we can find this polynomial indirectly as follows. Note that since $[\mathbb{Q}(\sqrt[5]{7}, i):\mathbb{Q}(i)] = 5$, Proposition 12.4 implies that the degree $m(x)$ is 5. But $x^5 - 7 \in \mathbb{Q}(i)[x]$ and $\sqrt[5]{7}^{\,5} - 7 = 0$. So by Proposition 10.3, $x^5 - 7$ is in the ideal $(m(x))$, i.e., $x^5 - 7 = m(x)f(x)$ for some $f(x) \in \mathbb{Q}(i)[x]$. Then by Proposition 8.1, $\deg (x^5 - 7) = \deg m(x) + \deg f(x)$, and hence $\deg f(x) = 0$, i.e., $f(x) = \phi_0$ for some $\phi_0 \in \mathbb{Q}(i)$. So $x^5 - 7 = m(x)\phi_0$ and since $m(x)$ is monic of degree 5, equating coefficients of x^5 gives $1 = \phi_0 1$. So $m(x) = x^5 - 7$ and therefore the minimum polynomial of $\sqrt[5]{7}$ over $\mathbb{Q}(i)$ is $x^5 - 7$. Note that we can then conclude that $x^5 - 7$ is irreducible over $\mathbb{Q}(i)$. $\diamond$

In Example 12.7, before we could apply Proposition 12.5, we had to check that $\mathbb{Q}(\sqrt[5]{7}, i)$ is a finite-dimensional extension of $\mathbb{Q}(i)$ and that $\mathbb{Q}(i)$ was a finite-dimensional extension of $\mathbb{Q}$. We can avoid checking this by applying the following bit of linear algebra.

Proposition 12.8   *If K is a finite-dimensional extension field of a field F and L is a subfield of K which contains F, then K is a finite-dimensional extension field of L and L is a finite-dimensional extension field of F.*

Proof   As a vector space over F, K has a finite basis $b_1, \ldots, b_n$. Then every element of K may be written as a linear combination of $b_1, \ldots, b_n$ and elements of F. Since $F \subseteq L$, this implies that every element of K may also be written as a linear combination of $b_1, \ldots, b_n$ and elements of L, i.e., that $\{b_1, \ldots, b_n\}$ is a spanning set for K as a vector space over L. Then by Proposition 12.1, $\{b_1, \ldots, b_n\}$ contains a (necessarily finite) basis for K as a vector space over L, and hence K is a finite-dimensional extension field of L. That L is finite-dimensional over F follows from Proposition 12.1 and the observation that L is a subspace of K over F. ❖

Example 12.9   Consider the polynomial $f(x) = x^6 + 21x^5 - 14x^3 - 7$. Since $f(0) = -7 < 0 < 1 = f(1)$, the intermediate value theorem implies that the equation $f(x) = 0$ has a solution $r \in \mathbb{R}$. We will find the dimension of $\mathbb{Q}(r, \zeta_3)$ as a vector space over $\mathbb{Q}(\zeta_3)$.

Clearly $f(x) \in \mathbb{Q}[x]$ and $f(x)$ is monic, and by hypothesis $f(r) = 0$. Furthermore, by Eisenstein's criterion (with prime 7), $f(x)$ is irreducible over $\mathbb{Q}$, and thus $f(x)$ is the minimum polynomial of r over $\mathbb{Q}$. So by Proposition 12.4, $[\mathbb{Q}(r):\mathbb{Q}] = 6$. Furthermore, if $m(x) = x^2 + x + 1$, then $m(x) \in \mathbb{Q}(r)[x]$ and $m(x)$ is monic, and by Example 10.5, $m(\zeta_3) = 0$. Since $r \in \mathbb{R}$, $\mathbb{Q}(r) \subseteq \mathbb{R}$, and thus neither ζ_3 nor ζ_3^2 is in $\mathbb{Q}(r)$. But we noted in Example 10.5, that ζ_3 and ζ_3^2 are the roots of $m(x)$ and thus by Proposition 8.12, $m(x)$ is irreducible in $\mathbb{Q}(r)[x]$. It follows that $m(x)$ is the minimum polynomial of ζ_3 over $\mathbb{Q}(r)$, and hence by Proposition 10.4, that $[\mathbb{Q}(r, \zeta_3):\mathbb{Q}(r)] = 2$. Then by Proposition 12.5,

$$[\mathbb{Q}(r, \zeta_3):\mathbb{Q}] = [\mathbb{Q}(r, \zeta_3):\mathbb{Q}(r)][\mathbb{Q}(r):\mathbb{Q}] = 2 \cdot 6 = 12.$$

Now by Proposition 12.8, $\mathbb{Q}(r, \zeta_3)$ is a finite-dimensional extension of $\mathbb{Q}(\zeta_3)$ and $\mathbb{Q}(\zeta_3)$ is a finite-dimensional extension of $\mathbb{Q}$. So by Proposition 12.5, $[\mathbb{Q}(r, \zeta_3):\mathbb{Q}] = [\mathbb{Q}(r, \zeta_3):\mathbb{Q}(\zeta_3)][\mathbb{Q}(\zeta_3):\mathbb{Q}]$. But we have seen in Example 10.5 that $[\mathbb{Q}(\zeta_3):\mathbb{Q}] = 2$, and thus $12 = [\mathbb{Q}(r, \zeta_3):\mathbb{Q}(\zeta_3)] \cdot 2$. It follows that $[\mathbb{Q}(r, \zeta_3):\mathbb{Q}(\zeta_3)] = 6$.

Note that an argument similar to that in Example 12.7 can now be used to show that $f(x)$ is the minimum polynomial of r over $\mathbb{Q}(\zeta_3)$ (see Exercise 46). ❖

Note that the techniques used in Examples 12.7 and 12.9 can even be used sometimes to find dimensions of the form $[\mathbb{Q}(a, b):\mathbb{Q}]$, where neither $[\mathbb{Q}(a, b):\mathbb{Q}(a)]$ nor $[\mathbb{Q}(a, b):\mathbb{Q}(b)]$ can be determined directly (cf. Exercises 32–38).

The point of view taken in Examples 12.7 and 12.9 can be used in general to show that finite-dimensional extensions consist entirely of algebraic elements and that the degree of each element divides the dimension of the extension (Proposition 12.10). We will need this result in Chapter 14 to show that finite-dimensional extensions have single generators (Proposition 14.4). As well, it is the key to showing that not every angle can be trisected by straightedge and compass alone, a problem posed, but never solved, by the mathematicians of ancient Greece (the details are given in Appendix D).

Proposition 12.10 *If K is a finite-dimensional extension field of a field F and $k \in K$, then k is algebraic over F and $[k:F]$ divides $[K:F]$.*

Proof Let $[K:F] = n$, and consider the set $\{1, k, k^2, \ldots, k^n\}$. Either two of its elements are the same or it has $n + 1$ elements. In both cases, it is linearly dependent, and hence there exist $\alpha_0, \ldots, \alpha_n \in F$, not all 0, such that $\alpha_0 1 + \alpha_1 k + \cdots + \alpha_n k^n = 0$. Then $q(k) = 0$, where $0 \neq q(x) = \alpha_0 + \alpha_1 x + \cdots + \alpha_n x^n \in F[x]$, and hence k is algebraic over F. By Proposition 12.8, K is a finite-dimensional extension field of $F(k)$ and $F(k)$ is a finite-dimensional extension field of F, and hence by Proposition 12.5, $[K:F] = [K:F(k)][F(k):F]$. Since k is algebraic over F, Proposition 12.4 implies that $[k:F] = [F(k):F]$, and thus $[k:F]$ divides $[K:F]$. ❖

We noted that Proposition 12.10 will be used in Chapter 14 and Appendix D. Other applications are given in some of the exercises that follow.

Exercises

Exercises 1–4 concern a complex solution r of the equation $x^6 - 3x^2 + 3 = 0$. In each exercise, express the given elements in $\mathbb{Q}(r)$ in the form

$$\alpha_0 + \alpha_1 r + \alpha_2 r^2 + \alpha_3 r^3 + \alpha_4 r^4 + \alpha_5 r^5.$$

1. $(r^2)^{-1}$ **2.** $(r^3 - r + 1)(r^5 + r)^{-1}$

3. $(r^5 - 2r^4 + r - 3)(2r^3 - 1)^{-1}$ **4.** $(7 - 3r^3 - 2r^5)(1 + r^3)^{-1}(1 - 4r^2)$

In Exercises 5–11, show that a is algebraic over $\mathbb{Q}$ and find $[\mathbb{Q}(a):\mathbb{Q}]$.

5. $a = i$ **6.** $a = 5 + 3i$ **7.** $a = 3 - i\sqrt{7}$

8. $a = \zeta_7$ (*Hint:* See Exercise 20 in Chapter 11.)

9. $a = \zeta_5$ (*Hint:* See Exercise 19 in Chapter 11.)

10. a is a nonreal solution of $x^6 - 1 = 0$

11. a is a solution of $3x^6 - 12x^5 + 6x^3 + 18x - 6 = 0$

In Exercises 12–17, determine the dimension of the given extension field over $\mathbb{Q}$ and find a basis for it over $\mathbb{Q}$.

12. $\mathbb{Q}(i)$ **13.** $\mathbb{Q}(\sqrt[5]{13})$ **14.** $\mathbb{Q}(\sqrt[4]{7}, i)$

15. $\mathbb{Q}(\sqrt[3]{6}, i)$ **16.** $\mathbb{Q}(1 - \sqrt[5]{7}, 2 + 3i)$ **17.** $\mathbb{Q}(2 + \sqrt[7]{12}, 3 - i)$

18. Show that $\sqrt[3]{2} \notin \mathbb{Q}(\sqrt{2})$.

19. Show that $\sqrt[12]{13} \notin \mathbb{Q}(\sqrt[10]{11})$.

20. If p and q are prime and $m > n$, show that $\sqrt[m]{p} \notin \mathbb{Q}(\sqrt[n]{q})$.

21. Show that $\sqrt[3]{2} \notin \mathbb{Q}(\sqrt[4]{5})$.

22. If p and q are prime and m does not divide n, show that $\sqrt[m]{p} \notin \mathbb{Q}(\sqrt[n]{q})$.

23. Show that $\sqrt[5]{12} \notin \mathbb{Q}(\sqrt[11]{10})$.

In Exercises 24–27, Use Exercise 20 to find the dimensions of the given vector space over $\mathbb{Q}$.

24. $\mathbb{Q}(\sqrt{2}, \sqrt[3]{2}, i)$ **25.** $\mathbb{Q}(\sqrt[3]{5}, \sqrt{3}, i)$

26. $\mathbb{Q}(\sqrt[3]{2}, 1 - 5i, \sqrt{3})$ **27.** $\mathbb{Q}(1 + 5i, 1 - 4\sqrt[3]{11}, 7 + 3\sqrt{13})$

In Exercises 28–31, find the dimension of the field K over its subfield F.

28. $K = \mathbb{Q}(\sqrt{2}, i), F = \mathbb{Q}(i)$ **29.** $K = \mathbb{Q}(\sqrt{11}, i), F = \mathbb{Q}(i)$

30. $K = \mathbb{Q}(\sqrt[3]{2}, \zeta_3), F = \mathbb{Q}(\zeta_3)$ **31.** $K = \mathbb{Q}(\sqrt{6}, \zeta_3), F = \mathbb{Q}(\zeta_3)$

32. Show that the dimension of $\mathbb{Q}(\sqrt[5]{3}, \zeta_5)$ as a vector space over $\mathbb{Q}$ is 20. (This exercise will be used in some of the exercises in Chapter 14.)

(*Hint:* First show that $[\mathbb{Q}(\sqrt[5]{3}) : \mathbb{Q}] = 5$;

then use Exercise 53 in Chapter 3 and Exercise 19 in Chapter 11 to show that $[\mathbb{Q}(\zeta_5) : \mathbb{Q}] = 4$;

then show that $[\mathbb{Q}(\sqrt[5]{3}, \zeta_5) : \mathbb{Q}] \geq 20$;

then use Exercise 53 in Chapter 3 to show that $[\mathbb{Q}(\sqrt[5]{3}, \zeta_5) : \mathbb{Q}(\zeta_5)] \leq 5$;

finally show that $[\mathbb{Q}(\sqrt[5]{3}, \zeta_5) : \mathbb{Q}] = 20$.)

In Exercises 33–38, use the technique of Exercise 32 to find the dimensions of the given extension field over its subfield $\mathbb{Q}$.

33. $\mathbb{Q}(\sqrt[5]{7}, \zeta_5)$ **34.** $\mathbb{Q}(\sqrt{11}, \zeta_7)$ **35.** $\mathbb{Q}(\sqrt[5]{7}, \sqrt[3]{5})$

36. $\mathbb{Q}(\sqrt[5]{7}, \sqrt[7]{3}, \sqrt[11]{5})$ **37.** $\mathbb{Q}(\sqrt[5]{7}, \sqrt[3]{5}, \zeta_5)$ **38.** $\mathbb{Q}(3 + 5\sqrt[7]{2}, 1 - \sqrt[11]{11}, 3\zeta_7)$

39. Find the dimension of $\mathbb{Q}(\sqrt[5]{7}, \sqrt[3]{5}, \zeta_5)$ as a vector space over $\mathbb{Q}(\zeta_5)$.

40. Find the dimension of $\mathbb{Q}(3 + 5\sqrt[4]{2}, 1 - \sqrt[5]{11}, 3\zeta_7)$ as a vector space over $\mathbb{Q}(\zeta_7)$.

41. Suppose that a and b are algebraic over a field F, of degrees m and n respectively. Show that if m and n have no common prime factor, then $F(a, b)$ has dimension mn as a vector space over F. (*Hint:* Generalize the technique of Exercise 32.)

42. Suppose that K is an extension field of a field L and that $\{a_1, \ldots, a_m\}$ spans K as a vector space over L. Suppose also that L is an extension field of a field F and that $\{b_1, \ldots, b_n\}$ spans L as a vector space over F. Show that $\{a_1b_1, \ldots, a_1b_n, a_2b_1, \ldots, a_mb_n\}$ spans K as a vector space over F.

43. Suppose that K is an extension field of a field L and that $\{a_1, \ldots, a_m\}$ is a linearly independent subset of K as a vector space over L. Suppose also that L is an extension field of a field F and that $\{b_1, \ldots, b_n\}$ is a linearly independent subset of L as a vector space over F. Show that $\{a_1b_1, \ldots, a_1b_n, a_2b_1, \ldots, a_mb_n\}$ is a linearly independent subset of K as a vector space over F.

44. Suppose that K is an extension field of a field F and that $k \in K$. Show that if there exists a positive integer n such that $\{1, \ldots, k^n\}$ is a linearly dependent subset of K as a vector space over F, then k is algebraic over F.

45. Suppose that F is a field and that r and s are algebraic over F. Show that there exist nonnegative integers m and n such that $\{1, \ldots, r^m, s, \ldots, sr^m, \ldots, s^nr^m\}$ is a basis of $F(r, s)$ as a vector space over F.

46. Let $f(x) = x^6 + 21x^5 - 14x^3 - 7 \in \mathbb{Q}(\zeta_3)[x]$ and let $r \in \mathbb{R}$ be a root of $f(x)$. Complete Example 12.9 by showing that $f(x)$ is the minimum polynomial of r over $\mathbb{Q}(\zeta_3)$.

47. Show that $x^6 - 13$ is the minimum polynomial of $\sqrt[6]{13}$ over $\mathbb{Q}(i)$.

In Exercises 48–51, find the minimum polynomial of t over F.

48. $t = \sqrt[7]{5}, F = \mathbb{Q}(i)$

49. $t = \sqrt[7]{5}, F = \mathbb{Q}(\zeta_3)$

50. $t = \sqrt[5]{7}, F = \mathbb{Q}(\sqrt[3]{5})$

51. $t = \sqrt[5]{7}, F = \mathbb{Q}(\sqrt[3]{5}, \zeta_3)$

52. Suppose that F is a field, that r is algebraic over F, and that s is algebraic over $F(r)$. Show that s is algebraic over F.

In Exercises 53 and 54, suppose that K is an extension field of a field F.

53. Show that if $\{a_1, a_2\}$ is a basis of K as a vector space over F, then for at least one a_i, $\{1, a_i\}$ is also a basis of K over F.

54. Show that if $\{a_1, a_2, a_3\}$ is a basis of K as a vector space over F, then for at least one a_i, $\{1, a_i, a_i^2\}$ is also a basis of K over F.

Exercises 55–57 give an alternate proof of Theorem 10.8. In these exercises, let F be a field contained in an integral domain D.

55. Prove that D is a vector space over F.

56. Suppose that as a vector space, D has finite dimension over F. Prove that D is a field. (*Hint:* Use the technique given in the proof of Proposition 12.10 to show that any element $0 \neq d \in D$ is algebraic over F. Then use this observation to find an inverse for d.)

57. Use Exercise 56 to give an alternate proof of Theorem 10.8.

13

Automorphisms of Fields

The crucial insight of Galois theory is that the solutions $r_1, \ldots, r_m$ of a polynomial equation $a_0 + a_1 x + \cdots + a_n x^n = 0$ are very closely linked to certain functions from $F(r_1, \ldots, r_m)$ to itself. In particular, if an equation is solvable by radicals, then this set of functions must have certain special properties. We will describe these properties in Chapter 27. In this chapter, we begin the process of determining these properties by specifying the special functions that are intimately connected with the solutions of polynomial equations.

First we recall two special kinds of functions. In general, a function $f: S \to T$ is a rule that assigns to each element s in S a unique element $f(s)$ in T. Recall that if every element of T arises in this way from some element of S, then f is said to be onto; and if f assigns different elements of T to different elements of S, then f is said to be one-to-one. Specifically we have the following.

DEFINITION

Let S and T be sets and let $f: S \to T$; S is called the **domain** of f; T is called the **codomain** of f. Then

   (i) f is **one-to-one** if $x \neq y$ in S implies that $f(x) \neq f(y)$ in T,

   (ii) f is **onto** if for all $t \in T$, there exists $s \in S$ such that $f(s) = t$.

To prove that a function is one-to-one, it is usually easier to prove the contrapositive of the condition given in the definition, i.e., that if $f(x) = f(y)$ in T, then $x = y$ in S. The typical proof then begins by assuming that $f(x) = f(y)$ for arbitrary elements x and y in domain S, continues by applying the function to create an equation in the codomain T, and finishes by manipulating the equation to show that $x = y$.

To prove that a function is onto can be more difficult since a direct proof requires reversing the action of the function. The typical proof begins by picking an arbitrary element t in the codomain T and then somehow finding an element s in the domain S such that $f(s) = t$.

To prove that a function f is <u>not</u> one-to-one, it is usually easiest to find two elements a and b in the domain S of f such that $a \neq b$ but $f(a) = f(b)$. Note that a proof involving arbitrary elements is not required; it is sufficient to specify two particular elements a and b with the required property.

Finally, to prove that a function f is <u>not</u> onto, it is necessary to prove the negation of the definition, i.e., that <u>there</u> <u>exists</u> w in the codomain of f such that <u>for</u> <u>all</u> s in the domain of f, $f(s) \neq w$. The element w should be specified, while the equation $f(s) \neq w$ must be proved for an arbitrary element s in the domain.

Example 13.1   Let $h:\mathbb{R} \to \mathbb{C}$ be the function $h(r) = r + 0i$.

If $h(r) = h(s)$, then $r + 0i = s + 0i$ and hence $r = s$ so that h is one-to-one.

To see that h is not onto, consider $0 + 1i \in \mathbb{C}$ (the codomain of H). If $h(r) = 0 + 1i$ for some $r \in \mathbb{R}$ (the domain of h), then $r + 0i = h(r) = 0 + 1i$ and hence, since the coefficients of i must be equal, $0 = 1$, a contradiction. Thus $h(r) \neq 0 + 1i$ for any $r \in \mathbb{R}$ and hence h is not onto. ❖

Example 13.2   Let $f:\mathbb{Z} \to \mathbb{Z}_7$ be the function $f(n) = [n]_7$.

Since $f(0) = [0]_7 = [7]_7 = f(7)$ and $0 \neq 7$ in $\mathbb{Z}$, f is not one-to-one.

To show that h is onto, we begin by picking an arbitrary element $Y \in \mathbb{Z}_7$ (the codomain of f). Then there exists $n \in \mathbb{Z}$ (the domain of f) such that $Y = [n]_7$, and $f(n) = [n]_7 = Y$. It follows that f is onto. ❖

Example 13.3   Let $f:\mathbb{C} \to \mathbb{C}$ be the function $f(z) = \bar{z}$ (i.e., $f(a + bi) = a - bi$).

If $f(z) = f(w)$, then $\bar{z} = \bar{w}$ and hence $z = \bar{\bar{z}} = \bar{\bar{w}} = w$; thus f is one-to-one.

To show that f is onto, we begin by picking an arbitrary element $c \in \mathbb{C}$ (the codomain of f). We then observe that $\bar{\bar{c}} = c$ and that $\bar{c} \in \mathbb{C}$ (the domain of f). That is, for any element c in the codomain of f, the element $\bar{c}$ in the domain of f is such that $f(\bar{c}) = \bar{\bar{c}} = c$, and hence f is onto. ❖

We are interested in sets with algebraic structure (fields and rings) and in functions that preserve algebraic structure. Specifically, we want to consider the following sorts of functions.

DEFINITION

Let R_1 and R_2 be rings. A **homomorphism** is a function $f:R_1 \to R_2$ that satisfies the conditions

   (*i*)   for all $a, b \in R_1, f(a + b) = f(a) + f(b)$ ("f preserves addition"),
   (*ii*)   for all $a, b \in R_1, f(ab) = f(a)f(b)$ ("f preserves multiplication");

an **isomorphism** is a homomorphism that is both one-to-one and onto; an **automorphism** is an isomorphism from a ring to itself, i.e., one for which $R_1 = R_2$.

Example 13.4   As in Example 13.1, let $h: \mathbb{R} \to \mathbb{C}$ be the function $h(r) = r + 0i$. Then $h(r + s) = (r + s) + 0i = (r + 0i) + (s + 0i) = h(r) + h(s)$; so h preserves addition. And $h(rs) = rs + 0i = (rs - 0 \cdot 0) + (r0 + 0s)i = (r + 0i)(s + 0i) = h(r)h(s)$; so h preserves multiplication. It follows that h is a homomorphism which, by Example 13.1, is one-to-one but not onto. ❖

Example 13.5   As in Example 13.2, let $f: \mathbb{Z} \to \mathbb{Z}_7$ be the function $f(n) = [n]_7$. Then $f(n + m) = [n + m]_7 = [n]_7 + [m]_7 = f(n) + f(m)$; so f preserves addition. And $f(nm) = [nm]_7 = [n]_7[m]_7 = f(n)f(m)$; so f preserves multiplication. It follows that f is a homomorphism which, by Example 13.2, is onto but not one-to-one. ❖

Example 13.6   As in Example 13.3, let $f: \mathbb{C} \to \mathbb{C}$ be the function $f(z) = \bar{z}$ (i.e., $f(a + bi) = a - bi$). Then by Proposition 3.3, $f(z + w) = \overline{z + w} = \bar{z} + \bar{w} = f(z) + f(w)$ and $f(zw) = \overline{zw} = \bar{z}\,\bar{w} = f(z)f(w)$; so f preserves addition and f preserves multiplication. It follows that f is a homomorphism. According to Example 13.3, f is both one-to-one and onto and hence f is an isomorphism. In fact, since the domain and codomain of f are both $\mathbb{C}$, f is an automorphism of $\mathbb{C}$. ❖

Example 13.7   By Example 10.5, $m(x) = x^2 + x + 1$ is the minimum polynomial of ζ_3 over $\mathbb{Q}$ and hence by Theorem 10.8, $\mathbb{Q}(\zeta_3) = \{a + b\zeta_3 \mid a, b \in \mathbb{Q}\}$. If $k: \mathbb{Q}(\zeta_3) \to \mathbb{Q}(\zeta_3)$ is the function $k(a + b\zeta_3) = a - b\zeta_3$, then

$$k((a + b\zeta_3) + (c + d\zeta_3)) = k((a + c) + (b + d)\zeta_3) = (a + c) - (b + d)\zeta_3$$
$$= (a - b\zeta_3) + (c - d\zeta_3) = k(a + b\zeta_3) + k(c + d\zeta_3);$$

so k preserves addition. However, since $\zeta_3^2 + \zeta_3 + 1 = 0$ by Example 10.5, $\zeta_3^2 = -1 - \zeta_3$, and hence

$$k(\zeta_3 \cdot \zeta_3) = k(\zeta_3^2) = k(-1 - \zeta_3) = -1 + \zeta_3$$
$$\neq -1 - \zeta_3 = \zeta_3^2 = (-\zeta_3)(-\zeta_3) = k(\zeta_3)k(\zeta_3);$$

so k does <u>not</u> preserve multiplication. It follows that k is not a homomorphism. ❖

The functions f and k of Examples 13.6 and 13.7 are defined similarly. The elements in their domains are both written in the form $a + b\Lambda$ for a fixed number Λ, and each function maps such an element to one of the form $a - b\Lambda$ for the same number Λ. Why is one a homomorphism and the other not? The answer to this question will be very important; so we will investigate it in some detail.

We first observe that nontrivial homomorphisms map 0 to 0 and 1 to 1.

Proposition 13.8   *Let f be a homomorphism from the ring R to the ring S. Then*

    (i) $f(0) = 0$; and

    (ii) if R and S are integral domains and $f(r) \neq 0$ for some $r \in R$, then $f(1) = 1$.

Proof   (*i*)   Since f preserves addition, $f(0) = f(0 + 0) = f(0) + f(0)$. Subtracting $f(0)$ from both sides, we have $0 = f(0)$.

(*ii*)   Since $f(1)f(r) = f(1r) = f(r) = 1f(r)$, $(f(1) - 1)f(r) = 0$. Then since $f(r) \neq 0$ and S is an integral domain, $f(1) - 1 = 0$, and therefore $f(1) = 1$. ∎

We next observe that if r is algebraic over F and if f is a homomorphism with domain $F(r)$ which leaves the elements of F unchanged, then the only possibilities for $f(r)$ are the roots of the minimum polynomial of r.

Proposition 13.9   *Suppose that K and L are extension fields of a field F and that $f: K \to L$ is a homomorphism such that $f(q) = q$ for all $q \in F$. Suppose further that $r \in K$ and that $m(x) = \mu_0 + \cdots + \mu_n x^n \in F[x]$.*

(*i*)   *If r is a root of $m(x)$, then $f(r)$ is also a root of $m(x)$.*

(*ii*)   *In particular, if $m(x)$ is the minimum polynomial of r over F, then $f(r)$ is a root of $m(x)$.*

Proof   (*i*)   Since each coefficient μ_i is in F, $f(\mu_i) = \mu_i$ for all i. Thus since f preserves addition and multiplication,

$$f(m(r)) = f(\mu_0 + \mu_1 r + \cdots + \mu_n r^n)$$
$$= f(\mu_0) + f(\mu_1)f(r) + \cdots + f(\mu_n)f(r)^n$$
$$= \mu_0 + \mu_1 f(r) + \cdots + \mu_n f(r)^n = m(f(r)).$$

Furthermore, by Proposition 13.8, $f(0) = 0$, and thus, if r is a root of $m(r)$, then $m(r) = 0$ and hence $0 = f(0) = f(m(r)) = m(f(r))$. So $f(r)$ is also a root of $m(x)$.

(*ii*)   Since $m(x)$ is the minimum polynomial of r, r is a root of $m(r)$, and hence by part (*i*), $f(r)$ is also a root of $m(x)$. ∎

Now, in light of Proposition 13.9, consider the functions f of Example 13.6 and k of Example 13.7.

We first note that for all $r \in \mathbb{R}$, $f(r) = \bar{r} = r$. Furthermore, if $m(x) = x^2 + 1$, then $m(i) = 0$, and hence according to Proposition 13.9(*i*), if f is a homomorphism, then $f(i)$ must also solve $m(x) = 0$, i.e., $f(i)$ must be either i or $-i$. Since f fixes $\mathbb{R}$ and $f(i) = \bar{i} = -i$, the pair i and $f(i)$ does not violate Proposition 13.9(*i*) and thus Proposition 13.9 does not preclude the possibility that f is a homomorphism (and indeed it is a homomorphism, as we saw in Example 13.6).

In the case of k, we note that for all $q \in \mathbb{Q}$, $k(q) = k(q + 0\zeta_3) = q - 0\zeta_3 = q$. And furthermore, if $m(x) = x^2 + x + 1$, then $m(\zeta_3) = 0$, so that if k is a homomorphism, Proposition 13.9(*i*) implies that $k(\zeta_3)$ must also solve $m(x) = 0$. By Example 10.5, the roots of $m(x)$ are ζ_3 and ζ_3^2, and neither of these roots can equal $-\zeta_3$ because $-\zeta_3$ is in the fourth quadrant, while ζ_3 is in the first quadrant and ζ_3^2 is in the third quadrant. However, $k(\zeta_3) = k(0 + \zeta_3) = 0 - \zeta_3 = -\zeta_3$, and thus the pair ζ_3 and $k(\zeta_3)$ violate Proposition 13.9(*i*). But then Proposition 13.9 implies that k <u>cannot</u> be a homomorphism (and indeed it is not, as we observed in Example 13.7).

As a guide to redefining k, note that f can be expressed $f(a + bi) = a + b(-i)$ where $-i$ is a root of $m(x)$. This suggests that the following change in the definition of k might create a homomorphism.

Example 13.10 By Example 10.5, $m(x) = x^2 + x + 1$ is the minimum polynomial of ζ_3 over $\mathbb{Q}$ and also of $\zeta_3{}^2$ over $\mathbb{Q}$. So by Theorem 10.8, $\mathbb{Q}(\zeta_3) = \{a + b\zeta_3 \mid a, b \in \mathbb{Q}\}$ and $\mathbb{Q}(\zeta_3{}^2) = \{a + b\zeta_3{}^2 \mid a, b \in \mathbb{Q}\}$.

In view of the previous discussion, we define $t : \mathbb{Q}(\zeta_3) \to \mathbb{Q}(\zeta_3{}^2)$ by letting $t(a + b\zeta_3) = a + b\zeta_3{}^2$. That is, we define t by replacing the root ζ_3 by the root $\zeta_3{}^2$. By definition of t,

$$
\begin{aligned}
t((a + b\zeta_3) + (c + d\zeta_3)) &= t((a + c) + (b + d)\zeta_3) = (a + c) + (b + d)\zeta_3{}^2 \\
&= (a + b\zeta_3{}^2) + (c + d\zeta_3{}^2) = t(a + b\zeta_3) + t(c + d\zeta_3);
\end{aligned}
$$

so t preserves addition. Furthermore, since $\zeta_3{}^2 + \zeta_3 + 1 = 0$, $\zeta_3{}^2 = -1 - \zeta_3$, and since $(\zeta_3{}^2)^2 + \zeta_3{}^2 + 1 = 0$, we have $(\zeta_3{}^2)^2 = -1 - \zeta_3{}^2$ as well. Thus

$$
\begin{aligned}
t((a + b\zeta_3)(c + d\zeta_3)) &= t(ac + (ad + bc)\zeta_3 + bd\zeta_3{}^2) \\
&= t((ac - bd) + (ad + bc - bd)\zeta_3) \\
&= (ac - bd) + (ad + bc - bd)\zeta_3{}^2 \\
&= ac + (ad + bc)\zeta_3{}^2 + bd(\zeta_3{}^2)^2 \\
&= (a + b\zeta_3{}^2)(c + d\zeta_3{}^2) = t(a + b\zeta_3)t(c + d\zeta_3);
\end{aligned}
$$

so t also preserves multiplication. Therefore, t is a homomorphism. Note that t also leaves elements of $\mathbb{Q}$ unchanged; for all $q \in \mathbb{Q}$, $t(q) = t(q + 0 \cdot \zeta_3) = q + 0 \cdot \zeta_3{}^2 = q$.

Furthermore, if $t(a + b\zeta_3) = t(c + d\zeta_3)$, then $a + b\zeta_3{}^2 = c + d\zeta_3{}^2$, and we previously observed that $[\zeta_3{}^2 : \mathbb{Q}] = 2$ so that by Lemma 12.2, $a = c$ and $b = d$. Therefore, $a + b\zeta_3 = c + d\zeta_3$ so that t is one-to-one. In fact, since $\mathbb{Q}(\zeta_3{}^2) = \{a + b\zeta_3{}^2 \mid a, b \in \mathbb{Q}\}$, t is obviously onto as well and thus is an isomorphism from $\mathbb{Q}(\zeta_3)$ to $\mathbb{Q}(\zeta_3{}^2)$. Note that $\mathbb{Q}(\zeta_3)$ is a subfield of $\mathbb{C}$ which contains both $\mathbb{Q}$ and $\zeta_3{}^2$ and hence $\mathbb{Q}(\zeta_3{}^2) \subseteq \mathbb{Q}(\zeta_3)$. Note as well that $\mathbb{Q}(\zeta_3{}^2)$ is a subfield of $\mathbb{C}$ which contains both $\mathbb{Q}$ and ζ_3 (because $\zeta_3 = \zeta_3{}^2\zeta_3{}^2$) and hence $\mathbb{Q}(\zeta_3{}^2) \supseteq \mathbb{Q}(\zeta_3)$. It follows that $\mathbb{Q}(\zeta_3{}^2) = \mathbb{Q}(\zeta_3)$ and hence that t is in fact an automorphism of $\mathbb{Q}(\zeta_3)$. ❖

The isomorphism of Example 13.10 was created by replacing the root ζ_3 of $m(x)$ in the expression $a + b\zeta_3$ by the root $\zeta_3{}^2$. Can this method always be used to define an isomorphism? The answer is yes, in the following general circumstances.

Proposition 13.11   *Suppose that F is a subfield of a field K, that $u \in K$ is algebraic over F with minimum polynomial $m(x) = \mu_0 + \cdots + \mu_{n-1}x^{n-1} + x^n \in F[x]$, and that $v \in K$ is also a root of $m(x)$. Define $f: F(u) \to K$ by letting $f(\alpha_0 + \cdots + \alpha_{n-1}u^{n-1}) = \alpha_0 + \cdots + \alpha_{n-1}v^{n-1}$. Then f is an isomorphism of $F(u)$ onto $F(v)$ such that $f(q) = q$ for all $q \in F$.*

Proof   Certainly $m(x)$ is the minimum polynomial of v over F and hence, by Theorem 10.8, any element of $F(v)$ may be written $\alpha_0 + \cdots + \alpha_{n-1}v^{n-1}$ for $\alpha_i \in F$. Then $\alpha_0 + \cdots + \alpha_{n-1}u^{n-1} \in F(u)$ and $f(\alpha_0 + \cdots + \alpha_{n-1}u^{n-1}) = \alpha_0 + \cdots + \alpha_{n-1}v^{n-1}$ so   that f is   onto $F(v)$.   If $f(\delta_0 + \cdots + \delta_{n-1}u^{n-1}) = f(\epsilon_0 + \cdots + \epsilon_{n-1}u^{n-1})$,   then $\delta_0 + \cdots + \delta_{n-1}v^{n-1} = \epsilon_0 + \cdots + \epsilon_{n-1}v^{n-1}$. Since $n = [v:F]$, $\delta_i = \epsilon_i$ for all i by Lemma 12.2, and hence $\delta_0 + \cdots + \delta_{n-1}u^{n-1} = \epsilon_0 + \cdots + \epsilon_{n-1}u^{n-1}$ so that f is one-to-one. To see that f is a homomorphism, let $\delta_0 + \cdots + \delta_{n-1}u^{n-1}, \epsilon_0 + \cdots + \epsilon_{n-1}u^{n-1} \in F(u)$. Since

$$f((\delta_0 + \cdots + \delta_{n-1}u^{n-1}) + (\epsilon_0 + \cdots + \epsilon_{n-1}u^{n-1}))$$
$$= f((\delta_0 + \epsilon_0) + \cdots + (\delta_{n-1} + \epsilon_{n-1})u^{n-1})$$
$$= (\delta_0 + \epsilon_0) + \cdots + (\delta_{n-1} + \epsilon_{n-1})v^{n-1}$$
$$= (\delta_0 + \cdots + \delta_{n-1}v^{n-1}) + (\epsilon_0 + \cdots + \epsilon_{n-1}v^{n-1})$$
$$= f(\delta_0 + \cdots + \delta_{n-1}u^{n-1}) + f(\epsilon_0 + \cdots + \epsilon_{n-1}u^{n-1}),$$

f preserves addition. To see that f preserves multiplication, form the polynomials $d(x) = \delta_0 + \cdots + \delta_{n-1}x^{n-1}$ and $e(x) = \epsilon_0 + \cdots + \epsilon_{n-1}x^{n-1}$ in $F[x]$. By the division algorithm (Theorem 8.3), there exist $q(x), r(x) \in F[x]$ such that $d(x)e(x) = m(x)q(x) + r(x)$ and either $r(x) = 0$ or $\deg r(x) < \deg m(x)$. In either case, $r(x) = \sigma_0 + \cdots + \sigma_{n-1}x^{n-1}$ for $\sigma_0, \ldots, \sigma_{n-1} \in F$ (some or possibly all the σ_i being 0). Then $r(u) = d(u)e(u) - m(u)q(u) = d(u)e(u) - 0 \cdot q(u) = d(u)e(u)$, and similarly $r(v) = d(v)e(v)$. So

$$f((\delta_0 + \cdots + \delta_{n-1}u^{n-1})(\epsilon_0 + \cdots + \epsilon_{n-1}u^{n-1})) = f(d(u)e(u))$$
$$= f(r(u)) = f(\sigma_0 + \cdots + \sigma_{n-1}u^{n-1}) = \sigma_0 + \cdots + \sigma_{n-1}v^{n-1}$$
$$= r(v) = d(v)e(v) = (\delta_0 + \cdots + \delta_{n-1}v^{n-1})(\epsilon_0 + \cdots \epsilon_{n-1}v^{n-1})$$
$$= f(\delta_0 + \cdots + \delta_{n-1}u^{n-1})f(\epsilon_0 + \cdots + \epsilon_{n-1}u^{n-1}),$$

and hence f preserves multiplication. Finally note that for any $q \in F$, $f(q) = f(q + 0 \cdot u + \cdots + 0 \cdot u^{n-1}) = q + 0 \cdot v + \cdots 0 \cdot v^{n-1} = q$. ∎

Example 13.12   By Proposition 3.8, the complex roots of $m(x) = x^3 - 2$ are $\sqrt[3]{2}$, $\sqrt[3]{2}\zeta_3$, and $\sqrt[3]{2}\zeta_3^2$. Clearly $m(x) \in \mathbb{Q}[x]$ and $m(x)$ is monic, and since none of these roots are in $\mathbb{Q}$, $m(x)$ is irreducible in $\mathbb{Q}[x]$ by Proposition 8.12. So $m(x)$ is the minimum polynomial of $\sqrt[3]{2}$ over $\mathbb{Q}$, and hence by Theorem 10.8, $\mathbb{Q}(\sqrt[3]{2}) = \{\alpha_0 + \alpha_1\sqrt[3]{2} + \alpha_2(\sqrt[3]{2})^2 \mid \alpha_i \in \mathbb{Q}\}$. Similarly $\mathbb{Q}(\sqrt[3]{2}\zeta_3) = \{\alpha_0 + \alpha_1\sqrt[3]{2}\zeta_3 + \alpha_2(\sqrt[3]{2}\zeta_3)^2 \mid \alpha_i \in \mathbb{Q}\}$. If we define

$f: \mathbb{Q}(\sqrt[3]{2}) \rightarrow \mathbb{Q}(\sqrt[3]{2}\zeta_3)$ by letting $f(\alpha_0 + \alpha_1\sqrt[3]{2} + \alpha_2\sqrt[3]{2})^2) = \alpha_0 + \alpha_1\sqrt[3]{2}\zeta_3 + \alpha_2(\sqrt[3]{2}\zeta_3)^2$, then since $\sqrt[3]{2}\zeta_3$ is a root of $m(x)$, Proposition 13.11 assures us that f is an isomorphism such that $f(q) = q$ for all $q \in \mathbb{Q}$. ❖

Exercises

In Exercises 1–12, determine whether $f(x)$ is (1) one-to-one and (2) onto.

1. $f: \mathbb{R} \rightarrow \mathbb{R}, f(x) = x^2$

2. $f: \mathbb{R} \rightarrow \mathbb{R}, f(x) = 4x - \pi$

3. $f: \mathbb{Z} \rightarrow \mathbb{Z}, f(n) = 2n$

4. $f: \mathbb{Z}_6 \rightarrow \mathbb{Z}_{12}, f([n]_6) = 2[n]_{12}$

5. $f: \mathbb{Q}(\sqrt{3}, i) \rightarrow \mathbb{Q}(\sqrt{3}, i), f(z) = z^3$

6. $f: \mathbb{C} \rightarrow \mathbb{R}, f(r(\cos\Theta + i\sin\Theta)) = r$

7. $f: \mathbb{Z}_8 \rightarrow \mathbb{Z}, f([n]_8) = n$ for $0 \leq n < 8$

8. $f: \mathbb{Q}(\sqrt{3}) \rightarrow \mathbb{Q}(\sqrt{3}), f(a_0 + a_1\sqrt{3}) = a_1 + a_0\sqrt{3}$

9. $f: \mathbb{Q}(\zeta_3) \rightarrow \mathbb{Q}(\zeta_3), f(a_0 + a_1\zeta_3) = a_0a_1\zeta_3$

10. $f: \mathbb{Q}(\sqrt{3}, i) \rightarrow \mathbb{Q}(\sqrt{3}, i),$
$$f(a_0 + a_1\sqrt{3} + b_0i + b_1i\sqrt{3}) = a_0 - a_1\sqrt{3} - b_0i - b_1i\sqrt{3}$$

11. $f: \mathbb{Q}(\sqrt{3}, i) \rightarrow \mathbb{Q}(\sqrt{3}, i),$
$$f(a_0 + a_1\sqrt{3} + b_0i + b_1i\sqrt{3}) = a_0 - a_1\sqrt{3} - b_0i + b_1i\sqrt{3}$$

12. $f: \mathbb{R} \rightarrow \mathbb{C}, f(x) = \dfrac{1}{\sqrt{1 + x^2}} + i\dfrac{x}{\sqrt{1 + x^2}}$

In Exercises 13–25, determine whether $f(x)$ is a homomorphism, an isomorphism, or an automorphism. Give a proof or a counterexample, whichever is appropriate.

13. $f: \mathbb{Z} \rightarrow \mathbb{R}, f(n) = \pi n$

14. $f: \mathbb{Z} \rightarrow \mathbb{Z}_{12}, f(n) = [n]_{12}$

15. $f: \mathbb{Z} \rightarrow \mathbb{Z}_{14}, f(n) = [14n]_{14}$

16. $f: \mathbb{Z} \rightarrow \mathbb{Z}_4, f(n) = [n - 6]_4$

17. $f: \mathbb{Z} \rightarrow \mathbb{Z}_{12}, f(n) = [n - 6]_{12}$

18. $f: \mathbb{R} \rightarrow M_2(\mathbb{R}), f(x) = \begin{pmatrix} x & 0 \\ 0 & x \end{pmatrix}$

19. $f: \mathbb{R} \rightarrow M_2(\mathbb{R}), f(x) = \begin{pmatrix} 0 & x \\ x & 0 \end{pmatrix}$

20. $f: M_2(\mathbb{R}) \rightarrow M_2(\mathbb{R}), f\begin{pmatrix} a & b \\ c & d \end{pmatrix} = \begin{pmatrix} a & 0 \\ 0 & d \end{pmatrix}$

21. $f: M_2(\mathbb{R}) \rightarrow M_2(\mathbb{R}), f\begin{pmatrix} a & b \\ c & d \end{pmatrix} = \begin{pmatrix} a & b \\ 0 & 0 \end{pmatrix}$

22. $f: \mathbb{Q}(\sqrt{3}) \rightarrow \mathbb{Q}(\sqrt{3}), f(a_0 + a_1\sqrt{3}) = a_0 - a_1\sqrt{3}$

23. $f: \mathbb{Q}(\sqrt{3}) \rightarrow \mathbb{Q}(\sqrt{3}), f(a_0 + a_1\sqrt{3}) = a_1 + a_0\sqrt{3}$

24. $f: \mathbb{Q}(\sqrt{3}, i) \rightarrow \mathbb{Q}(\sqrt{3}, i),$
$$f(a_0 + a_1\sqrt{3} + b_0i + b_1i\sqrt{3}) = a_0 - a_1\sqrt{3} - b_0i - b_1i\sqrt{3}$$

25. $f: \mathbb{Q}(\sqrt{3}, i) \rightarrow \mathbb{Q}(\sqrt{3}, i),$
$$f(a_0 + a_1\sqrt{3} + b_0i + b_1i\sqrt{3}) = a_0 - a_1\sqrt{3} - b_0i + b_1i\sqrt{3}$$

In Exercises 26–31, find an isomorphism $f:K \to L$ which is not the identity function. (Remember to justify your answer.)

26. $K = \mathbb{Q}(\sqrt{7})$, $L = \mathbb{Q}(-\sqrt{7})$ 27. $K = \mathbb{Q}(\zeta_5)$, $L = \mathbb{Q}(\zeta_5^2)$

28. $K = \mathbb{Q}(\zeta_5^2)$, $L = \mathbb{Q}(\zeta_5^3)$ 29. $K = \mathbb{Q}(\sqrt[3]{11})$, $L = \mathbb{Q}(\sqrt[3]{11}\zeta_3)$

30. $K = \mathbb{Q}(\sqrt[3]{7})$, $L = \mathbb{Q}(\sqrt[3]{7}\zeta_3^2)$ 31. $K = \mathbb{Q}(\sqrt[4]{7})$, $L = \mathbb{Q}(i\sqrt[4]{7})$

32. Define $f:\mathbb{Q}[x] \to \mathbb{Q}[x]$ by letting $f(\alpha_0 + \cdots + \alpha_n x^n) = \alpha_1 x + \cdots + \alpha_n x^n$. Determine whether f is a homomorphism. (Remember to justify your answer.)

33. Suppose that K is an extension field of a field F, that u and v in K are algebraic over F, and that $f:F(u) \to F(v)$ is a homomorphism such that $f(q) = q$ for all $q \in F$. Show that if $f(u) = v$, then u and v have the same minimum polynomial.

34. Suppose that R and S are rings and that $f:R \to S$ is a homomorphism. Show that $f(-\rho) = -f(\rho)$ for all $\rho \in R$.

35. Show that if K and L are fields and if $f:K \to L$ is a homomorphism which is not the zero homomorphism, then f is one-to-one.

36. Suppose that K and L are fields and that $f:K \to L$ is a homomorphism which is not the zero homomorphism. Show that if $0 \neq \rho \in K$, then $f(\rho) \neq 0$ and $f(\rho^{-1}) = f(\rho)^{-1}$.

In Exercises 37–39, suppose that K and L are extension fields of a field F, that $f:K \to L$ is a homomorphism such that $f(q) = q$ for all $q \in F$, that $r \in K$, and that $m(x) = \mu_0 + \cdots + \mu_n x^n \in F[x]$.

37. Show that f is one-to-one.

38. Show that r solves $m(x) = 0$ if and only if $f(r)$ solves $m(x) = 0$.

39. Show that $m(x)$ is the minimum polynomial of r over F if and only if $m(x)$ is the minimum polynomial of $f(r)$ over F.

40. Suppose that K is an extension field of a field F and that $u \in K$ is algebraic over F of degree n. Suppose further that $v \in K$ is a root of a polynomial $p(x) \in F[x]$ which is of degree n and has u as a root. Define $f:F(u) \to K$ by letting $f(\alpha_0 + \cdots + \alpha_{n-1}u^{n-1}) = \alpha_0 + \cdots + \alpha_{n-1}v^{n-1}$. Show that F is an isomorphism of $F(u)$ onto $F(v)$.

41. Let $p(x) = (x^2 - 3)(x^3 - 2)$. Does there exist a homomorphism $f:\mathbb{Q}^{p(x)} \to \mathbb{Q}^{p(x)}$ such that $f(\sqrt{3}) = \sqrt[3]{2}$ and $f(q) = q$ for all $q \in \mathbb{Q}$? Why?

42. Find $v \in \mathbb{R}$ and $p(x) \in \mathbb{Q}[x]$ such that $p(v) = 0$ and $p(\sqrt[5]{7}) = 0$ but the following function $f:\mathbb{Q}(\sqrt[5]{7}) \to \mathbb{Q}(v)$ is not a homomorphism:

$$f(\alpha_0 + \alpha_1\sqrt[5]{7} + \alpha_2\sqrt[5]{7}^2 + \alpha_3\sqrt[5]{7}^3 + \alpha_4\sqrt[5]{7}^4) =$$
$$\alpha_0 + \alpha_1 v + \alpha_2 v^2 + \alpha_3 v^3 + \alpha_4 v^4.$$

(Remember to justify your answer.)

43. Find a field F and a function $s: F \to F$ such that s is one-to-one and onto and preserves addition but s does not preserve multiplication.

Evariste Galois

Evariste Galois was born in Bourg-la-Reine, France, (near Paris) on October 25, 1811, and died in Paris, France, on May 31, 1832. The details concerning his death are not known; Rothman's account[1] is given below. Galois's father was Nicholas-Gabriel Galois who directed a school with about sixty boarders, and his mother was Adelaide-Marie Demente. Both were politically liberal with no known inclination to mathematics; Nicholas-Gabriel was active in local politics and had a talent for composing rhymed couplets. Adelaide-Marie supervised Evariste's education until he was twelve, ensuring that he had a solid classical education and understood her own skeptical views on religion.

In 1823, Galois entered the Collège Louis-le-Grand in Paris as a boarder. His performance was neither exceptionally good nor exceptionally bad. Although he won a prize and three mentions for his work, he was required to repeat his third year because of poor work in rhetoric. His introduction to mathematics came in February 1827, when he took the preparatory course from H. J. Vernier from whom he constantly received accolades for his work. Apparently his enthusiasm for mathematics cut into the time allotted to his other work because his other marks began to decline, especially in rhetoric. Vernier encouraged Galois to work more systematically, but Galois did not take his teacher's advice. In 1828, he attempted the entrance examination to the École Polytechnique a year early and failed because his background in basic mathematics was so weak. Throughout his life, Galois had difficulty dealing with rejection, and indeed he viewed his failure on this examination as a denial of justice.

He did not give up, however, and enrolled in Louis-Paul-Emile Richard's advanced course. Richard saw Galois's talent and his encouragement led to the publication of Galois's first small paper, "Proof of a theorem on periodic continued fractions." Then in the spring of 1829, he submitted a paper on solvability of equations of prime degree to the Academy, a paper for which Cauchy was to act as referee.

Simultaneously, a tragedy was unfolding in Bourg-la-Reine. Galois's father was the victim of a number of maliciously forged epigrams directed at his own family, and he committed suicide. Evariste was shaken and it is hard to discount the effect his father's suicide had on his later actions. The first such effect may have taken place just a few days later when he took the Polytechnique entrance examination for the second and final time. The examination was oral and its details are unknown. It is known that Galois worked almost entirely in his head, and there is an unverified tradition that he became so enraged at the stupidity of an examiner that he threw an eraser at him. Whatever the case, he failed the examination and became even more embittered.

These failures did not close the door to further education; they merely closed the door to the upper story. Since the Baccalaureate examinations were not required for entrance to the Polytechnique, Galois had not planned to take them. But now he reconsidered, took these examinations, passed, and early in 1830, entered the less prestigious École Normale. At the same time, on January 18, 1830, Cauchy was, in his own words, "supposed to present today to the Academy first a report on the work of the young Galois"[2] and a second memoir of his own. He was ill, however, and at the next session a week later, he did not present Galois's work. The reason is unknown. It is possible that he persuaded Galois to combine his work into a single memoir to be submitted for a prize. For this is exactly what Galois did, presenting his entry to Fourier (as secretary) in February 1830. However, when Fourier died in April, Galois's memoir was not found among his papers and hence was not even judged. Galois was convinced that the loss was not an accident but he seems not to have dwelt too long on it. For April saw the publication of one of his papers on the theory of equations and June saw the publication of another two.

He desperately wanted to join the July revolution of 1830, but the director of the École Normale locked the students in the school's grounds and Galois was unable to scale the wall. The revolution successful, Galois joined the Society of the Friends of the People, one of the most extreme of the republican societies. In December, he wrote a letter to the *Gazette des Écoles* denouncing the actions of the director of the École Normale and was summarily expelled. He joined the Artillery of the National Guard, composed almost entirely of republicans, and attempted to organize a private class in mathematics. At the first meeting, there were forty students but their number dwindled and the class did not last long. In January 1831, he submitted a third version of his memoir to the Academy, this time with Poisson as referee. This paper was lost and was eventually published. It presents the rudiments of Galois theory, including extension fields, groups of automorphisms, and primitive elements.

On May 9, 1831, at a party celebrating the acquittal of nineteen republicans on charges of conspiracy, Galois raised a glass and a dagger. In the words of Alexandre Dumas, "All I could perceive was that there was a threat and that the name of Louis-Phillipe had been mentioned; the intention was made clear by the open knife."[3] Galois was arrested but insisted that he had said "To Louis-Phillipe, *if he betrays*"[4] and the jury acquitted him with minimal discussion.

Shortly after, in July, Galois appeared, heavily armed, in the uniform of the now disbanded Artillery Guard. He was arrested, and in October, after already spending three months in prison, he was finally charged and sentenced to an additional six months. At the same time, he learned that his memoir had been rejected by the Academy. Poisson commented in his report: "We have made every effort to understand M. Galois's proofs. His argument is neither sufficiently clear nor sufficiently developed to allow us to judge its rigor; it is not even possible for us to give an idea of this paper. The author claims that the propositions contained in his manuscript are part of a general theory which has rich application. Often different parts of a theory clarify each other and can be more easily understood when taken together than when taken in isolation. One should rather wait to form a more definite opinion, therefore, until the author publishes a more complete account of his work."[5] Galois considered this encouraging rejection the last straw in his strained dealings with the Academy and resolved to have nothing more to do with them. He collected his manuscripts and with the help of a friend planned to publish them privately.

On March 16, 1832, Galois, along with other prisoners, was transferred to the pension Sieur Faultrier to prevent his being exposed to a cholera epidemic that was sweeping Paris. There is scant information about his activities in the last two months of his life. He was due to be released on April 29, and on May 25, he wrote a letter alluding to a broken love affair. It is certain that he committed himself to a duel, and on May 29, the evening before the duel, he wrote, "I beg patriots, my friends, not to reproach me for dying otherwise than for my country. I die the victim of an infamous coquette and her two dupes. It is in a miserable piece of slander that I end my life. Oh! Why die for something so little, so contemptible?"[6] On May 30, he fought the duel; on May 31, he died; and at least two thousand republicans attended his funeral on June 2. He also wrote on May 30, "Please remember me since fate did not give me enough of a life to be remembered by my country."[7] But in spite of all the turmoil of his short life, he worked constantly at his mathematics, and so in this final judgment, he was wrong. Not only his country, but the entire mathematical world remembers and reveres his work.

According to Dumas, the other duelist was Pescheux d'Herbinville, another republican, and it appears that the "infamous coquette" was Stéphanie-Félicie Poterin du Motel, daughter of the resident physician at the Sieur Faultrier. On the backs of one of his papers, Galois had made copies of parts of her letters to him. Included there are the words, "Please let us break up this affair. . . . I do not have the wit to follow . . . a correspondence of this nature . . . but I will try to have enough to converse with you as I did before anything happened. . . . Do not think about those things which did not exist and which never have existed. . . . Mademoiselle Stéphanie D, 14 May 183-"[8]

References

1. Rothman, Tony. "Genius & Biographers: The Fictionalization of Evariste Galois," in the *American Mathematical Monthly* **89** (1982), 84–106.
2. *Ibid.*, p. 88.
3. *Ibid.*, p. 92.
4. *Ibid.*, p. 93.
5. *Ibid.*, p. 96.
6. *Ibid.*, p. 97.
7. *Ibid.*, p. 98.
8. *Ibid.*, p. 101.

14

Counting Automorphisms

According to Proposition 13.9, if K is an extension field of the field F, then homomorphisms $f: K \to K$ which leave F unchanged take roots of polynomials to roots of the same polynomials. So if we know how these homomorphisms behave, we also know how the roots of polynomials behave. In fact, as we will see in Chapter 27, when these homomorphisms are determined by equations that are solvable by radicals, they possess very special properties. The precise sets of functions we wish to consider are the following.

DEFINITION

Let K be an extension field of a field F. The set of all automorphisms f of K such that $f(q) = q$ for all $q \in F$ is denoted **Gal K/F.** That is, Gal K/F consists of all functions $f: K \to K$ which satisfy the following conditions:

    (i) f preserves addition,

    (ii) f preserves multiplication,

    (iii) f is one-to-one,

    (iv) f is onto,

    (v) if $q \in F$, then $f(q) = q$.

The function f of Examples 13.3 and 13.5 is an automorphism of $\mathbb{C}$ for which $f(r) = r$ for all $r \in \mathbb{R}$ and therefore $f \in$ Gal $\mathbb{C}/\mathbb{R}$. It is easy to see that the identity function $\iota(k) = k$ for all $k \in K$ is always in Gal K/F; sometimes (Example 14.10) this set contains only the identity function.

We are interested in extension fields $F^{p(x)}$, and thus in the function sets Gal $F^{p(x)}/F$, for polynomials $p(x) \in F[x]$ for which the equation $p(x) = 0$ is solvable by radicals. In the succeeding chapters, we will address the special properties of the sets Gal $F^{p(x)}/F$ determined by these polynomials. In this chapter, we want to consider the sets Gal $F^{p(x)}/F$ in general for subfields F of $\mathbb{C}$. In particular, we will show (Proposition 14.8) that for these fields the number of functions in the Gal $F^{p(x)}/F$ is precisely the dimension of $F^{p(x)}$ as a vector space over F. The proof will rely on two properties of subfields F of $\mathbb{C}$: first (Proposition 14.2) that an irreducible polynomial in $F[x]$ of degree n has n distinct roots, and second (Proposition 14.3) that any extension field of F is generated by a single element. In fact, both these properties will hold in many fields which are not subfields of $\mathbb{C}$. However, we are primarily interested in extensions $\mathbb{Q}^{p(x)}$ of $\mathbb{Q}$ and restricting ourselves to $\mathbb{C}$ allows us to avoid some technical details and some special cases.

One important property of $\mathbb{C}$ is that polynomials with complex coefficients have a complete set of complex roots (the fundamental theorem of algebra—Theorem 3.11). In the language of polynomials this means that every polynomial in $\mathbb{C}[x]$ can be written as a product of linear factors.

Proposition 14.1   *Suppose that $p(x)$ is a nonconstant polynomial in $\mathbb{C}[x]$. Then there exist a positive integer k, a nonzero element d in $\mathbb{C}$, k distinct complex numbers $c_1, \ldots, c_k$, and k positive integers $e_1, \ldots, e_k$ such that*

$$p(x) = d(x - c_1)^{e_1}(x - c_1)^{e_2} \cdots (x - c_k)^{e_k}.$$

Proof   First divide $p(x)$ by the coefficient of its highest power of x to obtain $p(x) = dm(x)$ for $d \in \mathbb{C}$ and a monic polynomial $m(x) \in \mathbb{C}[x]$. By Theorem 3.11, $m(x)$ has a complex root r_1 and by Proposition 8.5, $x - r_1$ divides $m(x)$. That is, there exists a monic polynomial $m_1(x) \in \mathbb{C}[x]$ such that $p(x) = d(x - r_1)m_1(x)$. Continuing in this fashion, we obtain $r_1, \ldots, r_n \in \mathbb{C}$ such that $p(x) = d(x - r_1) \cdots (x - r_n)$. Collecting equal factors, we have positive integers $e_1, \ldots, e_k$ and distinct complex numbers $c_1, \ldots, c_k$ such that $p(x) = d(x - c_1)^{e_1} \cdots (x - c_n)^{e_k}$. ∎

Our strategy for counting the number of elements in Gal $F^{p(x)}/F$ is to show that $F^{p(x)} = F(w)$ for some w, that if $m(x)$ is the minimum polynomial of w over F, then the number of elements in Gal $F(w)/F =$ Gal $F^{p(x)}/F$ is the number of roots of $m(x)$, and that the number of roots of $m(x)$ is deg $m(x)$. We will then be able to conclude from Proposition 12.4 that Gal $F^{p(x)}/F$ has $[F^{p(x)} : F]$ elements.

We first show that if $m(x)$ is irreducible, then it has deg $m(x)$ distinct roots. The proof uses the decomposition of Proposition 14.1 and the derivative! Actually no calculus is involved. The derivative of a polynomial may be defined formally and the various rules for differentiating polynomials may be derived from this formal definition (see Exercises 42 and 43).

Proposition 14.2   *Let F be a subfield of $\mathbb{C}$ and suppose that $m(x)$ is an irreducible polynomial in $F[x]$ of degree n. Then $m(x)$ has exactly n distinct complex roots and hence there exist a complex number d and n distinct complex numbers c_j such that $m(x) = d(x - c_1) \cdots (x - c_n)$.*

Proof   By Proposition 14.1, there exist a positive integer k, a complex number d, k distinct complex numbers $c_1, \ldots, c_k$, and k positive integers $e_1, \ldots, e_k$ such that $m(x) = d(x - c_1)^{e_1}(x - c_2)^{e_2} \cdots (x - c_k)^{e_k}$. Clearly each c_i is a root of $m(x)$. Furthermore, if r is any root of $m(x)$, then $0 = m(r) = d(r - c_1)^{e_1}(r - c_2)^{e_2} \cdots (r - c_k)^{e_k}$, and since $\mathbb{C}$ is an integral domain, $r = c_i$ for some $i = 1, \ldots, k$. So there are exactly k distinct complex roots of $m(x)$, and thus if $k = n$, then $m(x)$ has exactly n distinct complex roots. But by Proposition 8.2, $n = e_1 + \cdots + e_k$, and hence if $e_i = 1$ for all i, then $k = n$ and $m(x) = d(x - c_1) \cdots (x - c_n)$.

We will show that $e_1 = 1$; the other cases follow from a similar argument. Let $g(x) = (x - c_2)^{e_2} \cdots (x - c_k)^{e_k}$ so that $m(x) = d(x - c_1)^{e_1}g(x)$ and note that since $m(x) \in F[x]$, its derivative $m'(x) \in F[x]$ as well. Using the power rule and the product rule, we have $m'(x) = de_1(x - c_1)^{e_1 - 1}g(x) + d(x - c_1)^{e_1}g'(x)$. But $\deg m'(x) < \deg m(x)$, and hence $m(x)$ cannot divide $m'(x)$. Thus the greatest common divisor of $m(x)$ and $m'(x)$ in $F[x]$ must be 1 because $m(x)$ is irreducible in $F[x]$. Then by Proposition 8.9, 1 is also the greatest common divisor of $m(x)$ and $m'(x)$ in $F(c_1, \ldots, c_k)[x]$. But if $e_1 > 1$, then $e_1 - 1 > 0$ and hence $x - c_1$ divides $m'(x)$ in $F(c_1, \ldots, c_k)[x]$. Since $x - c_1$ also divides $m(x)$ in $F(c_1, \ldots, c_k)[x]$, the greatest common division of $m(x)$ and $m'(x)$ in $F(c_1, \ldots, c_k)[x]$ cannot be 1. This is a contradiction and hence $e_1 = 1$. ∎

We next show (Proposition 14.4) that any finite-dimensional extension of F, and thus in particular any extension of the form $F^{p(x)}$, has a single generator. Before presenting the abstract proof, we consider the following special case.

Example 14.3   We will show that $\mathbb{Q}(\sqrt[4]{3}, i) = \mathbb{Q}(\sqrt[4]{3} + i)$. Clearly $\mathbb{Q}(\sqrt[4]{3}, i)$ is a subfield of $\mathbb{C}$ which contains both $\mathbb{Q}$ and $\sqrt[4]{3} + i$, and thus, since $\mathbb{Q}(\sqrt[4]{3} + i)$ is the smallest such subfield, $\mathbb{Q}(\sqrt[4]{3}, i) \supseteq \mathbb{Q}(\sqrt[4]{3} + i)$. Also $\mathbb{Q}(\sqrt[4]{3} + i)$ is a subfield of $\mathbb{C}$ which contains $\mathbb{Q}$ and thus, to show that $\mathbb{Q}(\sqrt[4]{3}, i) \subseteq \mathbb{Q}(\sqrt[4]{3} + i)$, it suffices to show that $\mathbb{Q}(\sqrt[4]{3} + i)$ contains $\sqrt[4]{3}$ and i. We will show this by showing that the polynomial $x - i$ is in $\mathbb{Q}(\sqrt[4]{3} + i)[x]$. Then i will be in $\mathbb{Q}(\sqrt[4]{3} + i)$ and hence so will $\sqrt[4]{3}$. We first observe that it is not difficult to show that $V(x) = x^4 - 3$ is the minimum polynomial of $\sqrt[4]{3}$ over $\mathbb{Q}$ and that $C(x) = x^2 + 1$ is the minimum polynomial of i over $\mathbb{Q}$. Then the polynomial $f(x) = V(\sqrt[4]{3} + i - x)$ is in $\mathbb{Q}(\sqrt[4]{3} + i)[x]$, and we have $f(i) = V(\sqrt[4]{3}) = 0$, while $f(-i) = V(\sqrt[4]{3} + 2i) \neq 0$. But $\pm i$ are the complex roots of $C(x)$, and hence if $d(x)$ is the greatest common divisor of $f(x)$ and $C(x)$ in $\mathbb{Q}(\sqrt[4]{3} + i)[x]$, then the only complex root of $d(x)$ is i. So by Proposition 14.1, $d(x) = (x - 1)^r$ for some $r \geq 1$. But if $r > 1$, then $(x - i)^r$ does not divide $(x - i)(x + i) = x^2 + i = C(x)$, and hence r must equal 1. So $x - i = d(x) \in \mathbb{Q}(\sqrt[4]{3} + i)[x]$, and therefore $i \in \mathbb{Q}(\sqrt[4]{3} + i)$. It follows that as well, $\sqrt[4]{3} = (\sqrt[4]{3} + i) - i \in \mathbb{Q}(\sqrt[4]{3} + i)$, and hence that $\mathbb{Q}(\sqrt[4]{3}, i) \subseteq \mathbb{Q}(\sqrt[4]{3} + i)$. ❖

A similar argument works in the general case.

Proposition 14.4   *If F and K are subfields of $\mathbb{C}$ and if K is a finite-dimensional extension of F, then there exists $w \in K$ such that $K = F(w)$.*

Proof   (The element w is called a <u>primitive element</u> of K over F.) Since K is a finite-dimensional extension of F, there exist $c_1, \ldots, c_n \in K$ such that $K = F(c_1, \ldots, c_n)$. We proceed by induction on n.

(i)   If $n = 1$, then $F(c_1, \ldots, c_n) = F(c_1)$.

(ii)   Suppose that $c_1, \ldots, c_{n+1} \in \mathbb{C}$ and that there exists $v \in F(c_1, \ldots, c_n)$ such that $F(c_1, \ldots, c_n) = F(v)$. Then

$$K = F(c_1, \ldots, c_{n+1}) = F(c_1, \ldots, c_n)(c_{n+1}) = F(v)(c_{n+1}) = F(v, c_{n+1})$$

and hence it suffices to find $w \in K$ such that $F(v, c_{n+1}) = F(w)$. Since $v, c_{n+1} \in K$, Proposition 12.10 implies that v and c_{n+1} are algebraic over F. Let $V(x) \in F[x]$ be the minimum polynomial of v over F and let $a_1 = v, a_2, \ldots, a_k$ be its distinct complex roots; let $C(x) \in F[x]$ be the minimum polynomial of c_{n+1} over F and let $b_1 = c_{n+1}, b_2, \ldots, b_l$ be its distinct complex roots. Since F is a subfield of $\mathbb{C}$, F is infinite and hence there exists $u \in F$ such that $a_i + ub_j \neq a_1 + ub_1$ for all $i = 1, \ldots, k$ and all $j = 2, \ldots, l$. Let $w = a_1 + ub_1 = v + uc_{n+1}$. Since $F(v, c_{n+1})$ is a subfield of $\mathbb{C}$ which contains both F and w, $F(w) \subseteq F(v, c_{n+1})$. To see that $F(w) \supseteq F(v, c_{n+1})$, consider the polynomial $f(x) = V(w - ux) \in F(w)[x]$. Since $f(c_{n+1}) = V(w - uc_{n+1}) = V(v) = 0$, c_{n+1} is a root of both $C(x)$ and $f(x)$. However, the other roots of $C(x)$ are $b_2, \ldots, b_l$, and if $f(b_j) = 0$ for some $j = 2, \ldots, l$, then $V(W - ub_j) = 0$ so that $w - ub_j = a_i$ for some $i = 1, \ldots, k$. This contradicts our choice of w, and thus the only common root of $C(x)$ and $f(x)$ is c_{n+1}. We conclude that if $d(x)$ is the greatest common divisor of $C(x)$ and $f(x)$ in $F(w)[x]$, then $d(x)$ has only c_{n+1} as a root, and hence by Proposition 14.1, for some $r \geq 1, d(x) = (x - c_{n+1})^r$. But by Proposition 14.2, since $C(x)$ is irreducible, if $r > 1, (x - c_{n+1})^r$ does not divide $C(x)$ and therefore $d(x) = x - c_{n+1}$. So $x - c_{n+1} = d(x) \in F(w)[x]$ and   hence $c_{n+1} \in F(w)$. Then $v = w - uc_{n+1} \in F(w)$ as   well,   and   thus $F(w) \supseteq F(v, c_{n+1})$. It follows that $K = F(v, c_{n+1}) = F(w)$. ∎

In some cases, it is easy to find a primitive element.

Example 14.5   It is easy to show that the minimum polynomial of $\sqrt{2}$ over $\mathbb{Q}$ is $x^2 - 2$, with roots $a_1 = \sqrt{2}$ and $a_2 = -\sqrt{2}$, and that the minimum polynomial of $\sqrt{5}$ over $\mathbb{Q}$ is $x^2 - 5$, with roots $b_1 = \sqrt{5}$ and $b_2 = -\sqrt{5}$. If $u = 1$, then

$$a_1 + ub_2 = \sqrt{2} + (1)(-\sqrt{5}) \neq \sqrt{2} + (1)(\sqrt{5}) = a_1 + ub_1,$$

and

$$a_2 + ub_2 = -\sqrt{2} + (1)(-\sqrt{5}) \neq \sqrt{2} + (1)(\sqrt{5}) = a_1 + ub_1$$

so that by the proof of Proposition 14.4, $w = a_1 + ub_1 = \sqrt{2} + \sqrt{5}$ is a primitive element of $\mathbb{Q}(\sqrt{2}, \sqrt{5})$, i.e., $\mathbb{Q}(\sqrt{2}, \sqrt{5}) = \mathbb{Q}(\sqrt{2} + \sqrt{5})$. ❖

Note that it is not always the case that $F(a, b) = F(a + b)$.

Example 14.6   For instance, since $\sqrt{5}$ is real, $\mathbb{Q}(i + (\sqrt{5} - i)) = \mathbb{Q}(\sqrt{5}) \subseteq \mathbb{R}$, and thus since $i \in \mathbb{Q}(i, \sqrt{5} - i)$, $\mathbb{Q}(i, \sqrt{5} - i) \neq \mathbb{Q}(i + (\sqrt{5} - i))$. The easiest way to find a primitive element for $\mathbb{Q}(i, \sqrt{5} - i)$ is to observe that $\mathbb{Q}(i, \sqrt{5}) = \mathbb{Q}(i, \sqrt{5} - i)$ and then apply an argument similar to that in Example 14.5 to show that $\mathbb{Q}(i, \sqrt{5}) = \mathbb{Q}(i + \sqrt{5})$ (see Exercise 34). ❖

The difficulty in applying the method given in the proof of Proposition 14.4 lies of course in finding the minimum polynomials and their roots.

Example 14.7   Consider $\mathbb{Q}(\sqrt{7}, i - \sqrt{5})$. The minimum polynomial, $x^2 - 7$, of $\sqrt{7}$ has roots $a_1 = \sqrt{7}$ and $a_2 = -\sqrt{7}$. To find a primitive element of $\mathbb{Q}(\sqrt{7}, i - \sqrt{5})$, we also need to determine the minimum polynomial, $m(x) \in \mathbb{Q}[x]$, of $i - \sqrt{5}$ over $\mathbb{Q}$. Note that by Corollary 9.11 and Proposition 10.3, $m(x)$ is the monic polynomial of minimal degree in the ideal $I = \{g(x) \in \mathbb{Q}[x] \mid g(i - \sqrt{5}) = 0\}$. So let $g(x)$ be a nonzero polynomial in I. Since by Proposition 3.12 roots of polynomials with real coefficients occur in conjugate pairs, $-i - \sqrt{5}$ must also be a root of $g(x)$, and hence by Proposition 8.3, $(x - (i - \sqrt{5}))(x - (-i - \sqrt{5})) = x^2 + 2\sqrt{5}x + 6$ must divide $g(x)$. Then deg $g(x) \geq 2$. But if deg $g(x) = 2$, then $c(x^2 + 2\sqrt{5}x + 6) = g(x) \in \mathbb{Q}[x]$ so that on the one hand $c \in \mathbb{Q}$ while on the other $2c\sqrt{5} \in \mathbb{Q}$. This implies that $\sqrt{5} \in \mathbb{Q}$, a contradiction of Proposition 1.18, and hence in fact deg $g(x) > 2$. Suppose that deg $g(x) = 3$. Then for some complex numbers c and r, $g(x) = c(x - r)(x^2 + 2\sqrt{5}x + 6)$, and hence $cx^3 + c(2\sqrt{5} - r)x^2 + c(6 - 2r\sqrt{5})x - 6cr = g(x) \in \mathbb{Q}[x]$ so that c, $6cr$, and $c(2\sqrt{5} - r)$ are all in $\mathbb{Q}$. This implies that $\sqrt{5} \in \mathbb{Q}$, a contradiction of Proposition 1.18. So deg $g(x) \geq 4$.

But we can construct a fourth degree polynomial in the ideal I as follows. Suppose that $b, c \in \mathbb{C}$ are such that

$$(x^2 + 2\sqrt{5}x + 6)(x^2 + bx + c)$$
$$= x^4 + (b + 2\sqrt{5})x^3 + (c + 2b\sqrt{5} + 6)x^2 + (2c\sqrt{5} + 6b)x + 6c \in \mathbb{Q}[x].$$

Then, since $6c \in \mathbb{Q}$, $c \in \mathbb{Q}$, and since $b + 2\sqrt{5} \in \mathbb{Q}$, $b = q - 2\sqrt{5}$ for some $q \in \mathbb{Q}$. Then $2c\sqrt{5} + 6(q - 2\sqrt{5}) \in \mathbb{Q}$, hence $(2c - 12)\sqrt{5} \in \mathbb{Q}$, and therefore $c = 6$. Also, $c + 2\sqrt{5}(q - 2\sqrt{5}) + 6 \in \mathbb{Q}$, hence $2\sqrt{5}q \in \mathbb{Q}$, and therefore $q = 0$ so that $b = -2\sqrt{5}$. Then $x^4 - 8x + 36 = (x^2 + 2\sqrt{5}x + 6)(x^2 - 2\sqrt{5}x + 6) \in I$, and since it is monic and has minimal degree in I, we must have $m(x) = x^4 - 8x + 36$ by Corollary 9.11 and Proposition 10.3. Applying the quadratic formula to $x^2 + 2\sqrt{5}x + 6$ and $x^2 - 2\sqrt{5}x + 6$, we find that the roots of $m(x)$ are

$$b_1 = i - \sqrt{5}, \qquad b_2 = -i - \sqrt{5}, \qquad b_3 = i + \sqrt{5}, \qquad b_4 = -i + \sqrt{5}.$$

If $u = 1$, then

$$a_1 + ub_2 = \sqrt{7} + (1)(-i - \sqrt{5}) \neq \sqrt{7} + (1)(i - \sqrt{5}) = a_1 + ub_1,$$
$$a_1 + ub_3 = \sqrt{7} + (1)(i + \sqrt{5}) \neq \sqrt{7} + (1)(i - \sqrt{5}) = a_1 + ub_1,$$
$$a_1 + ub_4 = \sqrt{7} + (1)(-i + \sqrt{5}) \neq \sqrt{7} + (1)(i - \sqrt{5}) = a_1 + ub_1,$$
$$a_2 + ub_2 = -\sqrt{7} + (1)(-i - \sqrt{5}) \neq \sqrt{7} + (1)(i - \sqrt{5}) = a_1 + ub_1,$$
$$a_2 + ub_3 = -\sqrt{7} + (1)(i + \sqrt{5}) \neq \sqrt{7} + (1)(i - \sqrt{5}) = a_1 + ub_1,$$
$$a_2 + ub_4 = -\sqrt{7} + (1)(-i + \sqrt{5}) \neq \sqrt{7} + (1)(i - \sqrt{5}) = a_1 + ub_1$$

because in each case equality would imply that $i \in \mathbb{R}$. Then according to the proof of Proposition 14.4, $\mathbb{Q}(\sqrt{7}, i - \sqrt{5}) = \mathbb{Q}(\sqrt{7} + i - \sqrt{5})$. ❖

We can now prove what we previously asserted: namely that for subfields F of $\mathbb{Q}$, the number of functions in the Gal $F^{p(x)}/F$ is precisely the dimension of $F^{p(x)}$ as a vector space over F.

Proposition 14.8   *If F is a subfield of $\mathbb{C}$ and $p(x) \in F[x]$, then the number of elements in Gal $F^{p(x)}/F$ is exactly $[F^{p(x)}:F]$.*

Proof   By Proposition 14.4, $F^{p(x)} = F(u)$ for some $u \in F^{p(x)}$. Let $m(x)$ be the minimum polynomial of u over F and let $n = \deg m(x)$. Note that by Theorem 10.8, $F(u) = \{\alpha_0 + \cdots + \alpha_{n-1}u^{n-1} \mid \alpha_i \in F\}$. By Proposition 14.2, $m(x)$ has exactly n distinct roots and by Proposition 12.4, $n = [F^{p(x)}:F]$. Thus it suffices to show that Gal $F^{p(x)}/F$ consists precisely of the functions $f(\alpha_0 + \cdots + \alpha_{n-1}u^{n-1}) = \alpha_0 + \cdots + \alpha_{n-1}v^{n-1}$, where v is a root of $m(x)$. To see that all functions in Gal $F^{p(x)}/F$ are of this form, note that for any $\phi \in$ Gal $F^{p(x)}/F$, $\phi(u)$ solves $m(x) = 0$ by Proposition 13.9, and since ϕ preserves addition and multiplication and for all $q \in F$, $\phi(q) = q$,

$$\phi(\alpha_0 + \cdots + \alpha_{n-1}u^{n-1}) = \phi(\alpha_0) + \cdots + \phi(\alpha_{n-1})\phi(u)^{n-1}$$
$$= \alpha_0 + \cdots + \alpha_{n-1}\phi(u)^{n-1}.$$

So $\phi(a_0 + \cdots + a_{n-1}u^{n-1}) = a_0 + \cdots + a_{n-1}v^{n-1}$ for some root v of $m(x)$. Conversely, Proposition 13.11 shows that for any root v of $m(x)$, the function $f: F(u) \to F(v)$ defined by letting $f(\alpha_0 + \cdots + \alpha_{n-1}u^{n-1}) = \alpha_0 + \cdots + \alpha_{n-1}v^{n-1}$ is an isomorphism such that for all $q \in F, f(q) = q$, and hence to see that

$f \in \text{Gal } F^{p(x)}/F$, it suffices to show that $F^{p(x)} = F(v)$. But if $c_1, \ldots, c_m$ are the roots of $p(x)$, then by Proposition 13.9 for each i, there exists j such that $f(c_i) = c_j$. Then $F(c_1, \ldots, c_m) = F^{p(x)}$ is a subfield of $\mathbb{C}$ which contains F and $\{f(c_1), \ldots, f(c_m)\}$, and therefore $F(f(c_1), \ldots, f(c_m)) \subseteq F^{p(x)}$. Since $u \in F^{p(x)} = F(c_1, \ldots, c_m)$, Proposition 12.5 implies that u can be expressed as a linear combination of products of $c_1, \ldots, c_m$ and hence that $f(u)$ can be expressed as a linear combination of products of $f(c_1), \ldots, f(c_m)$. So since $v = f(u)$, $v \in F(f(c_1), \ldots, f(c_m))$ and hence $v \in F^{p(x)}$. Then $F^{p(x)}$ is a subfield of $\mathbb{C}$ which contains F and v, and thus $F^{p(x)} \supseteq F(v)$. But by Proposition 12.4, $[F^{p(x)}:F] = [F(u):F] = n$, and since v is also a root of $m(x), n = [F(v):F]$ as well. So $F^{p(x)} = F(v)$ by Proposition 12.1. ∎

It can be difficult to find primitive elements, and for this reason, to construct the functions in $\text{Gal } F^{p(x)}/F$, it is usually easiest to iterate Proposition 13.11 in the following manner. Note the crucial role which Proposition 14.8 plays at the end of the construction.

Example 14.9   Let $p(x) = x^3 - 7 \in \mathbb{Q}[x]$ and recall that by Proposition 3.8, the roots of $p(x)$ are $\sqrt[3]{7}$, $\sqrt[3]{7}\zeta_3$, and $\sqrt[3]{7}\zeta_3^2$. Thus since $\mathbb{Q}(\sqrt[3]{7}, \zeta_3)$ is a subfield of $\mathbb{C}$, since $\mathbb{Q} \subseteq \mathbb{Q}(\sqrt[3]{7}, \zeta_3)$, and since $\sqrt[3]{7}, \sqrt[3]{7}\zeta_3, \sqrt[3]{7}\zeta_3^2 \in \mathbb{Q}(\sqrt[3]{7}, \zeta_3)$, $\mathbb{Q}(\sqrt[3]{7}, \zeta_3) \supseteq \mathbb{Q}^{p(x)}$. Similarly since $\mathbb{Q}^{p(x)}$ is a subfield of $\mathbb{C}$, since $\mathbb{Q} \subseteq \mathbb{Q}^{p(x)}$, and since $\sqrt[3]{7} \in \mathbb{Q}^{p(x)}$ and $\zeta_3 = (\sqrt[3]{7})^{-1}(\sqrt[3]{7}\zeta_3) \in \mathbb{Q}^{p(x)}$, $\mathbb{Q}(\sqrt[3]{7}, \zeta_3) \subseteq \mathbb{Q}^{p(x)}$. So $\mathbb{Q}^{p(x)} = \mathbb{Q}(\sqrt[3]{7}, \zeta_3)$ and hence finding all the functions in $\text{Gal } \mathbb{Q}^{p(x)}/\mathbb{Q}$ is equivalent to finding all the functions in $\text{Gal } \mathbb{Q}(\sqrt[3]{7}, \zeta_3)/\mathbb{Q}$.

By Eisenstein's criterion with prime 7, $p(x)$ is irreducible in $\mathbb{Q}[x]$, and since $p(x)$ is clearly monic and $p(\sqrt[3]{7}) = 0$, $p(x)$ is the minimum polynomial of $\sqrt[3]{7}$ over $\mathbb{Q}$. Thus by Theorem 10.8, $\mathbb{Q}(\sqrt[3]{7}) = \{a_0 + a_1\sqrt[3]{7} + a_2(\sqrt[3]{7})^2 | a_i \in \mathbb{Q}\}$. Since $\sqrt{-1 - 4 \cdot 1 \cdot 1} \notin (\mathbb{Q}\sqrt[3]{7})$, $m(x) = x^2 + x + 1$ is irreducible in $\mathbb{Q}(\sqrt[3]{7})[x]$ by Proposition 8.12, and thus, since $m(x)$ is monic and since $m(\zeta_3) = 0$ (Example 10.5), $m(x)$ is the minimum polynomial of ζ_3 over $\mathbb{Q}(\sqrt[3]{7})$. Therefore by Theorem 10.8,

$$\begin{aligned}\mathbb{Q}(\sqrt[3]{7}, \zeta_3) &= \{A_0 + A_1\zeta_3 | A_i \in \mathbb{Q}(\sqrt[3]{7})\} \\ &= \{(\alpha_0 + \alpha_1\sqrt[3]{7} + \alpha_2(\sqrt[3]{7})^2) \\ &\quad + (\beta_0 + \beta_1\sqrt[3]{7} + \beta_2(\sqrt[3]{7})^2)\zeta_3 | \alpha_i, \beta_j \in \mathbb{Q}\}.\end{aligned}$$

Now suppose that $f \in \text{Gal } \mathbb{Q}(\sqrt[3]{7}, \zeta_3)/\mathbb{Q}$. Since f preserves addition and multiplication and leaves elements of $\mathbb{Q}$ unchanged,

$$\begin{aligned}&f((\alpha_0 + \alpha_1\sqrt[3]{7} + \alpha_2(\sqrt[3]{7})^2) + (\beta_0 + \beta_1\sqrt[3]{7} + \beta_2(\sqrt[3]{7})^2)\zeta_3) \\ &\quad = (f(\alpha_0) + f(\alpha_1)f(\sqrt[3]{7}) + f(\alpha_2)(f(\sqrt[3]{7}))^2) \\ &\qquad + (f(\beta_0) + f(\beta_1)f(\sqrt[3]{7}) + f(\beta_2)(f(\sqrt[3]{7}))^2)f(\zeta_3) \\ &\quad = (\alpha_0 + \alpha_1 f(\sqrt[3]{7}) + \alpha_2(f\sqrt[3]{7}))^2) + (\beta_0 + \beta_1 f(\sqrt[3]{7}) + \beta_2(f(\sqrt[3]{7}))^2)f(\zeta_3).\end{aligned}$$

Furthermore, Proposition 13.9 implies that $f(\sqrt[3]{7})$ is a root of $x^3 - 7$ and that $f(\zeta_3)$ is a root of $x^2 + x + 1$. That is, $f(\sqrt[3]{7}) = \sqrt[3]{7}, \sqrt[3]{7}\zeta_3, \sqrt[3]{7}\zeta_3^2$ and $f(\zeta_3) = \zeta_3, \zeta_3^2$. Thus isomorphisms satisfying the previous equation may be succinctly described by combining the equation with the following table of possible substitutions for $f(\sqrt[3]{7})$ and $f(\zeta_3)$:

	f_1	f_2	f_3	f_4	f_5	f_6
$\sqrt[3]{7} \to$	$\sqrt[3]{7}$	$\sqrt[3]{7}$	$\sqrt[3]{7}\zeta_3$	$\sqrt[3]{7}\zeta_3$	$\sqrt[3]{7}\zeta_3^2$	$\sqrt[3]{7}\zeta_3^2$
$\zeta_3 \to$	ζ_3	ζ_3^2	ζ_3	ζ_3^2	ζ_3	ζ_3^2

Thus far, we have shown that Gal $\mathbb{Q}(\sqrt[3]{7}, \zeta_3)/\mathbb{Q} \subseteq \{f_1, f_2, f_3, f_4, f_5, f_6\}$. But by Propositions 12.4 and 12.5, $[\mathbb{Q}(\sqrt[3]{7}, \zeta_3):\mathbb{Q}] = [\mathbb{Q}(\sqrt[3]{7}, \zeta_3):\mathbb{Q}(\sqrt[3]{7})] \cdot [\mathbb{Q}\sqrt[3]{7}):\mathbb{Q}] = 2 \cdot 3 = 6$. So since $\mathbb{Q}(\sqrt[3]{7}, \zeta_3) = \mathbb{Q}^{p(x)}$, we may apply Proposition 14.8 and conclude that Gal $\mathbb{Q}(\sqrt[3]{7}, \zeta_3)/\mathbb{Q}$ contains exactly six functions. It follows that Gal $\mathbb{Q}(\sqrt[3]{7}, \zeta_3)/\mathbb{Q} = \{f_1, f_2, f_3, f_4, f_5, f_6\}$.

Note that is easy to recover a formula for f_i from the equation and the table. Just substitute the appropriate entry from the table into the equation. For example, to recover f_4, substitute in the equation $\sqrt[3]{7}\zeta_3$ for $f_4(\sqrt[3]{7})$ and ζ_3^2 for $f_4(\zeta_3)$:

$$f_4((\alpha_0 + \alpha_1\sqrt[3]{7} + \alpha_2(\sqrt[3]{7})^2) + (\beta_0 + \beta_1\sqrt[3]{7} + \beta_2(\sqrt[3]{7})^2)\zeta_3)$$
$$= (\alpha_0 + \alpha_1\sqrt[3]{7}\zeta_3 + \alpha_2(\sqrt[3]{7}\zeta_3)^2) + (\beta_0 + \beta_1\sqrt[3]{7}\zeta_3 + \beta_2(\sqrt[3]{7}\zeta_3)^2)\zeta_3^2. \; \clubsuit$$

The method used to construct Gal $\mathbb{Q}(\sqrt[3]{7}, \zeta_3)/\mathbb{Q}$ in Example 14.9 can be used generally. For instance, in the following example, we use it to show that Gal K/F need not always have $[K:F]$ functions.

Example 14.10 As in Example 14.9, $\mathbb{Q}(\sqrt[3]{7}) = \{\alpha_0 + \alpha_1\sqrt[3]{7} + \alpha_2(\sqrt[3]{7})^2 \,|\, \alpha_i \in \mathbb{Q}\}$. If $f \in$ Gal $\mathbb{Q}(\sqrt[3]{7})/\mathbb{Q}$, then f preserves addition and multiplication and leaves elements of $\mathbb{Q}$ unchanged and hence $f(\alpha_0 + \alpha_1\sqrt[3]{7} + \alpha_2(\sqrt[3]{7})^2) = \alpha_0 + \alpha_1 f(\sqrt[3]{7}) + \alpha_2(f(\sqrt[3]{7}))^2$. Thus f is completely determined by this equation and $f(\sqrt[3]{7})$. Since $\sqrt[3]{7}$ solves $x^3 - 7 = 0, f(\sqrt[3]{7})$ also solves $x^3 - 7 = 0$ by Proposition 13.9, and hence by Proposition 3.8, $f(\sqrt[3]{7}) = \sqrt[3]{7}, \sqrt[3]{7}\zeta_3$, or $\sqrt[3]{7}\zeta_3^2$. Therefore there are <u>at most</u> three functions in Gal $\mathbb{Q}(\sqrt[3]{7})/\mathbb{Q}$. They are determined by the equation $f(\alpha_0 + \alpha_1\sqrt[3]{7} + \alpha_2(\sqrt[3]{7})^2) = \alpha_0 + \alpha_1 f(\sqrt[3]{7}) + \alpha_2(f(\sqrt[3]{7}))^2$ and the following table of possible substitutions:

	f_1	f_2	f_3
$\sqrt[3]{7} \to$	$\sqrt[3]{7}$	$\sqrt[3]{7}\zeta_3$	$\sqrt[3]{7}\zeta_3^2$

However, functions in Gal $\mathbb{Q}(\sqrt[3]{7})/\mathbb{Q}$ must take elements in $\mathbb{Q}(\sqrt[3]{7})$ to $\mathbb{Q}(\sqrt[3]{7})$, and neither $\sqrt[3]{7}\zeta_3$ nor $\sqrt[3]{7}\zeta_3^2$ is an element of $\mathbb{Q}(\sqrt[3]{7})$. So neither f_2 nor f_3 can be in Gal $\mathbb{Q}(\sqrt[3]{7})/\mathbb{Q}$. But f_1 is the identity function on $\mathbb{Q}(\sqrt[3]{7})$, and certainly the identity function preserves addition and multiplication, is one-to-one and onto, and leaves elements of $\mathbb{Q}$ unchanged. So f_1 is in Gal $\mathbb{Q}(\sqrt[3]{7})/\mathbb{Q}$, and therefore Gal $\mathbb{Q}(\sqrt[3]{7})/\mathbb{Q} = \{f_1\}$. Since $[\mathbb{Q}(\sqrt[3]{7}):\mathbb{Q}] = 3$, the number of functions in Gal $\mathbb{Q}(\sqrt[3]{7})/\mathbb{Q}$ is <u>not</u> equal to $[\mathbb{Q}(\sqrt[3]{7}):\mathbb{Q}]$. Note that therefore, according to Proposition 14.8, $\mathbb{Q}(\sqrt[3]{7}) \neq \mathbb{Q}^{p(x)}$ for any $p(x) \in \mathbb{Q}[x]$. ❖

Exercises

In Exercises 1–6, write the given polynomial in the form given in Proposition 14.1.

1. $x^2 + 4x + 4$ **2.** $x^2 + x + 1$ **3.** $5x^3 - 25$

4. $3x^3 - 7$ **5.** $2x^5 - 12x^4 + 18x^3$ **6.** $x^3 + 3x^2 - 4$

In Exercises 7–18, determine the number of elements in the given set Gal K/F. (*Hint:* In Exercises 11–14, use the method of Exercise 32 in Chapter 12; in Exercises 15–18, use Exercise 22 in Chapter 12.)

7. Gal $\mathbb{Q}(\sqrt[4]{3}, i)/\mathbb{Q}$ **8.** Gal $\mathbb{Q}(\sqrt[4]{7}, i)/\mathbb{Q}$

9. Gal $\mathbb{Q}(\sqrt[3]{2}, \zeta_3)/\mathbb{Q}(\sqrt[3]{2})$ **10.** Gal $\mathbb{Q}(\sqrt[3]{7}, \zeta_3)/\mathbb{Q}(\zeta_3)$

11. Gal $\mathbb{Q}(\sqrt[5]{3}, \zeta_5)/\mathbb{Q}$ **12.** Gal $\mathbb{Q}(\sqrt[5]{7}, \zeta_5)/\mathbb{Q}$

13. Gal $\mathbb{Q}(\sqrt[5]{11}, \zeta_5)/\mathbb{Q}(\sqrt[5]{11})$ **14.** Gal $\mathbb{Q}(\sqrt[5]{11}, \zeta_5)/\mathbb{Q}(\zeta_5)$

15. Gal $\mathbb{Q}(\sqrt{5}, \sqrt[3]{11}, \zeta_3)/\mathbb{Q}$ **16.** Gal $\mathbb{Q}(\sqrt{5}, \sqrt[3]{11}, \zeta_3)/\mathbb{Q}(\sqrt{5})$

17. Gal $\mathbb{Q}(\sqrt{5}, \sqrt[3]{11}, \zeta_3)/\mathbb{Q}(\zeta_3)$ **18.** Gal $\mathbb{Q}(\sqrt{5}, \sqrt[3]{11}, \zeta_3)/\mathbb{Q}(\sqrt{5}, \zeta_3)$

In Exercises 19–24, use the method of Example 14.9 to determine all the functions in Gal $\mathbb{Q}^{p(x)}/\mathbb{Q}$ for the given $p(x) \in \mathbb{Q}[x]$.

19. $p(x) = x^2 + 11$ **20.** $p(x) = x^2 + 3x + 1$ **21.** $p(x) = x^2 + 3x + 4$

22. $p(x) = x^3 - 5$ **23.** $p(x) = x^3 + 11$ **24.** $p(x) = x^4 - 7$

In Exercises 25–33, use the method of Examples 14.9 and 14.10 to determine all the functions in Gal K/F.

25. Gal $\mathbb{Q}(\sqrt{11})/\mathbb{Q}$ **26.** Gal $\mathbb{Q}(\sqrt{13})/\mathbb{Q}$ **27.** Gal $\mathbb{Q}(\sqrt[4]{3}, \zeta_4)/\mathbb{Q}$

28. Gal $\mathbb{Q}(\sqrt[3]{13}, \zeta_3)/\mathbb{Q}$ **29.** Gal $\mathbb{Q}(\sqrt[3]{11})/\mathbb{Q}$ **30.** Gal $\mathbb{Q}(\sqrt[5]{19})/\mathbb{Q}$

31. Gal $\mathbb{Q}(\sqrt[5]{7}, \zeta_5)/\mathbb{Q}$ **32.** Gal $\mathbb{Q}(\sqrt[7]{5})/\mathbb{Q}$ **33.** Gal $\mathbb{Q}(\sqrt[7]{13}, \zeta_7)/\mathbb{Q}$

In Exercises 34 and 35, use the method of Example 14.3 to prove the given equality.

34. $\mathbb{Q}(i, \sqrt{5}) = \mathbb{Q}(i + \sqrt{5})$ **35.** $\mathbb{Q}(\zeta_3, \sqrt[5]{7}) = \mathbb{Q}(\zeta_3 + \sqrt[5]{7})$.

In Exercises 36–41, find $w \in F$ such that $F = \mathbb{Q}(w)$.

36. $F = \mathbb{Q}(\sqrt{3}, \sqrt{11})$ **37.** $F = \mathbb{Q}(\sqrt{7}, \zeta_3)$ **38.** $F = \mathbb{Q}(i, \sqrt{11} + i)$

39. $F = \mathbb{Q}(\sqrt{5}, \sqrt{3} + i)$ **40.** $F = \mathbb{Q}(i, \sqrt{11}, \sqrt{3})$ **41.** $F = \mathbb{Q}(i, \zeta_3, \sqrt{7})$

Exercises 42 and 43 derive the familiar power and product rules for the derivative in the following general setting. Let F be a field. For any $p(x) = a_0 + a_1 x + a_2 x^2 + \cdots + a_n x^n \in F[x]$, the derivative of $p(x)$, denoted $p'(x)$, is the polynomial $p'(x) = a_1 + 2a_2 x + 3a_3 x^2 + \cdots + n a_n x^{n-1}$. Prove the assertions in Exercises 42 and 43 by using this formal definition rather than by appealing to results from calculus.

42. If $p(x) = d(x + c)^n$, show that $p'(x) = nd(x + c)^{n-1}$.

(*Hint:* Recall the binomial theorem (Exercise 44 in Chapter 1): for any positive integer n,

$$(x + c)^n = \sum_{r=0}^{n} \binom{n}{r} x^r c^{n-r}, \text{ where } \binom{n}{r} = \frac{n!}{r!(n-r)!}.)$$

43. For $p(x), q(x) \in F[x]$, let $(pq)(x) = p(x)q(x)$. Show that $(pq)'(x) = p'(x)q(x) + p(x)q'(x)$.

44. Suppose that F is a subfield of $\mathbb{C}$ and that a and b are in $\mathbb{C}$. Show that there are an infinite number of elements $u \in F$ such that $F(a,b) = F(a + ub)$.

Exercises 45 and 46 concern the following situation. Suppose that F is a subfield of $\mathbb{C}$ and that $p(x)$ is a polynomial in $F[x]$ for which there are n functions in Gal $F^{p(x)}/F$. Suppose further that $u \in F^{p(x)}$ and that its minimum polynomial $m(x)$ over F has at least n complex roots.

45. Show that $m(x)$ has exactly n complex roots.

46. Show that $F(u) = F^{p(x)}$.

47. Suppose that F is a subfield of $\mathbb{C}$ and that $r \in \mathbb{C}$ is algebraic over F of degree n. Show that Gal $F(r)/F$ has at most n elements.

48. Suppose that F, L, and K are all subfields of $\mathbb{C}$ and that $F \subseteq L \subseteq K$. Suppose further that $f(x)$ is irreducible in $F[x]$ and that $g(x)$ is in $L[x]$. Show that if there is exactly one element c in K which is a root of both $f(x)$ and $g(x)$, then this element c is in L.

49. Suppose that F is a subfield of $\mathbb{C}$ and that $p(x)$ is a polynomial in $F[x]$. Suppose further that $m(x)$ is the minimum polynomial of an element u in $F^{p(x)}$. Show that all the complex roots of $m(x)$ are in $F^{p(x)}$. (*Hint:* Proving this requires Proposition 27.3 and some sophisticated reasoning.)

Exercises 50 and 51 concern the following. Let $p(x)$ be a nonzero polynomial in $\mathbb{Q}[x]$, let $d(x)$ be the greatest common divisor of $p(x)$ and $p'(x)$, and suppose that $p(x) = s(x)d(x)$.

50. Show that $s(x)$ is a polynomial having the same roots as $p(x)$.

51. If $n = \deg s(x)$, show that $s(x)$ has exactly n distinct complex roots.

52. The proof of Proposition 14.1 is technically an induction proof. Write the proof as an induction proof; clearly delineate both steps of the induction; clearly state the induction hypothesis.

HISTORICAL NOTE

Richard Dedekind

Julius Wilhelm Richard Dedekind was born in Brunswick, Germany, on October 6, 1831, and died in Brunswick on February 12, 1916. Both his father and his mother came from distinguished families; together they included physicians, lawyers, professors, and an imperial postmaster. Dedekind was the youngest of four children and, like his second sister, Julie, never married. She was a respected writer, and they lived in the same house for most of their lives.

Dedekind's first interests were chemistry and physics, but he soon turned to mathematics because he felt that the other subjects lacked sufficiently logical structure. In 1849–1850, he gave private lessons in mathematics, and then at Easter 1850, he matriculated at the University of Göttingen. The next winter, he attended Gauss's lecture on the method of least squares, and fifty years later still remembered the lecture vividly, recalling how his interest had constantly increased as the lecture progressed. He completed his doctoral work under Gauss in 1852 but, not satisfied with the breadth of his knowledge, spent two more years attending lectures. In 1854, he became a university lecturer but still attended lectures by his contemporary Riemann and Gauss's successor Dirichlet. His own lectures, to a small audience, introduced Galois theory for probably the first time and replaced the specific notion of permutation group with the general notion of abstract group. In 1858, he went to Switzerland to teach at the Polytechnikum in Zurich (now the Eidgenössische Technische Hochschule), and then in 1862, he obtained a position at the Polytechnikum in Brunswick and returned home. He spent the rest of his life there in the company of his brother and sister, traveling occasionally and for the most part quite healthy both physically and intellectually.

He is best known for his construction of the real numbers from the rational numbers by means of "Dedekind cuts." But certainly his influence extends far beyond that construction. In 1871, he published an edition of Dirichlet's lectures on number theory for which he wrote a supplement establishing the theory of algebraic number fields, giving the general definition of ideal, and showing how these ideals could be written as products of prime ideals. This work contained the seeds of Emmy Noether's great achievements fifty years later. And in 1897, he introduced the notion of an abstract lattice, an idea which in the twentieth century has led to the development of a vast and powerful theory. Dedekind's influence has been enormous; it is difficult to overestimate the degree to which his ideas and philosophy permeate contemporary mathematics.

PART 3

Elementary Group Theory

15

Groups

We are interested in solutions of polynomial equations $p(x) = 0$. The results of Chapters 13 and 14 show that the functions in Gal $F^{p(x)}/F$ are intricately related to the roots of $p(x)$ and thus that one way of learning about the solutions of polynomial equations is to study the structure of the sets Gal $F^{p(x)}/F$. In our earlier work, we exploited algebraic structure by looking at rings and fields. The sets Gal $F^{p(x)}/F$ are unlikely to fit into either of these classes because they possess only one natural operation (composition) rather than two. Does this single operation obey any of the field axioms involving only addition or multiplication? As we show next, composition is indeed closed and associative and has an identity, and with respect to it, all elements have inverses.

Proposition 15.1   *If K is an extension field of a field F, then the operation of composition of function in* Gal K/F *satisfies the following conditions.*

(i)   *If $f, g \in$ Gal K/F, then $f \circ g \in$ Gal K/F.*   (*closure*)

(ii)   *If $f, g, h \in$ Gal K/F, then $f \circ (g \circ h) = (f \circ g) \circ h$.*   (*associativity*)

(iii)   *There exists $\iota \in$ Gal K/F such that for all $f \in$ Gal K/F,*
        $f \circ \iota = f = \iota \circ f$.   (*existence of an identity*)

(iv)   *For all $f \in$ Gal K/F, there exists $t \in$ Gal K/F such that*
        $f \circ t = \iota = t \circ f$.   (*existence of inverses*)

Proof   (i)   To show that Gal K/F is closed, we must show that $f \circ g$ is a homomorphism from K to K which is one-to-one and onto and leaves F unchanged. Since for $a, b \in K$ and $c \in F$,

$$f(g(a + b)) = f(g(a) + g(b)) = f(g(a)) + f(g(b)),$$
$$f(g(ab)) = f(g(a)g(b)) = f(g(a))f(g(b)),$$
$$f(g(c)) = f(c) = c,$$

$f \circ g$ is a homomorphism and leaves F unchanged. If $a, b \in K$ and $f \circ g(a) = f \circ g(b)$, then $g(a) = g(b)$ because f is one-to-one and $a = b$ because g is one-to-one; so

$f \circ g$ is one-to-one. Furthermore, since f is onto, there exists $r \in K$ such that $f(r) = a$ and since g is onto, there exists $s \in K$ such that $g(s) = r$. Then $f \circ g(s) = a$ and hence $f \circ g$ is onto. It follows that $f \circ g \in \mathrm{Gal}\, K/F$.

(*ii*) For associativity, note that for any $a \in K$,

$$(f \circ (g \circ h))(a) = f(g \circ h(a)) = f(g(h(a))) = f \circ g(h(a)) = ((f \circ g) \circ h)(a),$$

and hence $f \circ (g \circ h) = (f \circ g) \circ h$.

(*iii*) The function ι is of course the function on K defined by letting $\iota(k) = k$. It is easy to check (see Exercise 19) that ι is a homomorphism from K to K which is one-to-one and onto and which leaves F unchanged, i.e., that $\iota \in \mathrm{Gal}\, K/F$. Furthermore, for any $f \in \mathrm{Gal}\, K/F$ and any $k \in K$, $\iota(f(k)) = f(k) = f(\iota(k))$, and thus $f \circ \iota = f = \iota \circ f$. So ι is an identify function with respect to composition.

(*iv*) For inverses, note that since f is one-to-one and onto, it has an inverse function t, i.e., there exists a function $t : K \to K$ such that $f \circ t = \iota = t \circ f$ (t may be defined by letting $t(k)$ be the unique element j of K such that $f(j) = k$). So, in particular, for all $k \in K$, $f(t(k)) = k$ and $t(f(k)) = k$. We need to show that $t \in \mathrm{Gal}\, K/F$. If $t(a) = t(b)$ for $a, b \in K$, then $a = f(t(a)) = f(t(b)) = b$ and hence t is one-to-one. If $k \in K$, then $f(k) \in K$ and $t(f(k)) = k$ so that t is onto. If $c \in F$, then $f(c) = c$ and thus $t(c) = t(f(c)) = c$; so t leaves F unchanged. Finally, if $a, b \in K$, then

$$t(a + b) = t(f[t(a)] + f[t(b)]) = t(f[t(a) + t(b)]) = t(a) + t(b),$$
$$t(a\,b) = t(f[t(a)]f[t(b)]) = t(f[t(a)t(b)]) = t(a)t(b),$$

and therefore t is a homomorphism. We conclude that $t \in \mathrm{Gal}\, K/F$. ∎

There is of course one important axiom missing from the list in Proposition 15.1, viz., commutativity. And indeed, as the next example shows, composition of functions in $\mathrm{Gal}\, K/F$ may not be commutative.

Example 15.2 Let $p(x) = x^3 - 7 \in \mathbb{Q}[x]$ and let f_4 and f_5 be the functions in $\mathrm{Gal}\, \mathbb{Q}^{p(x)}/\mathbb{Q}$ determined in Example 14.9. Then $f_4(\sqrt[3]{7}) = \sqrt[3]{7}\zeta_3$, $f_4(\zeta_3) = \zeta_3{}^2$, and $f_5(\sqrt[3]{7}) = \sqrt[3]{7}\zeta_3{}^2$, $f_5(\zeta_3) = \zeta_3$. Hence since both f_4 and f_5 preserve multiplication,

$$f_5 \circ f_4(\sqrt[3]{7}) = f_5(f_4(\sqrt[3]{7})) = f_5(\sqrt[3]{7}\zeta_3) = f_5(\sqrt[3]{7})f_5(\zeta_3) = \sqrt[3]{7}\zeta_3{}^2\zeta_3 = \sqrt[3]{7},$$
$$f_4 \circ f_5(\sqrt[3]{7}) = f_4(\sqrt[3]{7}\zeta_3{}^2) = f_4(\sqrt[3]{7})f_4(\zeta_3)^2 = \sqrt[3]{7}\zeta_3(\zeta_3{}^2)^2 = \sqrt[3]{7}\zeta_3{}^2.$$

Thus $f_5 \circ f_4 \ne f_4 \circ f_5$. ❖

Algebraic structures, such as $\mathrm{Gal}\, K/F$, which have a single operation which is closed and associative and has an identity and with respect to which all elements have inverses, form perhaps the single most important class of algebraic structures, and, as with fields and rings, are singled out and given their own name.

DEFINITION

A **group** is a set G with a binary operation · satisfying the following axioms.

G1. *CLOSURE: For all x, $y \in G$, $x \cdot y \in G$.*

G2. *ASSOCIATIVITY: For all x, y, $z \in G$, $x \cdot (y \cdot z) = (x \cdot y) \cdot z$.*

G3. *EXISTENCE OF AN IDENTITY: There exists $e \in G$ such that for all $x \in G$,*
$x \cdot e = x = e \cdot x$.

G4. *EXISTENCE OF INVERSES: For all $0 \neq x \in G$, there exists $z \in G$ such that*
$x \cdot z = e = z \cdot x$.

An **Abelian** (or **commutative**) **group** is a group G which satisfies the following <u>additional</u> axiom

G5. *COMMUTATIVITY: For all x, $y \in G$, $x \cdot y = y \cdot x$.*

According to Proposition 15.1, Gal K/F is a group with respect to composition of functions. It is called the Galois group of K over F.

DEFINITION

Let K be an extension field or a field F. The **Galois group** of K over F is the set Gal K/F with composition of functions as the operation.

Groups are ubiquitous and it is important (1) to understand the conventions used in studying them and (2) to have a collection of familiar examples. Many well-known examples arise inside the complex numbers and illustrate some of the conventions that must be adopted. In particular, the complex numbers have <u>two</u> natural operations, and thus there might be confusion about which operation is being used. In such cases, the operation can be specified. For example, $(\mathbb{Q}, +)$ refers to the rational numbers with the operation addition, whereas $(\mathbb{Q}, \cdot)$ refers to the rational numbers with the operation multiplication. Note that there is no actual ambiguity in either case. For $(\mathbb{Q}, +)$ is an Abelian group, and $(\mathbb{Q}, \cdot)$ is not a group because 0 has no multiplicative inverse. The <u>group</u> $\mathbb{Q}$ must therefore mean $(\mathbb{Q}, +)$. On the other hand, if $\mathbb{Q}^{\times}$ denotes the nonzero rational numbers, then $(\mathbb{Q}^{\times}, \cdot)$ is an Abelian group while $(\mathbb{Q}^{\times}, +)$ is not a group because it is not closed ($-1 + 1 = 0 \notin \mathbb{Q}^{\times}$), and hence the <u>group</u> $\mathbb{Q}^{\times}$ must mean $(\mathbb{Q}^{\times}, \cdot)$. Similarly, $\mathbb{C}$, $\mathbb{R}$, and $\mathbb{Z}$ are all Abelian groups with respect to addition, while $\mathbb{C}^{\times}$ and $\mathbb{R}^{\times}$ are both Abelian groups with respect to multiplication and $\mathbb{Z}^{\times}$ is not a group with respect to either operation. In summary: WHEN WE REFER TO THE GROUPS $\mathbb{Z}$, $\mathbb{Q}$, $\mathbb{R}$, AND $\mathbb{C}$, THE GROUP OPERATION WE HAVE IN MIND MUST BE $+$, AND WHEN WE REFER TO THE GROUPS $\mathbb{Q}^{\times}$, $\mathbb{R}^{\times}$, AND $\mathbb{C}^{\times}$, IT MUST BE ·.

In some of the previous examples, the group operation is addition $(+)$ and in others it is multiplication $(\cdot)$. In accordance with the convention for rings, we usually write multiplication $x \cdot y$ as juxtaposition xy, while addition $x + y$ is always written as $x + y$. Note the following convention: AN ARBITRARY (POSSIBLY NONCOMMUTATIVE) GROUP IS USUALLY WRITTEN MULTIPLICATIVELY WITH IDENTITY e, WHILE AN ARBITRARY ABELIAN GROUP IS USUALLY WRITTEN ADDITIVELY WITH IDENTITY 0.

Some other examples of groups are the following.

Example 15.3 For any positive integer n, recall that we can form the set $\mathbb{Z}_n$ of integers mod n as follows (for details, see Example 4.1). For $i \in \mathbb{Z}$, let $[i]_n = \{i + kn \,|\, k \in \mathbb{Z}\}$. Then $[i]_n = [j]_n$ if and only if n divides $j - i$, and hence

$$\mathbb{Z}_n = \{[i]_n \,|\, i \in \mathbb{Z}\} = \{[0]_n, [1]_n, \ldots, [n-1]_n\}.$$

We can add and multiply in $\mathbb{Z}_n$ by using the rules

$$[i]_n + [j]_n = [i + j]_n, \qquad \text{and} \qquad [i]_n[j]_n = [ij]_n.$$

Then $\mathbb{Z}_n$ is an Abelian group with respect to addition.

Similarly, the set $M_2(\mathbb{R})$ of two-by-two matrices with real entries is also an Abelian group with respect to addition (see Example 7.3).

In fact, in general, if R is a ring, then $(R, +)$ is always an Abelian group. ❖

Example 15.4 It is not difficult to check that for any integer n, the set $(n) = \{kn \,|\, k \in \mathbb{Z}\}$ of all multiples of n is an Abelian group with respect to addition.

In general, if I is an ideal of a commutative ring R, then $(I, +)$ is always an Abelian group. ❖

Example 15.5 It is not difficult to check that the set $\{\zeta_n, (\zeta_n)^2, (\zeta_n)^3, \ldots, (\zeta_n)^{n-1}, 1\}$ of complex nth roots of 1 is an Abelian group with respect to multiplication of complex numbers (Exercise 1). ❖

Example 15.6 As noted above, $(\mathbb{C}^\times, \cdot)$, $(\mathbb{R}^\times, \cdot)$ and $(\mathbb{Q}^\times, \cdot)$ are all Abelian groups. If p is prime, then the same will be true for the set $\mathbb{Z}_p$ of integers mod p defined in Examples 4.1 and 15.3. That is, if $\mathbb{Z}_p^\times$ denotes the set of nonzero elements in $\mathbb{Z}_p$, then $(\mathbb{Z}_p^\times, \cdot)$ is an Abelian group (by Example 4.1 and Proposition 4.3).

In general, let F be a field and consider $(F^\times, \cdot)$, the nonzero elements of F with multiplication as the operation. Then the field axioms imply that $(F^\times, \cdot)$ is closed, associative, and commutative, contains an identity, and has inverses; so $(F^\times, \cdot)$ is always an Abelian group. ❖

Example 15.7 According to Example 4.5, with respect to multiplication, the set $\mathbb{H}^\times$ of nonzero quaternions is closed and associative, contains an identity, and has inverses; so $(\mathbb{H}^\times, \cdot)$ is a group. Since $ik = -j \neq j = ki$, $(\mathbb{H}^\times, \cdot)$ is not an Abelian group. ❖

Example 15.8   The set $M_2(\mathbb{R})$ of two-by-two matrices with real entries is of course not a group with respect to multiplication because the zero matrix has no inverse. In this case, the set of nonzero matrices also fails to be a group because it also contains matrices (the singular ones) which have no multiplicative inverses. However, if we restrict ourselves to the set $M_2(\mathbb{R})^{ns}$ of nonsingular matrices, we do have a group. That is, $(M_2(\mathbb{R})^{ns}, \cdot)$ is a group that is not an Abelian group (Exercise 20). ❖

We will give more examples in the exercises and in the next chapter. However, we first note that—as in the case of fields (Proposition 5.1) and rings (Proposition 7.4)—there are several elementary properties which hold in all groups. Some of the most useful are the following.

Proposition 15.9   *Let G be a group. Then*

> (i)   *the identity of G is unique:*
> *for any $x \in G$, if $gx = g$ for all $g \in G$, or if $xg = g$ for all $g \in G$, then $x = e$;*
>
> (ii)   *inverses are unique in G:*
> *if $a, b, c \in G$ are such that $ab = e = ca$, then $b = c$;*
> *for any $a \in G$, we use a^{-1} (or $-a$ if G is written additively) to denote the unique inverse of a;*
>
> (iii)   *for all $g \in G$, $(g^{-1})^{-1} = g$;*
> (iv)   *for all $x, y \in G$, $(xy)^{-1} = y^{-1}x^{-1}$;*
> (v)   *if $xy = xz$ for $x, y, z \in G$, then $y = z$ (left cancellation);*
> (vi)   *if $xy = zy$ for $x, y, z \in G$, then $x = z$ (right cancellation).*

Proof   (i)   Suppose that $x \in G$ and for all $g \in G$, $gx = g$. Then $e = ex = x$. An analogous argument works for the second hypothesis.

(ii)   Suppose that $a, b, c \in G$ are such that $ab = e = ca$. Then by associativity, $b = eb = (ca)b = c(ab) = ce = c$.

(iii)   We have $(g^{-1})^{-1} = (g^{-1})^{-1}e = (g^{-1})^{-1}(g^{-1}g) = ((g^{-1})^{-1}g^{-1})g = eg = g$.

(iv)   We have

$$(xy)^{-1} = (xy)^{-1}e = (xy)^{-1}(x(yy^{-1})x^{-1})$$
$$= ((xy)^{-1}(xy))y^{-1}x^{-1} = ey^{-1}x^{-1} = y^{-1}x^{-1}.$$

(v)   If $xy = xz$, then $y = x^{-1}xy = x^{-1}xz = z$.
(vi)   The proof is similar to that of (v). ∎

NOTE THAT IN NONCOMMUTATIVE GROUPS, THE ORDER IN PROPOSITION 15.9 (*iv*) IS <u>VERY</u> IMPORTANT. While it is certainly true that $(xy)^{-1} = y^{-1}x^{-1} = x^{-1}y^{-1}$ in any Abelian group, it may happen in a non-Abelian group that $(xy)^{-1} \neq x^{-1}y^{-1}$. Permutation groups, which will be introduced in the next chapter, provide many examples of such behavior (cf. in particular Example 16.3).

There are of course many examples of groups other than those given here or in the next chapter. Note in particular the quaternion group of Exercises 22 and 23 and the dihedral groups of Exercises 35 and 36. Exercises 26 and 27 show how to create even more groups by forming products of known groups. We will investigate subgroups in Chapter 18, and yet another method of creating groups will be given in Chapter 23.

In principle, the group operation of a finite group can always be completely described by using a multiplication table. In practice, of course, such an approach is useful only for groups with a sufficiently small number of elements. The following examples illustrate the construction and use of such tables.

Example 15.10   By Example 15.3, $\mathbb{Z}_4$ is an additive Abelian group with respect to $+$. Its addition table is the following.

$+$	$[0]_4$	$[1]_4$	$[2]_4$	$[3]_4$
$[0]_4$	$[0]_4$	$[1]_4$	$[2]_4$	$[3]_4$
$[1]_4$	$[1]_4$	$[2]_4$	$[3]_4$	$[0]_4$
$[2]_4$	$[2]_4$	$[3]_4$	$[0]_4$	$[1]_4$
$[3]_4$	$[3]_4$	$[0]_4$	$[1]_4$	$[2]_4$

Certain properties of the group correspond in a very direct fashion to properties of the table. In particular, note the following:

(1)   Since $[0]_4$ is the identity, the column below it is identical to the column under the $+$; for the same reason, the row to the right of $[0]_4$ is identical to the row to the right of the $+$.

(2)   Since $+$ is commutative, the table is symmetric about its diagonal.

(3)   The existence of inverses means that the identity occurs symmetrically in each row and each column.

(4)   In the body of the table, each row contains each element exactly once, as does each column. (The corresponding general property of groups is given in Exercises 33 and 34.)

(5)   When a row and a column intersect in the identity, the elements determining the row and the column are an inverse pair.

Observation (5) allows the inverse of any element to be found by reading the table backward. For instance, to find the inverse of $[1]_4$, find the identity $[0]_4$ in the row to the right of $[1]_4$ and then go to the top of the column containing the $[0]_4$. The element $[3]_4$, at the top of this column, is the inverse of $[1]_4$. ❖

Example 15.11   Let G be the set $\{a, b, c, d\}$. Define an operation on G by using the following table.

·	a	b	c	d
a	a	b	c	d
b	b	a	d	c
c	c	d	a	d
d	d	c	d	a

Since the entries in the body of the table all are drawn from G, · is closed. Since the table is symmetric about its diagonal, · is commutative. Since the column below a is identical to the column below · and the row to the right of a is identical to the row to the right of ·, a is an identity for ·. And since a occurs symmetrically in each row and each column, each element has an inverse. However, since the column below c does not contain b, the operation determined by the table cannot be a group operation (cf. Exercise 34). ❖

Example 15.12   Let G be the set $\{a, b, c, d\}$. Define an operation on G by using the following table.

·	a	b	c	d
a	a	b	c	d
b	b	a	d	c
c	c	d	a	b
d	d	c	b	a

By the same reasoning as in Example 15.10, we can conclude by inspecting the table that · is closed and commutative and has a as an identity and that each element has an inverse. Thus to show that $(G, ·)$ is a group, it remains only to show that · is associative. Since this is not readily apparent from the table, it must be checked in some other way, the most direct being to check individually the validity of all the equations $x(yz) = (xy)z$. We leave it to the reader to do this. (See Exercise 25.) Note that if any of x, y, z is the identity a, then the equation certainly holds and thus it suffices to check only those equations that do not involve a. Another approach is given in Exercise 29. ❖

Exercises

1. Show that the set $\{\zeta_n, (\zeta_n)^2, (\zeta_n)^3, \ldots, (\zeta_n)^{n-1}, 1\}$ of complex nth roots of 1 is an Abelian group with respect to multiplication.

In Exercises 2–18, determine whether the given set is a group. In each case, give a proof or counterexample, whichever is appropriate.

2. $(\mathbb{Z}, -)$

3. $((9), -)$

4. $((9), \cdot)$

5. $\{q \in \mathbb{Q} \mid q > 0\}$ with respect to addition

6. $\{q \in \mathbb{Q} \mid q > 0\}$ with respect to multiplication

7. $\{r \in \mathbb{R} \mid |r| \leq 1\}$ with respect to multiplication

8. $\{r \in \mathbb{R} \mid |r| = 1\}$ with respect to multiplication

9. $\{z \in \mathbb{C} \mid |z| = 1\}$ with respect to multiplication

10. $\{[i]_7 \in \mathbb{Z}_7 \mid [i]_7 \neq [7]_7\}$ with respect to multiplication

11. $\{[i]_{12} \in \mathbb{Z}_{12} \mid [i]_{12} \neq [0]_{12}\}$ with respect to multiplication

12. $\{f: \mathbb{R} \to \mathbb{R} \mid f$ is one-to-one and onto$\}$ with respect to composition

13. $\{f: \mathbb{R} \to \mathbb{R} \mid f(x) = kx$ for some $0 < k \in \mathbb{R}\}$ with respect to composition

14. $\{f: \mathbb{R} \to \mathbb{R} \mid f(x + y) = f(x) + f(y)\}$ with respect to composition

15. $\{f: \mathbb{C} \to \mathbb{C} \mid f$ is one-to-one and onto$\}$ with respect to composition

16. $\{f: \mathbb{Q}(\sqrt{3}) \to \mathbb{Q}(\sqrt{3}) \mid f$ is one-to-one and onto and $f(x + y) = f(x) + f(y)\}$ with respect to composition

17. $\{f: \mathbb{Q}(\sqrt{3}) \to \mathbb{Q}(\sqrt{3}) \mid f$ is one-to-one and onto and $f(xy) = f(x)f(y)\}$ with respect to composition

18. $\{f: \mathbb{Q}(\sqrt[3]{5}, \zeta_3) \to \mathbb{Q}(\sqrt[3]{5}, \zeta_3) \mid f$ is one-to-one and onto, $f(x + y) = f(x) + f(y)$, and $f(xy) = f(x)f(y)\}$ with respect to composition

19. Complete the proof of Proposition 15.1 by showing in detail that the identity function $\iota: K \to K$, $\iota(k) = k$, is in Gal K/F.

20. Show that the set $M_2(\mathbb{R})^{ns}$ of nonsingular matrices is a noncommutative group with respect to multiplication of matrices.

21. Prove that if $xx = x$ in a group, then $x = e$.

Exercises 22 and 23 concern the subset $Q = \{\pm 1, \pm i, \pm j, \pm k\}$ of the quaternions $\mathbb{H}$ (Example 4.5). (This group will appear regularly in subsequent exercise sets.)

22. Construct the multiplication table of $Q = \{\pm 1, \pm i, \pm j, \pm k\}$.

23. Show that $Q = \{\pm 1, \pm i, \pm j, \pm k\}$ is a noncommutative group with respect to multiplication.

24. Show that the set of all differentiable functions $f: \mathbb{R} \to \mathbb{R}$ is an Abelian group with respect to addition of functions (cf. Exercise 3 in Chapter 7).

25. Show that the operation determined by the multiplication table given in Example 15.12 is associative.

Exercises 26–29 concern the following construction (which will occur regularly in subsequent exercise sets). For groups G and H, the set $G \times H$ ("G cross H") is defined to be

$$G \times H = \{(g, h) \mid g \in G, h \in H\}.$$

Define a binary operation on $G \times H$ by coordinatewise multiplication:

$$(a, b)(x, y) = (ax, by).$$

26. Show that $G \times H$ is a group.

27. Show that if G and H are Abelian groups, then so is $G \times H$.

28. Construct the addition table of $\mathbb{Z}_2 \times \mathbb{Z}_2$.

29. Use Exercise 27 to show that the operation defined in Example 15.12 is a group operation.

In Exercises 30–32, determine whether the set $S = \{a, b, c, d, e, f\}$ is a group with respect to the operation determined by the given table. Remember to justify your answer.

30.

·	a	b	c	d	e	f
a	a	b	c	d	e	f
b	b	c	d	e	f	a
c	c	d	e	f	a	b
d	d	e	f	a	b	c
e	e	f	a	b	c	d
f	f	a	b	c	d	e

31.

·	a	b	c	d	e	f
a	a	b	c	d	e	f
b	b	d	f	a	c	e
c	c	f	d	e	a	b
d	d	a	e	c	d	a
e	e	c	a	d	b	f
f	f	e	b	a	f	c

32.

·	*a*	*b*	*c*	*d*	*e*	*f*
a	*a*	*b*	*c*	*d*	*e*	*f*
b	*b*	*c*	*a*	*e*	*f*	*d*
c	*c*	*a*	*b*	*f*	*d*	*e*
d	*d*	*e*	*f*	*a*	*b*	*c*
e	*e*	*f*	*d*	*b*	*c*	*a*
f	*f*	*d*	*e*	*c*	*a*	*b*

33. If G is a group and $x, y \in G$, show that there exists a unique $z \in G$ such that $xz = y$.

34. Explain how Exercise 33 shows that each row and each column of the multiplication table of a group contains each element of the group exactly once.

Exercises 35 and 36 concern the following construction. Any of the rigid motions of the plane defined in Figure 15.1 map a square centered at the origin to itself:

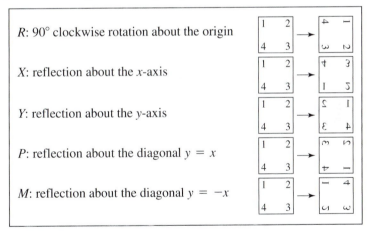

R: 90° clockwise rotation about the origin

X: reflection about the x-axis

Y: reflection about the y-axis

P: reflection about the diagonal $y = x$

M: reflection about the diagonal $y = -x$

FIGURE 15.1

Multiply any two of these motions by following one with the other. For example, RY denotes a reflection about the y-axis followed by a 90° clockwise rotation, and X^2 denotes one reflection about the x-axis followed by another. Some combinations of these motions are the same as others. For example, RY is the same as YXM or RPR or P itself.

35. Describe all possible products of the motions defined above. (There are eight distinct products.)

36. Show that the set of all possible products found in Exercise 35 is a group with respect to the given multiplication. This group is called the <u>dihedral group</u> for $n = 4$ and is denoted D_4; its identity is usually denoted by I.

37. Use the method of Exercises 35 and 36 and an <u>equilateral triangle</u> centered at the origin to construct the dihedral group for $n = 3$ (denoted by D_3).

38. Generalize Exercises 26 and 27 to a finite number of groups $G_1, \ldots, G_n$.

39. Prove that any set G with a binary operation $\lozenge$ satisfying the given postulates is a group.

    (*i*)   If $x, y \in G$, then $x \lozenge y \in G$.
    (*ii*) For all $x, y, z \in G$, $(x \lozenge y) \lozenge z = x \lozenge (y \lozenge z)$.
    (*iii*) There exists $e \in G$ such that $e \lozenge x = x$ for all $x \in G$.
    (*iv*) For all $x \in G$, there exists $y \in G$ such that $y \lozenge x = e$.

40. Prove that any set G with a binary operation $\lozenge$ satisfying the given postulates is a group.

    (*i*)   If $x, y \in G$, then $x \lozenge y \in G$.
    (*ii*) For all $x, y, z \in G$, $(x \lozenge y) \lozenge z = x \lozenge (y \lozenge z)$.
    (*iii*) For all $a, b \in G$, the equations $x \lozenge a = b$ and $a \lozenge y = b$ both have solutions x and y in G.

41. Prove that the following postulates describe an Abelian group.

    (*i*)   If $x, y \in G$, then $x \cdot y \in G$.
    (*ii*) For all $x, y, z \in G$, $(x \cdot y) \cdot z = x \cdot (z \cdot y)$.
    (*iii*) There exists $e \in G$ such that $e \cdot x = x$ for all $x \in G$.
    (*iv*) For all $x \in G$, there exists $y \in G$ such that $y \cdot x = e$.

42. Prove that any <u>finite</u> set G with an operation $*$ satisfying the following postulates is a group.

    (*i*)   If $x, y \in G$, then $x * y \in G$.
    (*ii*) For all $x, y, z \in G$, $(x * y) * z = x * (y * z)$.
    (*iii*) If $a, b, c \in G$ are such that $a * b = a * c$, then $b = c$.
    (*iv*) If $a, b, c \in G$ are such that $b * a = c * a$, then $b = c$.

HISTORICAL NOTE

Walther von Dyck

Walther Franz Anton von Dyck was born in Munich, Germany, on December 6, 1856, and died in Munich on November 5, 1934. His father was a painter and the director of the Munich Kunstgewerbeschule; his mother was Marie Royko. Walther began his studies in Munich and then went to Berlin and Leipzig, where he became a lecturer and assistant to Felix Klein in 1882. He returned to Munich in 1884 to become a professor at the Munich Polytechnikum, where he remained for the rest of his career. He married Auguste Müller in 1886, with whom he had two daughters.

He was strongly influenced by Klein, Cayley, and Hamilton. His 1882 paper "Gruppentheoretische Studien" began with a quote from Cayley: "A group is defined by means of the laws of combination of its symbols."[1] In this paper he introduced abstract groups in full generality. In his words, his intention was "to define a group of discrete operations, which are applied to a certain object, while abstracting from any special form of representation of the single objects and supposing the operations to be given only by those properties that are essential for the formation of the group."[2]

Von Dyck devoted himself to the Polytechnikum and to his home city of Munich as well as to mathematics. He was appointed director of the Polytechnikum in 1900 and oversaw its rise to university status as a Technische Hochschule. He was rector twice (1903–1906 and 1919–1925) and in this position carried out a major building expansion. He was also a leader in the establishment and early development of Munich's Deutsches Museum, and as a member of the Bayerische Akademie der Wissenschaften, he organized the publication of the works of Kepler.

References

1. van der Waerden, B. L. *A History of Algebra.* New York: Springer-Verlag, 1985, p. 152.
2. *Ibid.*, p. 152.

16

Permutation Groups

We began the previous chapter by showing that Gal $F^{p(x)}/F$ is a group with respect to composition of functions. In this chapter, we will introduce another large and important class of groups which also arises by using composition. These groups will consist of functions called permutations.

> **DEFINITION**
>
> A **permutation** of a set S is a function $f: S \to S$ which is both one-to-one and onto.

For example, the functions in Gal $F^{p(x)}/F$ are permutations of $F^{p(x)}$. Indeed we can view these functions more simply as permutations of the sets of roots of $p(x)$. This is why groups are so extremely useful in the study of field extensions.

Proposition 16.1 *Let F be a subfield of $\mathbb{C}$, let $p(x) \in F[x]$, and let $a_1, \cdots, a_n$ be the roots of $p(x)$ in $\mathbb{C}$. If $f \in$ Gal $F^{p(x)}/F$, let f_r denote f restricted to the set $\{a_1, \cdots, a_n\}$ of roots of $p(x)$, i.e., let $f_r: \{a_1, \cdots, a_n\} \to \{a_1, \cdots, a_n\}$ be the function $f_r(a_i) = f(a_i)$. Then f_r is a permutation of $\{a_1, \ldots, a_n\}$.*

Proof Note that by Proposition 13.9, for all $i = 1, \ldots, n$, $f_r(a_i)$ is a member of $\{a_1, \cdots, a_n\}$ so that f_r is indeed a function from $\{a_1, \cdots, a_n\}$ to itself. It remains to show that f_r is one-to-one and onto. If $f_r(a_i) = f_r(a_j)$, then $f(a_i) = f(a_j)$ and, since f is one-to-one, $a_i = a_j$. Thus, f_r is one-to-one. To see that f_r is onto, note that by Proposition 15.1, f has an inverse f^{-1} in Gal $F^{p(x)}/F$. Then for each a_i, Proposition 13.9 implies that $f^{-1}(a_i) \in \{a_1, \cdots, a_n\}$, and $f_r(f^{-1}(a_i)) = f(f^{-1}(a_i)) = a_i$. It follows that f_r is onto and hence that f_r is a permutation of $\{a_1, \cdots, a_n\}$. ∎

We can use general permutations to construct groups by observing that the set of all permutations of a given set forms a group with respect to composition. Since the resulting group is not Abelian for sets with more than two elements, this class of groups provides many examples of noncommutative groups.

Proposition 16.2 *Let S be a set. The set of permutations of S is a group with respect to composition of functions. If S has at least three elements, then the group of permutations is not Abelian.*

Proof The proof that the permutations of S form a group is similar to, but simpler than, the proof of Proposition 15.1 and is therefore relegated to Exercise 24. So suppose that a, b, and c are distinct elements of S. Then it is easy to see that the functions of $\sigma : S \to S$ and $\tau : S \to S$ defined by letting

$$\sigma(x) = \begin{cases} b, & \text{if } x = a, \\ c, & \text{if } x = b, \\ a, & \text{if } x = b, \\ x, & \text{otherwise}, \end{cases} \quad \text{and} \quad \tau(x) = \begin{cases} b, & \text{if } x = a, \\ a, & \text{if } x = b, \\ x, & \text{otherwise}, \end{cases}$$

are permutations of S. But $\sigma \circ \tau(a) = \sigma(b) = c$, while $\tau \circ \sigma(a) = \tau(b) = a$, and thus $\sigma \circ \tau \neq \tau \circ \sigma$ and hence the group of permutations of S is not Abelian. ∎

For any positive integer n, we let S_n denote the set of permutations of a set with n elements. For notational convenience, we usually use the set $\{1, \ldots, n\}$. By Proposition 16.2, for $n \geq 3$, S_n is a noncommutative group. For small n, the elements of this group may be expressed in a variety of ways, several of which are described in the following example.

Example 16.3 Consider S_4. Each element of S_4 is a function from $\{1, 2, 3, 4\}$ to $\{1, 2, 3, 4\}$ and thus will be completely determined by its action on the elements of the domain $\{1, 2, 3, 4\}$. This action may be pictured by using lines to show where the function takes each element of the domain. (See Figure 16.1.)

FIGURE 16.1

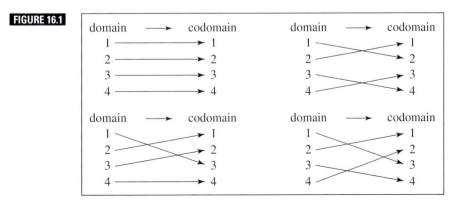

This notation is quite cumbersome, even for S_4; for larger values of n, it is completely impractical. One way to simplify it is to use 2×4 matrices:

$$\begin{pmatrix} 1 & 2 & 3 & 4 \\ \sigma(1) & \sigma(2) & \sigma(3) & \sigma(4) \end{pmatrix}.$$

In this notation, the elements of S_4 described in Figure 16.1 with arrows are written:

$$\begin{pmatrix} 1 & 2 & 3 & 4 \\ 1 & 2 & 3 & 4 \end{pmatrix} \quad \begin{pmatrix} 1 & 2 & 3 & 4 \\ 2 & 1 & 4 & 3 \end{pmatrix}.$$

$$\begin{pmatrix} 1 & 2 & 3 & 4 \\ 3 & 1 & 2 & 4 \end{pmatrix} \quad \begin{pmatrix} 1 & 2 & 3 & 4 \\ 3 & 1 & 4 & 2 \end{pmatrix}$$

It is easy to list all the elements of S_4 in this notation. In the first row of the following list, we place all permutations that leave 1 fixed, the first two permutations being those that also fix 2, the second pair being those that take 2 to 3, and the third pair being those that take 2 to 4. If 1 is fixed, then 2 can only go to itself or to 3 or 4, and hence these six permutations are the only ones that fix 1. We continue by listing, in the same fashion, all the permutations that take 1 to 2 in the second row, all those that take 1 to 3 in the third row, and all those that take 1 to 4 in the fourth row. In this way, we can systematically list all the elements of S_4:

$$\begin{pmatrix} 1 & 2 & 3 & 4 \\ 1 & 2 & 3 & 4 \end{pmatrix}, \begin{pmatrix} 1 & 2 & 3 & 4 \\ 1 & 2 & 4 & 3 \end{pmatrix}, \begin{pmatrix} 1 & 2 & 3 & 4 \\ 1 & 3 & 2 & 4 \end{pmatrix}, \begin{pmatrix} 1 & 2 & 3 & 4 \\ 1 & 3 & 4 & 2 \end{pmatrix}, \begin{pmatrix} 1 & 2 & 3 & 4 \\ 1 & 4 & 2 & 3 \end{pmatrix}, \begin{pmatrix} 1 & 2 & 3 & 4 \\ 1 & 4 & 3 & 2 \end{pmatrix},$$

$$\begin{pmatrix} 1 & 2 & 3 & 4 \\ 2 & 1 & 3 & 4 \end{pmatrix}, \begin{pmatrix} 1 & 2 & 3 & 4 \\ 2 & 1 & 4 & 3 \end{pmatrix}, \begin{pmatrix} 1 & 2 & 3 & 4 \\ 2 & 3 & 1 & 4 \end{pmatrix}, \begin{pmatrix} 1 & 2 & 3 & 4 \\ 2 & 3 & 4 & 1 \end{pmatrix}, \begin{pmatrix} 1 & 2 & 3 & 4 \\ 2 & 4 & 1 & 3 \end{pmatrix}, \begin{pmatrix} 1 & 2 & 3 & 4 \\ 2 & 4 & 3 & 1 \end{pmatrix},$$

$$\begin{pmatrix} 1 & 2 & 3 & 4 \\ 3 & 1 & 2 & 4 \end{pmatrix}, \begin{pmatrix} 1 & 2 & 3 & 4 \\ 3 & 1 & 4 & 2 \end{pmatrix}, \begin{pmatrix} 1 & 2 & 3 & 4 \\ 3 & 2 & 1 & 4 \end{pmatrix}, \begin{pmatrix} 1 & 2 & 3 & 4 \\ 3 & 2 & 4 & 1 \end{pmatrix}, \begin{pmatrix} 1 & 2 & 3 & 4 \\ 3 & 4 & 1 & 2 \end{pmatrix}, \begin{pmatrix} 1 & 2 & 3 & 4 \\ 3 & 4 & 2 & 1 \end{pmatrix},$$

$$\begin{pmatrix} 1 & 2 & 3 & 4 \\ 4 & 1 & 2 & 3 \end{pmatrix}, \begin{pmatrix} 1 & 2 & 3 & 4 \\ 4 & 1 & 3 & 2 \end{pmatrix}, \begin{pmatrix} 1 & 2 & 3 & 4 \\ 4 & 2 & 1 & 3 \end{pmatrix}, \begin{pmatrix} 1 & 2 & 3 & 4 \\ 4 & 2 & 3 & 1 \end{pmatrix}, \begin{pmatrix} 1 & 2 & 3 & 4 \\ 4 & 3 & 1 & 2 \end{pmatrix}, \begin{pmatrix} 1 & 2 & 3 & 4 \\ 4 & 3 & 2 & 1 \end{pmatrix}.$$

While useful for small values of n, this notation is still too cumbersome for large values, even if they are small in an absolute sense. We will therefore adopt what is known as <u>cycle notation</u>. In this notation, the identity permutation is denoted by $\iota : \iota(a) = a$ for all $a \in \{1, \ldots, n\}$. Any other permutation σ moves some elements of $\{1, \ldots, n\}$ but may leave others fixed. The elements that are fixed are ignored and the elements that are moved are written in a series of lists, the function moving one number in a particular list to the next number in the same list with the understanding that the last number moves to the first number (it is from this property that the notion gets its name—cf. Figure 16.2). Each list is called a k-cycle, k being the number of elements in the list. For instance, in cycle notation, the four elements we described first with lines and then with matrices are written as follows; the pictures are included to illustrate how the cycles should be interpreted.

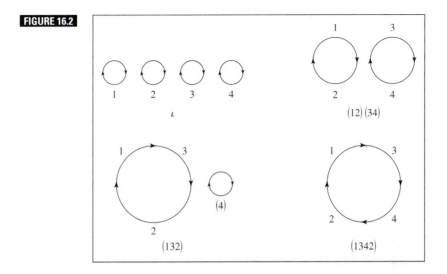

FIGURE 16.2

The elements of S_4, which we wrote above as matrices, then become the following in cycle notation.

ι,   (34),   (23),   (234),   (243),   (24),
(12),   (12)(34),   (123),   (1234),   (1243),   (124),
(132),   (1342),   (13),   (134),   (13)(24), (1324),
(1432),   (142),   (143),   (14),   (1423),   (14)(23).

Note that there are several ways of writing a cycle, e.g., $(123) = (231) = (312)$. A way to remove this ambiguity is to write each cycle with the smallest number at the beginning. We will adopt this convention whenever possible.

Whatever notation is used, permutations are functions and are multiplied by using composition of functions. Hence to multiply two permutations in cycle notation, start on the right, feed in each number of the underlying set, and follow it through each cycle. For example, if $\sigma = (1342)$ and $\tau = (13)$, then $(1342)(13)$ will be the permutation $\sigma \circ \tau$ written in cycle notation. The action of $(1342)(13)$ on $\{1, 2, 3, 4\}$ may therefore be determined as follows:

τ takes 1 to 3 and then σ takes 3 to 4, i.e., $\sigma \circ \tau(1) = \sigma(\tau(1)) = \sigma(3) = 4$;
τ takes 4 to 4 and then σ takes 4 to 2, i.e., $\sigma \circ \tau(4) = \sigma(\tau(4)) = \sigma(4) = 2$;
τ takes 2 to 2 and then σ takes 2 to 1, i.e., $\sigma \circ \tau(2) = \sigma(\tau(2)) = \sigma(2) = 1$;
τ takes 3 to 1 and then σ takes 1 to 3, i.e., $\sigma \circ \tau(3) = \sigma(\tau(3)) = \sigma(1) = 3$.

So $(1342)(13) = (142)$. Similarly, $(13)(1342) = (234)$ and $(1234)(13) = (14)(23)$.

This method of computation makes inverses of permutations written in cycle notation easy to determine; the cycles of the inverse permutation are those of the original permutation written backward. For example, it is easy to see that $(124)(421) = i$ and hence that $(124)^{-1} = (421)$. Since we conventionally start cycles with the smallest number, we preferably write $(124)^{-1} = (142)$. Similarly, $(14)^{-1} = (14)$ and $(1324)^{-1} = (1423)$.

Note that according to Proposition 16.2, not all permutations in S_4 commute. For instance, $(123)(143) = (14)(23) \neq (12)(34) = (143)(123)$, and $(23)(134) = (1234) \neq (1324) = (134)(23)$.

So S_4 is not commutative and thus provides an example of how important the order of the factors are in Proposition 15.9 (*iv*). For example, in S_4,

$$((134)(24))^{-1} = (1243) = (24)(143) = (24)^{-1}(134)^{-1},$$

while

$$((134)(24))^{-1} = (1243) \neq (1423) = (134)^{-1}(24)^{-1}.$$

Note that this does not <u>always</u> happen. For instance, it is not difficult to check that $(34)(12) = (12)(34)$ so that $[(12)(34)]^{-1} = (34)(12) = (12)(34)$. ❖

Example 16.4   We can easily list all the elements of S_3 in cycle notation as follows. The only permutations in S_3 that fix a number are the identity and the 2-cycles, while the permutations that move all three numbers are 3-cycles. So

$$S_3 = \{\iota, (12), (13), (23), (123), (132)\}.$$

Note that, in agreement with Proposition 16.2, S_3 is not Abelian: $(12)(23) = (123) \neq (132) = (23)(12)$ and $(12)(123) = (23) \neq (13) = (123)(12)$. ❖

Similar computations show that the noncommutativity exhibited at the end of Examples 16.3 and 16.4 is not unusual, that in fact permutations rarely commute. In spite of this, some permutations do commute. For example, it is easy to check that (135) and (2467) commute in S_7: $(135)(2467) = (2467)(135)$. Why do these particular permutations commute? It is because they don't interfere with each other, i.e., because they move entirely different sets of elements. The next result (Proposition 16.5) shows that this behavior is typical.

DEFINITION

Two cycles $(a_1 a_2 \cdots a_m)$ and $(b_1 b_2 \cdots b_k)$ in S_n are **disjoint** if $\{a_1, a_2, \cdots, a_m\} \cap \{b_1, b_2, \cdots, b_k\} = \varnothing$.

Note that the definition of cycle notation implies that every permutation can be written as a product of disjoint cycles (cf. Exercise 32).

Proposition 16.5   *Disjoint cycles in S_n commute.*

Proof   Let $\sigma = (a_1 a_2 \cdots a_m)$ and $\tau = (b_1 b_2 \cdots b_k)$ be disjoint cycles in S_n. If $x \neq a_i$ for any i and $x \neq b_j$ for any j, then $\sigma(x) = x$ and $\tau(x) = x$ and hence

$$\sigma \circ \tau(x) = x = \tau \circ \sigma(x).$$

If $x = a_i$ for some i, then $\sigma(x) = a_j$ for some j. However, since neither a_i nor a_j is in $\{b_1, b_2, \cdots, b_k\}$, $\tau(x) = x$ and $\tau(\sigma(x)) = \sigma(x)$. Thus

$$\sigma \circ \tau(x) = \sigma(\tau(x)) = \sigma(x) = \tau(\sigma(x)) = \tau \circ \sigma(x).$$

Similarly, if $x = b_j$ for some j, then $\sigma \circ \tau(x) = \tau(x) = \tau \circ \sigma(x)$. We conclude that $\sigma \circ \tau(x) = \tau \circ \sigma(x)$ for all x and hence that σ and τ commute. ■

Finally, we note that it is easy to count the number of elements in S_n.

Proposition 16.6   *The group S_n has $n!$ elements.*

Proof   The elements of S_n permute the set $\{1, \ldots, n\}$. We may construct all these permutations σ as follows. There are n choices for $\sigma(1)$. Once this choice has been made, there are $n - 1$ remaining choices for $\sigma(2)$. Then there are $n - 2$ remaining choices for $\sigma(3)$, and so forth. This process results in $n \cdot (n - 1) \cdot (n - 2) \cdots 2 \cdot 1 = n!$ different permutations σ. ■

Exercises

In Exercises 1–6, write the given product in S_6 as a product of disjoint cycles.

1. $(135)(2536)$ **2.** $(152)(2456)(162)$ **3.** $(13642)(15246)(25)$

4. $(256)(346)(16234)$ **5.** $(3456)(2345)(1234)$ **6.** $(12)(364)(352)(16523)$

In Exercises 7–12, find the inverse of the given element of S_5. Write the inverse as a product of disjoint cycles.

7. (12345) **8.** $(245)(13)$ **9.** (135)

10. $(12)(34)$ **11.** $(12)(345)$ **12.** $(134)(1542)$

In Exercises 13–23, write the given product in S_6 as a product of disjoint cycles.

13. $(135)(2536)^{-1}$ **14.** $(152)^{-1}(2456)(162)^{-1}$

15. $(134)^{-1}(26)(56)(1345)^{-1}$ **16.** $(2634)(234)^{-1}(162)^{-1}$

17. $(15234)^{-1}(12)^{-1}(3645)^{-1}$ **18.** $(13642)(15246)^{-1}(25)$

19. $[(256)(346)]^{-1}(16234)^{-1}$ **20.** $[(3654)(1436)]^{-1}[(16)(23)(14)]^{-1}$

21. $(156)[(346)(12356)]^{-1}(16234)^{-1}$

22. $(254)[(14326)(1245)]^{-1}[(26534)(12)(145)]^{-1}$

23. $(12)(34)[(2435)(456)(234)]^{-1}[(56)(12)(145)]^{-1}(123)^{-1}(234)$

24. Prove the first part of Proposition 16.2 in detail, that the set of permutations of a set S is a group with respect to composition of functions.

25. Find two distinct permutations in S_3 which commute but which are not disjoint and neither of which is the identity.

26. Find two distinct permutations in S_4 which commute but which are not disjoint and neither of which is the identity.

27. Let G be the group of all permutations of $\mathbb{C}$. Find two elements of G which do not commute.

28. Show that every cycle can be written as a product of 2-cycles.

29. Show that every permutation in S_n can be written as a product of 2-cycles.

30. Show that if $\sigma \in S_n$ can be written as a product of an even number of 2-cycles, then σ can be written as a product of 3-cycles.

31. Find an integer $n \geq 3$ and a permutation $\sigma \in S_n$ such that σ cannot be written as a product of 3-cycles.

32. We have observed that any permutation can be written as a product of disjoint cycles. Show that this decomposition is unique in the sense that any decomposition of a permutation as a product of a finite number of disjoint cycles always involves the same cycles.

33. Show the set $S = \{a, b, c, d, e, f\}$ is a group with respect to the operation determined by the table. Is S an Abelian group?

·	a	b	c	d	e	f
a	a	b	c	d	e	f
b	b	a	f	e	d	c
c	c	e	a	f	b	d
d	d	f	e	a	c	b
e	e	c	d	b	f	a
f	f	d	b	c	a	e

34. Find all the permutations in S_5.

35. For $\sigma \in S_n$, let $\text{Ch}(\sigma)$ denote the number of elements of $\{1, \ldots, n\}$ which are left fixed by σ. So, for example, if $\sigma = (124) \in S_5$, then $\text{Ch}(\sigma) = 2$. If $\sigma, \tau \in S_n$, show that $\text{Ch}(\sigma\tau\sigma^{-1}) = \text{Ch}(\tau)$.

36. Prove that there is no permutation $\sigma \in S_6$ such that $\sigma(135)\sigma^{-1} = (15)(23)$. (*Hint:* Exercise 35.)

17

Group Homomorphisms

According to Proposition 16.1, restricting a function in Gal $F^{p(x)}/F$ to the set of roots of $p(x)$ creates a permutation of the set of roots. This association is based on the intimate relation between functions in Gal $F^{p(x)}/F$ and the roots of $p(x)$ and allows knowledge of the permutation group of the roots to be applied to Gal $F^{p(x)}/F$. In this chapter, we want to make this association precise. Specifically, we will show that this association defines a one-to-one operation-preserving function from Gal $F^{p(x)}/F$ to the permutation group of the roots (Proposition 17.5).

We begin by adopting the following definitions. ("One-to-one" and "onto" are discussed in detail at the beginning of Chapter 13.)

DEFINITION

A function f from a set S to a set T is **one-to-one** if $x \neq y$ in S implies that $f(x) \neq f(y)$ in Y, and **onto** if for all $t \in T$, there exists $s \in S$ such that $f(s) = t$. If G and H are groups, then a function $f: G \rightarrow H$ is a **homomorphism** if for all $x, y \in G, f(xy) = f(x)f(y)$. An **isomorphism** is a homomorphism which is both one-to-one and onto. The groups G and H are said to be **isomorphic** if there exists an isomorphism $f: G \rightarrow H$.

The equation $f(xy) = f(x)f(y)$ in the definition of a homomorphism is frequently described by saying that the function f "preserves the group operation" or "preserves multiplication" or (when the operation is written additively) "preserves addition."

Note that we have already used the above terms in connection with rings and fields. If the meaning is not clear from the context, usually the term "group homomorphism" or "ring homomorphism" or "field homomorphism" is used rather than just "homomorphism." The same prefixes may of course be added to the other terms.

Note also that by Exercise 35, if G and H are isomorphic, then H and G are also isomorphic so that the order in which G and H appear in the phrase "G and H are isomorphic" makes no difference.

Example 17.1   By Example 15.3 $(\mathbb{Z}, +)$ is a group, and by Example 15.4 $((5), +)$ is a group. If $f : (\mathbb{Z}, +) \to ((5), +)$ is the function $f(n) = 5n$, then

$$f(n + m) = 5(n + m) = 5n + 5m = f(n) + f(m)$$

and hence f is a homomorphism. If $f(n) = f(m)$, then $5n = 5m$ and hence $n = m$. So f is one-to-one. If $y \in (5)$, then $y = 5n$ for some $n \in \mathbb{Z}$ and $f(n) = 5n = y$. So f is also onto, and hence f is an isomorphism. If, on the other hand, $g : (\mathbb{Z}, +) \to ((5), +)$ is the function $g(n) = 20n$, then a similar argument shows that g is a one-to-one homomorphism but since $g(n) \neq 5$ for any $n \in \mathbb{Z}$, g is not onto. If $h : (\mathbb{Z}, +) \to ((5), +)$ is the function $h(n) = 5n + 20$, then

$$h(1 + 2) = h(3) = 35 \neq 55 = 25 + 30 = h(1) + h(2)$$

and hence h is not even a homomorphism. ❖

Example 17.2   Let $f : (\mathbb{R}, +) \to (\mathbb{R}^{\times}, \cdot)$ be the function $f(x) = e^x$. Since

$$f(x + y) = e^{x+y} = e^x e^y = f(x)f(y),$$

f is a homomorphism. We know from calculus that f is one-to-one. However, since $f(x) = e^x > 0$ for all $x \in \mathbb{R}$, f is not onto. ❖

Example 17.3   If k is a positive integer and $f : (\mathbb{Z}, +) \to (\mathbb{Z}_k, +)$ is the function $f(n) = [n]_k$, then

$$f(n + m) = [n + m]_k = [n]_k + [m]_k = f(n) + f(m),$$

and hence f is a homomorphism. Since $f(0) = [0]_k = f(k)$ and $0 \neq k$, f is not one-to-one. If $y \in \mathbb{Z}_k$, then $y = [n]_k$ for some $n \in \mathbb{Z}$ and $f(n) = [n]_k = y$; so f is onto.

Suppose that k is even and let $g : (\mathbb{Z}, +) \to (\mathbb{Z}_k, +)$ be the function $g(n) = [2n]_k$. Then

$$g(n + m) = [2(n + m)]_k = [2n + 2m]_k = [2n]_k + [2m]_k = g(n) + g(m),$$

and hence g is a homomorphism. Since $g(0) = [0]_k = g(k)$ and $0 \neq k$, g is not one-to-one. Since k is even, $2n + kj$ is even for all $n, j \in \mathbb{Z}$, and hence the set $[2n]_k$ must consist entirely of even integers. Then $g(n) \neq [1]_k$ for any $n \in \mathbb{Z}$, and hence g cannot be onto. ❖

The point of view taken in the next example has far-reaching implications (see, for instance, Exercise 42, Cayley's theorem).

Example 17.4 Consider the group $(\mathbb{Z}_4, +)$. We will construct a one-to-one homomorphism from the <u>group</u> $\mathbb{Z}_4$ into the group of permutations of the <u>set</u> $\mathbb{Z}_4$. The group of permutations of the set $\mathbb{Z}_4$ is just S_4; the nature of the elements permuted does not affect the permutations. So define a function $f_{[1]_4}\colon \mathbb{Z}_4 \to \mathbb{Z}_4$ by letting $f_{[1]_4}([k]_4) = [1]_4 + [k]_4$. If $f_{[1]_4}([k]_4) = f_{[1]_4}([j]_4)$, then $[1]_4 + [k]_4 = [1]_4 + [j]_4$ and hence $[k]_4 = [j]_4$; thus $f_{[1]_4}$ is one-to-one. Furthermore, for any $[k]_4 \in \mathbb{Z}_4$, $-[1]_4 + [k]_4 \in \mathbb{Z}_4$ and $f_{[1]_4}(-[1]_4 + [k]_4) = [k]_4$; hence $f_{[1]_4}$ is onto. We conclude that $f_{[1]_4}$ is a permutation of $\mathbb{Z}_4$. (If we identify $[0]_4 \sim 1$, $[1]_4 \sim 2$, $[2]_4 \sim 3$, and $[3]_4 \sim 4$, then $f_{[1]_4}$ is just the cycle (1234).) Similarly for $i = 0, 2, 3$, we may define permutations $f_{[i]_4}$ by letting $f_{[i]_4}([k]_4) = [i]_4 + [k]_4$. Notice what happens when we compose these permutations:

$$f_{[i]_4} \circ f_{[j]_4}([k]_4) = f_{[i]_4}([j]_4 + [k]_4) = [i]_4 + [j]_4 + [k]_4 = f_{[i]_4 + [j]_4}([k]_4).$$

In summary, with each $[i]_4 \in \mathbb{Z}_4$, we have associated a permutation $f_{[i]_4}$ in such a way that $f_{[i]_4} \circ f_{[j]_4} = f_{[i]_4 + [j]_4}$. Define $F\colon \mathbb{Z}_4 \to S_4$ by letting $F([i]_4) = f_{[i]_4}$. Then F is a homomorphism because

$$F([i]_4 + [j]_4) = f_{[i]_4 + [j]_4} = f_{[i]_4} \circ f_{[j]_4} = F([i]_4) \circ F([j]_4).$$

If $F([i]_4) = F([j]_4)$, then $f_{[i]_4} = f_{[j]_4}$ and hence

$$[i]_4 = [i]_4 + [0]_4 = f_{[i]_4}([0]_4) = f_{[j]_4}([0]_4) = [j]_4 + [0]_4 = [j]_4.$$

Therefore, F is one-to-one. Note that $\mathbb{Z}_4$ has four elements while S_4 has $4! = 24$ elements; so F cannot be onto.

In summary, we have constructed a one-to-one homomorphism F from the <u>group</u> $\mathbb{Z}_4$ into the group of permutations of the <u>set</u> $\mathbb{Z}_4$. ❖

One way of using the point of view taken in Example 17.4 is to show that restricting a function in $\mathrm{Gal}\ F^{p(x)}/F$ to the set of roots of $p(x)$ is a one-to-one homomorphism from the Galois group $\mathrm{Gal}\ F^{p(x)}/F$ to the permutation group of the roots of $p(x)$.

Proposition 17.5 *Let F be a subfield of $\mathbb{C}$, let $p(x) \in F[x]$, and let $a_1, \ldots, a_n \in \mathbb{C}$ be the distinct solutions of $p(x) = 0$. Define $T\colon \mathrm{Gal}\ F^{p(x)}/F \to S_n$ by letting $T(f)$ be f restricted to $\{a_1, \ldots, a_n\}$. Then T is a one-to-one homomorphism.*

Proof Let $f \in \mathrm{Gal}\ F^{p(x)}/F$. By Proposition 16.1, the function $f_r\colon \{a_1, \ldots, a_n\} \to \{a_1, \ldots, a_n\}$, defined by letting $f_r(a_i) = f(a_i)$, is a permutation of $\{a_1, \ldots, a_n\}$. By hypothesis, $T(f) = f_r$, and thus, since $\{a_1, \ldots, a_n\}$ has n elements, T maps the function in $\mathrm{Gal}\ F^{p(x)}/F$ to permutations in S_n. It remains to show that T is one-to-one and a homomorphism. If $f, g \in \mathrm{Gal}\ F^{p(x)}/F$, then $f_r \circ g_r(a_i) = f(g(a_i)) = (f \circ g)_r(a_i)$, hence $T(f \circ g) = (f \circ g)_r = f_r \circ g_r = T(f) \circ T(g)$, and therefore T is a homomorphism. To see that T is one-to-one, suppose that $T(f) = T(g)$. Then

$f(a_i) = f_r(a_i) = g_r(a_i) = g(a_i)$ for all i. But as in the proof of Proposition 14.8, Proposition 12.5 implies that every element of $F^{p(x)}$ can be expressed as a linear combination of products of $a_1, \ldots, a_n$, and thus, since both f and g leave elements of F unchanged, $f(w) = g(w)$ for all $w \in F^{p(x)}$, i.e., $f = g$. We conclude that T is a one-to-one homomorphism. ∎

To use the intuition gained by Proposition 17.5 to the fullest, we need a way of writing $T(f)$ in cycle notation. The following two examples show how to do this.

Example 17.6 Let $p(x) = x^3 - 7 \in \mathbb{Q}[x]$. Then Gal $\mathbb{Q}^{p(x)}/\mathbb{Q} = \{f_1, f_2, f_3, f_4, f_5, f_6\}$, where the functions f_i are determined as follows. According to Example 14.9,

$$\mathbb{Q}^{p(x)} = \mathbb{Q}(\sqrt[3]{7}, \zeta_3)$$
$$= \{(a_0 + a_1\sqrt[3]{7} + a_2(\sqrt[3]{7})^2) + (b_0 + b_1\sqrt[3]{7} + b_2(\sqrt[3]{7})^2)\zeta_3 \mid a_i, b_j \in \mathbb{Q}\},$$

and if $f \in$ Gal $\mathbb{Q}^{p(x)}/\mathbb{Q}$, then

$$f((a_0 + a_1\sqrt[3]{7} + a_2(\sqrt[3]{7})^2) + (b_0 + b_1\sqrt[3]{7} + b_2(\sqrt[3]{7})^2)\zeta_3)$$
$$= (a_0 + a_1 f(\sqrt[3]{7}) + a_2(f(\sqrt[3]{7}))^2) + (b_0 + b_1 f(\sqrt[3]{7}) + b_2(f(\sqrt[3]{7}))^2)f(\zeta_3).$$

and the possible substitutions for $f(\sqrt[3]{7})$ and $f(\zeta_3)$ in this equation are

	f_1	f_2	f_3	f_4	f_5	f_6
$\sqrt[3]{7} \rightarrow$	$\sqrt[3]{7}$	$\sqrt[3]{7}$	$\sqrt[3]{7}\zeta_3$	$\sqrt[3]{7}\zeta_3$	$\sqrt[3]{7}\zeta_3^2$	$\sqrt[3]{7}\zeta_3^2$
$\zeta_3 \rightarrow$	ζ_3	ζ_3^2	ζ_3	ζ_3^2	ζ_3	ζ_3^2

Let T be the function defined in Proposition 17.5. We want to express the permutations $T(f_i)$ in cycle notation. Since $p(x)$ has three roots, $\sqrt[3]{7}$, $\sqrt[3]{7}\zeta_3$, and $\sqrt[3]{7}\zeta_3^2$, the codomain of T is S_3. To use cycle notation, we associate $\sqrt[3]{7} \sim 1$, $\sqrt[3]{7}\zeta_3 \sim 2$, and $\sqrt[3]{7}\zeta_3^2 \sim 3$. To find $T(f_i)$, we need to determine what this permutation does to 1, 2, and 3, i.e., we need to determine what f_i does to $\sqrt[3]{7}$, $\sqrt[3]{7}\zeta_3$, and $\sqrt[3]{7}\zeta_3^2$. Since f_i preserves multiplication, the preceding table determines this action. For example,

$$f_4(\sqrt[3]{7}) = \sqrt[3]{7}\zeta_3,$$
$$f_4(\sqrt[3]{7}\zeta_3) = f_4(\sqrt[3]{7})f_4(\zeta_3) = \sqrt[3]{7}\zeta_3\zeta_3^2 = \sqrt[3]{7},$$
$$f_4(\sqrt[3]{7}\zeta_3^2) = f_4(\sqrt[3]{7})f_4(\zeta_3)^2 = \sqrt[3]{7}\zeta_3(\zeta_3^2)^2 = \sqrt[3]{7}\zeta_3^2.$$

So f_4 maps 1 to 2, 2 to 1, and 3 to 3, i.e., $T(f_4) = (12)$. In a similar fashion,

$$T(f_1) = \iota, \qquad T(f_2) = (23), \qquad T(f_3) = (123),$$
$$T(f_5) = (132), \qquad T(f_6) = (13).$$

Since permutations are much easier to manipulate than field automorphisms, this is a very advantageous way to view the Galois group Gal $\mathbb{Q}^{p(x)}/\mathbb{Q}$.

As we noted in Example 16.4, $S_3 = \{\iota, (12), (13), (23), (123), (132)\}$, and hence our preceding work shows that T is onto. ❖

Although the function T determined in Example 17.6 is onto, this need not always be the case, as the next example shows.

Example 17.7 Let $p(x) = x^4 - 1 \in \mathbb{Q}[x]$. Since $\zeta_4 = i$, Proposition 3.8 implies that the four complex roots of $p(x)$ are $\pm 1, \pm i$, and hence that $\mathbb{Q}^{p(x)} = \mathbb{Q}(\pm 1, \pm i)$. It is easy to see that $\mathbb{Q}(\pm 1, \pm i) = \mathbb{Q}(i)$. Furthermore, if $m(x) = x^2 + 1$, then $m(x) \in \mathbb{Q}[x]$, $m(x)$ is monic, $m(x) = 0$, and since the roots of $m(x)$ are $\pm i \notin \mathbb{Q}$, $m(x)$ is irreducible by Proposition 8.12, and hence $m(x)$ is the minimum polynomial of i over $\mathbb{Q}$. So by Theorem 10.8, $\mathbb{Q}^{p(x)} = \mathbb{Q}(i) = \{a_0 + a_1 i \mid a_0, a_1 \in \mathbb{Q}\}$. If $f \in \text{Gal } \mathbb{Q}^{p(x)}/\mathbb{Q}$, then $f(a_0 + a_1 i) = a_0 + a_1 f(i)$ and by Proposition 13.9, $f(i) = \pm i$. It follows that Gal $\mathbb{Q}^{p(x)}/\mathbb{Q} \subseteq \{f_1, f_2\}$, where

$$f_1(a_0 + a_1 i) = a_0 + a_1 i, \quad \text{and} \quad f_2(a_0 + a_1 i) = a_0 - a_1 i.$$

Since $[\mathbb{Q}^{p(x)}:\mathbb{Q}] = 2$, Proposition 14.8 implies that Gal $\mathbb{Q}^{p(x)}/\mathbb{Q} = \{f_1, f_2\}$. Since $p(x)$ has four distinct roots, the function T of Proposition 17.5 maps Gal $\mathbb{Q}^{p(x)}/\mathbb{Q}$ into S_4. And since S_4 has $4! = 24$ elements by Proposition 16.6, T cannot be onto.

We can describe T explicitly as follows. Associate $1 \sim 1$, $-1 \sim 2$, $i \sim 3$, and $-i \sim 4$. Since f_1 is the identity on $\mathbb{Q}^{p(x)}$, $T(f_1) = \iota$. Since $f_2(1) = 1$, $f_2(-1) = -1$, $f_2(i) = -i$, and $f_2(-i) = i$, $T(f_2) = (34)$. ❖

Exercises

In Exercises 1–13, determine whether the given function is (1) one-to-one, (2) onto, and (3) a homomorphism. In each case, give a proof or a counterexample, whichever is appropriate.

1. $f: (\mathbb{Z}, +) \to (\mathbb{R}, +), f(n) = 5n$

2. $f: (\mathbb{Z}, +) \to (\mathbb{R}, +), f(n) = 5n + 5$

3. $f: (\mathbb{Z}, +) \to (\mathbb{R}^{\times}, \cdot), f(n) = 5^n$

4. $f: (\mathbb{Z}, +) \to (\mathbb{Z}_{12}, +), f(n) = [3n]_{12}$

5. $f: (\mathbb{Z}, +) \to (M_2(\mathbb{R}), +), f(n) = \begin{pmatrix} n & 0 \\ 0 & n \end{pmatrix}$

6. $f: (\mathbb{Z}, +) \to (M_2(\mathbb{R}), +), f(n) = \begin{pmatrix} \pi n & -n \\ 0 & 3n \end{pmatrix}$

7. $f: (M_2(\mathbb{R}), +) \to (\mathbb{R}, +), f\begin{pmatrix} a & b \\ c & d \end{pmatrix} = \det\begin{pmatrix} a & b \\ c & d \end{pmatrix}$

8. $f: S_{11} \to (\mathbb{Z}, +), f(\sigma) = \sigma(1)$

9. $f: S_7 \to (\mathbb{Z}_7, +), f(\sigma) = [\sigma(1)]_7$

10. $f: S_3 \to (\mathbb{Z}_6, +), f(\sigma) = [\sigma(1)]_6$

11. $f: (\mathbb{C}^{\times}, \cdot) \to (\mathbb{C}^{\times}, \cdot), f(z) = z^3$

12. $f: (\mathbb{C}^{\times}, \cdot) \to (\mathbb{C}^{\times}, \cdot), f(r(\cos\phi + \sin\phi)) = \cos\phi + i \sin\phi$

13. $f: (\mathbb{C}^{\times}, \cdot) \to (\mathbb{R}^{\times}, \cdot), f(z) = |z|$.

14. Prove that the set $G \times G = \{(g, h) \mid g, h \in G\}$ is a group with respect to coordinatewise multiplication: $(a, b)(c, d) = (ac, bd)$.

Let G be a group and recall from Exercise 26 in Chapter 15 or Exercise 14 above that the set $G \times G = \{(g, h) \mid g, h \in G\}$ is a group with respect to coordinatewise multiplication: $(a, b)(c, d) = (ac, bd)$. In Exercises 15–18, determine whether the given function is a homomorphism. Give a proof or a counterexample, whichever is appropriate.

15. $h: G \rightarrow G \times G$, $h(x) = (x, x)$

16. $\alpha: G \rightarrow G \times G$, $\alpha(x) = (x, e)$

17. $\mu: G \times G$, $\mu(x, y) = xy$

18. $\pi_1: G \times G \rightarrow G$, $\pi_1(x, y) = x$

19. If G is an Abelian group, show that the function $f: G \rightarrow G$ defined by letting $f(g) = g^{-1}$ is an isomorphism.

20. Find a group G such that the function for which $f: G \rightarrow G$ defined by letting $f(g) = g^{-1}$ is not an isomorphism.

21. Define an isomorphism $T: (\mathbb{R}^+, \cdot) \rightarrow (\mathbb{R}, +)$, where $\mathbb{R}^+ = \{r \in \mathbb{R} \mid r > 0\}$. (Remember to prove that the function is indeed an isomorphism.)

According to Exercise 23 in Chapter 15, the set Δ, of differentiable functions $f: \mathbb{R} \rightarrow \mathbb{R}$ is a group with respect to addition of functions. Exercises 22–24 concern the function $D: \Delta \rightarrow \Delta$ defined by letting $D(f) = f'$.

22. Show that D is a homomorphism.

23. Is D one-to-one? Remember to justify your answer.

24. Is D onto? Remember to justify your answer.

In Exercises 25–28 construct a one-to-one homomorphism from the group A to the group B. Prove that the function you define is a one-to-one homomorphism.

25. $A = (\mathbb{Z}_6, +)$, $B = S_6$

26. $A = (\mathbb{Z}_{11}, +)$, $B = S_{11}$

27. $A = S_3$, $B = S_6$

28. $A = S_4$, $B = S_{24}$

Exercises 29–34 involve a polynomial $p(x)$ and a positive integer n. Use the method of Example 17.6 to construct a function $T: \text{Gal } \mathbb{Q}^{p(x)}/\mathbb{Q} \rightarrow S_n$. Explain why T is, or is not, onto.

29. $p(x) = x^2 + x + 1$, $n = 2$

30. $p(x) = x^3 - 5$, $n = 3$

31. $p(x) = x^3 - 11$, $n = 3$

32. $p(x) = x^4 - 7$, $n = 4$

33. $p(x) = x^4 - 25$, $n = 4$

34. $p(x) = x^5 - 7$, $n = 5$

35. Show that if a group G is isomorphic to a group H, then H is also isomorphic to G. That is, show that if $f: G \rightarrow H$ is an isomorphism, then the inverse function $f^{-1}: H \rightarrow G$ is also an isomorphism.

Exercises 36–38 concern the polynomial $p(x) = x^5 - 11 \in \mathbb{Q}[x]$, the real number $r = \sqrt[5]{11}$, and the group $G = \text{Gal } \mathbb{Q}^{p(x)}/\mathbb{Q}$.

36. Use the statement of Proposition 17.5 to define concisely a one-to-one homomorphism $T: G \rightarrow S_5$.

37. If $f \in G$ is such that $f(r) = r\zeta_5$ and $f(\zeta_5) = (\zeta_5)^3$, write $T(f)$ in cycle notation.

38. If $g \in G$ is such that $g(r) = r(\zeta_5)^2$ and $g(\zeta_5) = (\zeta_5)^4$, write $T(g)$ in cycle notation.

39. Suppose that G is a <u>finite</u> group and that $f: G \rightarrow G$ is a homomorphism. Show that f is one-to-one if and only if it is onto.

40. Find an example of a group G and a one-to-one homomorphism $f: G \rightarrow G$ which is not onto.

41. Find an example of a group G and an onto homomorphism $f: G \rightarrow G$ which is not one-to-one.

42. Prove Cayley's theorem: For any group $(G, \cdot)$, there exists a one-to-one homomorphism from the <u>group</u> $(G, \cdot)$ into the group of permutations of the <u>set</u> G. (*Hint:* Use the method of Example 17.4.)

HISTORICAL NOTE

Arthur Cayley

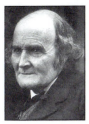

Arthur Cayley was born in Richmond, Surrey, England, on August 16, 1821, and died in Cambridge, England, on January 26, 1895. His father, Henry Cayley, was English but had settled in St. Petersburg, Russia, as a merchant. His mother, Maria Antonia Doughty, may have been Russian but was probably English. Arthur was born during a short visit to England but spent his first eight years in Russia. On returning to England in 1829, he attended a private school, whose master, observing his exceptional mathematical abilities, recommended that he go to the University of Cambridge rather than, as originally intended, into business. He did indeed enter Cambridge when he was seventeen and spent most of his undergraduate years studying mathematics. His recreational pursuits were confined to novel reading and mountain climbing, both of which he continued throughout his life. He graduated as Senior Wrangler and won the first Smith's prize, and remained at Cambridge on a fellowship. However, because the fellowship lasted for only four years, he had to choose a profession and decided on the law. He was called to the bar in 1849 and made a specialty of conveyancing. Altogether he spent fourteen years as a lawyer and during that time jealously reserved a portion of his time for mathematics. This resulted in almost three hundred papers, incorporating some of his best and most original work.

In 1863, he became the first Sadlerian professor of pure mathematics at Cambridge, giving up a lucrative legal practice for a modest salary. In exchange, of course, he obtained the time to devote himself entirely to what he wanted to do most, mathematics. He immediately married Susan Moline of Greenwich and settled in Cambridge. He published an enormous number of mathematical papers and because of his legal and administrative experience was frequently called upon to help in the administration of the university. He was editing his own mathematical papers for publication when he died on January 26, 1895.

Cayley's prodigious output is indicative of his mathematical legacy. He introduced matrices into linear algebra and championed vectors over quaternions. In his own words, "I have the highest admiration for the notion of a quaternion; but, as I consider the full moon more beautiful than any moonlit view, so I regard the notion of a quaternion as far more beautiful than any of its applications. As another illustration, I compare a quaternion formula to a pocket-map—a capital thing to put in one's pocket, but which for use must be unfolded: the formula, to be understood, must be translated into coordinates."[1] He was one of the founders of the theory of abstract groups and his eponymous theorem concerning the representation of any group as a group of permutations appeared in his 1854 paper "On the theory of groups, as depending on the symbolic equation $\Theta^n = 1$."

Reference   1. MacFarlane, Alexander. *Ten British Mathematicians,* New York: John Wiley and Sons. 1916, p. 71.

18

Subgroups

According to Proposition 17.5, if $p(x)$ has n distinct roots, then the Galois group Gal $F^{p(x)}/F$ may be embedded in S_n by means of a function that preserves the group operation. According to Example 17.7, this function need not be onto. However, since the function preserves the operation on Gal $F^{p(x)}/F$, the image of the function should be a group with respect to the operation on S_n. And indeed this will be the case.

We will show below that in general for any homomorphism $f: G \to H$, the image of G in H will always be a group with respect to the operation of H and that all the elements of G which f maps to the identity in H will also form a group with respect to the operation of G. Such subsets of groups are called subgroups.

> **DEFINITION**
>
> Let $(G, \cdot)$ be a group. A subset S of G which is a group with respect to $\cdot$ is a **subgroup** of G.

Obviously the group $(\mathbb{Z}, +)$ is a subgroup of the groups $(\mathbb{Q}, +)$, $(\mathbb{R}, +)$, and $(\mathbb{C}, +)$; and the group $(\mathbb{Q}^\times, \cdot)$ is a subgroup of the groups $(\mathbb{R}^\times, \cdot)$ and $(\mathbb{C}^\times, \cdot)$. Example 15.4 shows that for any $n \in \mathbb{Z}$, the set (n) of all multiples of n is a subgroup of $\mathbb{Z}$.

We first note that subgroups inherit commutativity.

Proposition 18.1 *A subgroup of an Abelian group is Abelian.*

Proof Suppose that S is a subgroup of an Abelian group G, and let $x, y \in S$. Then $x, y \in G$ and hence $xy = yx$. ∎

We next note that to prove that a given subset of a group is a subgroup, we do not have to check all the group axioms.

Proposition 18.2   *Let S be a subset of a group G. Then S is a subgroup of G if and only if S satisfies the following conditions:*

(i) $S \neq \varnothing$;

(ii)   *if $x, y \in S$, then $xy \in S$;*

(iii)   *if $x \in S$ and x^{-1} is the inverse of x in G, then $x^{-1} \in S$.*

Proof   The proof is similar to that of Proposition 4.2. If S is a subgroup, then it certainly is a group and hence it satisfies (i), (ii) and (iii). On the other hand, suppose that S satisfies the three conditions. Then by (i), the operation on S is closed. If $x, y, z \in S$, then $x, y, z \in G$ as well, and hence $x(yz) = (xy)z$ so that the operation is associative. By (i), S has an element x; by (ii), $x^{-1} \in S$; and so by (iii), $e = xx^{-1} \in S$. But every element x of S is also in G and e is the identity for G so that $xe = x = ex$. So e is also an identity for S. Finally, by (iii), every element in S has an inverse, and therefore S is a subgroup of G. ∎

Proposition 18.2 shows easily that every group has at least two subgroups.

Proposition 18.3   *For any group G, both $\{e\}$ and G itself are subgroups of G.*

Proof   Since $e \in \{e\}$, $\{e\} \neq \varnothing$; since $ee = e \in \{e\}$, $\{e\}$ is closed; and $e^{-1} = e \in \{e\}$. So by Proposition 18.2, $\{e\}$ is a subgroup of G. Since $e \in G$, $G \neq \varnothing$; by definition, G is closed and every element of G has an inverse in G. So by Proposition 18.2, G is a subgroup of itself. ∎

Example 18.4   Consider the subset $S = \{a + b\sqrt[4]{7} \,|\, a, b \in \mathbb{Q}\}$ of $\mathbb{R}$. Since $0 = 0 + 0\sqrt[4]{7} \in S$, $S \neq \varnothing$. If $a + b\sqrt[4]{7}, c + d\sqrt[4]{7} \in S$, then $(a + b\sqrt[4]{7}) + (c + d\sqrt[4]{7}) = (a + c) + (b + d)\sqrt[4]{7} \in S$ because $a + c$, $b + d \in \mathbb{Q}$, and $-(a + b\sqrt[4]{7}) = (-a) + (-b)\sqrt[4]{7} \in S$ because $-a, -b \in \mathbb{Q}$. So by Proposition 18.2, S is a subgroup of $(\mathbb{R}, +)$. ❖

Example 18.5   Let G be the group of permutations of $\mathbb{Z}$, and for each $n \in \mathbb{Z}$, let $f_n : \mathbb{Z} \rightarrow \mathbb{Z}$ be the function $f_n(x) = x + n$. If $f_n(x) = f_n(y)$, then $x + n = y + n$ and hence $x = y$ so that f_n is one-to-one. If $y \in \mathbb{Z}$, then $y - n \in \mathbb{Z}$ and $f_n(y - n) = y - n + n = y$ so that f_n is onto. It follows that each f_n is a permutation of $\mathbb{Z}$, and hence that $S = \{f_n \,|\, n \in \mathbb{Z}\}$ is a subset of G. We will show that S is a subgroup of G. Since $f_0 \in S$, $S \neq \varnothing$. Suppose that $f_n, f_m \in S$. Then for all $x \in \mathbb{Z}$, $f_n \circ f_m(x) = f_n(f_m(x)) = f_n(x + m) = x + m + n = f_{m+n}(x)$, and hence $f_n \circ f_m = f_{m+n} \in S$. Finally, since $f_n \circ f_{-n}(x) = x - n + n = x = x + n - n = f_{-n} \circ f_n(x)$, $f_n \circ f_{-n} = \iota = f_{-n} \circ f_n$, and thus $f_n^{-1} = f_{-n} \in S$. So by Proposition 18.2, S is a subgroup of G.

Note that since $f_n \circ f_m = f_{m+n} = f_{n+m} = f_m \circ f_n$, S is Abelian. However, it is easy to see that the function $g : \mathbb{Z} \rightarrow \mathbb{Z}$, defined by letting $g(x) = -x$, is in

G, and $f_1 \circ g(2) = f_1(g(2)) = -2 + 1 = -1 \neq -3 = g(f_1(2)) = g \circ f_1(2)$. So $f_1 \circ g \neq g \circ f_1$ and thus G is not Abelian. ❖

Determining whether a finite subset of a group is a subgroup is easier than determining whether an infinite subset is a subgroup. To prove this as well as many other results, we need the following notation. Let G be a group, let $x \in G$, and let $n \in \mathbb{Z}$. If G is written multiplicatively with identity e, then

$$\text{for } n > 0, x^n = x \cdots x \ (n \text{ times}),$$
$$\text{for } n = 0, x^n = x^0 = e, \text{ and}$$
$$\text{for } n < 0, x^n = x^{-1} \cdots x^{-1} \ (-n \text{ times}).$$

If G is written additively with identity 0, then

$$\text{for } n > 0, nx = x + \cdots + x \ (n \text{ times}),$$
$$\text{for } n = 0, nx = 0x = 0, \text{ and}$$
$$\text{for } n < 0, nx = (-x) + \cdots + (-x) \ (-n \text{ times}).$$

For use in the sequel, we isolate part of the proof in the following lemma.

Lemma 18.6   *If G is finite group and $x \in G$ is such that $\{x, x^2, x^3, \ldots\}$ is finite, then there exists a positive integer k such that $x^k = e$.*

Proof   Since $\{x, x^2, x^3, \ldots\}$ is finite, $x^n = x^m$ for some integers $m > n > 1$. Then $m - n$ is a positive integer, and $ex^n = x^n = x^m = x^{m-n} x^n$ so that by Proposition 15.9 (*vi*), $e = x^{m-n}$. ∎

Proposition 18.7   *Let G be a group and let S be a finite subset of G. Then S is a subgroup of G if and only if S satisfies.*
  (i) $S \neq \emptyset$;
  (ii)   if $x, y \in S$, then $xy \in S$.

Proof   By Proposition 18.2, this result will be proved if we can show that for any finite subset S of G satisfying (*i*) and (*ii*), $x^{-1} \in S$ whenever $x \in S$ and x^{-1} is the inverse of x in G. To this end, let $x \in S$. Since $\{x, x^2, x^3, \ldots\} \subseteq S$, $\{x, x^2, x^3, \ldots\}$ is finite, and hence by Lemma 18.6, $x^k = e$ for some positive integer k. If $k = 1$, then $e = x$ and hence $x^{-1} = e^{-1} = e = x \in S$. Otherwise $k > 1$ and hence $x^{k-1} \in S$; since $x^{k-1}x = e = xx^{k-1}$, $x^{-1} = x^{k-1} \in S$. ∎

Example 18.8   Consider the group $(\mathbb{Z}, +)$. The subset $\{0, 1, 2, 3, \ldots\}$ is nonempty and closed but is not a subgroup. Thus Proposition 18.7 may <u>not</u> be true for infinite subsets. ❖

Example 18.9 Obviously, for $n > 0$, the set $U_n = \{\zeta_n, (\zeta_n)^2, \ldots, (\zeta_n)^n\}$ of complex nth roots of 1 (cf. Example 15.5) is a nonempty subset of $\mathbb{C}^\times$. As well, if $(\zeta_n)^i, (\zeta_n)^j \in U_n$, then either $1 \leq i + j \leq n$ or $n < i + j \leq 2n$, and in the former case, $(\zeta_n)^i(\zeta_n)^j = (\zeta_n)^{i+j} \in U_n$, and in the latter case, since $\zeta_n{}^n = 1$, $(\zeta_n)^i(\zeta_n)^j = (\zeta_n)^{i+j} = (\zeta_n)^{i+j}(\zeta_n{}^n)^{-1} = (\zeta_n)^{i+j-n} \in U_n$. So U_n is multiplicatively closed. Therefore, since U_n is finite, we may apply Proposition 18.7 and conclude that U_n is a subgroup of $(\mathbb{C}^\times, \cdot)$. Note that while U_n is finite, the group $(\mathbb{C}^\times, \cdot)$ is infinite. ❖

Example 18.10 If S is a subgroup of $\mathbb{Z}_8$ and $[1]_8 \in S$, then $S = \mathbb{Z}_8$ because any element of $\mathbb{Z}_8$ may be obtained as $[1]_8$ added to itself a finite number of times. Furthermore, if $[k]_8 \in S$ for k odd, then $[k]_8 = [1]_8$, $[k]_8 = [3]_8$, $[k]_8 = [5]_8$, or $[k]_8 = [7]_8$, and since

$$[3]_8 + [3]_8 + [3]_8 = [9]_8 = [1]_8,$$
$$[5]_8 + [5]_8 + [5]_8 + [5]_8 + [5]_8 = [25]_8 = [1]_8,$$
$$[7]_8 + [7]_8 + [7]_8 + [7]_8 + [7]_8 + [7]_8 + [7]_8 = [49]_8 = [1]_8,$$

$[1]_8 \in S$ in all cases, i.e., $S = \mathbb{Z}_8$ in all cases. Since $[0]_8$ is an element of any subgroup of $\mathbb{Z}_8$, the only remaining candidates for subgroups of $\mathbb{Z}_8$ are its following subsets:

$$\{[0]_8\}, \quad \{[0]_8, [2]_8\}, \quad \{[0]_8, [2]_8, [4]_8\}, \quad \{[0]_8, [2]_8, [4]_8, [6]_8\},$$
$$\{[0]_8, [4]_8\}, \quad \{[0]_8, [4]_8, [6]_8, \quad \{[0]_8, [6]_8\}, \quad \mathbb{Z}_8.$$

But since $[2]_8 + [2]_8 + [2]_8 = [6]_8$, neither $\{[0]_8, [2]_8\}$ nor $\{[0]_8, [2]_8, [4]_8\}$ are subgroups, and since $[6]_8 + [6]_8 + [6]_8 = [2]_8$, $\{[0]_8, [6]_8\}$ is not a subgroup. By Proposition 18.3, $\{[0]_8\}$ and $\mathbb{Z}_8$ itself are subgroups of $\mathbb{Z}_8$. We will show that the remaining subsets are also subgroups. Clearly, $\{[0]_8, [2]_8, [4]_8, [6]_8\}$ is nonempty and the following table shows that it is closed with respect to addition.

$+$	$[0]_8$	$[2]_8$	$[4]_8$	$[6]_8$
$[0]_8$	$[0]_8$	$[2]_8$	$[4]_8$	$[6]_8$
$[2]_8$	$[2]_8$	$[4]_8$	$[6]_8$	$[0]_8$
$[4]_8$	$[4]_8$	$[6]_8$	$[0]_8$	$[2]_8$
$[6]_8$	$[6]_8$	$[0]_8$	$[2]_8$	$[4]_8$

Hence by Proposition 18.7, $\{[0]_8, [2]_8, [4]_8, [6]_8\}$ is a subgroup of $\mathbb{Z}_8$. Similarly, $\{[0]_8, [4]_8\}$ is a subgroup of $\mathbb{Z}_8$. We conclude that $\{[0]_8\}$, $\{[0]_8, [4]_8\}$, $\{[0]_8, [2]_8, [4]_8, [6]_8\}$, and $\mathbb{Z}_8$ are the only subgroups of $\mathbb{Z}_8$. Figure 18.1 summarizes the situation. ❖

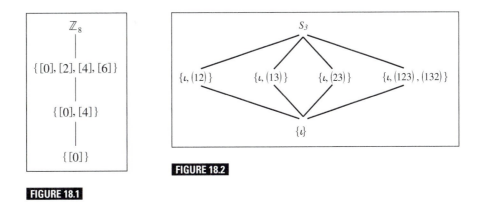

FIGURE 18.1

FIGURE 18.2

Example 18.11   Arguments similar to those in Example 18.10 show that the subgroups of S_3 are $\{\iota\}$, $\{\iota, (12)\}$, $\{\iota, (13)\}$, $\{\iota, (23)\}$, $\{\iota, (123), (132)\}$, and S_3. (An easier argument using Lagrange's theorem will be given in Example 21.11.) Figure 18.2 summarizes the situation. ❖

Figures 18.1 and 18.2 are examples of <u>subgroup lattices</u>. Such figures display the subgroups of a group with downward lines to indicate containment and can be very useful for understanding a group's structure.

We claimed at the beginning of this chapter that if $f: G \to H$ is a homomorphism of groups, then the set of functional values of f is a subgroup of H. Not only is this the case but the set of elements which f maps to the identity in H also forms a subgroup of G. With these claims in mind, we make the following definitions.

DEFINITION

Let G and H be groups and let $f: G \to H$ be a homomorphism. The **homomorphic image** of f is the subset $f(G) = \{f(x) \mid x \in G\}$ of H. The **kernel** of f is the subset $\ker f = \{x \in G \mid f(x) = e_H\}$ of G. (See Figure 18.3.)

FIGURE 18.3

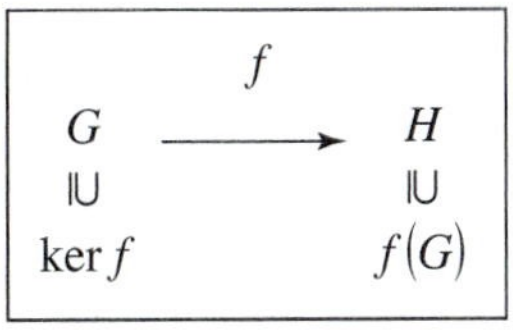

$$G \xrightarrow{\quad f \quad} H$$
$$\text{IU} \qquad\qquad \text{IU}$$
$$\ker f \qquad\quad f(G)$$

It is important to note that the kernel of f, $\ker f$, resides in G, whereas the homomorphic image of f, $f(G)$, resides in H.

As noted above, both the homomorphic image and the kernel are in fact subgroups of H and of G, respectively. To prove this requires some technical knowledge about the behavior of homomorphisms.

Lemma 18.12   *Let G and H be groups with identities e_G and e_H, respectively, and let $f: G \to H$ be a homomorphism. Then*

    (i) $f(e_G) = e_H$;
    (ii) for all $x \in G, f(x^{-1}) = f(x)^{-1}$.

Proof   (i)   Since $f(e_G)e_H = f(e_G) = f(e_G e_G) = f(e_G)f(e_G)$, Proposition 15.9($v$) implies that $f(e_G) = e_H$.

    (ii) By   (i), $f(x^{-1}) = e_H f(x^{-1}) = f(x)^{-1}f(x)f(x^{-1}) = f(x)^{-1}f(xx^{-1}) = f(x)^{-1}f(e_G) = f(x)^{-1}e_H = f(x)^{-1}$. ∎

Proposition 18.13   *Let G and H be groups and suppose that $f: G \to H$ is a homomorphism. Then $\ker f$ is a subgroup of G and $f(G)$ is a subgroup of H. As well, if G is Abelian, then $f(G)$ is Abelian.*

Proof   Since $f(e_G) = e_H$ by Lemma 18.12 (i), $\ker f \neq \emptyset$. If $x, y \in \ker f$, then $f(xy) = f(x)f(y) = e_H e_H = e_H$, and hence $xy \in \ker f$. Furthermore, if $x \in \ker f$, then by Lemma 18.12 (ii), $f(x^{-1}) = f(x)^{-1} = e_H{}^{-1} = e_H$, and hence $x^{-1} \in \ker f$. We conclude that $\ker f$ is a subgroup of G by Proposition 18.2.

    Since $e_G \in G, f(e_G) \in f(G)$, and thus $f(G) \neq \emptyset$. If $X, Y \in f(G)$, then there exist $x, y \in G$ such $f(x) = X$ and $f(y) = Y$, and hence $XY = f(x)f(y) = f(xy) \in f(G)$ because f is a homomorphism, and $X^{-1} = f(x)^{-1} = f(x^{-1}) \in f(G)$ by Lemma 18.12 (ii). So by Proposition 18.2, $f(G)$ is a subgroup of H.

    If G is Abelian and $X, Y \in f(G)$, then there exist $x, y \in G$ such $f(x) = X$ and $f(y) = Y$. And, since G is Abelian, $XY = f(x)f(y) = f(xy) = f(yx) = f(y)f(x) = YX$, and therefore $f(G)$ is also Abelian. ∎

One immediate consequence of Proposition 18.13 is that isomorphisms preserve commutativity.

Corollary 18.14   *If G and H are isomorphic groups and G is Abelian, then H is also Abelian.*

Proof   Since G and H are isomorphic, there is an isomorphism $f: G \to H$. Since f is onto, $f(G) = H$, and thus by Proposition 18.13, since G is Abelian, H is Abelian. ∎

A second consequence of Proposition 18.13 concerns the function T, defined in Proposition 17.5.

Corollary 18.15   *Let F be a subfield of $\mathbb{C}$, let $p(x) \in F[x]$, and let $c_1, \cdots, c_n \in \mathbb{C}$ be the distinct solutions of $p(x) = 0$. Then there exists a one-to-one homomorphism T of $\operatorname{Gal} F^{p(x)}/F$ onto a subgroup of S_n.*

Proof Apply Proposition 18.13 to Proposition 17.5. ∎

Another way of stating Corollary 18.15 is to say that Gal $F^{p(x)}/F$ is isomorphic to a subgroup of S_n. This says that Gal $F^{p(x)}/F$ is group-theoretically indistinguishable from a subgroup of S_n and thus that the study of permutation groups is the key to the study of Gal $F^{p(x)}/F$.

We conclude this section with a very important observation about the groups Gal K/F, viz., that a chain of extension fields determines a chain of Galois subgroups. Specifically, we have the following. (See Figure 18.4.)

FIGURE 18.4

$$
\begin{array}{cc}
F & \text{Gal } K/F \\
\cap| & |\cup \\
L & \text{Gal } K/L \\
\cap| & |\cup \\
K & \text{Gal } K/K = \{\iota\}
\end{array}
$$

Proposition 18.16 *Suppose that L is an extension field of a field F and that K is an extension field of L. Then* Gal K/L *is a subgroup of* Gal K/F *which contains the subgroup* Gal $K/K = \{\iota\}$.

Proof If $f \in$ Gal K/L, then f leaves elements of L unchanged, and since $F \subseteq L$, f therefore leaves elements of F unchanged. Hence $f \in$ Gal K/F and thus Gal $K/L \subseteq$ Gal K/F. But by Proposition 15.1, Gal K/L and Gal K/F are both groups with respect to composition. In particular, Gal K/L is a subset of the group Gal K/F which satisfies (*i*), (*ii*), and (*iii*) of Proposition 18.2 and hence in a subgroup of Gal K/F. By the same argument, Gal K/K is a subgroup of Gal K/F. Furthermore, if $\phi \in$ Gal K/K, then $\phi(k) = k$ for all $k \in K$, and hence $\phi = \iota$. So Gal $K/K = \{\iota\}$. ∎

It follows from Proposition 18.16 that a chain of extension fields sets up a corresponding chain of Galois groups. We will be particularly interested in chains of this type formed by successive algebraic extensions.

Example 18.17 If F is a subfield of $\mathbb{C}$ and if $r_1, \ldots, r_m \in \mathbb{C}$, then we may form the fields $F_0 = F, F_1 = F_0(r_1), F_2 = F_1(r_2), \ldots, F_m = F_{m-1}(r_m)$. This chain of fields gives rise to the chain of Galois groups shown in Figure 18.5. ❖

FIGURE 18.5

$$
\begin{array}{cc}
F & \text{Gal } F_m/F \\
\cap| & |\cup \\
F_1 & \text{Gal } F_m/F_1 \\
\cap| & |\cup \\
\vdots & \vdots \\
\cap| & |\cup \\
F_m & \text{Gal } F_m/F_m = \{\iota\}
\end{array}
$$

Exercises

In Exercises 1–12, determine whether the subset S is a subgroup of the group G.

1. $G = (\mathbb{Z}_{12}, +)$, $S = \{[0]_{12}, [3]_{12}, [6]_{12}, [9]_{12}\}$

2. $G = (\mathbb{Z}_{12}, +)$, $S = \{[0]_{12}, [2]_{12}, [4]_{12}, [6]_{12}, [8]_{12}, [10]_{12}\}$

3. $G = (\mathbb{Q}, +)$, $S = \{4n \mid n \in \mathbb{Z}\}$

4. $G = (\mathbb{Q}^{\times}, \cdot)$, $S = \{4n \mid 0 \neq n \in \mathbb{Z}\}$

5. $G = (\mathbb{Q}^{\times}, \cdot)$, $S = \{2n \mid 0 \neq n \in \mathbb{Z}\}$

6. $G = (\mathbb{R}^{\times}, \cdot)$, $S = \{-1, 1\}$

7. $G = (\mathbb{C}^{\times}, \cdot)$, $S = \{z \in \mathbb{C} \mid |z| = 1\}$

8. $G = (M_2(\mathbb{R}), +)$, $S = \{A \in M_2(\mathbb{R}) \mid \det(A) = 0\}$

9. $G = S_5$, $S = \{\iota, (12), (34)\}$ **10.** $G = S_5$, $S = \{\iota, (12)(34)\}$

11. $G = S_6$, $S = \{\iota, (123), (456)\}$ **12.** $G = S_5$, $S = \{\iota, (253), (235)\}$

13. Prove or give a counterexample: If A and B are subgroups of G, then $A \cap B$ is a subgroup of G.

14. Prove or give a counterexample: If A and B are subgroups of G, then $A \cup B$ is a subgroup of G.

15. Prove or give a counterexample: For any group G, $S = \{g \in G \mid g^{-1} = g\}$ is a subgroup of G.

Let G be a group. Exercises 16–21 concern the set $G \times G = \{(g, h) \mid g, h \in G\}$ and the operation $(a, b)(c, d) = (ac, bd)$.

16. Prove that $G \times G$ is a group (cf. Exercise 26 in Chapter 15).

In Exercises 17–21, assume that $G \times G$ is a group and determine whether the given subset of $G \times G$ is a subgroup of $G \times G$. Give a proof or a counterexample, whichever is appropriate.

17. $\{(e, e)\}$ **18.** $\{(x, e) \mid x \in G\}$ **19.** $\{(x, x) \mid x \in G\}$

20. $\{(x, x^{-1}) \mid x \in G\}$ **21.** $\{(x^2, x) \mid x \in G\}$

In Exercises 22 and 23, show that if G is an Abelian group, then the given subset of G is a subgroup of G.

22. $\{x \in G \mid x^2 = e\}$ **23.** $\{x \in G \mid x^n = e \text{ for some } n \in \mathbb{N}\}$

In Exercises 24–29, construct the subgroup lattice of the given group. Remember to justify your answer.

24. S_2 **25.** $(\mathbb{Z}_2, +)$ **26.** $(\mathbb{Z}_6, +)$

27. $(\mathbb{Z}_{11}, +)$ **28.** $(\mathbb{Z}_5^{\times}, \cdot)$ **29.** $(\mathbb{Z}_7^{\times}, \cdot)$

Recall from Exercise 26 in Chapter 15 that the set $G \times H = \{(g, h) \mid g \in G,$ $h \in H\}$ is a group with respect to the operation $(a, b)(x, y) = (ax, by)$. Exercises 30 and 31 concern groups of this form.

30. Let A be a subgroup of a group G and let B be a subgroup of a group H. Show that $A \times B$ is a subgroup of $G \times H$.

31. Find a group G and a subgroup N of $G \times G$ such that for all subgroups L and M of G, $N \neq L \times M$.

32. Let G be any group and let $Z(G) = \{g \in G \mid xg = gx \text{ for all } x \in G\}$. Prove that $Z(G)$ is a subgroup of G ($Z(G)$ is called the <u>center</u> of G).

33. Let G and H be groups, let $f: G \to H$ be a homomorphism, and let $y \in H$. Show that $\{x \in G \mid f(x) = y\}$ is a subgroup of G if and only if $y = e_H$.

34. Let G and H be groups, let $f: G \to H$ be a homomorphism, and let N be a subgroup of H. Show that $\{x \in G \mid f(x) \in N\}$ is a subgroup of G.

35. Prove the following variant of Proposition 18.1 (cf. Exercise 18 in Chapter 9): A subset S of a group G is a subgroup of G if and only if S satisfies

    (i) $S \neq \varnothing$;

    (ii) if $x, y \in S$, then $xy^{-1} \in S$.

36. A subgroup S of a group G is <u>proper</u> if $S \neq G$. Find an example of a group G, a proper subgroup S of G, and an isomorphism $f: S \to G$.

19

Subgroups Generated by Subsets

Determining whether the polynomial equation $p(x) = 0$ is solvable by radicals depends on the internal structure of the extension field $F^{p(x)}$. If we are to understand the ways in which the Galois group Gal $F^{p(x)}/F$ reflects this structure, we will need to have some general understanding of the internal structure of an arbitrary group. One important part of this internal structure is the manner in which subsets generate subgroups.

Our objective in this chapter is to investigate how subgroups are generated by subsets. Our considerations will allow us to prove (Proposition 19.9) that if a subset of S_5 contains both a 2-cycle and a 5-cycle, then the only subgroup which contains it is S_5 itself.

This result is not only interesting in its own right but will also turn out to be essential in our construction of polynomial equations which are not solvable by radicals.

We first note that any subset of a group has a smallest subgroup containing it and that this smallest subgroup is easy to describe.

Proposition 19.1   *If X is a nonempty subset of a group G and H is the set of all finite products $x_1 x_2 \cdots x_n$ where either $x_i \in X$ or $x_i^{-1} \in X$, then H is a subgroup of G. Furthermore, any subgroup of G which contains X also contains H.*

Proof   Clearly $X \subseteq H$ and hence $H \neq \emptyset$. Let $z, w \in H$. Then $z = c_1 c_2 \cdots c_n$, and $w = d_1 d_2 \cdots d_m$, where $c_1 \in X$ or $c_i^{-1} \in X$ and $d_j \in X$ or $d_j^{-1} \in X$. Then $zw = c_1 c_2 \cdots c_n d_1 d_2 \cdots d_m$, and hence zw is also in H. Therefore, H is closed. By Proposition 15.9, $(s^{-1})^{-1} = s$ for all $s \in S$, and $z^{-1} = c_n^{-1} \cdots c_2^{-1} c_1^{-1}$. Therefore z^{-1} is also in H, and we conclude that H is a subgroup by Proposition 18.2.

Suppose that W is a subgroup of G which contains X, and let $z \in H$. Then $z = c_1 c_2 \cdots c_n$, where $c_i \in X$ or $c_i^{-1} \in X$, and since W contains X, $z = c_1 c_2 \cdots c_n$, where $c_i \in W$ or $c_i^{-1} \in W$. So since W is a subgroup, $z = c_1 c_2 \cdots c_n \in W$. Thus $H \subseteq W$. ∎

DEFINITION

Let G be a group and let X be a nonempty subset of G. The subgroup of G consisting of all finite products $x_1 x_2 \cdots x_n$ where either $x_i \in X$ or $x_i^{-1} \in X$ is called the **subgroup generated** by X. This subgroup is denoted by $S(X)$.

We will abbreviate the notation $S(X)$ when it is convenient. Specifically, if $x_1, \ldots, x_n \in G$, then $S(x_1, \ldots, x_n)$ will mean $S(\{x_1, \ldots, x_n\})$.

Groups generated by a single element are particularly important; so we use the following nomenclature to single them out. Note that by definition, if x is an element of a group G, then $S(x) = \{x^n \mid n \in \mathbb{Z}\}$ (or, if G is written additively, $S(x) = \{nx \mid n \in \mathbb{Z}\}$).

DEFINITION

A group G is **cyclic** if G is generated by one of its elements, i.e., if there exists $x \in G$ such that $G = S(x)$.

We will investigate cyclic groups in more detail in Chapter 21. For now, we present the following example.

Example 19.2   For any positive integer n, since $\mathbb{Z}_n = \{k[1]_n \mid k \in \mathbb{Z}\} = S([1]_n)$, $\mathbb{Z}_n$ is a cyclic group with generator $[1]_n$. And similarly, the group $\{1, \zeta_n, \zeta_n^2, \ldots, \zeta_n^{n-1}\}$ of Example 15.5 is cyclic with generator ζ_n because $\{1, \zeta_n, \zeta_n^2, \ldots, \zeta_n^{n-1}\} = S(\zeta_n)$. Note that not every cyclic group is finite; according to Exercise 24, $\mathbb{Z}$ is a cyclic group. ❖

For a given subset X of a group G, how can we determine $S(X)$, the subgroup generated by X? By Proposition 19.1, any subgroup which contains X must contain $S(X)$, and conversely $S(X)$ contains all elements of G which can be written as products of elements of X and inverses of elements of X. Thus, to determine $S(X)$, first find a candidate H by combining elements of X and inverses of elements of X. Then show

(a)   that H is a subgroup which contains X and

(b)   that every element of H can be written in terms of elements of X and inverses of elements of X.

Then by (a), $H \supseteq S(X)$, and by (b), $H \subseteq S(X)$, and hence $H = S(X)$.

Example 19.3   Consider the subset $X = \{(124), (12)\}$ of S_4. Calculate:

$$(12)^2 = \iota, \qquad (124)^2 = (142), \qquad (124)^3 = \iota,$$
$$(12)(124) = (24), \qquad (12)(124)(12) = (142),$$
$$(124)(12) = (14), \qquad (124)(12)(124) = (12).$$

Then, since every other permutation in S_4 moves 3 but no product of (124) and (12) moves 3, all other products of (124) and (12) should be in $H = \{\iota, (12), (14), (24), (124), (142)\}$. So we conjecture that H is $S(X)$, the subgroup generated by X.

(*a*)   Obviously H is nonempty. The multiplication table for H is

·	ι	(12)	(14)	(24)	(124)	(142)
ι	ι	(12)	(14)	(24)	(124)	(142)
(12)	(12)	ι	(142)	(124)	(24)	(14)
(14)	(14)	(124)	ι	(142)	(12)	(24)
(24)	(24)	(142)	(124)	ι	(14)	(12)
(124)	(124)	(14)	(24)	(12)	(142)	ι
(142)	(142)	(24)	(12)	(14)	ι	(124)

and hence H is multiplicatively closed so that by Proposition 18.5, H is a subgroup of S_4. Thus since $H \supseteq X$, H must contain $S(X)$.

(*b*)   Conversely, observe that each element of H may be written as a product of elements of X or inverses of elements of X:

$$\iota = (124)(124)^{-1}, \qquad (14) = (124)(12),$$
$$(24) = (12)(124), \qquad (142) = (124)^2,$$

and hence H must be contained $S(X)$. It follows that H is the subgroup of S_4 generated by X. ❖

Example 19.4   Consider the subset $X = \{4, 10, -18\}$ of $\mathbb{Z}$. Note that $\mathbb{Z}$ is a group with respect to addition so that $S(X)$ consists of all <u>sums</u> $x_1 + x_2 + \cdots + x_n$, where either $x_i \in X$ or $-x_i \in X$. Since $4 + 4 + 4 + (-10) = 3 \cdot 4 - 10 = 2, 2 \in S(X)$, and certainly every element of X is a multiple of 2. So we conjecture that $H = \{2k \,|\, k \in \mathbb{Z}\}$ is the subgroup, $S(X)$, generated by X.

(*a*)   Clearly H contains X, and we showed in Example 15.4 that H is a subgroup of $\mathbb{Z}$. So $H \supseteq S(X)$.

(*b*)   Conversely, observe that for any $k \in \mathbb{Z}, 2k = 3k \cdot 4 - k \cdot 10$ and hence that $H \subseteq S(X)$. It follows that H is the subgroup of $\mathbb{Z}$ generated by X.

Note that we have shown that $S(X) = S(4, 10, -18) = S(2)$ and hence that $S(X)$ is cyclic. ❖

Example 19.5   Consider the subset $X = \{\sqrt{2}, \sqrt{5}\}$ of $\mathbb{R}$. Since the group operation on $\mathbb{R}$ is $+$, certainly any real number of the form $m\sqrt{2} + n\sqrt{5}$, for integers m and n, must be in the subgroup, $S(X)$, generated by X. We will show that $S(X) = \{m\sqrt{2} + n\sqrt{5} \mid m, n \in \mathbb{Z}\}$.

(a)   Let $H = \{m\sqrt{2} + n\sqrt{5} \mid m, n \in \mathbb{Z}\}$, and observe that since $0 = 0 \cdot \sqrt{2} + 0 \cdot \sqrt{5} \in H$, $H \neq \varnothing$. Furthermore, for m, n, a, $b \in \mathbb{Z}$, $m + n$, $a + b \in \mathbb{Z}$, and hence

$$(m\sqrt{2} + n\sqrt{5}) + (a\sqrt{2} + b\sqrt{5}) = (m + a)\sqrt{2} + (n + b)\sqrt{5} \in H.$$

And as well, $-m, -n \in \mathbb{Z}$ so that

$$-m\sqrt{2} + n\sqrt{5}) = (-m)\sqrt{2} + (-n)\sqrt{5} \in H.$$

It then follows from Proposition 18.2 that H is a subgroup of $\mathbb{R}$, and thus, since H contains X, H must contain $S(X)$.

(b)   On the other hand, as we previously observed, any real number of the form $m\sqrt{2} + n\sqrt{5}$ must be in $S(X)$. So H is the subgroup of $\mathbb{R}$ generated by S. ❖

Note that in Example 19.4, the <u>infinite</u> group $H = \{2k \mid k \in \mathbb{Z}\}$ is generated by the <u>finite</u> set $\{4, 10, -18\}$. So small subsets need not generate small subgroups. In fact, as previously noted, we will show (Proposition 19.9) that S_5 may be generated by only two elements. In the course of proving this, we will derive some general results about subsets which generate S_n. In the literature, 2-cycles are frequently called <u>transpositions</u>; so to ensure familiarity with this terminology, we use it to phrase the following results.

Proposition 19.6   *Every permutation in S_n is a product of transpositions.*

Proof   Since every permutation may be written as a product of cycles, it suffices to show that every cycle is a product of transpositions. But if $(i_1 i_2 \ldots i_k)$ is a k-cycle in S_n, then $(i_1 i_2 \ldots i_k) = (i_1 i_2)(i_2 i_3) \cdots (i_{k-1} i_k)$. ∎

Corollary 19.7   *The group S_n is generated by its transpositions. That is, if X is the set of transpositions in S_n, then $S_n = S(X)$.*

Proof   Use Propositions 19.1 and 19.6. ∎

Proposition 19.8   *If $(i_1 i_2 \ldots i_n)$ is an n-cycle in the group S_n, then S_n itself is the subgroup generated by $(i_1 i_2 \ldots i_n)$ and the transposition $(i_1 i_2)$.*

Proof By Corollary 19.7, it suffices to show that $S((i_1i_2\ldots i_n),(i_1i_2))$ contains all the transpositons in S_n. That is, it suffices to show that every element of S_n can be written as a product of $(i_1i_2\ldots i_n)$, (i_1i_2), $(i_1i_2\ldots i_n)^{-1}$, and $(i_1i_2)^{-1}$. We have $(i_2i_3) = (i_1i_2\ldots i_n)(i_1i_2)(i_1i_2\ldots i_n)^{-1}$, and $(i_3i_4) = (i_1i_2\ldots i_n)(i_2i_3)(i_1i_2\ldots i_n)^{-1}$. And in general, every transposition of the form $(i_{k-1}i_k)$, for $1 < k \le n$, can be written as such a product. Furthermore, $(i_1i_3) = (i_1i_2)(i_2i_3)(i_1i_2)$, and $(i_1i_4) = (i_1i_3)(i_3i_4)(i_1i_3)$. So in general, every transportation of the form (i_1i_k) for $1 < k \le n$, can also be written as such a product. However, if (ab) is any transposition in S_n, then $(ab) = (i_1a)(i_1b)(i_1a)$ and hence (ab) can also be written as such a product. So by Corollary 19.7, S_n is the subgroup generated by $\{(i_1i_2\ldots i_n),(i_1i_2)\}$. ∎

As noted previously, the following stronger result is true for S_5.

Proposition 19.9 *The group S_5 is generated by any subset containing a transposition and a 5-cycle.*

Proof Let $\sigma \in S_5$ be a 5-cycle and let $\tau \in S_5$ be a transposition. We can write σ and τ so that they begin with the same number: $\tau = (ax)$ and $\sigma = (abcde)$. If $x = b$, then $S(\sigma,\tau) = S_5$ by Proposition 19.8. If $x = c$, then $(acebd) = (abcde)^2 \in S(\sigma,\tau)$ and hence $S(\sigma,\tau) = S_5$, again by Proposition 19.8. But $(adbec) = (abcde)^3 \in S(\sigma,\tau)$, $(aedcb) = (abcde)^4 \in S(\sigma,\tau)$, and hence by Proposition 19.8, $S(\sigma,\tau) = S_5$ as well when $x = d$ or $x = e$. ∎

Proposition 19.9 does not hold for S_4 (Exercise 31). However, it will hold for groups S_n, where n is prime (Exercise 33). Even when n is not prime, the technique used in the proof can sometimes be applied to show that a particular transposition and n-cycle generate S_n.

Example 19.10 Suppose that X is the subset $\{(14),(12345678)\}$ of S_8. Then $(14725836) = (12345678)^3 \in S(X)$ and hence by Proposition 19.8, $S(X) = S_8$. ❖

Exercises

In Exercises 1–6, write the given permutation as a product of transpositions.

1. (165372) **2.** $(1472)(395)$ **3.** $(1546)(2357)$

4. $(12)(1567)(23)$ **5.** $(173)(365)(254)$ **6.** $(259)[(1865)2875)]^{-1}$

In Exercises 7–18, determine the subgroup generated by the subset X of the group G. Remember to justify your answer.

7. $X = \{[1]_7,[2]_7\}$, $G = (\mathbb{Z}_7,+)$ **8.** $X = \{[1]_7,[2]_7\}$, $G = (\mathbb{Z}_7{}^\times,\cdot)$

9. $X = \{6, 18, 27\}, G = (\mathbb{Z}, +)$ 10. $X = \{-12, 20, -16\}, G = (\mathbb{Z}, +)$

11. $X = \{2, \sqrt{3}\}, G = (\mathbb{R}, +)$ 12. $X = \{6, \sqrt{12}\}, G = (\mathbb{R}, +)$

13. $X = \{(124), (23)\}, G = S_{11}$ 14. $X = \{(25), (12345)\}, G = S_5$

15. $X = \{(34), (14253)\}, G = S_5$ 16. $X = \{(17), (1753624)\}, G = S_7$

17. $X = \left\{ \begin{pmatrix} \pi & 0 \\ 0 & \pi \end{pmatrix} \right\}, G = (M_2(\mathbb{R}), +)$ 18. $X = \{(17), (1753624)\}, G = S_{12}$

Exercises 19–23 concern the subset $Q = \{\pm 1, \pm i, \pm j, \pm k\}$ of the quaternions, $\mathbb{H}$. Recall from Exercise 23 in Chapter 15 that Q is a group with respect to multiplication. In Exercises 19–21, determine the subgroups of Q generated by the following subsets of Q. Remember to justify your answer.

19. $\{-1\}$ 20. $\{i\}$ 21. $\{i, -j\}$

22. Show that if $H \neq Q$ is a subgroup of Q, then H is cyclic.

23. Show that Q is not cyclic.

24. Show that $\mathbb{Z}$ is cyclic.

25. Show that S_3 is not cyclic.

26. Prove that S_8 is generated by (14325876) and (12).

27. Prove that S_8 is generated by (16854723) and (46).

28. Let G be an Abelian group and let $g, h \in G$ be such that $g^2 = e$ and $h^3 = e$. Show that the subgroup generated by $\{g, h\}$ is $\{e, h, h^2, g, gh, gh^2\}$.

29. Show that the dihedral group D_4 of Exercise 36 in Chapter 15 is generated by its two elements, R and Y, and that $R^4 = I$, $Y^2 = I$, and $RY = YR^3$.

30. Prove that S_7 is generated by any set containing a transposition and a 7-cycle.

31. Show that Proposition 19.9 does not hold in general by finding a 4-cycle and a transposition in S_4 which do <u>not</u> generate S_4.

32. Prove directly without using Proposition 19.8 that S_6 is generated by (134256) and (24).

33. For any prime p, show that the group S_p is generated by any set containing a transposition and a p-cycle.

20

Cosets

We noted in Examples 18.16 and 18.17 that a chain of extension fields gives rise to a corresponding chain of Galois groups. Since a polynomial equation is solvable by radicals provided that it gives rise to a particular chain of extension fields, the conditions we will impose on the Galois group corresponding to solvability by radicals will concern a chain of subgroups within the Galois group. With this in mind, we turn our attention to the behavior of subgroups. We will be interested especially in relationships between the existence of special kinds of subgroups and restrictions on the group itself.

We begin by noticing that translations of subgroups behave in a particularly tractable way. In general, let S and T be subsets of a group G and let $x \in G$. We then form new subsets of G by letting

$$ST = \{st \mid s \in S, t \in T\},$$
$$xS = \{xs \mid s \in S\},$$
$$Sx = \{sx \mid s \in S\}.$$

DEFINITION

Let S be a subgroup of a group G. A set of the form xS for some $x \in G$ is called a **left coset** of S; a set of the form Sx is called a **right coset** of S. We denote the collection of left cosets of S by

$$G/S = \{xS \mid x \in G\}.$$

The number of elements in G/S is called the **index** of S in G and is denoted by $[G:S]$.

Of course, if the operation on G is addition, the left cosets of S are written

$$x + S = \{x + s \mid s \in S\},$$

and

$$G/S = \{x + S \mid x \in G\}.$$

Example 20.1   Let S be the subset $\{[0]_6, [2]_6, [4]_6\}$ of $\mathbb{Z}_6$. Clearly, $S \neq \emptyset$, and the following table shows that S is additively closed:

+	$[0]_6$	$[2]_6$	$[4]_6$
$[0]_6$	$[0]_6$	$[2]_6$	$[4]_6$
$[2]_6$	$[2]_6$	$[4]_6$	$[0]_6$
$[4]_6$	$[4]_6$	$[0]_6$	$[2]_6$

So by Proposition 18.7, S is a subgroup of $\mathbb{Z}_6$. The left cosets of S are the following.

$$
\begin{aligned}
[0]_6 + S &= \{[0]_6, [2]_6, [4]_6\} = S, \\
[1]_6 + S &= \{[1]_6, [3]_6, [5]_6\}, \\
[2]_6 + S &= \{[2]_6, [4]_6, [0]_6\} = S, \\
[3]_6 + S &= \{[3]_6, [5]_6, [1]_6\} = [1]_6 + S, \\
[4]_6 + S &= \{[4]_6, [0]_6, [2]_6\} = S, \\
[5]_6 + S &= \{[5]_6, [1]_6, [3]_6\} = [1]_6 + S;
\end{aligned}
$$

thus $\mathbb{Z}_6/S = \{S, [1]_6 + S\}$, and the index of S in $\mathbb{Z}_6$ is 2.

Note that for all $[i]_6, [j]_6 \in \mathbb{Z}_6$.

(*i*)   either $[i]_6 + S = [j]_6 + S$ or $([i]_6 + S) \cap ([j]_6 + S) = \emptyset$.

Also, if $[i]_6 + S = [1]_6 + S$, then $[i]_6 = [1]_6, [i]_6 = [3]_6$, or $[i]_6 = [5]_6$, i.e., $-[1]_6 + [i]_6 \in S$; checking the other cases in the same way shows that for all $[i]_6, [j]_6 \in \mathbb{Z}_6$,

(*ii*) $[i]_6 + S = [j]_6 + S$ if and only if $-[j]_6 + [i]_6 \in S$.

Clearly

(*iii*) $\mathbb{Z}_6$ is the disjoint union of its distinct cosets S and $[1]_6 + S$:

	$\mathbb{Z}_6$:		
S:	$[0]_6$	$[2]_6$	$[4]_6$
$[1]_6 + S$:	$[1]_6$	$[3]_6$	$[5]_6$

(*iv*)   each coset $[i]_6 + S$ has the same number of elements (three) as S.

Furthermore,

$$([1]_6 + S) + ([2]_6 + S) = \{[1]_6, [3]_6, [5]_6\} + \{[2]_6, [4]_6 [0]_6\}.$$
$$= \{[1]_6 + [2]_6, [1]_6 + [4]_6, [1]_6 + [0]_6, [3] + [2]_6, [3]_6 + [4]_6,$$
$$[3]_6 + [0]_6, [5]_6 + [2]_6, [5]_6 + [4]_6, [5]_6 + [0]_6\}$$
$$= \{[3]_6, [5]_6, [1]_6\} = [3]_6 + S = ([1]_6 + [2]_6) + S;$$

and checking the other cases in a similar fashion shows that for all $[i]_6, [j]_6 \in \mathbb{Z}_6$.

(*v*) $([i]_6 + S) + ([j]_6 + S) = ([i]_6 + [j]_6) + S.$ ❖

Example 20.2   For any positive integer n, Example 15.4 implies that $(n) = \{kn \mid k \in \mathbb{Z}\}$ is a subgroup of $\mathbb{Z}$, and since $\mathbb{Z}$ is written additively, the left cosets of (n) in $\mathbb{Z}$ are the sets $i + (n) = \{i + kn \mid k \in \mathbb{Z}\} = [i]_n$. Thus

$$\mathbb{Z}/(n) = \{(n), 1 + (n), 2 + (n), \dots, n - 1 + (n)\}$$
$$= \{[0]_n, [1]_n, [2]_n, \dots, [n - 1]_n\} = \mathbb{Z}_n,$$

and since $\mathbb{Z}/(n)$ has n elements, the index of (n) in $\mathbb{Z}$ is n.

Note that as in Example 20.1, for $i, j \in \mathbb{Z}$,

(*i*)   either $i + (n) = j + (n)$ or $(i + (n)) \cap (j + (n)) = \varnothing$;
(*ii*) $i + (n) = j + (n)$ if and only if $-j + i \in (n)$;
(*iii*) $\mathbb{Z}$ is the disjoint union of the elements of $\mathbb{Z}/(n)$;
(*iv*)   each coset has an infinite number of elements, as does (n):
(*v*) $(i + (n)) + (j + (n)) = i + j + (n).$ ❖

Example 20.3   Let H be the subset $\{\iota, (12)\}$ of S_3. Clearly, $H \neq \varnothing$, and the following multiplication table shows that H is closed:

$\cdot$	ι	(12)
ι	ι	(12)
(12)	(12)	ι

So H is a subgroup of S_3 by Proposition 18.7. Its left cosets are the following:

$$
\begin{aligned}
\iota H &= \{\iota, (12)\} &&= H, \\
(12)H &= \{(12), \iota\} &&= H, \\
(13)H &= \{(13), (123)\}, \\
(23)H &= \{(23), (132)\}, \\
(123)H &= \{(123), (13)\} = (13)H, \\
(132)H &= \{(132), (23)\} = (23)H.
\end{aligned}
$$

So $S_3/H = \{H, (13)H, (23)H\}$, and the index of H in S_3 is 3.

Note that, as in Examples 20.1 and 20.2, for all $\sigma, \tau \in S_3$,

(*i*)   either $\sigma H = \tau H$ or $(\sigma H) \cap (\tau H) = \varnothing$;

(*ii*) $\sigma H = \tau H$ if and only if $\tau^{-1}\sigma \in H$;

(*iii*) S_3 is the disjoint union of its distinct cosets H, $(13)H$, and $(23)H$:

	S_3:	
H:	ι	(12)
$(13)H$:	(13)	(123)
$(23)H$:	(23)	(132)

(*iv*)   each coset σH has the same number of elements (two) as H.

However, the cosets in S_3/H do <u>not</u> all satisfy property (*v*). In particular,

$$
\begin{aligned}
(13)H(23)H &= \{(13), (123)\}\{(23), (132)\} \\
&= \{(13)(23), (13)(132), (123)(23), (123)(132)\} \\
&= \{(132), (23), (12), \iota\},
\end{aligned}
$$

and hence, $(13)H(23)H$ is not another left coset of H. Note that each left coset of H has two elements and hence, to see that $(13)H(23)H$ is not a left coset of H, it in fact suffices to observe that it contains more than two elements. ❖

The properties (*i*), (*ii*), and (*iii*) described in Examples 20.1, 20.2, and 20.3 hold in general. Property (*iv*) holds in general as well but is complicated to state for infinite subgroups. However, as noted in Example 20.3, property (*v*) does <u>not</u> always hold. The subgroups for which property (*v*) does hold are very important and will be studied in Chapter 23.

Proposition 20.4   *Let G be a group, S be a subgroup of G, and x, y $\in$ G. Then*

(*i*)   *either xS + yS or xS $\cap$ yS = $\varnothing$;*

(ii) $xS = yS$ *if and only if* $y^{-1}x \in S$;

(iii) G *is the disjoint union of the distinct cosets in* G/S;

(iv)   *the function* $f: S \to xS$, *defined by letting* $f(s) = xs$, *is one-to-one and onto; so if S is finite, then xS and S contain the same number of elements.*

Proof   *(i)*   Either $xS \cap yS = \emptyset$ or $xS \cap yS \neq \emptyset$. Suppose that $xS \cap yS \neq \emptyset$, and let $w \in xS \cap yS$. Then there exist $a, b \in S$ such that $xa = w = yb$. For any $s \in S$,

$$xs = x(aa^{-1})s = y(ba^{-1}s) \in yS.$$
$$ys = y(bb^{-1})s = x(ab^{-1}s) \in xS.$$

Therefore, $xS \subseteq yS$ and $xS \supseteq yS$, i.e., $xS = yS$.

(ii)   If $xS = yS$, then, since $e \in S$, there exists $s \in S$ such that $x = xe = ys$; hence $y^{-1}x \in S$. Conversely, suppose that $y^{-1}x = s \in S$. Then $x = ys \in xS \cap yS$, and hence $xS \cap yS \neq \emptyset$. By *(i)*, $xS = yS$.

(iii)   Since G contains each coset in G/S, G contains their union. To prove the reverse containment, let $x \in G$. Since S is a subgroup, $e \in S$ and hence $x = xe \in xS$. But xS is one of the cosets in G/S and hence xS is contained in the union of those cosets. Then x is contained in that union and therefore G is also contained in that union. By *(i)*, this union is a disjoint union.

(iv)   If $f(a) = f(b)$ for $a, b \in S$, then $xa = xb$. and by Lemma 15.9 *(v)*, $a = b$. So f is one-to-one. If $xs \in xS$, then $f(s) = xs$ and hence f is onto. If S is finite, then since there exists a one-to-one and onto function from S to xS, S and xS have the same number of elements. ∎

Example 20.5   Suppose that H is a subgroup of S_6 which contains (12345) and $(142)(36)$. (Note that $H \neq \{(12345), (142)(36)\}$; rather we are assuming merely that $H \supseteq \{(12345), (142)(36)\}$.) We will show that $(13)(24)H = (2536)H$ in two different ways, first by using Proposition 20.4 *(i)* and second by using Proposition 20.4 *(ii)*.

Since $(12345) \in H$, $(14532) = (13)(24)(12345) \in (13)(24)H$, and since $(142)(36) \in H$, $(14532) = (2536)(142)(36) \in (2536)H$ as well. So $(13)(24)H \cap (2536)H \neq \emptyset$ and thus $(13)(24)H = (2536)H$ by Proposition 20.4 *(i)*.

Alternatively, we may observe that

$$(2536)^{-1}(13)(24) = (2635)(13)(24) = (152463)$$
$$= (142)(36)(15432) = (142)(36)(12345)^{-1} \in H,$$

and hence $(13)(24)H = (2536)H$ by Proposition 20.4 *(ii)*. ❖

Cosets are generally useful. In Chapter 21 we will use cosets to prove Lagrange's theorem from which we will deduce that for p prime, $\mathbb{Z}_p$ is essentially the only group with p elements! In the chapter following that (Chapter 22), we will consider the general setting in which cosets arise.

While results from both these chapters will be necessary in Chapter 28 when we prove that certain fifth-degree polynomials are not solvable by radicals, they are on the surface something of a digression from our main task of finding conditions on the Galois group which reflect the solvability of the underlying polynomial by radicals. We will return to that task more directly in Chapter 25.

Exercises

In Exercises 1–9,

   (a)   show that S is a subgroup of the group G,

   (b)   find all left cosets of S in G,

   (c)   determine the index of S in G, and

   (d)   either prove that the product (or sum) of left cosets is another left coset or else find a pair of left cosets whose product (or sum) is not another left coset.

1. $S = \{[0]_6, [3]_6\}, G = (\mathbb{Z}_6, +)$ **2.** $S = \{[0]_8, [2]_8, [4]_8, [6]_8\}, G = (\mathbb{Z}_8, +)$

3. $S = \{[1]_7, [6]_7\}, G = (\mathbb{Z}_7{}^\times, \cdot)$ **4.** $S = \{[1]_{13}, [3]_{13}, [9]_{13}\}, G = (\mathbb{Z}_{13}{}^\times, \cdot)$

5. $S = \{\iota, (23)\}, G = S_4$ **6.** $S = \{\iota, (124), (142)\}, G = S_4$

7. $S = \{\iota, (12), (34), (12)(34)\}, G = S_4$

8. $S = \{\iota, (1234), (1432), (13)(24)\}, G = S_4$

9. $S = \{\iota, (123), (132), (124), (142), (134), (143), (234), (243), (12)(34),$
        $(13)(24), (14)(23)\}, G = S_4$

10. Suppose that S is a subgroup of S_5 which contains (123) and (34). Show that $(145)S = (1235)S$.

11. If S is a subgroup of S_5 which contains (1342), show that $(354)S = (1542)S$.

In Exercises 12–16, suppose that S is a subgroup of a group G and $x, y \in G$.

12. Show that $xS = S$ if and only if $x \in S$.

13. Use a counterexample to disprove the statement: If $xS = yS$, then $xy^{-1} \in S$.

14. Use a counterexample to disprove the statement: If $xy^{-1} \in S$, then $xS = yS$.

15. Prove the following statement or give a counterexample. Either $xS = Sy$ or $xS \cap Sy = \emptyset$.

16. Prove the following statement or give a counterexample. If S is finite, then Sy contains exactly the same number of elements as xS.

Exercises 17–20 prove Proposition 20.4 for right cosets. In these exercises, G is a group and S is a subgroup of G.

17. Prove that for all $x, y \in G$, either $Sx = Sy$ or $Sx \cap Sy = \emptyset$.

18. Prove that for all $x, y \in G$, $Sx = Sy$ if and only if $xy^{-1} \in S$.

19. Prove that G is the disjoint union of its distinct right cosets.

20. Prove that the function $f: S \to Sx$, defined by letting $f(s) = sx$, is one-to-one and onto; conclude that if S is finite, then Sx and S contain exactly the same number of elements.

In Exercises 21–24, n is a positive integer and $(n) = \{kn \mid k \in \mathbb{Z}\}$. Prove each assertion in detail and without using Proposition 20.4.

21. For all $i, j \in \mathbb{Z}$, either $i + (n) = j + (n)$ or $(i + (n)) \cap (j + (n)) = \emptyset$.

22. For all $i, j \in \mathbb{Z}$, $i + (n) = j + (n)$ if and only if $-j + i \in (n)$.

23. Prove that $\mathbb{Z}$ is the disjoint union of the elements of $\mathbb{Z}/(n)$.

24. For all $i, j \in \mathbb{Z}$, $(i + (n)) + (j + (n)) = i + j + (n)$.

21

Finite Groups and Lagrange's Theorem

In this chapter, we will be concerned primarily with finite groups, the main result being Lagrange's theorem (Proposition 21.10), an easy-to-prove yet powerful counting theorem for finite groups which involves the cosets of the group.

We will begin by investigating the idea of order. We will then prove Lagrange's theorem and use it to prove a very important and very surprising result about finite groups. Note that a group is a very general structure: a set, a single operation, four axioms. We have seen many dramatically different groups, from $\mathbb{Z}_n$ to S_n. And yet Lagrange's theorem can be used to prove that merely restricting the number of elements dramatically reduces the number of different groups. In fact, we will show that the only groups with a prime number of elements are the groups $\mathbb{Z}_p$, for primes p! (See Corollary 21.19.)

While Lagrange's theorem and Propositions 21.7 and 21.9 concerning orders of groups will be especially useful in determining equations that are not solvable by radicals, the characterization of groups of prime orders will not be directly involved in our study of solvability. It will, however, be very handy for creating examples.

DEFINITION

The **order** of a finite group is the number of elements which it contains; if the group has an infinite number of elements, it is said to be of **infinite order.** If x is an element of a group G, then the **order** of x is the least positive integer m such that $x^m = e$ (or, if G is written additively, such that $mx = 0$); if no such integer m exists, then x is said to have **infinite order.** The order of a finite group G is usually denoted $|G|$; the order of an element $x \in G$ is usually denoted $o(x)$.

Note that the same word, "order," is used for two ideas which at first glance may appear to be very different. When applied to an element of a group, "order" means the smallest power of the element which yields the identity, and when applied to the group itself, it means the number of elements in the underlying set. In fact, as the examples in this chapter indicate, these apparently dissimilar ideas are very closely connected. We will show (Proposition 21.14) that an element of finite order always generates a subgroup of the same order.

Example 21.1   Clearly the identity of any group is an element of order 1, and the subgroup consisting solely of the identity is a group of order 1. ❖

Example 21.2   The group $\mathbb{Z}_{12}$ contains twelve elements and therefore has order 12. For $[3]_{12} \in \mathbb{Z}_{12}$,

$$1[3]_{12} = [3]_{12} \neq [0]_{12},$$
$$2[3]_{12} = [3]_{12} + [3]_{12} = [6]_{12} \neq [0]_{12},$$
$$3[3]_{12} = [3]_{12} + [3]_{12} + [3]_{12} = [9]_{12} \neq [0]_{12},$$
$$4[3]_{12} = [3]_{12} + [3]_{12} + 3]_{12} + [3]_{12} = [12]_{12} = [0]_{12},$$

and hence the order of $[3]_{12}$ is 4. Note the subgroup generated by $\{[3]_{12}\}$ is $\{[0]_{12}, [3]_{12}, [6]_{12}, [9]_{12}\}$ which, since it contains four elements, also has order 4. By the same reasoning, $[6]_{12}$ has order 2 as does the subgroup it generates, and $[5]_{12}$ has order 12 as does the subgroup, $\mathbb{Z}_{12}$ itself, which it generates. ❖

Example 21.3   The group $(\mathbb{C}^{\times}, \cdot)$ is clearly of infinite order. Since $(2\zeta_6)^n = 2^n \zeta_6^n \neq 1$ for all $n > 0$, the element $2\zeta_6$ has infinite order. The subgroup it generates is

$$\{\ldots, (2\zeta_6)^{-3}, (2\zeta_6)^{-2}, (2\zeta_6)^{-1}, 1, (2\zeta_6), (2\zeta_6)^2, (2\zeta_6)^3, \ldots\}$$

which is also of infinite order. However, ζ_6 has order 6; for $(\zeta_6)^6 = 1$, but

$$\arg(\zeta_6) = \frac{2\pi}{6}, \qquad \arg((\zeta_6)^2) = \frac{4\pi}{6}, \qquad \arg((\zeta_6)^3) = \frac{6\pi}{6},$$
$$\arg((\zeta_6)^4) = \frac{8\pi}{6}, \qquad \arg((\zeta_6)^5) = \frac{10\pi}{6},$$

and thus $(\zeta_6)^n \neq 1$ for $1 \leq n \leq 5$. The subgroup which ζ_6 generates (cf. Example 18.9) is $\{1, \zeta_6, (\zeta_6)^2, (\zeta_6)^3, (\zeta_6)^4, (\zeta_6)^5\}$ which also has order 6. ❖

Example 21.4   Let F be a subfield of $\mathbb{C}$ and let $p(x) \in F[x]$. Then by Proposition 14.8, $|\operatorname{Gal} F^{p(x)}/F| = [F^{p(x)}:F]$. ❖

As previously mentioned, we will show below that the order of an element is the order of the subgroup which it generates. Note that Lemma 18.6 provides a less precise connection between the two ideas, viz., that any element of a finite group has finite order (cf. Corollary 21.16).

Note also that an element of finite order will have an infinite number of powers equaling the identity (for the nth power of an element x is the identity whenever n is a multiple of $o(x)$). We show next that the converse of this is also true (Proposition

21.5), and hence that if a prime power of an element is the identity, then the order of the element is precisely that prime (Corollary 21.6).

Proposition 21.5 *Let g be a group, let n,m be positive integers, and suppose that $g \in G$ has order n. Then $g^m = e$ if and only if n divides m.*

Proof If n divides m, then $m = nk$ and $g^m = g^{nk} = e^k = e$. On the other hand, by the division algorithm for the integers (Theorem 1.3) there always exist $q,r \in \mathbb{Z}$ such that $m = nq + r, 0 \leq r < n$, and if $g^m = e$, then $g^r = g^{m-nq} = g^m(g^n)^{-q} = e$. But n is the minimal positive power of g which equals e. Thus, $0 = r$ and hence n divides m. ∎

Corollary 21.6 *Let G be a group, let p be a prime, and let $e \neq g \in G$. If $g^p = e$, then g has order p.*

Proof Since $g^p = e$, g has finite order $n \leq p$, and since $g \neq e, n > 1$. If $n < p$, then by Proposition 21.6, n would divide p and hence p would not be prime. Thus $n = p$. ∎

Corollary 21.6 allows us to be very specific about the orders of certain elements of S_p for p prime.

Proposition 21.7 *If p is a prime, then the p-cycles are the only elements of order p in S_p.*

Proof If $\sigma = (a_1 a_2 \ldots a_k)$ is a k-cycle in S_p, then clearly $\sigma^k = \iota$. On the other hand, for any $0 < i < k$, $\sigma^i(a_1) = a_{i+1} \neq a_1$ and hence $\sigma^i \neq \iota$. Then the order of σ is k, i.e., the order of any cycle is its length. In particular, the order of any p-cycle is p. Suppose then that $\sigma \in S_p$ has order p and write σ as a product of disjoint cycles: $\sigma = \tau_1 \tau_2 \cdots \tau_r$. By Proposition 16.5, $\iota = \sigma^p = (\tau_1 \tau_2 \cdots \tau_r)^p = \tau_1^p \tau_2^p \cdots \tau_r^p$, and since the τ_i are disjoint, so are the τ_i^p. If $\tau_1^p \neq \iota$, then there exists x between 1 and p such that $\tau_1^p(x) \neq x$, and since the τ_i^p are disjoint, $\tau_i^p(x) = x$ for all $i > 1$. Then $\tau_1^p \tau_2^p \cdots \tau_r^p(x) = \tau_1^p(x) \neq x$, a contradiction, and thus $\tau_1^p = \iota$. So by Corollary 21.6, τ_1 has order p. But we showed before that the order of any cycle is its length, and hence τ_1 must have length p. Then τ_1 moves each element of $\{1, 2, \ldots, p\}$ and hence cannot be disjoint from any other cycle. It follows that $\sigma = \tau_1$ and hence that σ is a p-cycle. ∎

The preceding proof shows that k-cycles always have order k. However, if n is composite, the following example shows that S_n may contain elements of order n which are not n-cycles.

Example 21.8 If $\sigma = (123)(46) \in S_6$, then

$$\sigma^1 = (123)(46) \neq \iota, \qquad \sigma^2 = (132) \neq \iota, \qquad \sigma^3 = (46) \neq \iota,$$
$$\sigma^4 = (123) \neq \iota, \qquad \sigma^5 = (132)(46) \neq \iota, \qquad \sigma^6 = \iota,$$

and hence $o(\sigma) = 6$. So S_6 contains elements of order 6 which are not 6-cycles. ❖

Note also that one-to-one homomorphisms preserve order.

Proposition 21.9   *Suppose that G and H are groups and that $T: G \rightarrow H$ is a one-to-one homomorphism. If $g \in G$ has order n, then $T(g) \in H$ also has order n.*

Proof   Since T is a homomorphism, by Lemma 18.12, $T(g)^n = T(g^n) = T(e_G) = e_H$. Furthermore, if $0 < k < n$, then $g^k \neq e_G$, and since T is a homomorphism which is one-to-one, $T(g)^k = T(g^k) \neq T(e_G) = e_H$ by Lemma 18.12 again. It follows that $T(g)$ has order n. ∎

We now turn our attention to Lagrange's theorem and some of its consequences.

Proposition 21.10   **Lagrange's Theorem**

If G is a finite group and S is a subgroup of G, then the order of S times the index of S in G is the order of G, and hence the order of S divides the order of G.

Proof   By Proposition 20.4 (*iv*), the number of elements of each coset xS is the order of S, and by Proposition 20.4 (*iii*), G is the union of its disjoint cosets. Hence the order of G is the number of distinct cosets of S times the order of S, i.e., the order of S divides that of G. ∎

Note that Lagrange's theorem may be stated concisely by using the notation introduced above. In this notation, Lagrange's theorem says that for any subgroup S of a finite group G,

$$|S|[G:S] = |G|.$$

In general, Lagrange's theorem puts a substantial restriction on which subsets of a finite group may be subgroups. For this reason, it can be very helpful in finding subgroup lattices.

Example 21.11   (See Example 18.11). It is easy to use Proposition 18.7 to check that $\{\iota, (12)\}$, $\{\iota, (13)\}$, $\{\iota, (23)\}$, and $\{\iota, (123), (132)\}$ are all subgroups of S_3. But by Lagrange's theorem, a subgroup of S_3 can have only one, two, three, or six elements so that any nontrivial proper subgroup (i.e., one that is neither $\{\iota\}$ (nontrivial) nor S_3 itself (proper)) has either two or three elements. A proper subgroup containing a 3-cycle must contain its inverse, the other 3-cycle, and hence $\{\iota, (123), (132)\}$ is the only such subgroup. The product of two different 2-cycles is a 3-cycle and hence no proper subgroup can contain two different 2-cycles because such a subgroup would have to have more than three elements. It follows that the subgroups listed here are the only nontrivial proper subgroups of S_3. ❖

Recall from Chapter 19 that a group with a single generator is called a cyclic group. That is, a group of G is cyclic if it contains an element g such that $G = S(g)$. We saw many examples of cyclic groups in Example 19.2. An example of a non-cyclic group is S_3.

Example 21.12   It can be readily checked that in S_3 the powers of 2-cycles are again 2-cycles (or the identity) and the powers of 3-cycles are again 3-cycles (or the identity). Since every element of S_3 except the identity is a 2-cycle or a 3-cycle, this means that no single element of S_3 can generate S_3; thus S_3 is not cyclic. Note, however, that by Proposition 19.8, S_3 can be generated by two elements. ❖

An easier way to show that S_3 is not cyclic is to use the following result and Proposition 16.2.

Proposition 21.13   *Every cyclic group is Abelian.*

Proof   Suppose that G is cyclic with generator g and that $x, y \in G$. Then $x = g^n$ and $y = g^m$ for some $n, m \in \mathbb{Z}$. Hence $xy = g^n g^m = g^{n+m} = g^m g^n = yx$. ∎

We mentioned above that the subgroup which an element generates has the same order as the element (that is why the same term is used in both cases). We will now prove this and use it first, in conjunction with Lagrange's Theorem, to show that if p is prime, then any group of order p is cyclic (Corollary 21.17), and then to show that all groups of order p are isomorphic (Corollary 21.19).

Proposition 21.14   *Let G be a group and suppose that $g \in G$ has finite order m. Then $S(g) = \{e, g, g^2, \ldots, g^{m-1}\}$ and $S(g)$ has order m.*

Proof   Let $H = \{e, g, g^2, \ldots, g^{m-1}\}$.
   (a)   Clearly $S(g) = \{g^k \mid k \in \mathbb{Z}\} \supseteq H$.
   (b)   Conversely, if $k \in \mathbb{Z}$, then by the division algorithm (Theorem 1.3) there exist $q, r \in \mathbb{Z}$ such that $k = mq + r, 0 \le r < m$. Then $g^k = g^{mq+r} = (g^m)^q g^r = e^q g^r = g^r \in H$, and hence $S(g) \subseteq H$. It follows that $S(g) = H$.
   If the elements of H are not all distinct, then $g^i = g^j$ for some $0 \le i < j < m$, and hence $g^{j-i} = e$. However, since $0 < j - i < m$ and m is the order of g, this is impossible, and thus H has exactly m elements, i.e., $S(g)$ has order m. ∎

Corollary 21.15   *Let n be a positive integer and suppose that G is a cyclic group of order n. If g is a generator of G, then g has order n and $G = \{e, g, g^2, \ldots g^{n-1}\}$.*

Proof   Since $g \in G$, Lemma 18.6 implies that g has finite order m, and thus by Proposition 21.14, $S(g) = \{e, g, g^2, \ldots, g^{m-1}\}$ has m elements. But by hypothesis, $S(g) = G$, and hence $S(g)$ has n elements. So $m = n$ and $G = S(g) = \{e, g, g^2, \ldots, g^{n-1}\}$. ∎

Corollary 21.16   *The order of any element of a finite group is finite and divides the order of the group.*

Proof   By Lemma 18.6, the elements of a finite group are always of finite order. Then Proposition 21.14 says that the order of such an element is precisely the order of the subgroup it generates, and by Lagrange's theorem, the order of the subgroup divides the order of the group. ∎

Corollary 21.17   *Every finite group of prime order is cyclic and hence Abelian.*

Proof   Let G be a finite group of prime order p, let $e \neq g \in G$, and note that by Lemma 18.6, $o(g)$ is finite. Since $e \neq g$, $o(g) \neq 1$, and by Corollary 21.16, $o(g)$ divides p. So $o(g) = p$, and hence by Proposition 21.14, $S(g)$ has order p. Since G also has p elements, it follows that $S(g) = G$ so that G is cyclic and thus, by Proposition 21.13, is Abelian. ∎

That all cyclic groups of a given order are isomorphic follows from the characterization of cycle groups given in Corollary 21.15.

Proposition 21.18   *If n is a positive integer and if G is a cyclic group of order n, then G is isomorphic to $\mathbb{Z}_n$.*

Proof   By Corollary 21.15, G consists of the n distinct elements $e, g, g^2, \ldots, g^{n-1}$. Define $\Phi: G \to \mathbb{Z}_n$ by letting $\Phi(g^i) = [i]_n$ for $0 \leq i < n$, and note that the operation which makes $\mathbb{Z}_n$ a group is $+$. So to show Φ is a homomorphism, we must show that $\Phi(g^i g^j) = \Phi(g^i) + \Phi(g^j)$. But this follows from observing that for all $0 \leq i, j < n$,

$$\Phi(g^i g^j) = \left\{ \begin{array}{ll} \Phi(g^{i+j}) & \text{if } i + j < n \\ \Phi(g^{i+j-n}) & \text{if } i + j \geq n \end{array} \right\} = \left\{ \begin{array}{ll} [i+j]_n & \text{if } i + j < n \\ [i+j-n]_n & \text{if } i + j \geq n \end{array} \right\}$$
$$= [i+j]_n = [i]_n + [j]_n = \Phi(g^i) + \Phi(g^j).$$

Furthermore, since $\mathbb{Z}_n$ consists of the elements $[0]_n, \ldots, [n-1]_n$, Φ is certainly one-to-one and onto. So we can conclude that Φ is an isomorphism and thus that G is isomorphic to $\mathbb{Z}_n$. ∎

Corollary 21.19   *If p is a prime and G is a group of order p, then G is isomorphic to $\mathbb{Z}_p$.*

Proof   The corollary follows from Corollary 21.17 and Proposition 21.18. ∎

Note that in the remaining chapters we will be using some of the results from this chapter more frequently than others. Corollaries 21.6 and 21.17 and Proposition

21.10 will turn out to be especially useful, and Propositions 21.7 and 21.9 will be needed when we finally find a polynomial equation that is not solvable by radicals.

Exercises

In Exercises 1–12, find the order of the element x in the group G.

1. $G = (\mathbb{Z}_{18}, +), x = [6]_{18}$ **2.** $G = (\mathbb{Z}_{18}, +), x = [5]_{18}$

3. $G = (\mathbb{Z}_{18}, +), x = [12]_{18}$ **4.** $G = S_7, x = (1253)$

5. $G = S_7, x = (12)(357)$ **6.** $G = S_7, x = (12)(35)(47)$

7. $G = S_7, x = (127)(36)(15)$ **8.** $G = S_7, x = (1723645)$

9. $G = S_{12}, x = (1\,2\,3\,4)(5\,6\,7\,8\,9\,10)$

10. $G = S_{15}, x = (1\,2\,3\,4\,5\,6)(7\,8\,9\,10\,11\,12\,13\,14\,15)$

11. $G = (\mathbb{C}^\times, \cdot), x = \pi\zeta_5$ **12.** $G = (\mathbb{C}^\times, \cdot), x = 1 - i$

In Exercises 13–16, show that H is a subgroup of G and determine the order of H.

13. $G = (\mathbb{Z}_{18}, +), H = \{[0]_{18}, [3]_{18}, [6]_{18}, [9]_{18}, [12]_{18}, [15]_{18}\}$

14. $G = S_6, H = \{\iota, (16), (23), (16)(23)\}$

15. $G = S_6, H = \{\iota, (14), (145), (15), (45), (154)\}$

16. $G = \mathbb{C}, H = \mathbb{R}$

In Exercises 17–20, show that H is a subgroup of G and determine the index of H in G.

17. $G = (\mathbb{Z}_{18}, +), H = \{[0]_{18}, [9]_{18}\}$

18. $G = (\mathbb{Z}_{18}, +), H = \{[0]_{18}, [3]_{18}, [6]_{18}, [9]_{18}, [12]_{18}, [15]_{18}\}$

19. $G = S_6, H = \{\iota, (16), (23), (16)(23)\}$

20. $G = S_6, H = \{\iota, (14), (145), (15), (45), (154)\}$

In Exercises 21 and 22, use Lagrange's theorem to show that S is not a subgroup of G.

21. $G = (\mathbb{Z}_{12}, +), S = \{[0]_{12}, [6]_{12}, [8]_{12}, [2]_{12}, [10]_{12}\}$

22. $G = (\mathbb{Z}_{21}, +), S = \{[0]_{21}, [10]_{21}, [5]_{21}, [11]_{21}, [16]_{21}\}$

In Exercises 23 and 24, show that G cannot be isomorphic to H.

23. $G = (\mathbb{Z}_6, +), H = S_6$ **24.** $G = (\mathbb{Z}_6, +), H = S_3$

25. Show that S_n is not cyclic if $n \geq 3$.

26. Determine all the elements in S_5 of order 5.

27. Show that S_5 is generated by any set containing an element of order 5 and a transposition.

28. If G is a finite group of order n, prove that for all $g \in G$, $g^n = e$.

29. If G is a finite group of order n, prove that for all $g \in G$, the order of g is less than or equal to n.

By Example 18.9, for $n > 0$, $U_n = \{\zeta_n, (\zeta_n)^2, \ldots, (\zeta_n)^n\}$ is a subgroup of $(\mathbb{C}^\times, \cdot)$. In Exercises 30–33, show that the function T is a one-to-one homomorphism and determine the order of g and the order of $T(g)$.

30. $T: U_3 \to U_6$; $T((\zeta_3)^k) = (\zeta_6)^{2k}$; $g = \zeta_3$

31. $T: U_3 \to \mathbb{Z}_9$; $T((\zeta_3)^k) = [3k]_9$; $g = (\zeta_3)^2$

32. $T: U_6 \to \mathbb{Z}_{12}$; $T((\zeta_6)^k) = [2k]_{12}$; $g = \zeta_6$

33. $T: U_4 \to S_4$; $T((\zeta_4)^k) = (1234)^k$; $g = i$

34. Let G be an Abelian group and suppose that G has an element g of order 2 and an element h of order 5. If S is the subgroup of G generated by $\{g, h\}$, show that $S = \{e, h, h^2, h^3, h^4, g, gh, gh^2, gh^3, gh^4\}$.

In Exercises 35–39, construct the subgroup lattice of the given group.

35. $(\mathbb{Z}_{15}, +)$ 36. $(\mathbb{Z}_9, +)$ 37. $(\mathbb{Z}_{55}, +)$

38. $(\mathbb{Z}_{11}^\times, \cdot)$ 39. $(\mathbb{Z}_{13}^\times, \cdot)$

40. The subset $Q = \{\pm 1, \pm i, \pm j, \pm k\}$ of the quaternions is a noncommutative group with respect to multiplication of quaternions (Exercise 23 in Chapter 15). Construct the subgroup lattice of Q.

41. Show that $(\mathbb{Z}_5, +)$ is isomorphic to $\{1, \zeta_5, (\zeta_5)^2, (\zeta_5)^3, (\zeta_5)^4\}$.

42. Show that $(\mathbb{Z}, +)$ is isomorphic to $((8), +)$.

43. Explain why $(\mathbb{Z}, +)$ is not isomorphic to $(\mathbb{Z}, +) \times (\mathbb{Z}, +)$.

44. Suppose that G and H are groups and $f: G \to H$ is a homomorphism. Show that if $x \in G$ has order one, then $f(x)$ has order one.

45. Find a group G, a group H, a homomorphism $f: G \to H$, and an element $x \in G$ such that x and $f(x)$ have different orders.

In Exercises 46 and 47, let G be an Abelian group and let $X = \{g \in G \mid o(g) = 3\} \cup \{e\}$.

46. Show that X is a subgroup of G.

47. If G has order 25, what is the order of X?

48. Prove or give a counterexample: For any positive integer k, the only elements in S_k of order k are the k-cycles.

49. Prove or give a counterexample: For p prime and $0 < k < p$, the only elements of order k in S_p are the k-cycles.

50. Find a cyclic group G and an element $e \neq g \in G$ such that g does <u>not</u> generate G.

51. Suppose that G is a group all of whose elements have finite order. Show that a subset S of G is a subgroup of G if and only if S is nonempty and closed.

52. Find a group of order 4 which is not isomorphic to $(\mathbb{Z}_4, +)$. (*Hint:* Exercise 28 in Chapter 15.)

53. Show that a group with only two subgroups is cyclic.

54. Prove that every subgroup of a cyclic group is cyclic.

HISTORICAL NOTE

Joseph Louis Lagrange

Joseph Louis Lagrange was born Giuseppe Lodovico Lagrangia in Turin, Italy, on January 25, 1736, and died in Paris, France, on April 10, 1813. His great-grandfather, a Frenchman, had married an Italian and settled in Italy. His grandfather, father, and one of his brothers were all Treasurers of the Office of Public Works and Fortifications at Turin, and his mother was the daughter of a physician from a small town near Turin. In spite of their position, the family lived modestly and desired Joseph to become a lawyer. However, he found that he had a talent for the exact sciences and decided to pursue them early in his career. He initiated a correspondence with Euler in 1755, but published most of his early works, on the calculus of variation and planetary motion, in a local journal, *Miscellanea Taurinensia ou Mélanges de Turin.*

In 1766, Lagrange was asked to replace Euler in Berlin at an unusually high salary; he accepted and arrived in Berlin on October 27 after visits to Paris and London. During his time in Berlin, he continued to work at mechanics and differential and integral calculus and also began his research in the theory of equations and the theory of numbers. It was in 1770 that he proved that the order of a subgroup of S_n divides $n!$, the (preliminary) result from which Lagrange's theorem gets its name. The working conditions in Berlin were the equal of the financial compensation, and in 1767, he married Vittoria Conti. "My wife," he wrote, "who is one of my cousins and even lived for a long time with my family, is a very good housewife and has no pretensions at all."[1]

Vittoria died in 1783 after a long illness, and Frederick II, Lagrange's principal supporter in Berlin, died in 1786. Lagrange was courted by the princes in Italy but the French government made him the best offer and he moved to Paris in 1787. He kept a low political profile and the succession of French governments honored his contract for the rest of his life. In 1792, he married Renée-Françoise-Adélaïde Le Monnier and the second marriage, like the first, was a happy one. When he died in 1813, funeral ceremonies were held in Paris and in various universities in Italy. However, because Prussia had joined the coalition against Napoleon, there was no ceremony in Berlin.

Reference   1. Itard, Jean. "Lagrange, Joseph Louis." *Dictionary of Scientific Biography,* 16 vols. Gillespie, Charles Coulston, ed. New York: Charles Scribner's Sons, 1970, vol. VII, p. 503.

22 Equivalence Relations and Cauchy's Theorem

In Chapter 21, we saw how cosets could be used to determine the structure of certain finite groups. In this chapter, we will consider the general setting in which cosets arise. Expanding the notion of coset in this way will allow us to prove a special case of Cauchy's theorem, which in turn may be regarded as a partial converse of Corollary 22.7. Specifically, we will show (Proposition 22.9) that if 5 divides the order of a finite group, then the group has an element of order 5.

As noted at the end of Chapter 20, while this result will be necessary in Chapter 28 when we prove that certain fifth degree polynomials are not solvable by radicals, it is certainly on the surface something of a digression from our main task of finding conditions on the Galois group which reflect the solvability of the underlying polynomial by radicals. We will determine those conditions in Chapter 25.

We begin by noting that if H is a subgroup of a group G, then the relation that identifies elements of G if they generate the same left coset of H has properties in common with the equality relation.

Example 22.1 Let H be a subgroup of a group G. Define a binary relation on G by letting

$$a \sim b \quad \text{if and only if} \quad aH = bH,$$

That is, all the elements of a given left coset are related to themselves but to no other elements. Then this relationship shares the following properties with =.

(*i*) Each element is equivalent to itself, i.e., the relation is <u>reflexive</u>. This is clear because for any $a \in G$, $aH = aH$ and hence $a \sim a$.

(*ii*) The order in which elements are related does not matter, i.e., the relation is <u>symmetric</u>. For if $a \sim b$, then $aH = bH$ so that $bH = aH$ and hence $b \sim a$.

(*iii*) If one element is related to another and the second element is related to a third, then the first element is related to the third, i.e., the relation is transitive. To see this, merely observe that if $a \sim b$ and $b \sim c$, then $aH = bH$ and $bH = cH$ so that $aH = cH$ and hence $a \sim c$. ❖

In general, any relation that satisfies the three properties described in Example 22.1 is called an equivalence relation.

DEFINITION

Let S be a set. A binary relation $\sim$ on S is an **equivalence relation** if it satisfies the following conditions for all $a, b, c \in S$.

Reflexivity: $a \sim a$;

Symmetry: if $a \sim b$, then $b \sim a$;

Transitivity: if $a \sim b$ and $b \sim c$, then $a \sim c$.

Example 22.1 says that the relation defined there on a group G is an equivalence relation. We observed previously that equality on any set is also an equivalence relation. Some other examples follow, and examples are given in the exercises of binary relations that are not equivalence relations.

Example 22.2   Recall that for $z = a + bi \in \mathbb{C}$, the modulus of z is $|z| = |a + bi| = \sqrt{a^2 + b^2}$, and define a binary relation $\sim$ on $\mathbb{C}$ by letting

$$z \sim w \quad \text{if and only if} \quad |z| = |w|.$$

Since $|z| = |z|$, $z \sim z$ and hence $\sim$ is reflexive. If $z \sim w$, then $|z| = |w|$; hence $|w| = |z|$ and $w \sim z$. So $\sim$ is symmetric. If $z \sim w$ and $w \sim u$, then $|z| = |w|$ and $|w| = |u|$ and hence $|z| = |u|$, i.e., $z \sim u$. Thus $\sim$ is transitive.

Two complex numbers are related by $\sim$ if and only if they have the same modulus. Geometrically, this says that they are related by $\sim$ if and only if they both lie on the same circle in the complex plane centered at the origin. ❖

Example 22.3   For a positive integer n, define a binary relation on $\mathbb{Z}$ as follows:

$$a \equiv b \ (\text{mod } n) \quad \text{if and only if} \quad n \text{ divides } a - b.$$

Since $a - a = 0$, n divides $a - a$, and hence $\equiv$ is reflexive. If n divides $a - b$, then n divides $b - a$, and hence $\equiv$ is symmetric. Finally, if n divides both $a - b$ and $b - c$, then n divides $(a - b) + (b - c) = a - c$, and hence $\equiv$ is transitive. Thus, $\equiv$ is an equivalence relation on $\mathbb{Z}$. Note the following alternate definitions for the relation $\equiv$. They are easily seen to be equivalent to the one just given.

$$a \equiv b \ (\text{mod } n) \quad \text{if and only if} \quad -b + a \in (n);$$

$$a \equiv b \ (\text{mod } n) \quad \text{if and only if} \quad a + (n) = b + (n);$$

$$a \equiv b \ (\text{mod } n) \quad \text{if and only if} \quad [a]_n = [b]_n.$$

The second alternate definition above says that the relation $\equiv$ is just the relation $\sim$ described in Example 22.1 for the subgroup (n) of the group $\mathbb{Z}$. ❖

The equivalence relation $\sim$ defined on a group G in Example 22.1 has properties similar to those of equality on G. In fact, $\sim$ corresponds precisely to equality on G/H: two elements of G are equivalent with respect to $\sim$ if and only if the cosets of H that they generate are equal. Thus gathering together elements to form a coset is the same as gathering together all the elements that are equivalent to a generator of the coset. Specifically, for $a \in G$, $x \in aH$ if and only if $a^{-1}x \in H$; then by Proposition 20.4,

$$aH = \{x \in G \mid x \in aH\} = \{x \in G \mid a^{-1}x \in H\}$$
$$= \{x \in G \mid xH = aH\} = \{x \in G \mid x \sim a\}.$$

Thus another way of thinking of a coset is as the set of all the elements of G that are equivalent to a generator of the coset with respect to $\sim$. For a general equivalence relation, we describe this construction as follows.

DEFINITION

Let $\sim$ be an equivalence relation on a set S. For all $s \in S$, the **equivalence class** generated by s is the set

$$[s] = \{x \in S \mid x \sim s\}.$$

We denote the set of equivalence classes by

$$S/\sim = \{[s] \mid s \in S\}.$$

Example 22.4 Consider the binary relation on $\mathbb{C}$ defined in Example 22.2: $z \sim w$ if and only if $|z| = |w|$. The equivalence classes determined by this relation will be the sets $[z] = \{x \in \mathbb{C} \mid |x| = |z|\}$. Geometrically, $[z]$ will be all the points in the complex plane on the circle of radius $|z|$ centered at the origin. ❖

Example 22.5 Consider the binary relation on $\mathbb{Z}$ defined in Example 22.3: $a \equiv b \pmod{n}$ if and only if n divides $a - b$. The equivalence classes determined by this relation will be the sets $[b] = \{a \in \mathbb{Z} \mid n \text{ divides } a - b\}$. Since n divides $a - b$ if and only if $a = b + kn$ for some $k \in \mathbb{Z}$, $[b] = \{b + kn \mid k \in \mathbb{Z}\} = [b]_n$, and thus the equivalence classes of $\mathbb{Z}$ with respect to this relation are just the elements of $\mathbb{Z}_n$. ❖

Example 22.6 If G is a group and H is a subgroup of G, then by Example 22.1 the relation $\sim$ defined by letting $x \sim y$ if and only if $xH = yH$ is an equivalence relation on G. As noted above, for any $a \in G$, $aH = \{x \in G \mid x \sim a\}$, and therefore, $G/H = G/\sim$. Suppose for instance that $G = S_3$ and $H = \{\iota, (12)\}$. According to Example 20.3,

H is a subgroup of S_3, and hence the relation $\sim$ defined by letting $\sigma \sim \tau$ if and only if $\sigma H = \tau H$ is an equivalence relation on S_3. Since $\sigma H = \{\tau \in S_3 \mid \tau \sim \sigma\}$, we may think of $\sim$ as collecting the elements of S_3 into the left cosets of H in S_3:

S_3/H:	S_3:		$S_3/\sim$:
$\iota H = (12)H = H$	ι	(12)	$[\iota] = [(12)] = H$
$(13)H = (123)H$	(13)	(123)	$[(13)] = [(123)]$
$(23)H = (132)H$	(23)	(132)	$[(23)] = [(132)]$

❖

Equivalence classes consist of elements that are equivalent with respect to the given relation; the set of equivalence classes is the result of ignoring the distinction between equivalent elements of the original set. Thus equality on the set of equivalence classes should correspond to equivalence in the original set, and the original set should be the disjoint union of the elements of the distinct equivalence classes. The next result is the precise statement of these assertions. Note the similarity between this result and Proposition 20.4.

Proposition 22.7   *Let $\sim$ be an equivalence relation on a set S. Then for $a, b \in S$,*

   (i)   either $[a] = [b]$ or $[a] \cap [b] = \varnothing$;
   (ii) $S = \cup[a]$, where the union runs over one element from each class;
   (iii) $[a] = [b]$ if and only if $a \sim b$.

Proof   (i)   Suppose that $[a] \cap [b] \neq \varnothing$ and let $c \in [a] \cap [b]$. If $x \in [a]$, then $x \sim a$ and $c \sim a$. Then by symmetry, $a \sim c$, and hence by transitivity, $x \sim c$. But also $c \sim b$, and hence again by transitivity, $x \sim b$, i.e., $x \in [b]$. We conclude that $[a] \subseteq [b]$. A similar argument shows that $[b] \subseteq [a]$, and therefore that $[a] = [b]$.
   (ii)   Since the equivalence classes are subsets of S, $S \supseteq \cup[a]$. Conversely, for all $s \in S$, $s \in [s]$ by reflexivity so that every element of S is in some equivalence class, i.e., $S \subseteq \cup[a]$.
   (iii)   Note that by definition, $x \in [b]$ if and only if $x \sim b$ and by reflexivity, $a \in [a]$. Thus if $[a] = [b]$, then $a \in [b]$ and hence $a \sim b$. If, on the other hand, $a \sim b$, then $a \in [a] \cap [b]$ and hence by (i), $[a] = [b]$. ■

As mentioned at the beginning of the chapter, we want to use equivalence relations to prove a special case of Cauchy's theorem, viz., that any finite group whose order is divisible by 5 has an element of order 5. The proof relies on the following rather strange equivalence relation. (Note that this equivalence relation is of the type described in general in Exercise 27.)

Example 22.8   Let G be a group and let S be the set of ordered 5-tuples $(v_1, v_2, v_3, v_4, v_5)$ of G whose product is the identity:

$$S = \{(v_1, v_2, v_3, v_4, v_5) \mid v_1 v_2 v_3 v_4 v_5 = e\}.$$

If $(v_1, v_2, v_3, v_4, v_5) \in S$, then $v_1 v_2 v_3 v_4 v_5 = e$. Hence

$$v_5 v_1 v_2 v_3 v_4 = (v_5 v_1 v_2 v_3 v_4) v_5 v_5^{-1} = v_5 (v_1 v_2 v_3 v_4 v_5) v_5^{-1} = v_5 e v_5^{-1} = e,$$

and thus $(v_5, v_1, v_2, v_3, v_4) \in S$. Let $f\colon S \to S$ be the function $f((v_1, v_2, v_3, v_4, v_5)) = (v_5, v_1, V_2, v_3, v_4)$ and let $f^2 = f \circ f, f^3 = f \circ f \circ f$, etc. Using the notation $v = (v_1, v_2, v_3, v_4, v_5)$ for elements of S, we define a binary relation $\sim$ on S by letting

$$v \sim w \text{ if and only if there exists a positive integer } i \text{ such that } f^i(v) = w$$

That is, $v \sim w$ if and only if w is one of the following rearrangements of v:

$$(v_5, v_1, v_2, v_3, v_4), (v_4, v_5, v_1, v_2, v_3), (v_3, v_4, v_5, v_1, v_2), (v_2, v_3, v_4, v_5, v_1),$$
$$\text{or } (v_1, v_2, v_3, v_4, v_5).$$

Note that $f^5(v) = v$ and that in general, for $5 \geq i > 0$ and $a > 0, f^{i+5a}(v) = f^i(v)$.

We will show that $\sim$ is an equivalence relation on S. Since $f^5(v) = v, v \sim v$, and hence $\sim$ is reflexive. For symmetry, suppose that $v \sim w$. Then $f^i(v) = w$ for some i and, as noted above, we can assume that $0 < i \leq 5$. If $i = 5$, then $w = f^i(v) = v$ and since $\sim$ is reflexive, $w \sim v$. If $0 < i < 5$, then $5 - i > 0$ and $f^{5-i}(w) = f^{5-i}(f^i(v)) = f^5(v) = v$. So in this case as well, $w \sim v$, and hence $\sim$ is symmetric. For transitivity, suppose that $v \sim w$ and $w \sim u$. Then for positive integers i and j, $f^i(v) = w$ and $f^j(w) = u$, and hence $f^{j+i}(v) = f^j(f^i(v)) = f^j(w) = u$. Since $j + i$ is a positive integer, $v \sim u$, and hence $\sim$ is transitive. So $\sim$ is an equivalence relation. By the definition of $\sim$, its equivalence classes are the sets

$$[v] = \{f^i(v) \mid i > 0\} = \{f(v), f^2(v), f^3(v), f^4(v), f^5(v)\}$$
$$= \{(v_5, v_1, v_2, v_3, v_4), (v_4, v_5, v_1, v_2, v_3), (v_3, v_4, v_5, v_1, v_2),$$
$$(v_2, v_3, v_4, v_5, v_1), (v_1, v_2, v_3, v_4, v_5)\}.$$

Note that the elements in $[v]$ need not be distinct. For example, $[(e, e, e, e, e)]$ has only the one element (e, e, e, e, e). ❖

The equivalence relation defined in Example 22.8 may be used to prove the following special case of Cauchy's theorem.

Proposition 22.9   *Let G be a finite group of order m. If 5 divides m, then G has an element of order 5.*

Proof   **McKay**

Taking G as the group in Example 22.8, construct the set S and the equivalence relation $\sim$. Observe that picking v_1, v_2, v_3, and v_4 arbitrarily in G uniquely determines an element $(v_1, v_2, v_3, v_4, v_5) \in S$ because $v_1 v_2 v_3 v_4 v_5 = e$ implies that $v_5 = (v_1 v_2 v_3 v_4)^{-1}$.

This in turn implies that S contains m^4 elements. However, the elements of S may also be counted as follows. Each equivalence class $[v]$ contains between one and five elements, and if $v \in S$ and $f(v) = v$, then $[v]$ contains exactly one element, v. Now suppose that $[v]$ contains less than five elements. Then $f^i(v) = f^j(v)$ for some $5 \geq i > j > 0$, and $f^{5-i+j}(v) = f^{5-i}(f^j(v)) = f^{5-i}(f^i(v)) = f^{5-i+i}(v) = f^5(v) = v$. But $5 > 5 + (j - i) = 5 - i + j > 0$, and thus by Proposition 4.3, there exists $0 < k \leq 5$ such that $k(5 - i + j) = 1 + 5a$ for some $a > 0$. Then

$$f(v) = f^{1+5a}(v) = f^{k(5-i+j)}(v) = f^{5-i+j} \circ \cdots \circ f^{5-i+j}(v) = v,$$

and hence $[v] = \{v\}$. (Another way of looking at this is the following. If $v \in S$ and $v_1 = v_2 = v_3 = v_4 = v_5$, then v is the only element in the equivalence class $[v]$, and if, for example, $(v_5, v_1, v_2, v_3, v_4) = (v_3, v_4, v_5, v_1, v_2)$, then by equating corresponding coordinates, we have $v_5 = v_3 = v_1 = v_4 = v_2$, and hence $[(v_1, v_2, v_3, v_3, v_5)] = \{(v_1, v_2, v_3, v_4, v_5)\}$.) So each class $[v]$ contains five elements or contains one element. Let M be the number of equivalence classes containing five elements and let N be the number of equivalence classes containing one element. By Proposition 22.7 (i) and (ii), S has $N + 5M$ elements. So $m^4 = N + 5M$. But 5 divides m^4 by hypothesis and 5 obviously divides $5M$. So 5 divides $m^4 - 5M = N$. Clearly $(e, e, e, e, e) \in S$ and $[(e, e, e, e, e)]$ has only one element so that $N \neq 0$, and thus, since 5 divides N, $N > 1$. So there exists $v \in S$ such that $[v] \neq [(e, e, e, e, e)]$ and $[v]$ has one element. Then $f(v) = v$ and hence $v_5 = v_4 = v_3 = v_2 = v_1$. So there exists $e \neq g \in G$ such that $(g, g, g, g, g) = v \in S$. And since $g^5 = e$ by definition of S, g has order 5 by Corollary 21.6. ∎

Cauchy's theorem says that in general, if p is a prime dividing the order of a finite group, then the group has an element of order p. Proposition 22.9 is a special case of this result and a generalization of this argument may be used to prove the general theorem (Exercise 43).

Exercises

1. Prove Proposition 22.1 for right cosets: If H is a subgroup of a group G, then the binary relation on G defined by

$$x \sim y \quad \text{if and only if} \quad Hx = Hy$$

is an equivalence relation on G.

2. Let G be a group and define a binary relation on G as follows:

$$a \approx b \quad \text{if and only if} \quad \text{there exists } x \in G \text{ such that } x^{-1}ax = b.$$

Show that $\approx$ is an equivalence relation on G. This equivalence relation on G is called <u>conjugacy</u>. If $a \approx b$ in G, we say that a and b are <u>conjugate</u> in G.

3. Let G be a group and define a binary relation on G as follows:

$$x \cong y \quad \text{if and only if} \quad \text{there exist integers } n \text{ and } m \text{ such that } x^n = y \text{ and } y^m = x.$$

Show that $\cong$ is an equivalence relation on G.

4. If a and b are nonzero elements of a ring R with unit element, then a and b are <u>associates</u> if both a divides b and b divides a (i.e., if there exist $c, d \in R$ such that $ac = b$ and $a = bd$). Show that the relation $\cong$ defined by

$$a \cong b \quad \text{if and only if} \quad a \text{ and } b \text{ are associates}$$

is an equivalence relation on the set of nonzero elements of R.

5. Show that if H is a subgroup of a group G, then the equivalence classes of the equivalence relation $\sim$ defined in Exercise 1 are precisely the right cosets of H.

In Exercises 6–10, find the conjugacy classes of the given group. That is, find the equivalence classes of the conjugacy relation (Exercise 2) on the group.

6. $(\mathbb{Z}_6, +)$ **7.** $(\mathbb{Z}_{64}, +)$ **8.** S_3

9. S_4 **10.** $Q = \{\pm 1, \pm i, \pm j, \pm k\}$ (See Exercise 23 in Chapter 15.)

Exercises 11–15 concern the equivalence relation $\cong$ of Exercise 3; in each exercise, find the equivalence classes of the given group.

11. $(\mathbb{Z}_9, +)$ **12.** $(\mathbb{Z}_8, +)$ **13.** $(\mathbb{Z}_{36}, +)$

14. S_3 **15.** $Q = \{\pm 1, \pm i, \pm j, \pm k\}$ (See Exercise 23 in Chapter 15.)

In Exercises 16–21, find the equivalence classes of the associate relation (Exercise 4) on the nonzero elements of the given ring.

16. $\mathbb{Z}_6$ **17.** $\mathbb{Z}_7$ **18.** $\mathbb{Z}_9$

19. $\mathbb{Z}_8$ **20.** $\mathbb{Z}$ **21.** $\mathbb{Q}$

22. For any group G, define a relation $\sim$ on G by letting $x \sim y$ if and only if $xy = yx$. Determine whether $\sim$ is an equivalence relation on G.

23. Show that if G is a finite group whose order is divisible by 5, then G has a subgroup of order 5.

24. Let G be a finite group of order m. If 3 divides m, show that G has an element of order 3.

25. Let G be a finite group of order m. If 7 divides m, show that G has an element of order 7.

26. Let R be a ring with unit element. A <u>congruence relation</u> on R is an equivalence relation $\equiv$ which satisfies the additional condition:

$$\text{if } a \equiv b \text{ and } c \equiv d, \text{ then } a + c \equiv b + d \text{ and } ac \equiv bd.$$

Show that for any congruence relation $\equiv$ on R, $[0]$ is an ideal of R.

Exercises 27 and 28 concern the relation $\sim$ defined in the following way. Let S be a set and suppose that S is the disjoint union of the sets $A_1, \ldots, A_n$ (i.e., $A_i \cap A_j = \varnothing$ if $i \neq j$ and $S = \cup\, A_i$). Define a binary relation $\sim$ on S by letting

$$s \sim t \quad \text{if and only if} \quad A_i = A_j \text{ where } s \in A_i \text{ and } t \in A_j.$$

27. Show that $\sim$ is an equivalence relation on S.

28. Show that for all $s \in A_i$, $[s] = A_i$.

In Exercises 29–34, determine whether the given binary relation is reflexive, symmetric, or transitive. For each property, give a proof or a counterexample.

29. For $x, y \in \mathbb{C}$, $x \sim y \Leftrightarrow x \neq y$

30. For $x, y \in \mathbb{R}$, $x \sim y \Leftrightarrow x \leq y$

31. For $x, y \in \mathbb{R}$, $x \sim y \Leftrightarrow x < y$

32. For $x, y \in \mathbb{C}$, $x \sim y \Leftrightarrow |x| = |y|$

33. For $x, y \in \mathbb{R}$, $x \sim y \Leftrightarrow x \leq |y|$

34. For $x, y \in \mathbb{C}$, $x \sim y \Leftrightarrow x^2 + x = 2y + 1$.

Exercises 35–38 describe the construction of the rational numbers from the integers. Let $\mathbb{Z}^\times$ denote the set of nonzero integers and define a relation $\approx$ on $\mathbb{Z} \times \mathbb{Z}^\times$ as follows:

$$(m, n) \approx (x, y) \quad \text{if and only if} \quad my = nx.$$

35. Show that $\approx$ is an equivalence relation on $\mathbb{Z} \times \mathbb{Z}^\times$.

Let Q denote the set of equivalence classes determined by the relation $\approx$ and for $[(m, n)], [(x, y)] \in Q$, define

$$[(m, n)] + [(x, y)] = [(my + nx, ny)] \quad \text{and} \quad [(m, n)] \cdot [(x, y)] = [(mx, ny)].$$

36. Show that $+$ is a well-defined operation on Q.

37. Show that $\cdot$ is a well-defined operation on Q.

38. Show that $(Q, +, \cdot)$ is a field $((Q, +, \cdot)$ is the field of rational numbers).

Exercises 39–42 describe the construction of the field of quotients of an integral domain. For an integral domain D, let $D^\times$ denote the nonzero elements of D and define a relation $\approx$ on $D \times D^\times$ as follows:

$$(m, n) \approx (x, y) \quad \text{if and only if} \quad my = nx.$$

39. Show that $\approx$ is an equivalence relation on $D \times D^\times$.

Let Q denote the set of equivalence classes determined by the relation $\approx$ and for $[(m, n)], [(x, y)] \in Q$, define

$$[(m, n)] + [(x, y)] = [(my + nx, ny)] \quad \text{and} \quad [(m, n)] \cdot [(x, y)] = [(mx, ny)].$$

40. Show that $+$ is a well defined operation on Q.

41. Show that $\cdot$ is a well defined operation on Q.

42. Show that $(Q, +, \cdot)$ is a field. This field is called the field of quotients of D.

43. Prove Cauchy's theorem: If G is a group of order m and if p is a prime which divides m, then G has an element of order p. (*Hint:* Let S be the set of all ordered p-tuples $(g_1, g_2, \ldots, g_p)$ of G such that $g_1 g_2 \cdots g_p = e$, and let $f: S \to S$ be the function $f((g_1, g_2, \ldots, g_p)) = (g_p, g_1, g_2, \ldots, g_{p-1})$. Define $\approx$ on S by letting $g_1, g_2, \ldots, g_p \approx (h_1, h_2, \ldots, h_p)$ if and only if for some positive integer n, $(h_1, h_2, \ldots, h_p) = f^n((g_1, g_2, \ldots, g_p))$. Show that $\approx$ is an equivalence relation on S whose equivalence classes have either 1 element or p elements. Then proceed as in the proof of Proposition 22.9.)

HISTORICAL NOTE

Augustin-Louis Cauchy

Augustin-Louis Cauchy was born in Paris, France, on August 21, 1789, and died in Sceaux, France (near Paris), on May 22, 1857. His father, Louis-François Cauchy, a government administrator, had been born in Rouen and in 1787 had married Marie-Madeleine Desestre. Altogether, they had four children, the eldest being Augustin-Louis. During the French Revolution, the family escaped to the village of Arceuil, where they were neighbors of Laplace. Augustin-Louis received an excellent education, and in 1805, he was admitted to the École Polytechnique. At the time, it was not unusual for even the most promising students to work as engineers for several years after they graduated, and that is exactly what Cauchy did. His mathematical career began in earnest in 1811 when, at the urging of Lagrange, he solved a geometrical problem concerning the angles of a convex polygon which had been proposed two years earlier by Poinsot. Then, in 1812, he solved an older problem, proposed by Fermat, concerning polygonal numbers. And for the rest of his life, mathematics flowed from him continuously; he published an enormous amount of work in many disparate fields.

Cauchy was a royalist and so staunch a supporter of the Jesuits as to argue that they were hated and persecuted because of their virtue. So it was no surprise that in 1816, when Gaspard Monge was expelled from the Académie des Sciences for his republican activities, Cauchy was appointed (not elected) to succeed him. In 1818, he married Aloïse de Bure, the daughter of a publisher, with much pomp and ceremony, and as proof that his scientific reputation and royalist inclinations had not gone unnoticed, Louis XVIII and the entire royal family signed the marriage contract.

At the same time as he entered the Académie, he was promoted to professor at the École Polytechnique, where he immediately became deeply involved in efforts to rework the teaching of the calculus. Two results of his efforts were *Cour d'Analyse de l'École Royale Polytechnique* (1821) and *Résumé des Leçons Données à l'École Royale Polytechnique sur le Calcul Infinitésimal* (1823), both published by Aloïse's family. They were both landmarks in the theoretical development of the calculus, and the foundations they built still stand today. However, Cauchy's efforts at teaching were not as successful as his writing. His colleagues complained of the liberties he took with the syllabus (on which the examinations were based), and his students complained of the abstract nature of his presentation. Caught between his desire to present the theoretical foundations he was busily developing and the exercises and applications required by the syllabus, he opted for the former and upset the balance of the course. J. Bertrand commented, "When Cauchy replaced Poinsot, the students were divided. 'Poinsot did not teach us anything,' remarked the students who liked the new course; 'Cauchy will disenchant them with science forever,' said Poinsot who never bothered to hide his views. . . . Poinsot, it is true, did not teach many things in any given lecture; but what he did present was presented very well indeed! Cauchy, on the other hand, was forever going beyond bounds, and only a few, very gifted students could understand him."[1]

The July Revolution of 1830 precipitated a revolution in Cauchy's life as well. Louis-Philippe, the choice of the bourgeoisie, replaced Charles X, and Cauchy refused to take the oath of allegiance to the new king. Leaving his family, he journeyed first to Fribourg where he lived with the Jesuits, and then to the court of Charles X in exile in Prague. He continued publishing his mathematical results while he was away from France, but at a somewhat slower rate. In 1838, he returned to Paris, still a member of the Académie, but with no other academic position. So when one came vacant at the Bureau des Longitudes in the summer of 1839, he applied for it. He was duly elected, but since he still refused to take the oath of allegiance, he could neither receive payment nor attend meetings, and thus in reality still lacked a university position.

Even though he was now on the fringes of the French scientific community, his work continued unabated; in fact, since he was no longer teaching, his rate of publication even increased. He also continued his work on behalf of the Jesuits and for the Catholic alternative to public education, efforts which bore bitter fruit in 1843. In June of that year, he was one of three candidates for a vacant position at the public Collège de France. Since one of the other candidates, Liouville, deferred to him, and since the third candidate, Libri, was known as an incompetent mathematician and an embezzler, it was assumed that Cauchy would be elected to the position. However, Libri had recently become a very vocal opponent of the Jesuits and for this reason he, not Cauchy, won the election. And then, six months later, the Bureau des Longitudes, no longer able to tolerate holding Cauchy's position vacant, declared it truly vacant and elected Poinsot in his place. It was only with the establishment of the Second Republic in 1848 that the requirement of an oath of allegiance was repealed and Cauchy was able to resume his chair at the Sorbonne. And in 1852, when Napoleon III reestablished the oath, Cauchy still retained the chair by special exemption.

Between 1844 and 1846, Cauchy published a series of papers on substitution groups, one of which contained his proof that if a prime divides the order of a group, then the group contains an element of that prime order. As happened not infrequently, his work in this area was stimulated by a paper he was asked to referee. On March 17, 1845, he was asked to report on Joseph Bertrand's paper "Sur les nombres des valeurs que peut prendre une function quand on y permute les lettres qu'elle renferme." When he gave his report on the paper, he also reported on his own work, which went considerably beyond that of Bertrand and in fact included the theorem on elements of prime order. Somewhat later, in a paper in 1848, he interpreted the complex numbers as the quotient ring of $\mathbb{R}[x]/(x^2 + 1)$. He continued working right up to his death. Eighteen days before he died, he ended his last communication to the Académie with the words "C'est ce que j'expliquerai plus au long dans un prochain memoire."[2] ("I will furnish the details in my next report.")

While as a devout Catholic, Cauchy took a leading part in various charities, he seems to have been less successful in his personal relationships. Abel called him mad, infinitely Catholic, and bigoted; and Stendhal referred to him as "a veritable Jesuit in short frock."[3] And during an audience with the king of Sardinia in 1831, five times Cauchy responded to a question by saying "I expected your majesty would ask me this, so I have prepared to answer it,"[4] and then, retrieving a memoir from his pocket, he read it.

Cauchy was one of the most prolific mathematicians ever known. He founded the modern theories of complex numbers and elasticity; he made fundamental contributions to analysis and algebra, to partial differential equations and hydrodynamics. He published seven books and over eight hundred papers, and so it is not surprising that most of his work, although correct, was hastily written. In the words of one of his biographers, he was "the most superficial of the great mathematicians, the one who had a sure feeling for what was simple and fundamental without realizing it."[5]

References

1. Belhoste, Bruno. *Augustin-Louis Cachy: A Biography.* New York: Springer-Verlag, 1991, 73–74.
2. Fruedenthal, Hans. "Cauchy, Augustin-Louis." *Dictionary of Scientific Biography.* 16 vols. Charles Coulston, ed. New York: Charles Scribner's Sons, 1970, vol. III, p. 132.
3. Belhoste, Bruno, *op. cit.,* p. 139.
4. Fruedenthal, Hans, *op. cit.,* p. 133.
5. *Ibid.,* p. 135.

23 Normal Subgroups and Quotient Groups

We now return to our main task, that of finding conditions on the Galois group generated by a polynomial $p(x)$ which correspond to the equation $p(x) = 0$ being solvable by radicals. Such an equation is solvable by radicals provided that it gives rise to a particular chain of extension fields, and according to Proposition 18.16 and Example 18.17 such a chain of extension fields gives rise to a corresponding chain of Galois groups. We want to characterize this chain of subgroups group-theoretically. We will show in Chapters 25 and 27 that the left cosets of these subgroups always form Abelian groups.

The objective of this chapter is to determine when multiplication of cosets turns the set of cosets into a group (Proposition 23.6).

We saw in Chapter 20 (Example 20.3) that products of cosets need not always be cosets. We will see later (Proposition 23.6) that when they are, the set of cosets forms a group with respect to multiplication of cosets. We begin by noting several statements which are equivalent to products of cosets being cosets. The statements are all equally important, different applications requiring different phrasings of the same idea.

Proposition 23.1   *For a subgroup H of a group G, the following statements are equivalent.*

   (i)   *For all $x, y \in G$, there exists $g \in G$ such that $(xH)(yH) = gH$.*
   (ii)   *For all $x, y \in G$, $(xH)(yH) = (xy)H$.*
   (iii) *For all $x \in G$, $xHx^{-1} \subseteq H$.*
   (iv)   *For all $x \in G$, $xHx^{-1} = H$.*
   (v)   *For all $x \in G$, $xH = Hx$.*

Proof   We will show that $(i) \Rightarrow (ii)$, $(ii) \Rightarrow (iii)$, $(iii) \Rightarrow (iv)$, $(iv) \Rightarrow (v)$, and $(v) \Rightarrow (i)$. We may then conclude that all five statements are equivalent.

$(i) \Rightarrow (ii)$: Suppose that $(xH)(yH) = gH$. Since H is a subgroup, $e \in H$, and hence $xy = xeye \in (xH)(yH)$. So $xy \in gH$. But as well $xy = xye \in xyH$, and thus $xyH \cap gH \neq \varnothing$. So by Proposition 20.4 (i), $xyH = gH = (xH)(yH)$.

$(ii) \Rightarrow (iii)$: We must show that for any $x \in G$ and $s \in H$, $xsx^{-1} \in H$. Since H is a subgroup, $HH = H$ (cf. Exercise 16 below) and thus, by (ii), $xHx^{-1}H = xx^{-1}HH = eH = H$, and hence, since $s, e \in H$, $xsx^{-1}e \in H$.

$(iii) \Rightarrow (iv)$: Let $x \in G$. By (iii), $xHx^{-1} \subseteq H$ and $x^{-1}Hx = x^{-1}H(x^{-1})^{-1} \subseteq H$. From the second statement, $H = x(x^{-1}Hx)x^{-1} \subseteq xHx^{-1}$ and hence $xHx^{-1} = H$.

$(iv) \Rightarrow (v)$: Let $x \in G$. By (iv), $xHx^{-1} = H$ and hence $xH = xH(x^{-1}x) = (xHx^{-1})x = Hx$.

$(v) \Rightarrow (i)$: Let $x, y \in G$. Since $HH = H$, (v) implies that $(xH)(yH) = x(Hy)H = (xy)HH = (xy)H$. Letting $g = xy \in G$, we have $(xH)(yH) = (xy)H = gH$. ∎

The subgroups described in Examples 20.1 and 20.2 satisfy the equivalent conditions of Proposition 23.1 but the subgroup described in Example 20.3 does not. A subgroup H that satisfies these conditions satisfies condition (ii) in particular, and thus multiplication of cosets, $(xH)(yH) = (xy)H$, becomes a natural multiplication on the set G/H of left cosets. With this in mind, we single out for study the subgroups that satisfy the equivalent conditions of Proposition 23.1.

DEFINITION

Let G be a group. A subgroup N of G is **normal** if for all $x \in G$, $xNx^{-1} \subseteq N$. We use the notation $N \triangleleft G$ to indicate that N is a normal subgroup of G.

So if N is a subset of a group G, then by Proposition 18.1, $N \triangleleft G$ if and only if

(i) $N \neq \emptyset$,

(ii) if $x, y \in N$, then $xy \in N$,

(iii) if $x \in N$ and x^{-1} is the inverse of x in G, then $x^{-1} \in N$

(iv) for all $x \in G$, $xNx^{-1} \subseteq N$.

And by Proposition 18.7, if N is finite, condition (iii) is redundant.

Example 23.2   As noted in Example 21.11, $N = \{\iota, (123), (132)\}$ is a subgroup of S_3, and thus to show that $N \triangleleft S_3$, it suffices to show that for all $\sigma \in S_3$, $\sigma N \sigma^{-1} \subseteq N$. If $\sigma \in N$, then $\sigma \tau \sigma^{-1} \in N$ for all $\tau \in N$ because N is a subgroup, and hence $\sigma N \sigma^{-1} \subseteq N$ for all $\sigma \in N$. For the elements of S_3 not in N, we have

$$(12)N(12)^{-1} = (12)N(12) = \{\iota, (132), (123)\} = N,$$
$$(13)N(13)^{-1} = (13)N(13) = \{\iota, (132), (123)\} = N,$$
$$(23)N(23)^{-1} = (23)N(23) = \{\iota, (132), (123)\} = N,$$

and therefore $N \triangleleft S_3$. ❖

Example 23.3   We observed in Example 20.3 that $H = \{\iota, (12)\}$ is a subgroup of S_3 such that $(13)H(23)H$ is not a left coset. Thus by Proposition 23.1 (i), H is a subgroup of S_3 which is <u>not</u> normal.

In particular,

$$(123)H(123)^{-1} = \{(123)\,\iota(123)^{-1}, (123)(12)(123)^{-1}\} = \{\iota, (23)\} \not\subseteq H. \;❖$$

The next two results list some subgroups that are always normal.

Proposition 23.4   *Every subgroup of an Abelian group is normal.*

Proof   Proposition 23.1 (v) clearly holds for every subgroup of an Abelian group. ∎

Proposition 23.5   *If G and H are groups and $f: G \to H$ is a homomorphism, then $\ker f \lhd G$.*

Proof   By Proposition 18.13, $\ker f$ is a subgroup of G. To see that $\ker f$ is normal, let $x \in G$ and $a \in \ker f$. Then $f(a) = e_H$ and thus by Lemma 18.12, $f(xax^{-1}) = f(x)f(a)f(x)^{-1} = f(x)e_Hf(x)^{-1} = e_H$. Then $xax^{-1} \in \ker f$ and hence $\ker f \lhd G$. ∎

We will see in the next chapter that every normal subgroup is the kernel of some onto homomorphism (Proposition 24.1) and hence that Proposition 23.5 provides yet more characterizations of normal subgroups (Corollary 24.2).

We previously indicated that normal subgroups are important to us (and important in general) because if $N \lhd G$, then G/N is a group with respect to multiplication of cosets.

Proposition 23.6   *If G is a group and $N \lhd G$, then G/N is a group with respect to multiplication of cosets; if G is Abelian, then G/N is Abelian.*

Proof   By Proposition 23.1 (i), the product of two left cosets is another left coset and hence G/N is closed with respect to multiplication of cosets. The operation is associative because by Proposition 23.1, $(xNyN)(zN) = (xy)zN = x(yz)N = xN(yNzN)$. Since $eNxN = xN = xNeN$, $eN = N$ is an identity for G/N. If $xN \in G/N$, then $x^{-1}N \in G/N$ and $x^{-1}NxN = x^{-1}xN = eN = xx^{-1}N = xNx^{-1}N$; hence $x^{-1}N$ is an inverse of xN in G/N. If G is Abelian and $x, y \in G$, then $xNyN = xyN = yxN = yNxN$, and hence G/N is Abelian. ∎

DEFINITION

If $N \lhd G$, then the set $G/N = \{xN \mid x \in G\}$ with operation $xNyN = xyN$ is called the **quotient group** of G modulo N; G/N is frequently read "G mod N."

We are already very familiar with one quotient group.

Example 23.7 Let $n \in \mathbb{Z}$. It was noted in Example 20.2 that $k + (n) = [k]_n$ for all $k \in \mathbb{Z}$, and hence that $\mathbb{Z}/(n) = \mathbb{Z}_n$. Since $\mathbb{Z}$ is an additive group, the group operation on $\mathbb{Z}/(n)$ is $i + (n) + j + (n) = i + j + (n)$, and since $k + (n) = [k]_n$ for all k, this equation may be rewritten as $[i]_n + [j]_n = [i + j]_n$. Since this is precisely the operation in $\mathbb{Z}_n$, $\mathbb{Z}/(n)$ and $\mathbb{Z}_n$ are identical as groups as well as sets. ❖

In the finite case, Lagrange's theorem makes it easy to calculate the order of the quotient group.

Proposition 23.8 *Let G be a finite group of order m and suppose that N is a normal subgroup of G of order n. Then the order of G/N is m/n.*

Proof Proposition 23.8 follows immediately from Lagrange's theorem. ∎

Arguments based on Proposition 23.8 can make it easy to determine small quotient groups. For instance, here is an example of a quotient group formed from the noncommutative group S_3. Note that even though S_3 is not Abelian, the quotient group is Abelian.

Example 23.9 As noted in Example 23.2, $N = \{\iota, (123), (132)\}$ is a normal subgroup of S_3. Since $|N| = 3$ and $|S_3| = 6$, Proposition 23.9 implies that $|S_3/N| = 2$. And since 2 is prime, Corollary 21.17 implies that S_3/N is Abelian; in fact, by Corollary 21.19, S_3/N is isomorphic to $\mathbb{Z}_2$. We can use Proposition 20.4 to find the two elements in S_3/N. For since $\iota^{-1}(12) = (12) \notin N$, $(12)N \neq \iota N$ by Proposition 20.4 (*ii*) and thus $S_3/N = \{N, (12)N\}$. ❖

Exercises

In Exercises 1–5, determine whether the subset S of the group G is a subgroup of G; if it is a subgroup, determine whether it is normal.

1. $G = (\mathbb{Z}_{18}, +)$, $S = \{[0]_{18}, [3]_{18}, [9]_{18}, [15]_{18}, [6]_{18}, [12]_{18}\}$
2. $G = S_3$, $S = \{\iota, (23)\}$
3. $G = S_4$, $S = \{\iota, (12), (34), (12)(34)\}$
4. $G = S_4$, $S = \{\iota, (124), (142)\}$
5. $G = S_4$, $S = \{i, (12), (13), (23), (123), (132)\}$

As noted in Example 23.2, $N = \{\iota, (123), (132)\}$ is a normal subgroup of S_3. In Exercises 6–8, write the given element of S_3/N in the form σN for some $\sigma \in S_3$.

6. $[(12)N]^{-1}(23)N(12)N$
7. $(13)N[(123)N]^{-1}(12)N(23)N$
8. $[(23)N(13)N]^{-1}(123)N(12)N(23)N$

9. Show that every group G has two normal subgroups: $\{e\}$ and G itself.

10. Show that $(\mathbb{Z}, +) \lhd (\mathbb{Q}, +)$.

By Exercise 10, $\mathbb{Z} \lhd \mathbb{Q}$. In Exercises 11 and 12, write the given element of $\mathbb{Q}/\mathbb{Z}$ in the form $r + \mathbb{Z}$ for some $r \in \mathbb{Q}$.

11. $\left(\frac{8}{5} + \mathbb{Z}\right) - \left(\frac{3}{4} + \mathbb{Z}\right) + \left(\frac{107}{2} + \mathbb{Z}\right)$ **12.** $\left(\frac{75}{5} + \mathbb{Z}\right) + \left(\frac{9}{8} + \mathbb{Z}\right) - \left(\frac{1}{16} + \mathbb{Z}\right)$

By Exercise 23 in Chapter 15, the subset $Q = \{\pm 1, \pm i, \pm j, \pm k\}$ of the quaternions is a group with respect to multiplication. Exercises 13 and 14 concern the group Q.

13. List all the subgroups of Q (cf. Exercise 40 in Chapter 21).

14. Show that Q is a noncommutative group all of whose subgroups are normal.

15. Show that if N and M are normal subgroups of a group G, then $N \cap M$ is also a normal subgroup of G (cf. Exercise 13 in Chapter 18).

16. Show in detail that if G is a group and if H is a subgroup of G, then $HH = H$.

17. Let G and H be groups and suppose that $N \lhd G$ and $M \lhd H$. By Exercise 26 in Chapter 15, $G \times H$ is a group with respect to coordinatewise multiplication: $(a,b)(x,y) = (ab,xy)$. Show that $N \times M \lhd G \times H$.

18. Suppose that $N \lhd S_6$, which contains (14532). Determine whether $(2543)N\,(123)N = (1325)N$. (Remember to justify your answer.)

19. Prove Proposition 23.8 in detail: If G is a finite group of order m and N is a normal subgroup of G of order n, then the order of G/N is m/n.

20. If H is a subgroup of a group G, show that $xHx^{-1} = H$ for all $x \in H$.

21. Suppose that H is a subgroup of a group G. According to Exercise 20, $xHx^{-1} = H$ for all $x \in H$. Why does this <u>not</u> imply that H is a normal subgroup of G?

22. Suppose that G is a group of order 30, that $g \in G$ has order 6, and that $S(g)$, the subgroup of G generated by g, is normal in G. Show that $G/S(g)$ is Abelian.

23. Suppose that R and S are rings and that $f : R \to S$ is a ring homomorphism, and recall that ker $f = \{x \in R \mid f(x) = 0\}$. Show that ker f is an ideal of R.

24. Show that if N and M are normal subgroups of a group G, then $NM = \{nm \mid n \in N, m \in M\}$ is also a normal subgroup of G. (*Hint: $abxy = a(xx^{-1})bxy$.*)

In Exercises 25 and 26, let G be a group and let $Z(G) = \{g \in G \mid xg = gx$ for all $x \in G\}$ be the center of G (cf. Exercise 32 in Chapter 18).

25. Show that $Z(G) \lhd G$.

26. Show that if $G/Z(G)$ is cyclic, then G is Abelian.

27. Suppose that $f: G \to H$ is a homomorphism of groups and that $S \lhd H$. Show that $\{x \in G \mid f(x) \in S\} \lhd G$.

As noted in Example 15.3, if R is a ring, then $(R, +)$ is an Abelian group, and hence for any subgroup I of $(R, +)$, we may form the quotient group $R/I = \{x + I \mid x \in R\}$. Define a multiplication on R/I by letting

$$(x + I)(y + I) = xy + I.$$

(Note that this multiplication is <u>not</u> multiplication of the subsets $x + I$ and $y + I$ of R.) In Exercises 28 and 29, suppose that I is an ideal of R and consider R/I with this multiplication and the usual addition of cosets.

28. Show that R/I is a ring (called the <u>quotient ring</u> of I in R). (Remember to show that the multiplication is well defined—see Chapter 4.)

29. Show that the function $\pi: R \to R/I$ defined by letting $\pi(x) = x + I$ is a ring homomorphism such that ker $\pi = I$.

In Exercises 30–32, let G be a group and let $\equiv$ be a <u>congruence relation</u> on G. That is, let $\equiv$ be an equivalence relation on G which satisfies the additional condition:

$$\text{if } a \equiv b \text{ and } c \equiv d, \text{ then } ac \equiv bd.$$

30. Show that $[e] \lhd G$.

31. Show that with respect to the operation $[g] \cdot [h] = [gh]$, $G/\equiv$ is a group. (Remember to show that the operation is well defined—see Chapter 4.)

32. Show that $G/\equiv$ and $G/[e]$ are the same group.

33. State and prove Exercises 30–32 for rings (cf. Exercise 26 in Chapter 22).

34. Suppose that H is a subgroup of a group G. If H has index 2, show that $H \lhd G$.

HISTORICAL NOTE

Otto Hölder

Otto Ludwig Hölder was born in Stuttgart, Germany, on December 22, 1859, and died in Leipzig, Germany, on August 29, 1937. He came from a family of public officials and scholars, his father, also named Otto Hölder, being a teacher of French at the Stuttgart Polytechnikum. Otto (the son) began his university studies in engineering at the Polytechnikum, but in 1877, at the suggestion of a friend of his father's, he entered the University of Berlin to study mathematics.

In Berlin, he came under the influence of Weierstrass, who was somewhat cold but who responded correctly, coherently, and carefully to student questions, and

Kronecker, who was outgoing and friendly but tended to answer questions by turning the conversation to his own work rather than giving a thoughtful and considered
reply. Kronecker frequently invited students to his home but it was the analysis of
Weierstrass that caught Hölder's interest. Eventually in 1882, he developed a conti-
nuity condition for volume density and submitted this work for his dissertation at
Tübingen.

He then turned his attention to Fourier series of functions which he allowed to
be discontinuous and unbounded, and in 1884 submitted this work as his
Habilitationsschrift at Göttingen. He met Klein at Leipzig while he was working on
his Habilitation, and again in Göttingen in 1886 when Klein was lecturing on Galois
theory. He did not get on well with Klein but apparently through the earlier influence
of Kronecker and now that of Klein, Hölder's interests turned to algebra in general
and group theory in particular, a field in which he was to do some of his best work.

The mathematicians at Göttingen decided to offer Hölder an assistant professorship but the Ministry of Culture of the Prussian government refused to approve the
appointment because it was felt that Hölder had insufficient experience as a teacher.
Luckily, Tübingen soon informally made him a similar offer. He was quite prepared
to accept this second offer, but in the meantime, he had unluckily suffered a mental
collapse and had been admitted to a clinic in Erlangen. The mathematicians at
Tübingen discussed the matter earnestly and eventually in August 1889, on the
grounds that the offer had been made and accepted in good faith and in the hopes
that Hölder would recover quickly made the offer formal. Hölder accepted and was
well enough in June 1890 to give his inaugural address at Tübingen. This chain of
events turned out to be beneficial for all concerned; the next four years at Tübingen
were among Hölder's most mathematically fruitful.

In 1894, he was invited to succeed Minkowski at Königsberg. He accepted the
invitation but he was quite unhappy there and stayed only four years. In 1899 he
moved again, this time to succeed Sophus Lie at Leipzig, and the same year married
Helene Lautenschlager. Hölder found Leipzig much more to his liking, and lived
there happily for the rest of his life.

As noted above, his mathematical career began with work in differential equations and potential theory and he soon branched out into group theory and Galois
theory. Later he turned to the foundations of mathematics but attracted less attention
with this work.

He was one of the originators of the concept of quotient group, an idea that took a
surprisingly long time to develop. It was certainly implicit in the work of Galois
(1820s), and Dedekind probably understood the general concept as early as the
1850s (although his ideas remained for the most part unpublished until after his
death in 1916). Jordan's *Traité des Substitutions* (1870) used quotient groups in the
context of permutation groups and Dyck (1882) and Frobenius (1887) had the necessary machinery but failed to use it. It was left for Hölder in "Zurückführung einer

beliebigen alebraischen Gleichung auf eine Kette von Gleichungen" (1889) to state the definition of quotient group in its full generality. Jordan had used the notation $\frac{G}{H}$; Hölder introduced the name "quotient" and used the notation $G \,|\, N$. In his introduction, he stated that the concept was "a group-theoretic idea that has until now not been adequately appreciated."[1]

In the latter part of his life (1914–1923), Hölder concerned himself with philosophical questions, the result being the book *Die Mathematische Methode*. In it, he conceived of mathematics as a hierarchy in which old notions became the objects of mathematical discourse on which new notions were built. From this, he concluded that all of mathematics could never be encompassed by logical formalism, an idea that Gödel made precise in the next decade. Hölder also considered the definition of subset and decided that there can be no set of all subsets. From this, he concluded that Dedekind's construction of the real numbers was flawed and hence that, since the real numbers could not be constructed, their existence had to be postulated.

Reference   1. Quoted in Nicholson, Julia. "The Development and Understanding of the Concept of Quotient Group." *Historia Mathematica* **20**: (1993), p. 81.

24 The Homomorphism Theorem for Groups

The objective of this chapter is to show that if f is an onto homomorphism from the group G to the group H with kernel K, then H is isomorphic to G/K. This result goes by various names: the homomorphism theorem, the fundamental homomorphism theorem, the first isomorphism theorem, etc. In 1930, B. L. van der Waerden called it the homomorphism theorem for groups in his seminal work *Algebra*. This was the first book to make a unified presentation of abstract algebra and was based on lectures given by E. Artin and E. Noether and seminars conducted by Artin, O. Blaschke, O. Schreier, and van der Waerden himself. We will follow the lead of this august assembly and call it the homomorphism theorem for groups.

This result will play an essential role in the next chapter when we determine some key properties of Galois groups of radical extensions.

We begin by showing that for every normal subgroup N of a group G, the natural map from G to G/N $(x \mapsto xN)$ is an onto homomorphism (sometimes called the "canonical" onto homomorphism) whose kernel is N. Combining this result with the homomorphism theorem for groups establishes a one-to-one correspondence between onto homomorphisms and normal subgroups. In particular, the kernel of an onto homomorphism $f: G \to H$ is a normal subgroup K (Proposition 23.5) and thus the natural map from G to G/K is group-theoretically indistinguishable from f by the homomorphism theorem for groups.

Proposition 24.1   *Let G be a group, let $N \triangleleft G$, and define $\pi_N: G \to G/N$ by letting $\pi_N(x) = xN$. Then π_N is an onto homomorphism with kernel N.*

Proof   For $x, y \in G$, $\pi_N(xy) = (xy)N = (xN)(yN) = \pi_N(x)\pi_N(y)$, and hence π_N is a homomorphism. By definition of π_N, $\pi_N(x) = eN$ if and only if $xN = eN$, and by Proposition 20.4, $xN = eN$ if and only if $x \in N$. Thus, since eN is the identity of G/N, $N = \ker \pi_N$. Finally, if $xN \in G/N$, then $\pi_N(x) = xN$ and hence π_N is an onto homomorphism. ∎

As noted in Chapter 23, this result provides further characterizations of normal subgroups.

Corollary 24.2   *For a subset N of a group G, the following statements are equivalent.*

(*i*)   *N is a normal subgroup of G;*
(*ii*)   *N is the kernel of some onto homomorphism;*
(*iii*) *N is the kernel of some homomorphism.*

Proof   We have that $(i) \Rightarrow (ii)$ by Proposition 24.1, that $(ii) \Rightarrow (iii)$ a fortiori, and that $(iii) \Rightarrow (i)$ by Proposition 23.5. ∎

Example 24.3   Since $\mathbb{Q}$ is Abelian and $\mathbb{Z}$ is a subgroup of $\mathbb{Q}$, $\mathbb{Z} \triangleleft \mathbb{Q}$ by Proposition 23.4. Thus by Proposition 24.1, the function $f:\mathbb{Q} \to \mathbb{Q}/\mathbb{Z}$ defined by letting $f(q) = q\mathbb{Z}$ is an onto homomorphism with kernel $\mathbb{Z}$.

As well, we noted in Example 23.2 that $N = \{\iota, (123), (132)\}$ is a normal subgroup of S_3, and hence by Proposition 24.1, the function $p:S_3 \to S_3/N$ defined by letting $p(\sigma) = \sigma N$ is an onto homomorphism with kernel N. ❖

To prove the first homomorphism theorem for groups, we will use the following lemma. Its purpose is to give a quick, and frequently easy, method of showing that a homomorphism is one-to-one.

Lemma 24.4   *Let G and H be groups and let $f: G \to H$ be a homomorphism. Then f is one-to-one if and only if $\ker f \subseteq \{e_G\}$.*

Proof   Suppose that f is one-to-one. If $x \in \ker f$, then $f(x) = e_H = f(e_G)$ by Lemma 18.12, and hence $x = e_G$. Thus $\ker f \subseteq \{e_G\}$. Conversely, suppose that $\ker f \subseteq \{e_G\}$. If $f(x) = f(y)$, then by Lemma 18.12, $f(xy^{-1}) = f(x)f(y^{-1}) = f(x)f(y)^{-1} = e_H$, and thus $xy^{-1} \in \ker f$. Since $\ker f \subseteq \{e_G\}$, $xy^{-1} = e_G$, i.e., $x = y$, and therefore f is one-to-one. ∎

Example 24.5   Consider the function $f:\mathbb{Z} \to \mathbb{Z}$ defined by letting $f(n) = 2n$. Since $f(n + m) = 2(n + m) = 2n + 2m = f(n) + f(m)$, f is a homomorphism. If $n \in \ker f$, then $0 = f(n) = 2n$ and thus $n = 0$. So $\ker f \subseteq \{0\}$, and thus by Lemma 24.4, f is one-to-one.

On the other hand, consider $g:\mathbb{Z} \to \mathbb{Z}_6$, $g(n) = [2n]_6$. Since $g(n + m) = [2(n + m)]_6 = [2m]_6 + [2n]_6 = g(n) + g(m)$, g is a homomorphism. Since $g(3) = [6]_6 = [0]_6$, $3 \in \ker g$ and hence $\ker g \nsubseteq \{0\}$. So Lemma 24.4 implies that g is not one-to-one. ❖

Note that by Lemma 18.12, for all homomorphisms f, $e_G \in \ker f$, i.e., $\ker f \supseteq \{e_G\}$. Thus the test of Lemma 24.4 is sometimes phrased as follows: f is one-to-one if and only if $\ker f = \{e_G\}$.

Theorem 24.6   **Homomorphism Theorem for Groups**

Let G and H be groups and let $f:G \to H$ be an onto homorphism with kernel K. Then $K \lhd G$ and there exists an isomorphism $\phi:G/K \to H$ such that $\phi(\pi_K(x)) = f(x)$.

As indicated above, Theorem 24.6 says that F and π_K are group-theoretically indistinguishable, i.e., that there is no group-theoretic difference between f and π_K in Figure 24.1. The assertion that $\phi(\pi_K(x)) = f(x)$ is sometimes expressed by saying that the diagram in Figure 24.1 "commutes."

FIGURE 24.1

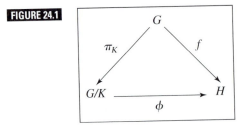

That ϕ is an isomorphism says that H is group-theoretically identical to G/K. Other theorems (given in the exercises) elaborate on the connections between the structure of H and that of G.

Proof   Define $\phi: G/K \to H$ by letting $\phi(xK) = f(x)$. Since it may happen that $xK = yK$ for $x \neq y$, we must first show that this definition is unambiguous. (Note that this amounts to showing that ϕ is well defined in the sense of Chapter 4.) Specifically, we must show that if $xK = yK$, then $f(x) = f(y)$. However, if $xK = yK$, then $y^{-1}x \in K$ by Proposition 20.4 (*ii*), and since $K = \ker f$, Lemma 18.12 implies that $e_H = f(y^{-1}x) = f(y^{-1})f(x) = f(y)^{-1}f(x)$. Hence $f(x) = f(y)$, and thus ϕ is well defined. If $xK, yK \in G/K$, then $\phi((xK)(yK)) = \phi((xy)K) = f(xy) = f(x)f(y) = \phi(xK)\phi(yK)$, and hence ϕ is a homomorphism. To see that ϕ is one-to-one, we will use Lemma 24.4. If $xK \in \ker \phi$, then $f(x) = \phi(xK) = e_H$ and hence $e_G^{-1}x = x \in \ker f = K$. Then by Proposition 20.4 (*ii*), $xK = e_G K$, and thus by Lemma 24.4, ϕ is one-to-one. To see that ϕ is onto, let $y \in H$. Since f is onto, there exists $x \in G$ such that $f(x) = y$, and then $xK \in G/K$ and $\phi(xK) = f(x) = y$. Hence ϕ is onto, and we conclude that ϕ is an isomorphism. Finally, note that by definition of π_K, $\phi(\pi_K(x)) = f(x)$. ∎

We will need Theorem 24.6 to prove Proposition 25.6. We conclude this chapter with two other applications.

Example 24.7   If $f:S_4 \to G$ is an onto homomorphism whose kernel has order 4, then G has order 6. For by Theorem 24.6, G is isomorphic to $S_4/(\ker f)$, and thus G has the same number of elements as $S_4/(\ker f)$. But by Proposition 16.6, the order of S_4 is 4!, and hence by Proposition 23.8, the order of $S_4/(\ker f)$ is $4!/4 = 6$. ❖

Example 24.8   Let $\mathbb{R}^+$ be the multiplicative group of positive real numbers, let $\mathbb{C}^\times$ be the multiplicative group of nonzero complex numbers, and let $U = \{z \in \mathbb{C}^\times \,|\, |z| = 1\}$ be the unit circle. We will use the first homomorphism theorem for groups to show that $U \lhd \mathbb{C}^\times$ and that $\mathbb{C}^\times/U$ is isomorphic to $\mathbb{R}^+$. By definition, $\varnothing \neq U \subseteq \mathbb{C}^\times$. If $z, w \in U$, then by Proposition 3.5, $|zw| = |z|\,|w| = 1 \cdot 1 = 1$ and hence $zw \in U$. Furthermore, $1 = |zz^{-1}| = |z|\,|z^{-1}| = 1|z^{-1}| = |z^{-1}|$ and hence $z^{-1} \in U$ as well. Thus by Proposition 18.1, U is a subgroup of $\mathbb{C}^\times$. Since $\mathbb{C}^\times$ is Abelian, Proposition 23.4 implies that $U \lhd \mathbb{C}^\times$.

   To apply the first homomorphism theorem for groups, we need to define an onto homomorphism $f \colon \mathbb{C}^\times \to \mathbb{R}^+$ with kernel U. That is, we want f to be such that $\{z \in \mathbb{C}^\times \,|\, |z| = 1\} = U = \ker f = \{z \in \mathbb{C}^\times \,|\, f(z) = 1\}$. So for $z \in \mathbb{C}^\times$, let $f(z) = |z|$. Then $f(zw) = |zw| = |z|\,|w| = f(z)f(w)$ and hence f is a homomorphism; if $r \in \mathbb{R}^+$, then $r \in \mathbb{C}^\times$ and $f(r) = |r| = r$ and hence f is an onto homomorphism. By definition of U and f, $U = \ker f$. Therefore by the first homomorphism theorem for groups, there exists an isomorphism $\phi \colon \mathbb{C}^\times/U \to \mathbb{R}^+$.

   Note that it is not really necessary to show separately that $U \lhd \mathbb{C}^\times$. For once we have shown that U is the kernel of f, then Theorem 24.6 (or Proposition 23.5) implies that $U \lhd \mathbb{C}^\times$. ❖

Exercises

In Exercises 1–4, find an onto homomorphism $f \colon G \to H$ and determine $\ker f$. (Remember to justify your answer.)

   1. $G = \mathbb{Z}, H = \mathbb{Z}/(5)$, where $(5) = \{5n \,|\, n \in \mathbb{Z}\}$

   2. $G = \mathbb{C}, H = \mathbb{C}/\mathbb{R}$

   3. $G = \{\pm 1, \pm i, \pm j, \pm k\}, H = G/\{\pm 1, \pm i\}$

   4. $G = \mathbb{C}^\times, H = \mathbb{C}^\times/\{\pm 1\}$

   5. Suppose that G is a group of order 36 and that $f \colon G \to H$ is an onto homomorphism whose kernel has order 9. Determine the order of H. (Remember to justify your answer.)

In Exercises 6–11, show that the function $f \colon G \to H$ is a group homomorphism; then use Lemma 24.4 to determine whether it is one-to-one.

   6. $G = \mathbb{Q}, H = \mathbb{R}, f(q) = \pi q$ **7.** $G = \mathbb{R}, H = \mathbb{C}, f(r) = ri$

   8. $G = \mathbb{C}^\times, H = \mathbb{C}^\times, f(x) = |z|$

9. $G = \mathbb{C}^\times, H = \mathbb{C}^\times, f(x) = \dfrac{z}{|z|}$

10. $G = \mathbb{Z}_2, H = \mathbb{Z}_4, f([n]_2) = [2n]_4$ (first show that f is well defined)

11. $G = \mathbb{Z}_4, H = \mathbb{Z}_2, f([n]_4) = [n]_2$ (first show that f is well defined)

12. Suppose that G is a group of order 35. Show that if $f: G \to H$ is a homomorphism with nontrivial kernel (i.e., ker $f \neq \{e\}$), then $f(G)$ is Abelian.

13. Suppose that p and q are distinct primes and that G is a group of order pq. Show that if $f: G \to H$ is an onto homomorphism which is not a one-to-one homomorphism, then H is Abelian.

In Exercises 14–16, let G and H be groups and let $N = \{(g, e_H) \in G \times H \,|\, g \in G\}$. (For the definition of $G \times H$, see Exercise 26 in Chapter 15.)

14. Show that $N \triangleleft G \times H$.

15. Show that N is isomorphic to G.

16. Show that $(G \times H)/N$ is isomorphic to H.

17. Suppose that G and H are groups and that $A \triangleleft G$ and $B \triangleleft H$. By Exercise 26 in Chapter 15, $G \times H$ is a group with respect to coordinatewise multiplication: $(a, b)(x, y) = (ab, xy)$. And by Exercise 17 in Chapter 23, $A \times B \triangleleft G \times H$. Show that $G/A \times H/B$ is isomorphic to $(G \times H)/(A \times B)$.

18. Show that there is no subgroup S of $\mathbb{Z}$ such that $\{29k \,|\, k \in \mathbb{Z}\} \subset S \subset \mathbb{Z}$. (*Note:* $\subset$ means <u>strict</u> containment: $[X \subset Y] \Leftrightarrow [X \subseteq Y$ and $X \neq Y]$.)

19. Let $f: G \to H$ be a homomorphism of groups. For all $y \in H$, let $f^{-1}(y) = \{g \in G \,|\, f(g) = y\}$. If $x \in G$ and $y \in H$ are such that $f(x) = y$, show that $f^{-1}(y) = \{kx \,|\, k \in \ker f\}$.

In Exercises 20–24, show that $f: G \to H$ is an onto homomorphism. Then construct an isomorphism $\phi: G/(\ker f) \to H$; prove directly (i.e., without using Theorem 24.6) that ϕ is a well-defined isomorphism.

20. $G = \mathbb{Z}, H = \mathbb{Z}_{17}, f(n) = [n]_{17}$

21. $G = \mathbb{Z}_9, H = \mathbb{Z}_3, f([n]_9) = [n]_3$ (show also that f is well defined)

22. $G = \mathbb{Z}_{24}, H = \mathbb{Z}_6, f([n]_{24}) = [n]_6$ (show also that f is well defined)

23. $G = S_3, H = \mathbb{Z}_2, f(\sigma) = \begin{cases} [1]_2 & \text{if } \sigma \text{ is a 2-cycle} \\ [0]_2 & \text{otherwise} \end{cases}$

24. $G = \mathbb{C}, H = \mathbb{C}, f(z) = \bar{z}$

25. (Correspondence theorem) Suppose that $f: G \to H$ is an onto group homomorphism with kernel K, and for any subgroup S of H, let

$$f^{-1}(S) = \{g \in G \mid f(g) \in S\}.$$

Show that for any subgroup S of H, $f^{-1}(S)$ is a subgroup of G, $K \lhd f^{-1}(S)$, and $f^{-1}(S)/K$ is isomorphic to S. Show also that $f^{-1}(S) \lhd G$ if and only if $S \lhd H$.

26. (Second isomorphism theorem) Suppose that $f: G \to H$ is an onto group homomorphism with kernel K. Suppose further that $N \lhd H$ and let $M = f^{-1}(N) = \{g \in G \mid f(g) \in N\}$. Show that $K \lhd M$, that $M \lhd G$, that G/M is isomorphic to H/N, and that G/M is isomorphic to $(G/K)/(M/K)$.

Exercises 27–30 prove the first isomorphism theorem. Let G be a group, let H be a subgroup of G, let N be a normal subgroup of G, and recall that $HN = \{hn \mid h \in H, n \in N\}$.

27. Prove that $H \cap N \lhd H$.

28. Prove that HN is a subgroup of G.

29. Prove that $N \lhd HN$.

30. Prove that $(HN)/N$ is isomorphic to $H/(H \cap N)$.

Exercises 31–33 concern the quotient rings constructed in Exercise 28 in Chapter 23. Let R be a ring and suppose that I is an ideal of R. According to Exercise 28 in Chapter 23, R/I is a ring with respect to the operations $[x + I] + [y + I] = [x + y] + I$ and $[x + I][y + I] = [xy] + I$. Exercise 31 is the homomorphism theorem for rings; Exercise 32 explains how to construct algebraic extensions by using quotient rings; Exercise 33 applies Exercise 32 to the complex numbers.

31. Let R and S be rings and let $f: R \to S$ be an onto ring homomorphism with kernel K. Then K is an ideal of R (by Exercise 23 in Chapter 23). Prove that there exists a ring isomorphism $\phi: R/K \to S$ such that $\phi(x + K) = f(x)$.

32. Let r be algebraic over a field F and let $m(x)$ in $F[x]$ be the minimum polynomial of r over F. Show that $F(r)$ is ring isomorphic to $F[x]/(m(x))$. (*Hint:* Define $T: F[x] \to F(r)$ by letting $T(p(x)) = p(r)$; then show that T is an onto ring homomorphism with kernel $(m(x))$ and apply Exercise 31.)

33. Show that $\mathbb{C}$ is isomorphic to $\mathbb{R}[x]/(x^2 + 1)$.

HISTORICAL NOTE

B. L. van der Waerden

Bartel Leendert van der Waerden was born in Amsterdam, the Netherlands, on February 27, 1903, and died in Zürich, Switzerland, on January 12, 1996. His father was a leftist Social Democrat who felt close to the communists but remained with the Social Democratic Party when the two groups split. Since his father taught mathematics, Bartel grew up in a household with many mathematics books. But they were locked away because his father wanted him to play outside! Bartel circumvented this prohibition by reading books from the local library and by creating his own mathematics. For instance, he learned of the cosine and from this found the other trigonometric functions; he defined the sine as "the square root of one minus the cosine squared."[1]

He finished high school early in 1919 and attended the University of Amsterdam from 1919 to 1924. While there, he became interested in algebraic geometry; in particular, he worried that the foundations of the subject were not secure. When he finished his studies at the university, he asked his father to support him for another year so that he could visit Göttingen which was at the time one of the most important mathematical institutions in the world. Because he had finished his schooling so quickly, his father agreed and Bartel headed off to Germany.

While at Göttingen, he met Hilbert whose home he visited often, and he also met the person who was to have such a decisive influence on his future career: Emmy Noether. Her previous work provided him with the tools he needed and she was able to direct him to the papers that were essential for him to continue his work.

In 1925, he returned to Holland, specifically to the marine base in Den Helder, to complete his mandatory year of military service. He was able to find enough time in between his military duties to complete his Ph.D. thesis on the foundations of algebraic geometry, and in 1926, he defended his thesis at the University of Amsterdam. His written thesis was quite short because it contained only the statements of his theorems. At the time, theses at the university could only be written in Dutch or Latin and he was working in German. So he published his proofs in German in *Mathematische Annalen* and merely summarized his work in Dutch.

Later in 1926, Emmy Noether's recommendation secured him a Rockefeller fellowship to visit Hamburg so that he could study with Artin, Hecke, and Schreier. And then, a year or so after that, he accepted a position in Groningen. The story of how he came to Groningen is rather curious. Among the books he had found in the library in Amsterdam when he was growing up was one on analytic geometry by Barrau, a professor at Groningen. Van der Waerden found "many theorems insufficiently proven, even insufficiently formulated,"[2] and wrote to Barrau, who responded by writing that, in van der Waerden's words, "should he [Barrau] leave

[Groningen], they would have to nominate van der Waerden as his successor."[3] And many years later that is exactly what they did.

In 1929, he returned to Göttingen as a visitor. And this time he met Camilla Rellich, the sister of the mathematician Franz Rellich. In Camilla's words, "We met in July and were married in September. Then we went to Groningen. After a while, Emmy Noether called and said 'Time to end the honeymoon; back to work again!'"[4] Van der Waerden did indeed go back to work but not at the expense of his marriage. He and Camilla remained together until his death some sixty-seven years later.

While van der Waerden was in Hamburg, Artin was supposed to be writing a book on algebra. He suggested that the two of them collaborate; so van der Waerden began to write. He finished the first chapter and then the second. Artin was very impressed with van der Waerden's work and suggested that he take over the whole project. The result was *Moderne Algebra,* one of the most influential mathematics books of the twentieth century. According to Saunders MacLane, "This beautiful and eloquent text served to transform the graduate teaching of algebra. . . . It formulated clearly and succinctly the conceptual and structural insights which Noether had expressed so forcefully. This was combined with the elegance and understanding with which Artin had lectured."[5]

In 1931, van der Waerden moved to Leipzig, where he met Heisenberg and became interested in physics. The result was another important book, this time on group theory and quantum mechanics.

One of his close friends in Leipzig was the philosopher Gadamer, who early in the war was able to use his influence with a former student, the wife of the then chief of police, to rescue van der Waerden from jail, where he had been taken because he was Dutch. With Heisenberg, van der Waerden talked only of science; with Gadamer, he talked politics. Once he attended a course that Gadamer was teaching on Plato. In van der Waerden's words, Gadamer "explained, as Plato shows in the *Republic,* that a dictator is necessarily antagonistic to a reasonable person and finally that a dictator necessarily destroys himself. At first he ruins his enemies, then his friends, and finally himself. There were certainly also Nazi students in the class, but they did not understand him."[6]

In 1943, he and his family were bombed out of Leipzig. While on the train to Dresden, they met one of van der Waerden's former students who suggested they join her family in Bischofswerda. They remained there for a year and then returned to Leipzig. But in 1945, the bombing intensified again and they went to Austria, near Graz, to stay with Camilla's mother. The war ended with them there, where as "displaced persons," they were returned to Holland.

Their situation in Holland was difficult. Van der Waerden had refused an offer to come to Utrecht during the war because it meant he would receive his appointment from the Nazi minister of public instruction and he did not want that. Nonetheless, since he had spent the war in Germany, he had a very difficult time finding work in

Holland. Eventually, when he was down to his last penny, Fruedenthal called and offered him a job with Shell. He readily accepted and spent the next couple of years solving optimization problems in industry.

In 1947, he journeyed to Baltimore to spend a year at Johns Hopkins, and the next year he was finally offered a position at the University of Amsterdam. However, it was still quite difficult for him in Holland. So when he received an offer from Zürich in 1951, he accepted and moved to Switzerland, where he spent the rest of his life.

When he was young, van der Waerden studied some of the history of mathematics, and when he was in Göttingen, he attended lectures on Greek mathematics given by Neugebauer. When he returned to Holland in 1951, he began the work in the history of mathematics which resulted in his books *Science Awakening* and somewhat later *Geometry and Algebra in Ancient Civilizations* and *A History of Algebra*.

Van der Waerden's mathematical career spanned several fields over many years. He wrote technical papers on algebraic geometry, wrote books on quantum mechanics, and applied his knowledge in industry. But very possibly the achievement for which he will be longest remembered is his seminal, and enormously influential, early work *Moderne Algebra*.

References

1. Dold-Samplonius, Yvonne. "Interview with Bartel Leendert van der Waerden." *Notices of the American Mathematical Society* **44** (1997): 313.
2. *Ibid.*, p. 315.
3. *Ibid.*, p. 315.
4. *Ibid.*, p. 316.
5. MacLane, Saunders. "Van der Waerden's Modern Algebra." *Notices of the American Mathematical Society* **44** (1997): p. 321.
6. Dold-Samplonius, *op. cit.*, p. 317.

PART 4

Polynomial Equations Not Solvable by Radicals

25 Galois Groups of Radical Extensions

We began our excursion into the study of groups in the hope that such a study could simplify and clarify the study of field extensions determined by the solutions of polynomial equations that are solvable by radicals. In this chapter, we return to consideration of such extensions by radicals and, in keeping with our avowed purpose, we determine conditions on the Galois group that many radical extensions must always satisfy. We will show in Chapter 28 that field extensions generated by the solutions of polynomial equations that are solvable by radicals also enjoy these properties.

Recall first that if F is a subfield of $\mathbb{C}$, if $p(x) \in F[x]$, and if the equation $p(x) = 0$ is solvable by radicals, then there exist nonzero complex numbers $r_1, \ldots, r_m$ and positive integers $k_1, \ldots, k_m$ such that

(a) $F^{p(x)} \subseteq F(r_1, \ldots, r_m)$,

(b) $(r_1)^{k_1} \in F$ and for $1 < i \leq m$, $(r_i)^{k_i} \in F(r_1, \ldots, r_{i-1})$.

If $F_0 = F$ and $F_i = F(r_1, \ldots, r_i)$ for each i, then by Example 18.15, the chain of fields

$$F_0 \subseteq F_1 \subseteq \cdots \subseteq F_{m-1} \subseteq F_m$$

in F_m gives rise to a chain of subgroups

$$\text{Gal } F_m/F_0 \supseteq \text{Gal } F_m/F_1 \supseteq \cdots \supseteq \text{Gal } F_m/F_{m-1} \supseteq \text{Gal } F_m/F_m = \{\iota\}$$

in $\text{Gal } F_m/F_0$. We want to investigate properties of this chain, properties that require that either $r_i = \zeta_{k_i}$ or ζ_{k_i} be already present when r_i is adjoined. This may not happen, as the following example shows.

Example 25.1   Consider $p(x) = (x^3 - 2)(x^5 - 7) \in \mathbb{Q}[x]$. To show that $p(x) = 0$ is solvable by radicals, we could pick the following:

$$r_1 = \sqrt[3]{2}, \qquad k_1 = 3, \qquad\qquad r_4 = \sqrt[5]{7}, \qquad k_4 = 5,$$
$$r_2 = \sqrt[3]{2}\,\zeta_3, \quad k_2 = 3, \qquad\qquad r_5 = \sqrt[5]{7}\zeta_5, \quad k_5 = 5,$$
$$r_3 = \sqrt[3]{2}\zeta_3^{\,2}, \quad k_3 = 3, \qquad\qquad r_6 = \sqrt[5]{7}\zeta_5^{\,2}, \quad k_6 = 5,$$
$$r_7 = \sqrt[5]{7}\zeta_5^{\,3}, \quad k_7 = 5,$$
$$r_8 = \sqrt[5]{7}\zeta_5^{\,4}, \quad k_8 = 5.$$

Then $\mathbb{Q}^{p(x)} = \mathbb{Q}(r_1, \ldots, r_8)$, $r_1^{k_1} = 2 \in \mathbb{Q}$, and it is easy to see that for all $i > 1$, $r_i^{k_i} \in \mathbb{Q}(r_1, \ldots, r_{i-1})$. However, $r_1 \neq \zeta_3 = \zeta_{k_1}$ and $\zeta_{k_1} \notin \mathbb{Q}$ and $r_2 \neq \zeta_3 = \zeta_{k_2}$ and $\zeta_{k_2} \notin \mathbb{Q}(r_1)$.

Since solvability by radicals requires only that $\mathbb{Q}^{p(x)}$ be <u>contained</u> in a radical extension of $\mathbb{Q}$ we can ensure that either $r_i = \zeta_{k_i}$ or ζ_{k_i} is already present when r_i is adjoined by augmenting the list of radicals as follows:

$$s_1 = \zeta_3, \qquad l_1 = 3, \qquad\qquad s_6 = \sqrt[5]{7}, \qquad l_6 = 5,$$
$$s_2 = \zeta_5, \qquad l_2 = 5, \qquad\qquad s_7 = \sqrt[5]{7}\zeta_5, \quad l_7 = 5,$$
$$s_3 = \sqrt[3]{2}, \qquad l_3 = 3, \qquad\qquad s_8 = \sqrt[5]{7}\zeta_5^{\,2}, \quad l_8 = 5,$$
$$s_4 = \sqrt[3]{2}\zeta_3, \quad l_4 = 3, \qquad\qquad s_9 = \sqrt[5]{7}\zeta_5^{\,3}, \quad l_9 = 5,$$
$$s_5 = \sqrt[3]{2}\zeta_3^{\,2}, \quad l_5 = 3, \qquad\qquad s_{10} = \sqrt[5]{7}\zeta_5^{\,4}, \quad l_{10} = 5.$$

Then certainly $\mathbb{Q}^{p(x)} \subseteq \mathbb{Q}(s_1, \ldots, s_{10})$, $s_1^{l_1} = 1 \in \mathbb{Q}$, and it is again easy to see that for all $i > 1$, $s_i^{l_i} \in \mathbb{Q}(s_1, \ldots, s_{i-1})$. In this case, as well, $s_1 = \zeta_{l_1}$, and for $i > 1$, either $s_i = \zeta_{l_i}$ or $\zeta_{l_i} \in \mathbb{Q}(s_1, \ldots, s_{i-1})$.

Of course our list is redundant; for $s_4, s_5 \in \mathbb{Q}(s_1, s_2, s_3)$ and $s_7, s_8, s_9, s_{10} \in \mathbb{Q}(s_1, s_2, s_3, s_6)$. So we could equally well choose

$$t_1 = \zeta_3, \qquad j_1 = 3, \qquad\qquad t_4 = \sqrt[5]{7}, \quad j_4 = 5.$$
$$t_2 = \zeta_5, \qquad j_2 = 5,$$
$$t_3 = \sqrt[3]{2}, \qquad j_3 = 3,$$

For in this case, as well, $\mathbb{Q}^{p(x)} \subseteq \mathbb{Q}(t_1, t_2, t_3, t_4)$; $t_1^{j_1} = 1 \in \mathbb{Q}$ and for all $i > 1$, $t_i^{j_i} \in \mathbb{Q}(t_1, \ldots, t_{i-1})$; and $t_1 = \zeta_{j_1}$ and for all $i > 1$, either $t_i = \zeta_{j_i}$ or $\zeta_{j_i} \in \mathbb{Q}(t_1, \ldots, t_{i-1})$. ❖

In general,

$$F_0^{\,p(x)} \subseteq F_0(r_1, \ldots, r_m) \subseteq F_0(\zeta_{k_1}, \ldots, \zeta_{k_m}, r_1, \ldots, r_m),$$

and hence, since for any positive integer j, $(\zeta_j)^j = 1 \in F_0 \subseteq F_{i-1}$, we may, by augmenting and rearranging our list of r_i if necessary, also assume that $r_1, \ldots, r_m$ satisfy the additional condition:

(c)   either $r_i = \zeta_{k_i}$ or $\zeta_{k_i} \in F_{i-1}$.

The main results of this chapter will thus be proved under the following hypotheses.

HYPOTHESES S

F_0 is a subfield of $\mathbb{C}$, $r_1, \ldots, r_m$ are nonzero complex numbers, $k_1, \ldots, k_m$ are positive integers, and for $1 \le i \le m$,

    (i) $F_i = F_0(r_1, \ldots, r_i)$,

    (ii) $(r_i)^{k_i} \in F_{i-1}$,

    (iii) either $r_i = \zeta_{k_i}$ or $\zeta_{k_i} \in F_{i-1}$.

Our objective is to show that with these hypotheses, $\mathrm{Gal}\, F_m/F_i$ is always a normal subgroup of $\mathrm{Gal}\, F_m/F_{i-1}$ and the resulting quotient group $(\mathrm{Gal}\, F_m/F_{i-1})/(\mathrm{Gal}\, F_m/F_i)$ is always Abelian.

The five parts of the proof may be summarized as follows.

(1)   (Lemma 25.2) Suppose that $f \in \mathrm{Gal}\, F_m/F_{i-1}$. By Propositions 3.8 and 13.9, $f(r_i) = r_i \zeta_{k_i}^{\,l}$ and since r_i, $\zeta_{k_i} \in F_i$ by hypotheses S, $f(r_i) \in F_i$. So by Theorem 10.8, $f(F_i) \subseteq F_i$.

(2)   (Proposition 25.3) Suppose that $f \in \mathrm{Gal}\, F_m/F_{i-1}$ and $g \in \mathrm{Gal}\, F_m/F_i$. Since $f^{-1} \in \mathrm{Gal}\, F_m/F_{i-1}$, Lemma 25.2 implies that $f^{-1}(x) \in F_i$ for all $x \in F_i$ and hence, since g leaves elements of F_i unchanged, $g(f^{-1}(x)) = f^{-1}(x)$. Then $f \circ g \circ f^{-1}(x) = f(f^{-1}(x)) = x$, and thus $f \circ g \circ f^{-1}$ also leaves elements of F_i unchanged, i.e., $f \circ g \circ f^{-1} \in \mathrm{Gal}\, F_m/F_i$. Therefore $\mathrm{Gal}\, F_m/F_i \vartriangleleft \mathrm{Gal}\, F_m/F_{i-1}$.

(3)   (Proposition 25.4) If σ, $\tau \in \mathrm{Gal}\, F_i/F_{i-1}$, then $\sigma(r_i) = r_i \zeta_{k_i}^{\,s}$ and $\tau(r_i) = r_i \zeta_{k_i}^{\,t}$ by Propositions 3.8 and 13.9. Thus, if $r_i = \zeta_{k_i}$, then $\sigma(\tau(r_i)) = \sigma(\zeta_{k_i}^{\,(t+1)}) = \zeta_{k_i}^{\,(s+1)(t+1)} = \tau(\zeta_{k_i}^{\,(s+1)}) = \tau(\sigma(r_i))$; and if $\zeta_{k_i} \in F_{i-1}$, then

$$\sigma(\tau(r_i)) = \sigma(r_i \zeta_{k_i}^{\,t}) = \sigma(r_i)\sigma(\zeta_{k_i})^t = r_i \zeta_{k_i}^{\,t+s} = \tau(r_i)\tau(\zeta_{k_i})^s = \tau(r_i \zeta_{k_i}^{\,s}) = \tau(\sigma(r_i)).$$

Hence, in either case, for $a_j \in F_{i-1}$,

$$\begin{aligned}
\sigma(\tau(a_0 + \cdots + a_d r_i^{\,d})) &= a_0 + \cdots + a_d \sigma(\tau(r_i))^d \\
&= a_0 + \cdots + a_d \tau(\sigma(r_i))^d = \tau(\sigma(a_0 + \cdots + a_d r_i^{\,d})),
\end{aligned}$$

and therefore by Theorem 10.8, $\mathrm{Gal}\, F_i/F_{i-1}$ is Abelian.

(4)   (Lemma 25.5) If $f \in \mathrm{Gal}\, F_m/F_{i-1}$, then by Lemma 25.2 we may define $f_R : F_i \to F_i$ by letting $f_R(z) = f(z)$. Since f preserves addition and multiplication, is one-to-one and leaves elements of F_0 unchanged, so does f_R. Since f^{-1} is also in $\mathrm{Gal}\, F_m/F_0$ and $f_R(f^{-1}(z)) = z$, f_R is onto. Thus $f_R \in \mathrm{Gal}\, F_i/F_{i-1}$ and we may define a function $R : \mathrm{Gal}\, F_m/F_{i-1} \to \mathrm{Gal}\, F_i/F_{i-1}$ by letting $R(f) = f_R$. By definition, R preserves composition, and since $R(f)$ is the identity if and only if f leaves every element of F_i fixed, $\ker R = \mathrm{Gal}\, F_m/F_i$.

(5)   (Proposition 25.6) Finally, since $R(\mathrm{Gal}\, F_m/F_{i-1})$ is a subgroup of $\mathrm{Gal}\, F_i/F_{i-1}$ and $\mathrm{Gal}\, F_i/F_{i-1}$ is Abelian, it follows that $R(\mathrm{Gal}\, F_m/F_{i-1})$ is Abelian. By

the homomorphism theorem for groups (Theorem 24.6), since ker $R = \text{Gal } F_m/F_i$, $(\text{Gal } F_m/F_{i-1})/(\text{Gal } F_m/F_i)$ is isomorphic to $R(\text{Gal } F_m/F_{i-1})$, and therefore $(\text{Gal } F_m/F_{i-1})/(\text{Gal } F_m/F_i)$ is Abelian.

The detailed arguments are as follows.

Lemma 25.2   *If **hypotheses S** hold, and if $f \in \text{Gal } F_m/F_{i-1}$, then $f(F_i) \subseteq F_i$.*

Proof   We must show that for all $b \in F_i, f(b) \in F_i$. Since $F_i = F_{i-1}(r_i)$, Theorem 10.8 implies that for some positive integer $d, b = a_0 + a_1 r_i + \cdots + a_d r_i^d$, where $a_j \in F_{i-1}$ for all j. Since $f \in \text{Gal } F_m/F_{i-1}, f$ leaves elements of F_{i-1} unchanged, and hence $f(a_j) = a_j$ for all j. To determine $f(r_i)$, note that if $p(x) = x^{k_i} - (r_i)^{k_i}$, then $p(x) \in F_{i-1}[x]$ and $p(r_i) = 0$. But since $r_i^{k_i} \in F_{i-1}$ and f leaves elements of F_{i-1} unchanged, Proposition 13.9 implies that $p(f(r_i)) = 0$ as well. Then by Proposition 3.8, $f(r_i) = r_i \zeta_{k_i}^l$ for some $0 \le l < k_i$, and by definition of $F_i, r_i \in F_i$. But since (by hypotheses S) $r_i = \zeta_{k_i}$ or $\zeta_{k_i} \in F_{i-1}, \zeta_{k_i}^l \in F_i$ as well, and we conclude that $f(r_i) = r_i \zeta_{k_i}^l \in F_i$. Therefore

$$f(b) = f(a_0 + a_1 r_i + \cdots + a_d r_i^d) = a_0 + a_1 f(r_i) + \cdots + a_d f(r_i)^d \in F_i. \quad \blacksquare$$

Proposition 25.3   *If **hypotheses S** hold, then $\text{Gal } F_m/F_i \triangleleft \text{Gal } F_m/F_{i-1}$.*

Proof   By Proposition 18.16, $\text{Gal } F_m/F_i$ is a subgroup of $\text{Gal } F_m/F_{i-1}$. To see that it is normal, let $g \in \text{Gal } F_m/F_i$ and $f \in \text{Gal } F_m/F_{i-1}$. It suffices to show that $f \circ g \circ f^{-1} \in \text{Gal } F_m/F_i$. And since $f \circ g \circ f^{-1} \in \text{Gal } F_m/F_{i-1}, f \circ g \circ f^{-1}$ is a one-to-one, onto homomorphism from F_m to itself, and hence it suffices merely to show that $f \circ g \circ f^{-1}$ leaves elements of F_i unchanged. But if $b \in F_i$, then $f^{-1}(b) \in F_i$ by Lemma 25.2 and hence $g(f^{-1}(b)) = f^{-1}(b)$. Then $f(g(f^{-1}(b))) = f(f^{-1}(b)) = b$, and therefore $f \circ g \circ f^{-1}$ leaves elements of F_i unchanged. $\quad \blacksquare$

Proposition 25.4   *If **hypotheses S** hold, then $\text{Gal } F_i/F_{i-1}$ is Abelian.*

Proof   Let $\sigma, \tau \in \text{Gal } F_i/F_{i-1}$ and let $b \in F_i$. We want to show that $\sigma \circ \tau(b) = \tau \circ \sigma(b)$. As in the proof of Lemma 25.2, Theorem 10.8 implies that for some positive integer $d, b = a_0 + a_1 r_i + \cdots + a_d r_i^d$, where $a_j \in F_{i-1}$ for all j. By Proposition 13.9, $\sigma(r_i)$ and $\tau(r_i)$ both solve $x^{k_i} - (r_i)^{k_i} = 0$; and hence by Proposition 3.8, $\sigma(r_i) = r_i(\zeta_{k_i})^s$ and $\tau(r_i) = r_i(\zeta_{k_i})^t$ for some $0 \le s, t < k_i$. If $r_i = \zeta_{k_i}$, then

$$\sigma \circ \tau(r_i) = \sigma \circ \tau(\zeta_{k_i}) = \sigma(\tau(\zeta_{k_i})) = \sigma(\zeta_{k_i}^{t+1})$$
$$= (\zeta_{k_i})^{(s+1)(t+1)} = \tau(\zeta_{k_i}^{s+1}) = \tau(\sigma(\zeta_{k_i})) = \tau \circ \sigma(\zeta_{k_i}) = \tau \circ \sigma(r_i).$$

Otherwise, by hypotheses S, $(\zeta_{k_i})^s, (\zeta_{k_i})^t \in F_{i-1}$, and thus, since σ and τ both fix F_{i-1},

$$\sigma \circ \tau(r_i) = \sigma(r_i(\zeta_{k_i})^t) = \sigma(r_i)\,\sigma((\zeta_{k_i})^t) = r_i(\zeta_{k_i})^s(\zeta_{k_i})^t = r_i(\zeta_{k_i})^{s+t}$$
$$= r_i(\zeta_{k_i})^t(\zeta_{k_i})^s = \tau(r_i)\tau((\zeta_{k_i})^s) = \tau(r_i(\zeta_{k_i})^s) = \tau \circ \sigma(r_1).$$

Therefore, in both cases,

$$\sigma \circ \tau(b) = a_0 + a_1\sigma \circ \tau(r_i) + \cdots + a_d(\sigma \circ \tau(r_i))^d$$
$$= a_0 + a_1\tau \circ \sigma(r_i) + \cdots + a_d(\tau \circ \sigma(r_i))^d = \tau \circ \sigma(b).$$

We conclude that $\sigma \circ \tau = \tau \circ \sigma$ and hence that Gal F_i/F_{i-1} is Abelian. ∎

We actually prove a more general version of Lemma 25.5 than the one described in the preceding summary. The generality introduced here will be needed in the sequel.

Lemma 25.5   *Suppose that F, L, and K are fields such that $F \subseteq L \subseteq K$ and for all $\tau \in$ Gal K/F, $\tau(L) \subseteq L$. Let $\sigma \in$ Gal K/F and for $v \in L$, let $\sigma_L(v) = \sigma(v)$. Then $\sigma_L \in$ Gal L/F and the function which takes $\sigma \in$ Gal K/F to $\sigma_L \in$ Gal L/F is a homomorphism with kernel Gal K/L.*

Proof   (The function σ_L is called the "restriction of σ to L.") By hypothesis, $\sigma(L) \subseteq L$ and hence $\sigma_L : L \to L$. Since σ preserves addition and multiplication, is one-to-one, and leaves elements of F unchanged, so does σ_L. To see that σ_L is onto, note that since $\sigma \in$ Gal K/F and Gal K/F is a group, $\sigma^{-1} \in$ Gal K/F. Thus by hypothesis, for any $v \in L$, $\sigma^{-1}(v) \in L$, and thus $\sigma_L(\sigma^{-1}(v)) = \sigma(\sigma^{-1}(v)) = v$. Therefore, σ_L is onto and hence $\sigma_L \in$ Gal L/F. Let $R :$ Gal $K/F \to$ Gal L/F be the function $R(\sigma) = \sigma_L$. By definition, $(\sigma \circ \tau)_L = \sigma_L \circ \tau_L$ for all $\sigma, \tau \in$ Gal K/F and hence $R(\sigma \circ \tau) = R(\sigma) \circ R(\tau)$ so that R is a homomorphism. If $R(\sigma) = \iota$, then $\sigma_L = \iota$, hence σ leaves elements of L unchanged, and thus $\sigma \in$ Gal K/L. Conversely, if $\sigma \in$ Gal K/L, then σ leaves elements of L unchanged and hence $R(\sigma) = \sigma_L = \iota$. We conclude that ker $R =$ Gal K/L. ∎

Proposition 25.6   *If hypotheses S hold, then (Gal $F_m/F_{i-1})/($Gal $F_m/F_i)$ is Abelian.*

Proof   If $\sigma \in$ Gal F_m/F_{i-1}, then by Lemma 25.2, $\sigma(F_i) \subseteq F_i$, and thus, by Lemma 25.5, restricting the domain of an element in Gal F_m/F_{i-1} to F_i defines a homomorphism $R :$ Gal $F_m/F_{i-1} \to$ Gal F_i/F_{i-1} with kernel Gal F_m/F_i. By Proposition 18.13, $R($Gal $F_m/F_{i-1})$ is a subgroup of Gal F_i/F_{i-1} and by Proposition 25.4, Gal F_i/F_{i-1} is Abelian. So by Proposition 18.1, $R($Gal $F_m/F_{i-1})$ is Abelian. But ker $R =$ Gal F_m/F_i, and thus, by the homomorphism theorem for groups (Theorem 24.6), (Gal $F_m/F_{i-1})/($Gal $F_m/F_i)$ is isomorphic to $R($Gal $F_m/F_{i-1})$. So the quotient group (Gal $F_m/F_{i-1})/($Gal $F_m/F_i)$ is a homomorphic image of the Abelian group $R($Gal $F_m/F_{i-1})$, and hence by Proposition 18.13, (Gal $F_m/F_{i-1})/($Gal $F_m/F_i)$ is Abelian. ∎

The properties of the Galois group described by the preceding results are summarized in Figure 25.1.

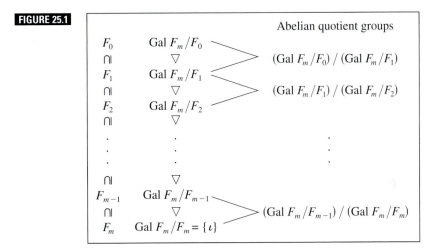

FIGURE 25.1

Example 25.7 Let $F_0 = \mathbb{Q}$ and let $K = \mathbb{Q}(\sqrt[5]{3}, \sqrt[5]{7}\zeta_5, \sqrt[5]{12})$. We may construct elements r_i and k_i which satisfy hypotheses S as follows. Let

$$r_1 = \zeta_5, k_1 = 5;$$
$$r_2 = \sqrt[5]{3}, k_2 = 5;$$
$$r_3 = \sqrt[5]{7}, k_3 = 5; \text{ and}$$
$$r_4 = \sqrt[5]{12}, k_4 = 5.$$

Then

$$r_1^{k_1} = 1 \in \mathbb{Q} = F_0 \text{ and } r_1 = \zeta_{k_1};$$
$$r_2^{k_2} = 3 \in \mathbb{Q} \subseteq F_0(r_1) \text{ and } \zeta_{k_2} = \zeta_5 \in F_0(r_1);$$
$$r_3^{k_3} = 7 \in \mathbb{Q} \subseteq F_0(r_1, r_2) \text{ and } \zeta_{k_3} = \zeta_5 \in F_0(r_1) \subseteq F_0(r_1, r_2); \text{ and}$$
$$r_4^{k_4} = 12 \in \mathbb{Q} \subseteq F_0(r_1, r_2, r_3) \text{ and } \zeta_{k_4} = \zeta_5 \in F_0(r_1) \subseteq F_0(r_1, r_2, r_3).$$

So the elements r_i and k_i satisfy hypotheses S, and hence by Propositions 25.3 and 25.6, the chain of fields

$$\mathbb{Q} = F_0 \subseteq F_1 = F_0(r_1) \subseteq F_2 = F_0(r_1, r_2)$$
$$\subseteq F_3 = F_0(r_1, r_2, r_3) \subseteq F_0(r_1, r_2, r_3, r_4) = K$$

gives rise to a chain of normal subgroups

$$\{\iota\} = \text{Gal } K/F_4 \lhd \text{Gal } K/F_3 \lhd \text{Gal } K/F_2 \lhd \text{Gal } K/F_1 \lhd \text{Gal } K/F_0,$$

each of whose quotient groups

$$(\text{Gal } K/F_0)/(\text{Gal } K/F_1), \qquad (\text{Gal } K/F_1)/(\text{Gal } K/F_2),$$
$$(\text{Gal } K/F_2)/(\text{Gal } K/F_3), \qquad (\text{Gal } K/F_3)/(\text{Gal } K/F_4),$$

is Abelian. ❖

Example 25.8   Let $F_0 = \mathbb{Q}(\zeta_3)$ and let $K = \mathbb{Q}(\zeta_3, \sqrt{7}, \sqrt[4]{11}, i, \sqrt[3]{6})$. We may construct elements r_i and k_i which satisfy hypotheses S as follows. Note that since ζ_3 is already in F_0, we do not have to adjoin it explicitly. Let

$$r_1 = \sqrt{7}, k_1 = 2;$$
$$r_2 = i, k_2 = 4;$$
$$r_3 = \sqrt[4]{11}, k_3 = 4; \text{ and}$$
$$r_4 = \sqrt[3]{6}, k_4 = 3.$$

Then, since $\zeta_2 = \cos(\pi) + i\sin(\pi) = -1$, we have

$$r_1{}^{k_1} = 7 \in \mathbb{Q} \subseteq F_0 \text{ and } \zeta_{k_1} = \zeta_2 = -1 \in \mathbb{Q} \subseteq F_0;$$
$$r_2{}^{k_2} = 1 \in \mathbb{Q} \subseteq F_0(r_1) \text{ and } r_2 = i = \zeta_4 = \zeta_{k_2};$$
$$r_3{}^{k_3} = 11 \in \mathbb{Q} \subseteq F_0(r_1, r_2) \text{ and } \zeta_{k_3} = \zeta_4 = i \in F_0(r_1, r_2); \text{ and}$$
$$r_4{}^{k_4} = 6 \in \mathbb{Q} \subseteq F_0(r_1, r_2, r_3) \text{ and } \zeta_{k_4} = \zeta_3 \in F_0 \subseteq F_0(r_1, r_2, r_3).$$

So the elements r_i and k_i satisfy hypotheses S, and hence, as in Example 25.7, Propositions 25.3 and 25.6 imply that the chain of fields

$$\mathbb{Q} = F_0 \subseteq F_1 = F_0(r_1) \subseteq F_2 = F_0(r_1, r_2)$$
$$\subseteq F_3 = F_0(r_1, r_2, r_3) \subseteq F_0(r_1, r_2, r_3, r_4) = K$$

gives rise to a chain of normal subgroups

$$\{\iota\} = \text{Gal } K/F_4 \triangleleft \text{Gal } K/F_3 \triangleleft \text{Gal } K/F_2 \triangleleft \text{Gal } K/F_1 \triangleleft \text{Gal } K/F_0,$$

each of whose quotient groups

$$(\text{Gal } K/F_0)/(\text{Gal } K/F_1), \qquad (\text{Gal } K/F_1)/(\text{Gal } K/F_2),$$
$$(\text{Gal } K/F_2)/(\text{Gal } K/F_3), \qquad (\text{Gal } K/F_3)/(\text{Gal } K/F_4),$$

is Abelian. ❖

Exercises

In Exercises 1–6, find a chain of fields

$$F = F_0 \subseteq F_1 \subseteq F_2 \subseteq \cdots \subseteq F_m = K$$

such that for all i, Gal $K/F_i \lhd$ Gal K/F_{i-1} and (Gal K/F_{i-1})/(Gal K/F_i) is Abelian.

1. $F = \mathbb{Q}(\zeta_3)$, $K = \mathbb{Q}(\sqrt[3]{5}, \zeta_3, \sqrt[3]{7})$

2. $F = \mathbb{Q}(\zeta_6)$, $K = \mathbb{Q}(\sqrt[6]{13}, \zeta_6, \sqrt[5]{5}, \sqrt[6]{-3})$

3. $F = \mathbb{Q}$, $K = \mathbb{Q}(\sqrt[3]{5}, \zeta_3, \sqrt[5]{-10}, \zeta_5, \sqrt[3]{6})$

4. $F = \mathbb{Q}$, $K = \mathbb{Q}(\sqrt[7]{5}, \zeta_4, \sqrt[9]{3}, \zeta_9, \sqrt[4]{-7}, \sqrt[7]{2}, \zeta_7)$

5. $F = \mathbb{Q}$, $K = \mathbb{Q}(\alpha, \beta, \zeta_3)$, where $\alpha, \beta \in \mathbb{C}$ are such that $\alpha^3 = 5$ and $\beta^3 = -1 + \alpha$

6. $F = \mathbb{Q}$, $K = \mathbb{Q}(\alpha, \zeta_9, \beta, \gamma, \zeta_3)$, where $\alpha, \beta, \gamma \in \mathbb{C}$ are such that $\alpha^3 = 2$, $\beta^9 = 1 - \alpha$, and $\gamma^3 = 5 + \beta$

7. Find nonzero complex numbers r_1, r_2 and positive integers k_1, k_2 such that $r_i{}^{k_i} \in \mathbb{Q}$ for $i = 1, 2$, but there exists a function $f \in$ Gal $\mathbb{Q}(r_1, r_2)/\mathbb{Q}$ such that $f(\mathbb{Q}(r_1)) \nsubseteq \mathbb{Q}(r_1)$. (Remember to justify your answer.)

8. Find nonzero complex numbers r_1, r_2, r_3 and positive integers k_1, k_2, k_3 such that $r_i{}^{k_i} \in \mathbb{Q}$ for $i = 1, 2, 3$, but there exists a function $f \in$ Gal $\mathbb{Q}(r_1, r_2, r_3)/\mathbb{Q}(r_1)$ such that $f(\mathbb{Q}(r_1, r_2)) \nsubseteq \mathbb{Q}(r_1, r_2)$. (Remember to justify your answer.)

In Exercises 9–12, for the fields $f \subset L \subset K$, find a homomorphism R: Gal $K/F \to$ Gal L/F with kernel Gal K/L.

9. $\mathbb{Q} \subseteq \mathbb{Q}(\zeta_3) \subseteq \mathbb{Q}(\zeta_3, \sqrt[3]{5})$

10. $\mathbb{Q} \subseteq \mathbb{Q}(i, \sqrt{2}) \subseteq \mathbb{Q}(i, \sqrt{2}, \zeta_3)$

11. $\mathbb{Q}(i) \subseteq \mathbb{Q}(i, \sqrt{2}) \subseteq \mathbb{Q}(i, \sqrt{2}, \zeta_3, \sqrt[3]{5})$

12. $\mathbb{Q} \subseteq \mathbb{Q}(i, \sqrt{3}) \subseteq \mathbb{Q}(i, \sqrt{3}, \sqrt[3]{5})$

In Exercises 13–16, suppose that F is a field, that $p(x) \in F[x]$, and that K is an extension field of F which contains $F^{p(x)}$.

13. Show that for all $\phi \in$ Gal K/F, $\phi(F^{p(x)}) = F^{p(x)}$.

14. Show that Gal $K/F^{p(x)} \lhd$ Gal K/F.

15. Find a field F and a polynomial $p(x) \in F[x]$ such that Gal $F^{p(x)}/F$ is not Abelian.

16. Find fields F and K and a polynomial $p(x) \in F[x]$ such that (Gal K/F)/ (Gal $K/F^{p(x)}$) is not Abelian.

17. If $p(x) = x^2 - 1 \in \mathbb{Q}[x]$, show that Gal $\mathbb{Q}^{p(x)}/\mathbb{Q}$ is Abelian.

18. Suppose that F is a subfield of $\mathbb{C}$ which contains ζ_n and let $c \in \mathbb{C}$. If $p(x) = x^n - c \in F[x]$, show that Gal $F^{p(x)}/F$ is Abelian.

26 Solvable Groups and Commutator Subgroups

We observed in Chapter 25 that the Galois group of an extension by radicals has a special chain of subgroups. Since such extensions determine whether a polynomial is solvable by radicals, abstract groups that have such chains of subgroups are called solvable.

DEFINITION

A group G is **solvable** if there exists a chain of subgroups
$$G = H_0 \supseteq H_1 \supseteq \cdots \supseteq H_k = \{e\}$$
such that for all $1 \leq i \leq k$,

(i) $H_i \lhd H_{i-1}$;

(ii) H_{i-1}/H_i is Abelian.

According to Propositions 25.3 and 25.6, extensions by radicals are solvable.

Proposition 26.1   *Suppose that F_0 is a subfield of $\mathbb{C}$, that $r_1, \ldots, r_m$ are nonzero complex numbers, that $k_1, \ldots, k_m$ are positive integers and that for $1 \leq i \leq m$,*

(i) $F_i = F_0(r_1, \ldots, r_i)$,

(ii) $(r_i)^{k_i} \in F_{i-1}$,

(iii) *either* $r_i = \zeta_{k_i}$ *or* $\zeta_{k_i} \in F_{i-1}$.

Then Gal F_m/F_0 *is a solvable group.* ∎

There are many other solvable groups. For example, all Abelian groups are solvable.

Proposition 26.2   *Every Abelian group is solvable.*

Proof   For the sequence of subgroups, pick $G = H_0 \supseteq H_1 = \{e\}$. By Propositions 18.3 and 23.4, $H_1 \lhd H_0$, and by Proposition 23.6, H_0/H_1 is Abelian. ∎

The Galois groups of Examples 25.7 and 25.8 are not Abelian but are solvable by Proposition 26.1. A simpler example of a solvable group which is not Abelian is S_3.

Example 26.3   By Example 23.2, $N = \{\iota, (123), (132)\} \lhd S_3$. By Propositions 16.6 and 23.8, the order of S_3/N is 2 and hence by Corollary 21.17, S_3/N is Abelian. It is easy to see that the set consisting solely of the identity is a normal subgroup of any group (cf. Proposition 18.3) and hence $\{\iota\} \lhd N$. By Propositon 23.8, $N/\{\iota\}$ has three elements and hence is Abelian by Corollary 21.17. Thus the chain $S_3 \supseteq N \supseteq \{\iota\}$ shows that S_3 is a solvable group. ❖

We will generalize Proposition 26.1 in the next chapter by showing that if $p(x)$ is solvable by radicals, then Gal $F^{p(x)}/F$ is a solvable group. Thus, to find an equation that is not solvable by radicals, it will suffice to find a polynomial whose Galois group is not solvable. So we need some nonsolvable groups. The rest of this chapter is devoted to showing that if $n \geq 5$, then S_n is not solvable. Note that by Proposition 14.2, a fifth-degree irreducible polynomial $p(x) \in \mathbb{Q}[x]$ has five distinct roots and hence, by Proposition 17.5, its Galois group may be embedded in S_5. If this embedding is onto, then the Galois group is not solvable and hence, by the results to be proved in Chapter 27, the equation $p(x) = 0$ is not solvable by radicals. Thus the reason that there are fifth-degree polynomials $p(x) \in \mathbb{Q}[x]$ such that $p(x) = 0$ is not solvable by radicals is the nonsolvability of S_5.

The definition of solvability is not very easy to use, and so to show that S_5 is not solvable we need a characterization of solvability that is more tractable. In particular, we want to avoid quotient groups. The next result tells us how to do this.

Proposition 26.4   *Let G be a group and let $N \lhd G$. Then G/N is Abelian if and only if $N \supseteq \{x^{-1}y^{-1}xy \,|\, x, y \in G\}$.*

Proof   If G/N is Abelian and $x, y \in G$, then

$$(x^{-1}y^{-1}xy)N = (x^{-1}N)(y^{-1}N)(xN)(yN)$$
$$= (x^{-1}N)(xN)(y^{-1}N)(yN) = (eN)(eN) = N,$$

and hence, since $e \in N$, $x^{-1}y^{-1}xy = (x^{-1}y^{-1}xy)e \in (x^{-1}y^{-1}xy)N = N$. Conversely, suppose that $x^{-1}y^{-1}xy \in N$ whenever $x, y \in G$. Then $xy = (yxx^{-1}y^{-1})xy = yx(x^{-1}y^{-1}xy) \in (yx)N$, and thus by Proposition 20.4 (i), $(xy)N = (yx)N$. Then $(xN)(yN) = (xy)N = (yx)N = (yN)(xN)$, and hence G/N is Abelian. ∎

DEFINITION

Let G be a group and let $x, y \in G$. The **commutator** of x and y is the element $x^{-1}y^{-1}xy$.

According to Proposition 26.4, the set of commutators determines when a quotient group is Abelian. It would be a happy circumstance if this set were always a group, but sometimes it is not (although examples are not easy to construct), and hence instead of using the set of commutators, we use the subgroup generated by this set. This subgroup is somewhat less complicated than it might be because the inverse of a commutator is always another commutator (if G is a group and $x, y \in G$, then $(x^{-1}y^{-1}xy)^{-1} = y^{-1}x^{-1}yx$ by Proposition 15.9) and thus the subgroup generated by the commutators consists entirely of products of commutators.

DEFINITION

Let G be a group. The set of all finite products of commutators of G is a subgroup of G, called the **commutator subgroup**; it is denoted by G' or $G^{(1)}$. The second commutator subgroup of G, $(G')'$, is denoted by $G^{(2)}$; in general, for $i \geq 2$, $G^{(i)} = (G^{(i-1)})'$.

Note that by Proposition 19.1, G' is the smallest subgroup which contains the commutators of G. So any subgroup which contains the set of commutators also contains G', and hence Proposition 26.4 may be rephrased.

Proposition 26.5   *Let G be a group and let $N \lhd G$. Then G/N is Abelian if and only if $N \supseteq G'$.*

In view of Propositions 23.6 and 26.5, it is not surprising that the commutator subgroups of an Abelian group are all trivial.

Proposition 26.6   *If G is Abelian, then $G^{(i)} = \{e\}$ for all i.*

Proof   If G is Abelian and $x, y \in G$, then $x^{-1}y^{-1}xy = x^{-1}xy^{-1}y = e$. So G', the set of all finite products of commutators of G is just $\{e\}$. If $i > 1$, then $G^{(i)}$ is a subgroup of $G' = \{e\}$, and hence, since $e \in G^{(i)}$, $G^{(i)} = \{e\}$ as well. ∎

Example 26.7   For the group S_3, we have

$$(12)^{-1}(13)^{-1}(12)(13) = ((13)^{-1}(12)^{-1}(13)(12))^{-1} = (123),$$
$$(12)^{-1}(23)^{-1}(12)(23) = ((23)^{-1}(12)^{-1}(23)(12))^{-1} = (132),$$
$$(13)^{-1}(23)^{-1}(13)(23) = ((23)^{-1}(13)^{-1}(23)(13))^{-1} = (123),$$

and thus $N = \{\iota, (123), (132)\} \subseteq S_3'$. However, all other nontrivial commutators of S_3 are of the form $\sigma^{-1}\tau^{-1}\sigma\tau$ where either $\sigma \in N$ or $\tau \in N$, and by Example 23.2, $N \lhd S_3$. So if $\sigma \in N$, then $\sigma^{-1}\tau^{-1}\sigma\tau = \sigma^{-1}(\tau^{-1}\sigma\tau) \in N$ and if $\tau \in N$, then $\tau^{-1} \in N$ and thus $\sigma^{-1}\tau^{-1}\sigma\tau = (\sigma^{-1}\tau^{-1}\sigma)\tau \in N$. We conclude that every commutator is in N and hence that $S_3' \subseteq N$. So $S_3' = N$.

Since N has three elements, it is Abelian by Corollary 21.17, and thus by Proposition 26.6, $S_3^{(2)} = N' = \{\iota\}$. Therefore $S_3^{(k)} = \{\iota\}$ for all $k \geq 2$. ❖

To phrase solvability in terms of commutator subgroups, we will need to know that all the commutator subgroups $G^{(i)}$ are normal in G.

Proposition 26.8 *For any group G, $G^{(i)} \lhd G$ for all $i \geq 1$.*

Proof It suffices to show that if N is a normal subgroup of a group H, then N' is also normal in H. For certainly G is a normal subgroup of itself, and thus applying this result i times shows that $G^{(i)} \lhd G$.

So suppose that $N \lhd H$. By definition N' is a subgroup of H. To see that N' is normal, let $a \in H$. If $x, y \in N$, then

$$a(x^{-1}y^{-1}xy)a^{-1} = (ax^{-1}a^{-1})(ay^{-1}a^{-1})(axa^{-1})(aya^{-1})$$
$$= (axa^{-1})^{-1}(aya^{-1})^{-1}(axa^{-1})(aya^{-1}).$$

Since $N \lhd H$, axa^{-1}, $aya^{-1} \in N$ and hence $a(x^{-1}y^{-1}xy)a^{-1}$ is a commutator of N. In summary, if c is a commutator of N, then aca^{-1} is also a commutator of N. If $z \in N'$, then $z = c_1 c_2 \cdots c_n$, where $c_1, c_2, \ldots, c_n$ are all commutators of N. Since

$$aza^{-1} = a(c_1 c_2 \cdots c_n)a^{-1} = (ac_1 a^{-1})(ac_2 a^{-1}) \cdots (ac_n a^{-1}),$$

aza^{-1} is also a product of commutators, and hence $aza^{-1} \in N'$. We conclude that $aN'a^{-1} \subseteq N'$ and hence that $N' \lhd H$. ∎

Proposition 26.8, together with Proposition 26.5, leads to the following characterization of solvability.

Proposition 26.9 *A group G is solvable if and only if $G^{(k)} = \{e\}$ for some positive integer k.*

Proof Suppose that G is solvable and let $G = H_0 \supseteq H_1 \supseteq \cdots \supseteq H_k = \{e\}$ be a chain of subgroups of G such that for all $1 \leq k \leq n$, $H_i \lhd H_{i-1}$ and H_{i-1}/H_i is Abelian. By Proposition 26.5, $H_{i-1}' \subseteq H_i$. Then $G^{(1)} = H_0' \subseteq H_1$, and $G^{(2)} \subseteq H_1' \subseteq H_2$. In general, $G^{(i)} \subseteq H_{i-1}' \subseteq H_i$. So since $H_k = \{e\}$, $G^{(k)} \subseteq \{e\}$. But since $G^{(k)}$ is a subgroup, $e \in G^{(k)}$ and therefore, $G^{(k)} = \{e\}$. Conversely, suppose that $G^{(k)} = \{e\}$ for some k, and consider the chain $G \supseteq G^{(1)} \supseteq G^{(2)} \supseteq \cdots \supseteq G^{(k)} = \{e\}$. By Proposition 26.8, for any $1 \leq i \leq k$, $G^{(i)} \lhd G$; so a fortiori $G^{(i)} \lhd G^{(i-1)}$. By Proposition 26.5, $G^{(i-1)}/G^{(i)}$ is Abelian, and therefore, G is solvable. ∎

The characterization of solvability given in Proposition 26.9 is extremely useful. We will use it to prove that homomorphic images of solvable groups are solvable (Proposition 26.10), and, as previously indicated, that for all $n \geq 5$, S_n is not solvable (Proposition 26.12). The method used to prove Proposition 26.10 is typical of the way Proposition 26.9 is used.

Proposition 26.10 *A homomorphic image of a solvable group is solvable.*

FIGURE 26.1

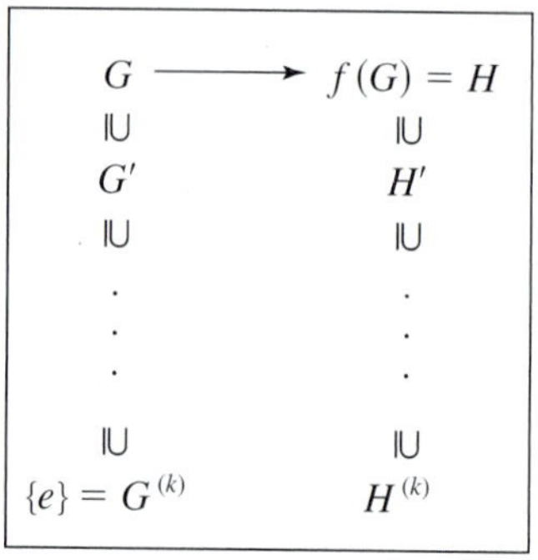

$$
\begin{array}{ccc}
G & \longrightarrow & f(G) = H \\
\cup\!| & & \cup\!| \\
G' & & H' \\
\cup\!| & & \cup\!| \\
\cdot & & \cdot \\
\cdot & & \cdot \\
\cdot & & \cdot \\
\cup\!| & & \cup\!| \\
\{e\} = G^{(k)} & & H^{(k)}
\end{array}
$$

Proof   Let G be a solvable group, let H be a group, and suppose that $f : G \to H$ is an onto homomorphism. We want to show that H is solvable. By Proposition 26.9, $G^{(k)} = \{e_G\}$ for some k; and by Proposition 26.9, it suffices to show that $H^{(k)} = \{e_H\}$ as well (Figure 26.1). Since f is onto, we know that $H \subseteq f(G)$. From this, we want to deduce that $H^k \subseteq f(G^{(k)}) = \{e_H\}$. We do this by showing that in general for any subgroup S of G, $f(S)' \subseteq f(S')$, and then applying this result k times.

We first prove

$$(*) \text{ that for any subgroup } S \text{ of } G, f(S)' \subseteq f(S').$$

Consider a commutator $z^{-1}w^{-1}zw$, where $z, w \in f(S)$. Since f is onto, $f(s) = z$ and $f(t) = w$ for some $s, t \in S$. Then $s^{-1}t^{-1}st \in S'$, and thus by Lemma 18.12, $z^{-1}w^{-1}zw = f(s)^{-1}f(t)^{-1}f(s)f(t) = f(s^{-1}t^{-1}st) \in f(S')$. But by Proposition 19.1 and the definition of $f(S)'$, $f(S)'$ is the smallest subgroup of H which contains all the commutators of $f(S)$. Since $f(S')$ is a subgroup of H by Proposition 18.13, $f(S)' \subseteq f(S')$.

Now since f is onto, $H \subseteq f(G)$ and by definition, $H' \subseteq f(G)'$; so by $(*)$, $H' \subseteq f(G)' \subseteq f(G')$. But similarly $H^{(2)} \subseteq f(G')'$ and hence, again by $(*)$, $H^{(2)} \subseteq f(G')' \subseteq f(G^{(2)})$. Continuing in the same fashion, eventually we have $H^{(k)} \subseteq f(G^{(k)}) = f(\{e_G\}) = \{e_H\}$. But $e_H \in H^{(k)}$ because $H^{(k)}$ is a subgroup, and therefore $H^{(k)} = \{e_H\}$. ∎

The proof that S_5 is not solvable depends similarly on an iterative process. In this case, however, we want to show that the iterations remain large.

Lemma 26.11   *Let $n \geq 5$ and suppose that N_1 and N_2 are normal subgroups of S_n such that $N_2 \subseteq N_1$ and N_1/N_2 is Abelian. If N_1 contains all the 3-cycles of S_n then so does N_2.*

Proof   Since $n \geq 5$, we have $(123), (145) \in S_n$. By hypothesis, $(123), (145) \in N_1$. Since N_1/N_2 is Abelian, N_2 contains all the commutators of N_1 by Proposition 26.4; in particular, $(135) = (123)^{-1}(145)^{-1}(123)(145) \in N_2$. Now let (abc) be a 3-cycle and let $\sigma \in S_n$ be such that $\sigma(1) = a$, $\sigma(3) = b$, $\sigma(5) = c$. We claim that $\sigma(135)\sigma^{-1} =$

(abc). Suppose that $k \in \{1, 2, \ldots, n\}$. If $k \notin \{a, b, c\}$, then, since σ is one-to-one, $\sigma^{-1}(k) \notin \{1, 3, 5\}$. Then $((135)\sigma^{-1})(k) = \sigma^{-1}(k)$ and thus $(\sigma(135)\sigma^{-1})(k) = \sigma(\sigma^{-1}(k)) = k$. Otherwise,

$$(\sigma(135)\sigma^{-1})(a) = (\sigma(135))(1) = \sigma(3) = b,$$
$$(\sigma(135)\sigma^{-1})(b) = (\sigma(135))(3) = \sigma(5) = c,$$
$$(\sigma(135)\sigma^{-1})(c) = (\sigma(135))(5) = \sigma(1) = a.$$

Thus $\sigma(135)\sigma^{-1} = (abc)$. Since $N_2 \lhd S_n$ and $(135) \in N_2$, we conclude that $(abc) \in N_2$, and thus that N_2 contains every 3-cycle. ∎

Proposition 26.12 *For $n \geq 5$, S_n is not solvable.*

Proof By Propositions 26.8 and 26.5, $S_n^{(k)}$ and $S_n^{(k-1)}$ are normal subgroups of S_n such that $S_n^{(k-1)}/S_n^{(k)}$ is Abelian. Since S_n contains all 3-cycles, k applications of Lemma 26.11 show that $S_n^{(k)}$ also contains all 3-cycles, and hence that $S_n^{(k)} \neq \{\iota\}$ for any k. By Proposition 26.9, S_n is not solvable. ∎

Two of the most famous results concerning solvable groups are connected with the British mathematician William Burnside (1852–1927). Burnside proved that any group of order $p^\alpha q^\beta$ is solvable, where p and q are primes, and conjectured that every group of odd order is solvable. In 1963, Walter Feit and John Thompson gave an affirmative answer to this conjecture.

Exercises

In Exercises 1–8, find the commutator of x and y in G.

1. $G = \mathbb{Z}_4$, $x = [2]_4$, $y = [3]_4$ **2.** $G = \mathbb{Z}_{12}$, $x = [7]_{12}$, $y = [10]_{12}$

3. $G = S_4$, $x = (12)$, $y = (34)$ **4.** $G = S_4$, $x = (1234)$, $y = (24)$

5. $G = \mathbb{C}^\times$, $x = i$, $y = 2 + 3i$ **6.** $G = S_4$, $x = (123)$, $y = (14)$

7. $G = S_4$, $x = (12)(34)$, $y = (234)$ **8.** $G = S_7$, $x = (1357)(246)$, $y = (15437)$

Exercises 9 and 10 concern the subset $S = \{\iota, (13), (14), (34), (134), (143)\}$ of S_4.

9. Show that S is a subgroup of S_4.

10. Use the <u>definition</u> to show that S is a solvable group.

Exercises 11 and 12 concern the quaternion group $Q = \{\pm 1, \pm i, \pm j, \pm k\}$ (cf. Exercise 23 in Chapter 15).

11. Show that the subset $Q = \{\pm 1, \pm i, \pm j, \pm k\}$ of the quaternions is a group with respect to multiplication.

12. Use the <u>definition</u> to show that Q is a solvable group.

13. Let $M = \left\{ \begin{pmatrix} a & b \\ 0 & c \end{pmatrix} \middle| a, b, c \in \mathbb{R}, a \neq 0 \neq c \right\}$. Use the <u>definition</u> to show that M is a solvable group with respect to multiplication of matrices.

In Exercises 14–25, use the <u>definition</u> to show that the given group is a solvable group. (For the definition of $G \times H$, see Exercise 25 in Chapter 15.)

14. $\mathbb{Z}_{64}$ 15. $\mathbb{Z}_{169}$ 16. S_2

17. $S_3 \times \mathbb{Z}_3$ 18. $S_3 \times \mathbb{Z}_{25}$ 19. $S_3 \times S_2$

20. S_4 21. Gal $\mathbb{Q}(\zeta_3)/\mathbb{Q}$ 22. Gal $\mathbb{Z}(\sqrt{5})/\mathbb{Q}$

23. Gal $\mathbb{Q}(\sqrt{5}, i)/\mathbb{Q}$ 24. Gal $\mathbb{Q}(\sqrt[3]{5}, \zeta_3)/\mathbb{Q}$ 25. Gal $\mathbb{Q}(\sqrt[5]{5}, \zeta_5)/\mathbb{Q}$

In Exercises 26–31, find all the commutator subgroups of the given group. (For the definition of $G \times H$, see Exercise 26 in Chapter 15.)

26. $\mathbb{Z}_{64}$ 27. $\mathbb{Z}_{169}$ 28. S_2

29. $S_3 \times \mathbb{Z}_3$ 30. $S_3 \times \mathbb{Z}_{25}$ 31. $S_3 \times S_2$

32. Let p and q be distinct primes. Show that a group of order pq which has a proper, nontrivial, normal subgroup is solvable. (A subgroup N of a group G is <u>proper</u> if $N \neq G$; N is <u>nontrivial</u> if $N \neq \{e\}$.)

33. Show that for groups G and H, $(G \times H)' = G' \times H'$. (For the definition of $G \times H$, see Exercise 26 in Chapter 15.)

34. Show that a group G is Abelian if and only if $G' = \{e\}$.

In Exercises 35–38, prove the given statement or show that it is false by giving a counterexample.

35. All groups of order 119 are solvable.

36. All groups of order 120 are solvable.

37. All groups of order 719 are solvable.

38. All groups of order 720 are solvable.

Exercises 39–43 concern the subset $M = \left\{ \begin{pmatrix} a & b \\ 0 & c \end{pmatrix} \middle| a, b, c \in \mathbb{R}, a \neq 0 \neq c \right\}$ of $M_3(\mathbb{R})$.

39. Show that M is a group with respect to multiplication of matrices.

40. Show that $M' = \left\{ \begin{pmatrix} 1 & r \\ 0 & 1 \end{pmatrix} \middle| r \in \mathbb{R} \right\}$.

41. Find $M^{(2)}$.

42. Show directly (with using Proposition 26.8) that $M' \triangleleft M$.

43. Show directly (without using Propositions 26.4 or 26.5) that M/M' is Abelian.

A group G is <u>metabelian</u> if it has a normal subgroup N such that both N and G/N are Abelian. Exercises 44–47 concern metabelian groups.

44. Show that S_3 is a metabelian group which is not Abelian.

45. Show that the quaternion group Q is a metabelian group which is not Abelian. (For the definition of Q, see Exercise 23 in Chapter 15.)

46. Show that the dihedral group D_4 is a metabelian group which is not Abelian. (For the definition of D_4, see Exercise 36 in Chapter 15.)

47. Show that a group G is metabelian if and only if $G^{(2)} = \{e\}$.

48. Prove or give a counterexample: If G is a solvable group and $N \lhd G$, then G/N is a solvable group.

49. Let $n \geq 3$ and suppose that N is a normal subgroup of S_n which contains a 3-cycle. Show that N contains all 3-cycles.

50. Show that a subgroup of a solvable group is solvable.

HISTORICAL NOTE

William Burnside

William Burnside was born in London, England, on July 2, 1852, and died in West Wickham, Kent, England, on August 21, 1927. He was the elder son of a merchant of Scottish ancestry and was left an orphan at the age of six. He won a mathematical scholarship at St. John's College, Cambridge, and began his studies there in October 1871. He was a rower of some repute throughout his undergraduate years and after leaving Cambridge regularly went fishing in Scotland during the holidays. In the mathematical tripos of 1875 he tied for second wrangler and subsequently came in first on the examination for the Smith's prizes. He became a lecturer at Pembroke College, and then, in 1885, he was appointed professor of mathematics at the Royal Naval College at Greenwich, where he spent the rest of his career. Soon after arriving at Greenwich, in 1886, he married Alexandrine Urqhart. At Greenwich, he taught ballistics, mechanics and heat, and dynamics, and developed a reputation as an excellent teacher. He was twice invited back to Pembroke College but declined both times because he did not want to become involved in the unavoidable administrative and social duties that would be required of him. Because of failing health, he retired from his post at Greenwich in 1919.

He published over one hundred and fifty papers, and his book *Theory of Groups* was the first English treatise on the subject and the first to develop it abstractly. He published his proof that every group of order $p^{\alpha}q^{\beta}$ is solvable in 1904 and was one of the principal originators of the theory of group representations. It was some of the results of this theory that led him to conjecture that every group of odd order is solvable.

27

Solvable Galois Groups

The objective of this chapter is to show that if $p(x) \in F[x]$ for a subfield F of $\mathbb{C}$, and if $p(x) = 0$ is solvable by radicals, then Gal $F^{p(x)}/F$ is a solvable group. In many simple cases, we already know this.

Example 27.1 Consider $p(x) = x^3 - 7 \in \mathbb{Q}[x]$. By Example 14.9, $\mathbb{Q}^{p(x)} = \mathbb{Q}(\zeta_3, \sqrt[3]{7})$, and if

$$r_1 = \zeta_3, k_1 = 3, \text{ and}$$
$$r_2 = \sqrt[3]{7}, k_2 = 3,$$

then $r_1^{k_1} = 1 \in \mathbb{Q}$ and $r_1 = \zeta_{k_1}$, and $r_2^{k_2} = 7 \in \mathbb{Q}(r_1)$ and $\zeta_{k_2} = \zeta_3 \in \mathbb{Q}(r_1)$. It then follows from Proposition 26.1 that Gal $\mathbb{Q}^{p(x)}/\mathbb{Q}$ is a solvable group. ❖

In general, whenever $F^{p(x)} = F(r_1, \ldots, r_m)$ for radicals $r_1, \ldots, r_m$ satisfying hypotheses S of Chapter 25, Gal $F^{p(x)}/F$ will be a solvable group by Proposition 26.1. Note that the definition of solvability by radicals does not guarantee <u>equality</u>; it only guarantees that $F^{p(x)}$ is <u>contained</u> in $F(r_1, \ldots, r_m)$. To deal with the general situation, we proceed as follows.

As we have observed, if $p(x) = 0$ is solvable by radicals, then $F^{p(x)} \subseteq F_m$ for an extension field F_m for which Gal F_m/F is a solvable group. Since the homomorphic image of a solvable group is solvable (Proposition 26.10), we can show that Gal $F^{p(x)}/F$ is solvable by showing that there exists an onto homomorphism from Gal F_m/F to Gal $F^{p(x)}/F$. But we already know (Lemma 25.5) that for some extensions $F \subseteq L \subseteq K$, restriction can define a homomorphism from Gal K/F to Gal L/F. We will show that there are always extensions F_m of F such that (1) F_m is an extension of F by radicals, (2) $F \subseteq F^{p(x)} \subseteq F_m$, and (3) the restriction function does indeed define an onto homomorphism from Gal F_m/F to Gal $F^{p(x)}/F$.

The difficult part of the proof will be showing that the restriction map is onto. This is not surprising because showing the map is onto amounts to showing that any function in Gal $F^{p(x)}/F$ can be extended to a function in Gal F_m/F. We will avoid dealing with extensions directly by using a counting argument, viz., by showing that the number of functions in the homomorphic image of the restriction map is the same as the number of functions in Gal $F^{p(x)}/F$. To count the functions, we will apply Propostion 14.8 to Gal F_m/F, and as shown in the following example, to do this may require extending F_m.

Example 27.2   Consider the field $\mathbb{Q}$ and let $r_1 = i$ and $k_1 = 2$, $r_2 = \sqrt{2}$ and $k_2 = 2$, and $r_3 = \sqrt[4]{1 + \sqrt{2}}$ and $k_3 = 4$. Then $r_1{}^{k_1} = -1 \in \mathbb{Q}$ and $\zeta_{k_1} = -1 \in \mathbb{Q}$; $r_2{}^{k_2} = 2 \in \mathbb{Q}(r_1)$ and $\zeta_{k_2} = -1 \in \mathbb{Q}(r_1)$; and $r_3{}^{k_3} = 1 + \sqrt{2} \in \mathbb{Q}(r_1, r_2)$ and $\zeta_{k_3} = i \in \mathbb{Q}(r_1, r_2)$. Thus $\mathbb{Q}$ and r_1, r_2, r_3 and k_1, k_2, k_3 satisfy hypotheses S of Chapter 25, and hence Gal $\mathbb{Q}(r_1, r_2, r_3)/\mathbb{Q}$ is a solvable group. To apply Proposition 14.8 to $\mathbb{Q}(r_1, r_2, r_3)$, we must have $\mathbb{Q}(r_1, r_2, r_3) = \mathbb{Q}^{p(x)}$ for some polynomial $p(x) \in \mathbb{Q}[x]$. However, the obvious choice for $p(x)$ is not in $\mathbb{Q}[x]$. For r_1 is a root of $x^2 + 1$, r_2 is a root of $x^2 - 2$, and r_3 is a root of $x^4 - (1 + \sqrt{2})$, but some of the coefficients of $p(x) = (x^2 + 1)(x^2 - 2)(x^4 - (1 + \sqrt{2}))$ are not in $\mathbb{Q}$ and hence $p(x) \notin \mathbb{Q}[x]$. The way out of this dilemma is to observe that for any $\phi \in$ Gal $\mathbb{Q}(r_1, r_2)/\mathbb{Q}$, either $\phi(1 + \sqrt{2}) = 1 + \sqrt{2}$ or $\phi(1 + \sqrt{2}) = 1 - \sqrt{2}$ (by Proposition 13.9) and that $(x^4 - (1 + \sqrt{2}))(x^4 - (1 - \sqrt{2})) = x^8 - 2x^4 - 1 \in \mathbb{Q}[x]$. With this in mind, let $r_4 \in \mathbb{C}$ be a fourth root of $1 - \sqrt{2}$ and let $k_4 = 4$. Then $r_4{}^{k_4} = 1 - \sqrt{2} \in \mathbb{Q}(r_1, r_2, r_3)$ and $\zeta_{k_4} = i \in \mathbb{Q}(r_1, r_2, r_3)$. So the elements r_1, r_2, r_3, r_4 and k_1, k_2, k_3, k_4 also satisfy hypotheses S of Chapter 25, and hence Gal $\mathbb{Q}(r_1, r_2, r_3, r_4)/\mathbb{Q}$ is also a solvable group. But in this case, we know that $\mathbb{Q}(r_1, r_2, r_3, r_4) = \mathbb{Q}^{g(x)}$, where $g(x)$ is the polynomial

$$g(x) = (x^2 + 1)(x^2 - 2)(x^4 - (1 + \sqrt{2}))(x^4 - (1 - \sqrt{2}))$$
$$= (x^2 + 1)(x^2 - 2)(x^8 - 2x^4 - 1) \in \mathbb{Q}[x]. \ \diamondsuit$$

To use the method described in Example 27.2 in general, we need to know that polynomials of the form $(x^{k_{j+1}} - \psi_1(r_{j+1})) \cdots (x^{k_{j+1}} - \psi_n(r_{j+1}))$, where $\{\psi_1, \ldots, \psi_n\} =$ Gal $F_0(r_1, \ldots, r_j)/F_0$, have all their coefficients in F_0. To prove this, we will need the following result.

Proposition 27.3   *Let F be a subfield of $\mathbb{C}$ and let $p(x) \in F[x]$. Then*

$$F = \{k \in F^{p(x)} \mid \sigma(k) = k \text{ for all } \sigma \in \text{Gal } F^{p(x)}/F\}.$$

Proof   Let $X = \{k \in F^{p(x)} \mid \sigma(k) = k \text{ for all } \sigma \in \text{Gal } F^{p(x)}/F\}$. It is not difficult to use Proposition 5.2 to check that X is a subfield of $F^{p(x)}$ (Exercise 13), and since $F \subseteq X \subseteq F^{p(x)}$, certainly $X^{p(x)} = F^{p(x)}$. By Proposition 14.8, $[F^{p(x)}:F] = |\text{Gal } F^{p(x)}/F|$ and $[F^{p(x)}:X] = [X^{p(x)}:X] = |\text{Gal } X^{p(x)}/X|$. But by definition of X, Gal $X^{p(x)}/X =$ Gal $F^{p(x)}/X =$ Gal $F^{p(x)}/F$ (Exercise 14), and therefore $[F^{p(x)}:X] = [F^{p(x)}:F]$. But Propositions 12.8 and 12.5 imply that $[F^{p(x)}:F] = [F^{p(x)}:X][X:F]$, and hence $[F^{p(x)}:X] = [F^{p(x)}:X][X:F]$. Then $[X:F] = 1$ and hence $X = F$. ■

Note that if the extension field in Proposition 27.3 is not of the form $F^{p(x)}$, then the conclusion may not hold.

Example 27.4   According to Example 14.10, Gal $\mathbb{Q}(\sqrt[3]{7})/\mathbb{Q} = \{\iota\}$ and hence

$$\{k \in \mathbb{Q}(\sqrt[3]{7}) \mid \sigma(k) = k \text{ for all } \sigma \in \text{Gal } \mathbb{Q}(\sqrt[3]{7})/\mathbb{Q}\} = \mathbb{Q}(\sqrt[3]{7}) \neq \mathbb{Q}.$$

However, according to Example 14.10, $\mathbb{Q}(\sqrt[3]{7}, \zeta_3) = \mathbb{Q}^{p(x)}$ for $p(x) = x^3 - 7 \in \mathbb{Q}[x]$, and hence Proposition 27.3 implies that

$$\{k \in \mathbb{Q}(\sqrt[3]{7}, \zeta_3) \mid \sigma(k) = k \text{ for all } \sigma \in \text{Gal } \mathbb{Q}(\sqrt[3]{7}, \zeta_3)/\mathbb{Q}\} = \mathbb{Q}. \text{ ❖}$$

We now want to show that a radical extension $F_0(r_1, \ldots, r_m)$ determined by hypotheses S is always contained in a radical extension $F_0(s_1, \ldots, s_\mu)$ which satisfies hypotheses S and which also can be written $F_0(s_1, \ldots, s_\mu) = F_0^{\pi(x)}$ for some polynomial $\pi(x) \in F_0[x]$. We will use Proposition 27.3 to show that the coefficients of the polynomial $\pi(x)$ that we will construct are all in F_0 (Lemma 27.7); for this proof, we will need the following computational lemma.

Lemma 27.5   *Let K be an extension field of a field F. For $p(x) = a_0 + \cdots + a_n x^n \in K[x]$ and $\sigma \in \text{Gal } K/F$, let $p(x)^\sigma = \sigma(a_0) + \cdots + \sigma(a_n)x^n$. Then for any $f(x), g(x) \in K[x]$ and $\sigma \in \text{Gal } K/F$, $(f(x)g(x))^\sigma = f(x)^\sigma g(x)^\sigma$.*

Proof   Let $f(x) = b_0 + \cdots + b_m x^m$ and $g(x) = c_0 + \cdots + c_n x^n$. Then

$$(f(x)g(x))^\sigma = \sum_{k=0}^{n+m} \sigma\left(\sum_{i+j=k} b_i c_j\right)x^k = \sum_{k=0}^{n+m} \sum_{i+j=k} \sigma(b_i)\sigma(c_j)x^k = f(x)^\sigma g(x)^\sigma. \text{ ∎}$$

Example 27.6   By Example 13.10, $\mathbb{Q}(\zeta_3) = \{a + b\zeta_3 \mid a, b \in \mathbb{Q}\}$ and letting $\sigma(a + b\zeta_3) = a + b\zeta_3^2$ defines an element $\sigma \in \text{Gal } \mathbb{Q}(\zeta_3)/\mathbb{Q}$. Consider $f(x) = x^2 + (2 - \zeta_3)$ and $g(x) = x^3 + (3 + \zeta_3)x^2 - 7x - \zeta_3$ in $\mathbb{Q}(\zeta_3)[x]$. Then in the notation of Lemma 27.5, $f(x)^\sigma = x^2 + (2 - \zeta_3^2)$ and $g(x)^\sigma = x^3 + (3 + \zeta_3^2)x^2 - 7x - \zeta_3^2$. Multiplying (and recalling from Example 13.10 that $\zeta_3^2 = -1 - \zeta_3$ and $(\zeta_3^2)^2 = -1 - \zeta_3^2$), we have

$$f(x)g(x) = x^5 + (3 + \zeta_3)x^4 - (5 + \zeta_3)x^3$$
$$+ (7 - \zeta_3)x^2 - (14 - 7\zeta_3)x - (1 + 3\zeta_3),$$
$$f(x)^\sigma g(x)^\sigma = x^5 + (3 + \zeta_3^2)x^4 - (5 + \zeta_3^2)x^3$$
$$+ (7 - \zeta_3^2)x^2 - (14 - 7\zeta_3^2)x - (1 + 3\zeta_3^2).$$

Note that in agreement with Lemma 27.5, $(f(x)g(x))^\sigma = f(x)^\sigma g(x)^\sigma$. ❖

Lemma 27.7   *Let F_0 be a subfield of $\mathbb{C}$, let $r_1, \ldots, r_m$ be nonzero complex numbers, let $k_1, \ldots, k_m$ be positive integers, let $F_i = F_0(r_1, \ldots, r_i)$, and suppose that*

(b) $(r_i)^{k_i} \in F_{i-1}$,

(c)   *either $r_i = \zeta_{k_i}$ or $\zeta_{k_i} \in F_{i-1}$.*

Then there exist a polynomial $h(x) \in F_0[x]$, nonzero complex numbers $s_1, \ldots, s_n$, and positive integers $l_1, \ldots, l_n$ such that if $L_0 = F_0$ and $L_i = L_0(s_1, \ldots, s_i)$, then

(A) $F_m \subseteq L_n$,

(B) $(s_i)^{l_i} \in L_{i-1}$,

(C) *either $s_i = \zeta_{l_i}$ or $\zeta_{l_i} \in L_{i-1}$*,

(D) $L_n = L_0^{h(x)}$.

Proof We prove the lemma by induction on m, the number of r_i.

(i) If $m = 1$, let $h(x) = x^{k_1} - (r_1)^{k_1}$, $n = 1$, $s_1 = r_1$, and $l_1 = k_1$. Then $h(x) \in F_0[x]$ and it is easy to see that (A), (B), (C), and (D) all hold.

(ii) Suppose the lemma holds for any pair of sequences $r_1, \ldots, r_m, k_1, \ldots, k_m$ of length m, and consider a pair of sequences $r_1, \ldots, r_{m+1}, k_1, \ldots, k_{m+1}$ of length $m + 1$. By the induction hypothesis applied to $r_1, \ldots, r_m$, there exist a polynomial $f(x) \in F_0[x]$, nonzero complex numbes $s_1, \ldots, s_n$, and positive integers $l_1, \ldots, l_n$ such that if $L_0 = F_0$ and $L_i = L_0(s_1, \ldots, s_i)$, then (A) $F_m \subseteq L_n$, (B) $(s_i)^{l_i} \in L_{i-1}$, (C) either $s_i = \zeta_{l_i}$ or $\zeta_{l_i} \in L_{i-1}$, and (D) $F_m \subseteq L_n = L_0^{f(x)}$.

Let $G = \mathrm{Gal}\, L_n/L_0$, and note that $(r_{m+1})^{k_{m+1}} \in F_m \subseteq L_n$. Since G is finite by Proposition 14.8, $\{\phi((r_{m+1})^{k_{m+1}}) \mid \phi \in G\} = \{t_1, \ldots, t_u\}$, where $t_i \neq t_j$ for $i \neq j$. Let

$$(x^{k_{m+1}} - t_1) \cdots (x^{k_{m+1}} - t_u) = g(x) = b_0 + \cdots + b_{w-1}x^{w-1} + x^w.$$

We will use Proposition 27.3 to show that $g(x) \in F_0[x]$. Suppose that $\sigma \in G$ and note that by Lemma 27.5 applied u times,

$$g(x)^\sigma = [(x^{k_{m+1}} - t_1) \cdots (x^{k_{m+1}} - t_u)]^\sigma = (x^{k_{m+1}} - t_1)^\sigma \cdots (x^{k_{m+1}} - t_u)^\sigma$$
$$= (x^{k_{m+1}} - \sigma(t_1)) \cdots (x^{k_{m+1}} - \sigma(t_u)).$$

For each i, there exists $\phi_i \in G$ such that $t_i = \phi_i((r_{m+1})^{k_{m+1}})$, and hence since $\sigma \circ \phi_i \in G$, $\sigma(t_i) = \sigma \circ \phi_i((r_{m+1})^{k_{m+1}}) \in \{t_1, \ldots, t_u\}$. That is, $\{\sigma(t_1), \ldots, \sigma(t_u)\} \subseteq \{t_1, \ldots, t_u\}$. But then since σ is one-to-one, $\{\sigma(t_1), \ldots, \sigma(t_u)\}$ has u elements and thus $\{\sigma(t_1), \ldots, \sigma(t_u)\} = \{t_1, \ldots, t_u\}$. So $g(x)^\sigma = g(x)$, i.e., $\sigma(b_j) = b_j$ for all j. Since σ was chosen arbitrarily in $G = \mathrm{Gal}\, L_0^{f(x)}/L_0$, Proposition 27.3 implies that $b_j \in F_0$ for all j, i.e., that $g(x) \in F_0[x]$. Then $h(x) = f(x)g(x)$ is in $F_0[x]$ as well. By hypothesis, either $r_{m+1} = \zeta_{k_{m+1}}$ or $\zeta_{k_{m+1}} \in F_m$. If $r_{m+1} = \zeta_{k_{m+1}}$, then $u = 1$ and $t_1 = 1$; in this case, let $s_{n+1} = \zeta_{k_{m+1}}$. Otherwise, $\zeta_{k_{m+1}} \in F_m \subseteq L_n$; in this case, for $i = 1, \ldots, u$, choose $s_{n+i} \in \mathbb{C}$ such that $(s_{n+i})^{k_{m+1}} = t_i$. In both cases, let $l_{n+i} = k_{m+1}$ for all $i = 1, \ldots, u$. Then (B) and (C) are clearly satisfied for the complex numbers $s_1, \ldots, s_{n+u}$ and the positive integers $l_1, \ldots, l_{n+u}$. Since G contains the identity function, $(r_{m+1})^{k_{m+1}} \in \{t_1, \ldots, t_u\}$ and hence, by Proposition 3.8, $r_{m+1} = (s_{n+i})(\zeta_{k_{m+1}})^v$ for some

$i = 1, \ldots, u$ and some $0 \leq v < k_{m+1}$. It follows that $r_{m+1} \in L_{n+u}$ and hence that $F_{m+1} = F_m(r_{m+1}) \subseteq L_n(r_{m+1}) \subseteq L_{n+u}$. So (A) holds. Furthermore, $L_{n+u} = L_0^{f(x)}(s_{n+1}, \ldots, s_{n+u})$, and hence Proposition 3.8 implies that in both cases $L_{n+u} = L_0^{h(x)}$, i.e., (D) holds as well. ∎

We now use the strategy described at the beginning of the chapter to show that equations that are solvable by radicals give rise to solvable Galois groups.

Theorem 27.8   *Let F be a subfield of $\mathbb{C}$ and suppose that $p(x) \in F[x]$. If the equation $p(x) = 0$ is solvable by radicals, then $\mathrm{Gal}\, F^{p(x)}/F$ is a solvable group.*

Proof   Since $p(x) = 0$ is solvable by radicals, there exist nonzero complex numbers $r_1, \ldots, r_m$ and positive integers $k_1, \ldots, k_m$ such that if $F_0 = F$ and $F_i = F(r_1, \ldots, r_i)$, then (a) $F^{p(x)} \subseteq F_m$ and (b) $(r_i)^{k_i} \in F_{i-1}$. As we observed in Chapter 25, we may also assume that (c) either $r_i = \zeta_{k_i}$ or $\zeta_{k_i} \in F_{i-1}$. Then by Lemma 27.7, there exist a polynomial $h(x) \in F[x]$, nonzero complex numbers $s_1, \ldots, s_n$, and positive integers $l_1, \ldots, l_n$ such that if $L_0 = F$ and $L_i = F(s_1, \ldots, s_i)$, then (A) $F^{p(x)} \subseteq F_m \subseteq L_n$, (B) $(s_i)^{l_i} \in L_{i-1}$, (C) either $s_i = \zeta_{l_i}$ or $\zeta_{l_i} \in L_{i-1}$, and (D) $L_n = F^{h(x)}$. By Proposition 26.1, $\mathrm{Gal}\, F^{h(x)}/F$ is a solvable group. We will show that $\mathrm{Gal}\, F^{p(x)}/F$ is a homomorphic image of $\mathrm{Gal}\, F^{h(x)}/F$ and hence by Proposition 26.10 that $\mathrm{Gal}\, F^{p(x)}/F$ is a solvable group.

Suppose that $\sigma \in \mathrm{Gal}\, F^{h(x)}/F$, and let $v_1, \ldots, v_d$ be the complex roots of $p(x)$. By Proposition 13.9, each $\sigma(v_i)$ is a root of $p(x)$ and hence

$$\sigma(F^{p(x)}) = \sigma(F(v_1, \ldots, v_d)) \subseteq F(\sigma(v_1), \ldots, \sigma(v_d)) \subseteq F^{p(x)}.$$

Thus by Lemma 25.5, restricting the domain of an element in $\mathrm{Gal}\, F^{h(x)}/F$ to $F^{p(x)}$ defines a homomorphism $R : \mathrm{Gal}\, F^{h(x)}/F \to \mathrm{Gal}\, F^{p(x)}/F$ with kernel $\mathrm{Gal}\, F^{h(x)}/F^{p(x)}$. We must show that R is onto. To this end, note that by the homomorphism theorem for groups (Theorem 24.6) and Proposition 23.8,

$$\left| R(\mathrm{Gal}\, F^{h(x)}/F) \right| = \left| \frac{\mathrm{Gal}\, F^{h(x)}/F}{\mathrm{Gal}\, F^{h(x)}/F^{p(x)}} \right| = \frac{\left| \mathrm{Gal}\, F^{h(x)}/F \right|}{\left| \mathrm{Gal}\, F^{h(x)}/F^{p(x)} \right|}.$$

But by Propositions 14.8, 12.7, and 12.5,

$$\left| \mathrm{Gal}\, F^{h(x)}/F \right| = [F^{h(x)} : F] = [F^{h(x)} : F^{p(x)}][F^{p(x)} : F]$$
$$= \left| \mathrm{Gal}\, F^{h(x)}/F^{p(x)} \right| \left| \mathrm{Gal}\, F^{p(x)}/F \right|.$$

It follows that $\left| R(\mathrm{Gal}\, F^{h(x)}/F) \right| = \left| \mathrm{Gal}\, F^{p(x)}/F \right|$ and hence that R is onto. We conclude that $\mathrm{Gal}\, F^{p(x)}/F$ is a solvable group by Proposition 26.10. ∎

Example 27.9   Consider $p(x) = x^2 + 4x + 5 \in \mathbb{Q}[x]$. By the quadratic formula, the roots of $p(x)$ are $-2 + i$ and $-2 - i$, and by Example 6.6, $p(x)$ is solvable by radicals over $\mathbb{Q}$. Thus by Theorem 27.8, $\mathrm{Gal}\, \mathbb{Q}(-2 + i, -2 - i)/\mathbb{Q}$ is a solvable group. ❖

Example 27.10 According to Example 2.5, the solutions of the polynomial equation $3x^4 + 12x^3 + 15x^2 - 9 = 0$ are $x = \dfrac{-1 \pm \sqrt{5}}{2}, \dfrac{-3 \pm \sqrt{-3}}{2}$, and according to Example 6.7, the given equation is solvable by radicals over $\mathbb{Q}$. Therefore by Theorem 27.8, the Galois group $\text{Gal } \mathbb{Q}\!\left(\dfrac{-1 \pm \sqrt{5}}{2}, \dfrac{-3 \pm \sqrt{-3}}{2}\right)\!\Big/\mathbb{Q}$ is a solvable group. ❖

Example 27.11 We observed in Example 6.4 that the solutions of $x^4 - 8x^3 + 22x^2 - 24x + 7 = 0$ are $x = 2 \pm \sqrt{1 + \sqrt{2}}, 2 \pm \sqrt{1 - \sqrt{2}}$ and that the given equation is solvable by radicals over $\mathbb{Q}$. Therefore by Theorem 27.8, the Galois group $\text{Gal } \mathbb{Q}(2 \pm \sqrt{1 + \sqrt{2}}, 2 \pm \sqrt{1 - \sqrt{2}})/\mathbb{Q}$ is a solvable group. ❖

The converse of Theorem 27.8 is also true. That is, if F is a subfield of $\mathbb{C}$, if $p(x) \in F[x]$, and if $\text{Gal } F^{p(x)}/F$ is a solvable group, then $p(x) = 0$ is solvable by radicals. We do not need this result, and its proof is substantially different from that of Theorem 27.8. Thus rather than include a detailed proof, we content ourselves with only the following outline.

Suppose that $\text{Gal } F^{p(x)}/F$ is a solvable group of order n, and let $L = F^{p(x)}(\zeta_n)$. Then $G = \text{Gal } L/F(\zeta_n)$ is not only solvable but in fact has a chain of subgroups.

$$G = H_1 \supseteq \cdots \supseteq H_m = \{\iota\}$$

such that $H_i \triangleleft H_{i-1}$ and H_{i-1}/H_i is cyclic of prime order. This chain of subgroups gives rise to a chain of extension fields.

$$F(\zeta_n) = K_1 \subseteq \cdots \subseteq K_m = L,$$

each of which is a radical extension of the previous one. Specifically, for all $i = 2, \ldots, m$, there exist $r_i \in K_i$ and $0 < k_i \in \mathbb{Z}$ such that $r_i^{k_i} \in K_{i-1}$ and $K_i = K_{i-1}(r_i)$. If we let $r_1 = \zeta_n$ and $k_1 = n$, then we have found complex numbers $r_1, \ldots, r_m$ and positive integers $k_1, \ldots, k_m$ such that $F^{p(x)} \subseteq F(r_1, \ldots, r_m)$, and $r_1^{k_1} = 1 \in F$ and for $i = 2, \ldots, m$, $r_i^{k_i} \in F(r_1, \ldots, r_{i-1})$. That is, we have found complex numbers and positive integers which show that the equation $p(x) = 0$ is solvable by radicals according to the definition.

Exercises

In Exercises 1–6, (*a*) use Proposition 26.1 to show that the given group is solvable, and (*b*) use Theorem 27.8 to show the same thing.

1. $\text{Gal } \mathbb{Q}^{p(x)}/\mathbb{Q}$, where $p(x) = x^2 + 6x + 2$
2. $\text{Gal } \mathbb{Q}^{p(x)}/\mathbb{Q}$, where $p(x) = x^3 + 6x^2 + 8x + 3$

3. Gal $\mathbb{Q}(\sqrt{11}, i)/\mathbb{Q}$

4. Gal $\mathbb{Q}(\sqrt{13}, i)/\mathbb{Q}$

5. Gal $\mathbb{Q}(\sqrt[3]{11}, \zeta_3)/\mathbb{Q}$

6. Gal $\mathbb{Q}(\sqrt[3]{13}, \zeta_3)/\mathbb{Q}$

In Exercises 7–12, determine whether $\mathbb{Q} = \{k \in K \mid \sigma(k) = k \text{ for all } \sigma \in \text{Gal } K/\mathbb{Q}\}$.

7. $K = \mathbb{Q}(i)$ **8.** $K = \mathbb{Q}(\sqrt[3]{5})$ **9.** $K = \mathbb{Q}(\zeta_3)$

10. $K = \mathbb{Q}(i, \zeta_3)$ **11.** $K = \mathbb{Q}\left(\sqrt[4]{3}, \sqrt{13}, i\right)$

12. $K = \mathbb{Q}\left(\sqrt{4 + \sqrt{5}}, \sqrt{4 - \sqrt{5}}\right)$

Suppose that K is a subfield of $\mathbb{C}$ and an extension field of F. Exercises 13–15 concern the following subset of K:

$$X = \{k \in K \mid \sigma(k) = k \text{ for all } \sigma \in \text{Gal } K/F\}$$

13. Show that X is a subfield of K.

14. Show that Gal $K/X =$ Gal K/F.

15. Show that X contains $\mathbb{Q}$.

In Exercises 16–21, show that $\sigma \in$ Gal $K/\mathbb{Q}$ and then for $p(x) \in K[x]$, find $p(x)^\sigma$. (*Hint:* For Exercises 19–21, see Example 14.9.)

16. $K = \mathbb{Q}(i), p(x) = x^2 + (2 - 3i), \sigma(a_0 + a_1 i) = a_0 - a_1 i$

17. $K = \mathbb{Q}(i), p(x) = (4 + i)x^3 - ix^2 + 11, \sigma(a_0 + a_1 i) = a_0 - a_1 i$

18. $K = \mathbb{Q}(\sqrt{3}), p(x) = \sqrt{3}x^4 + (3 + \sqrt{3})x^2 + (1 + \sqrt{3})x,$
$\sigma(a_0 + a_1\sqrt{3}) = a_0 - a_1\sqrt{3}$

19. $K = \mathbb{Q}(\sqrt[3]{7}, \zeta_3),$
$p(x) = (2\sqrt[3]{7} + (\sqrt[3]{7})^2 + \zeta_3)x + ((1 - 5(\sqrt[3]{7})^2 + (3 + 2\sqrt[3]{7})\zeta_3)$
$\sigma((a_0 + a_1\sqrt[3]{7} + a_2(\sqrt[3]{7})^2) + (b_0 + b_1\sqrt[3]{7} + b_2(\sqrt[3]{7})^2)\zeta_3)$
$\quad = (a_0 + a_1\sqrt[3]{7} + a_2(\sqrt[3]{7})^2) + (b_0 + b_1\sqrt[3]{7} + b_2(\sqrt[3]{7})^2)\zeta_3{}^2$

20. $K = \mathbb{Q}(\sqrt[3]{7}, \zeta_3),$
$p(x) = (2\sqrt[3]{7} + (3 - (\sqrt[3]{7})^2)\zeta_3)x^2 - x + ((\sqrt[3]{7})^2 + (3 + 2\sqrt[3]{7} + (\sqrt[3]{7})^2)\zeta_3)$
$\sigma((a_0 + a_1\sqrt[3]{7} + a_2(\sqrt[3]{7})^2) + (b_0 + b_1\sqrt[3]{} + 7 + b_2(\sqrt[3]{7})^2)\zeta_3)$
$\quad = (a_0 + a_1\sqrt[3]{7}\zeta_3 + a_2(\sqrt[3]{7}\zeta_3)^2) + (b_0 + b_1\sqrt[3]{7}\zeta_3 + b_2(\sqrt[3]{7}\zeta_3)^2)\zeta_3$

21. $K = \mathbb{Q}(\sqrt[3]{7}, \zeta_3),$
$p(x) = x^4 - ((2 + (\sqrt[3]{7})^2) + (\sqrt[3]{7})^2\zeta_3)x^3$
$\qquad\qquad\qquad + 13(\sqrt[3]{7})^2x^2 - x + (4\sqrt[3]{7} + 8(\sqrt[3]{7})^2\zeta_3)$
$\sigma((a_0 + a_1\sqrt[3]{7} + a_2(\sqrt[3]{7})^2) + (b_0 + b_1\sqrt[3]{7} + b_2(\sqrt[3]{7})^2)\zeta_3)$
$\quad = (a_0 + a_1\sqrt[3]{7}\zeta_3{}^2 + a_2(\sqrt[3]{7}\zeta_3{}^2)^2) + (b_0 + b_1\sqrt[3]{7}\zeta_3{}^2 + b_2(\sqrt[3]{7}\zeta_3{}^2)^2)\zeta_3{}^2$

In Exercises 22–27, let $p(x) = x^3 - 7 \in \mathbb{Q}[x]$, and let

$$r_1 = \zeta_3 \text{ and } k_1 = 3,$$
$$r_2 = \sqrt[3]{7} \text{ and } k_2 = 3,$$
$$r_3 = \sqrt{1 + \sqrt[3]{7}} \text{ and } k_3 = 2,$$
$$r_4 \in \mathbb{C} \text{ be such that } (r_4)^2 = 1 + \sqrt[3]{7}\zeta_3 \text{ and } k_4 = 2,$$
$$r_5 \in \mathbb{C} \text{ such that } (r_5)^2 = 1 + \sqrt[3]{7}\zeta_3^2 \text{ and } k_5 = 2.$$

According to Example 14.9, $\mathbb{Q}^{p(x)} = \mathbb{Q}(r_1, r_2)$, and Gal $\mathbb{Q}^{p(x)}/\mathbb{Q} = \{f_1, f_2, f_3, f_4, f_5, f_6\}$, where

	f_1	f_2	f_3	f_4	f_5	f_6
$\sqrt[3]{7} \rightarrow$	$\sqrt[3]{7}$	$\sqrt[3]{7}$	$\sqrt[3]{7}\zeta_3$	$\sqrt[3]{7}\zeta_3$	$\sqrt[3]{7}\zeta_3^2$	$\sqrt[3]{7}\zeta_3^2$
$\zeta_3 \rightarrow$	ζ_3	ζ_3^2	ζ_3	ζ_3^2	ζ_3	ζ_3^2

22. Show that $\mathbb{Q}$ and r_1, r_2, r_3 and k_1, k_2, k_3 satisfy hypotheses S of Chapter 25.

23. Show that $(x^3 - 7)(x^2 - (1 + \sqrt[3]{7})) \notin \mathbb{Q}[x]$.

24. Show that $\{f_i(1 + \sqrt[3]{7}) \mid f_i \in \text{Gal } \mathbb{Q}^{p(x)}/\mathbb{Q}\} = \{1 + \sqrt[3]{7}, 1 + \sqrt[3]{7}\zeta_3, 1 + \sqrt[3]{7}\zeta_3^2\}$.

25. Show that $(x^2 - (1 + \sqrt[3]{7}))(x^2 - (1 + \sqrt[3]{7}\zeta_3))(x^2 - (1 + \sqrt[3]{7}\zeta_3^2)) \in \mathbb{Q}[x]$.

26. Show that $\mathbb{Q}$ and r_1, r_2, r_3, r_4, r_5 and k_1, k_2, k_3, k_4, k_5 satisfy hypotheses S of Chapter 25.

27. Show that $\mathbb{Q}(r_1, r_2, r_3, r_4, r_5) = \mathbb{Q}^{\pi(x)}$ for $\pi(x) = (x^3 - 7)(x^6 - 3x^4 + 3x^2 - 8) \in \mathbb{Q}[x]$.

28 Polynomial Equations Not Solvable by Radicals

Our objective in this chapter is to find a polynomial equation that is not solvable by radicals.

By the contrapositive of Theorem 27.8, any polynomial $p(x) \in \mathbb{Q}[x]$ for which Gal $\mathbb{Q}^{p(x)}/\mathbb{Q}$ is not a solvable group is itself not solvable by radicals. But if Gal $\mathbb{Q}^{p(x)}/\mathbb{Q}$ is isomorphic to S_n for some $n \geq 5$, then by Propositions 26.12 and 26.10, Gal $\mathbb{Q}^{p(x)}/\mathbb{Q}$ is not solvable. To find a polynomial that is not solvable by radicals, we therefore need to find a polynomial $p(x) \in \mathbb{Q}[x]$ for which Gal $\mathbb{Q}^{p(x)}/\mathbb{Q}$ is isomorphic to S_n for some $n \geq 5$. Proposition 17.5 describes a one-to-one homomorphism from Gal $\mathbb{Q}^{p(x)}/\mathbb{Q}$ into the permutation group S_n, where n is the number of distinct roots of $p(x)$. This one-to-one homomorphism will be onto whenever any permutation of the roots generates an automorphism in the Galois group. For instance, we saw in Example 17.6 that every permutation of the three roots of $p(x) = x^3 - 7$ generates an automorphism in Gal $\mathbb{Q}^{p(x)}/\mathbb{Q}$ and hence that Gal $\mathbb{Q}^{p(x)}/\mathbb{Q}$ is isomorphic to the full permutation group S_3. We might reasonably expect such polynomials to occur frequently. And indeed the next result provides an entire class of polynomials with Galois groups isomorphic to S_5.

Lemma 28.1

Let $p(x)$ be an irreducible polynomial in $\mathbb{Q}[x]$ of degree 5, and suppose that $p(x)$ has three real roots and two nonreal roots. Then the Galois group $G = $ Gal $\mathbb{Q}^{p(x)}/\mathbb{Q}$ is isomorphic to S_5.

Proof Let r_1, r_2, r_3 be the real roots of $p(x)$ and let c_1, c_2 be its nonreal roots. By Proposition 14.2, the five solutions of $p(x) = 0$ are all distinct, and thus by Proposition 17.5, we may define a one-to-one homomorphism $T : G \to S_5$ by letting $T(\phi)$ be the restriction of ϕ to the five-element set $\{r_1, r_2, r_3, c_1, c_2\}$. We will show that $T(G)$ contains a 5-cycle and a transposition and hence, by Proposition 19.9, that $T(G)$ is all of S_5.

By Proposition 12.8, $[\mathbb{Q}^{p(x)}:\mathbb{Q}(r_1)]$ is finite; by Proposition 12.4, $[\mathbb{Q}(r_1):\mathbb{Q}]$ is finite; hence Proposition 12.5 implies that $[\mathbb{Q}^{p(x)}:\mathbb{Q}] = [\mathbb{Q}^{p(x)}:\mathbb{Q}(r_1)]\,[\mathbb{Q}(r_1):\mathbb{Q}]$. Furthermore, if $m(x) = \pi_5^{-1}p(x)$, where $p(x) = \pi_0 + \cdots + \pi_5 x^5$, then $m(x) \in \mathbb{Q}[x]$ is monic and irreducible and r_1 is a root of $m(x)$, and hence $m(x)$ is the minimum polynomial of r_1 over $\mathbb{Q}$. Therefore, $[\mathbb{Q}(r_1):\mathbb{Q}] = 5$ by Proposition 12.4, and thus 5 divides $[\mathbb{Q}^{p(x)}:\mathbb{Q}]$. But by Proposition 14.8, the order of G is $[\mathbb{Q}^{p(x)}:\mathbb{Q}]$ so that 5 divides the order of G. Then Proposition 22.9, implies that G has an element ϕ of order 5. Since T is a one-to-one homomorphism, Proposition 21.9 implies that $T(\phi)$ is an element of order 5 in $T(G)$ and is thus a 5-cycle in $T(G)$ by Proposition 21.7.

If $\psi: \mathbb{Q}^{p(x)} \to \mathbb{C}$ is the function $\psi(z) = \bar{z}$, then according to Example 13.6, ψ is a one-to-one homomorphism which fixes $\mathbb{Q}$. By definition of ψ, $\psi(r_1) = r_1$, $\psi(r_2) = r_2$, and $\psi(r_3) = r_3$, and since c_2 is the complex conjugate of c_1 by Proposition 3.12, $\psi(c_1) = c_2$ and $\psi(c_2) = c_1$. Hence $\psi(\mathbb{Q}^{p(x)}) = \psi(\mathbb{Q}(r_1, r_2, r_3, c_1, c_2)) = \mathbb{Q}(\psi(r_1),$ $\psi(r_2), \psi(r_3), \psi(c_1), \psi(c_2)) = \mathbb{Q}^{p(x)}$, and therefore $\psi \in G$. But by definition of T, $T(\psi)$ is the permutation of $\{r_1, r_2, r_3, c_1, c_2\}$ induced by ψ and hence is a transposition in $T(G)$.

By Propositions 17.5 and 18.13, T is an isomorphism of G onto $T(G)$, and by Proposition 19.9, since $T(G)$ contains both a 5-cycle and a transposition, $T(G) = S_5$. We conclude that G is isomorphic to S_5. ∎

Theorem 28.2 *Let $p(x)$ be an irreducible polynomial in $\mathbb{Q}[x]$ of degree 5, and suppose that $p(x)$ has three real roots and two nonreal roots. Then the equation $p(x) = 0$ is not solvable by radicals.*

Proof By Lemma 28.1, the Galois group Gal $\mathbb{Q}^{p(x)}/\mathbb{Q}$ is isomorphic to S_5 and thus if Gal $\mathbb{Q}^{p(x)}/\mathbb{Q}$ were a solvable group, then Proposition 26.10 would imply that S_5 is also solvable, a contradiction of Proposition 26.12. So Gal $\mathbb{Q}^{p(x)}/\mathbb{Q}$ is not a solvable group. But by the contrapositive of Theorem 27.8, any polynomial $p(x) \in \mathbb{Q}[x]$ for which Gal $\mathbb{Q}^{p(x)}/\mathbb{Q}$ is not a solvable group is not solvable by radicals. Therefore, $p(x)$ is not solvable by radicals. ∎

The only remaining question is whether polynomials of the type described in Theorem 28.2 exist. We give a specific one in the next example.

Example 28.3 Let $p(x) \in \mathbb{Q}[x]$ be the polynomial $4x^5 - 10x^2 + 5$. By Eisenstein's criterion with prime 5, $p(x)$ is irreducible. The derivative of $p(x)$ is $p'(x) = 20x^4 - 20x = 20x(x^3 - 1)$, and hence $p(x)$ is increasing for $x < 0$, decreasing for $0 < x < 1$, and increasing again for $1 < x$. Note that

$$p(-1) = -9 < 0, \quad p(0) = 5 > 0,$$
$$p(1) = -1 < 0, \quad p(2) = 93 > 0,$$

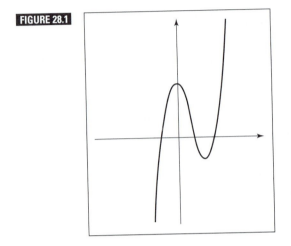

FIGURE 28.1

and hence the graph of $y = p(x)$ looks like Figure 28.1. By the intermediate value theorem of elementary calculus, $p(x) = 0$ has exactly three real solutions, and hence by Proposition 14.2, $p(x) = 0$ has two nonreal complex solutions. Therefore, by Theorem 28.2, the equation $p(x) = 0$ is not solvable by radicals. ❖

Exercises

In Exercises 1–9, determine whether the given polynomial equation is solvable by radicals over $\mathbb{Q}$.

1. $x^2 + 4x + 4 = 0$

2. $x^2 + 2x + 6 = 0$

3. $x^3 + 3x + 3 = 0$

4. $x^3 + 5x^2 + 5x = 0$

5. $2x^5 - 10x + 5 = 0$

6. $24x^5 - 30x^4 + 5 = 0$

7. $(x^2 + 2x + 2)(x^3 - 5) = 0$

8. $(x^4 + 3)(2x - 1) = 0$

9. $x^5 - 9x^3 + 3 = 0$

10. Suppose that $p(x) \in \mathbb{Q}[x]$ and that Gal $\mathbb{Q}^{p(x)}/\mathbb{Q}$ is isomorphic to S_8. Show that $p(x) = 0$ is not solvable by radicals.

11. Suppose that $f(x) \in \mathbb{Q}[x]$ and that Gal $\mathbb{Q}^{f(x)}/\mathbb{Q}$ is isomorphic to S_{10}. Show that $f(x) = 0$ is not solvable by radicals.

In Exercises 12–15, assume that Cauchy's theorem holds for any prime p (i.e., assume that if p divides the order of a finite group G, then G has an element of order p—see Exercise 43 in Chapter 22). Remember to justify your answer in each case, and note that your justification may have to include generalizing Proposition 19.9 (cf. Exercises 30 and 33 in Chapter 19).

12. Suppose that $p(x) \in \mathbb{Q}[x]$ is an irreducible polynomial of degree 7 with five real roots and two nonreal roots. Show that Gal $\mathbb{Q}^{p(x)}/\mathbb{Q}$ is isomorphic to S_7.

13. Suppose that $p(x) \in \mathbb{Q}[x]$ is an irreducible polynomial of degree 7 with five real roots and two nonreal roots. Show that $p(x) = 0$ is not solvable by radicals.

14. Find a polynomial $p(x)$ of degree 7 in $\mathbb{Q}[x]$ such that Gal $\mathbb{Q}^{p(x)}/\mathbb{Q}$ is isomorphic to S_7.

15. Find a polynomial $p(x)$ of degree 11 in $\mathbb{Q}[x]$ such that $p(x) = 0$ is not solvable by radicals.

HISTORICAL NOTE

Paolo Ruffini

Paolo Ruffini was born in Valentano, Italy, on September 22, 1765, and died in Modena, Italy, on May 10, 1822. Ruffini was the son of a physician and became a physician as well as a mathematician. While he was in his teens, his family moved from Valentano to Modena, where he spent the rest of his life. He attended the University of Modena, where he studied medicine, philosophy, literature, and mathematics, obtaining degrees in medicine, philosophy, and mathematics in 1788. He was given a position in mathematics soon afterward, and in 1791, he received his license to practice medicine. Napoleon's troops withdrew from Modena in 1796, and Ruffini was appointed, against his wishes, to the Junior Council of the Cisalpine Republic. After being relieved of these duties in 1798, he was asked to swear an oath of allegiance to the republic. He refused on religious grounds and was forced to give up teaching and all public life. He was not especially perturbed and continued his medical practice and his mathematical research. Eventually, in 1814, Napoleon fell, and Ruffini was appointed Rector of the restored university. It was an especially difficult time for the university, and Ruffini threw himself into his duties as Rector and returned to teaching both medicine and mathematics as well. He contracted typhus from a patient in 1817, but following his recovery, refused to moderate his activities. However, in 1819, his declining health forced him to cut back on his teaching, and then in April 1822, he contracted chronic pericarditis from a patient and died the next month.

It was during the self-enforced hiatus from his university duties (1798–1814) that Ruffini published his version of the theorem concerning equations not solvable by radicals. A preliminary demonstration appeared in "Teoria generale delle equazioni in cui si dimostra impossibile la soluzione algebrica dell equazioni generali di grado superiore al quarto" in 1799, and after many discussions with his colleagues, he presented a more detailed proof in "Riflessioni intorno alla soluzione delle equazioni algebriche generali" in 1813. Cauchy was alone among the leading mathematicians of the time to accord Ruffini the credit he deserved. It was not until Abel's independent proof in 1824 that Ruffini's proof was finally generally accepted.

PART 5

Finite Groups

29 Finite External Direct Products of Groups

Our overall goal in the remaining chapters is to describe all groups with less than sixteen elements. Specifically, for each integer $0 < n < 16$, we want to find a collection of groups such that every group with n elements is isomorphic to one of the groups in the collection.

For instance, consider groups with four elements. One group with four elements is, of course, $\mathbb{Z}_4$. Another group can be formed by combining $\mathbb{Z}_2$ with itself as follows. Form $\mathbb{Z}_2 \times \mathbb{Z}_2 = \{([0]_2, [0]_2), ([1]_2, [0]_2), ([0]_2, [1]_2), ([1]_2, [1]_2)\}$ with componentwise addition: $([a]_2, [b]_2) + ([k]_2, [l]_2) = ([a]_2 + [k]_2, [b]_2 + [l]_2)$. It is easy to see that $\mathbb{Z}_2 \times \mathbb{Z}_2$ is a group (see Exercise 26 in Chapter 15). Furthermore, since $2([1]_2, [0]_2) = ([0]_2, [0]_2) = 2([0]_2, [1]_2) = 2([1]_2, [1]_2)$, the elements of $\mathbb{Z}_2 \times \mathbb{Z}_2$ have either order 1 (the element $([0]_2, [0]_2)$) or order 2 (the remaining elements). But $[1]_4 \in \mathbb{Z}_4$ has order 4, and thus if $T : \mathbb{Z}_4 \to \mathbb{Z}_2 \times \mathbb{Z}_2$ is an isomorphism, then $T([1]_4)$ has order 4 in $\mathbb{Z}_2 \times \mathbb{Z}_2$ by Proposition 21.9. Since $\mathbb{Z}_2 \times \mathbb{Z}_2$ has no such element, $\mathbb{Z}_2 \times \mathbb{Z}_2$ cannot be isomorphic to $\mathbb{Z}_4$. We have thus found two distinct groups of order 4. We will see in Chapter 34 that every group of order 4 is Abelian and in Chapter 32 that there are no other ways of constructing Abelian groups of order 4.

In general, we want to combine groups $G_1, \ldots, G_n$ by forming the set of n-tuples $(g_1, \ldots, g_n)$, where $g_i \in G_i$, with the group operation defined coordinate by coordinate. We first show that the result is always a group.

Proposition 29.1   Let $G_1, \ldots, G_n$ be groups and let $H = \{(g_1, \ldots, g_n) \mid g_i \in G_i\}$. With respect to the operation $(a_1, \ldots, a_n)(b_1, \ldots, b_n) = (a_1b_1, \ldots, a_nb_n)$, H is a group.

Proof   Since each G_i is a group, each product $(a_1, \ldots, a_n)(b_1, \ldots, b_n) = (a_1b_1, \ldots, a_nb_n) \in H$, and hence H is closed. If $(a_1, \ldots, a_n)$, $(b_1, \ldots, b_n)$, $(c_1, \ldots, c_n) \in H$, then since each G_i is associative,

$$(a_1, \ldots, a_n)((b_1, \ldots, b_n)(c_1, \ldots, c_n)) = (a_1(b_1c_1), \ldots, a_n(b_nc_n))$$
$$= ((a_1b_1)c_1, \ldots, (a_nb_n)c_n) = ((a_1, \ldots, a_n)(b_1, \ldots, b_n))(c_1, \ldots, c_n),$$

and hence H is associative. Furthermore, if e_i represents the identity of G_i, then $(a_1, \ldots, a_n)(e_1, \ldots, e_n) = (a_1, \ldots, a_n) = (e_1, \ldots, e_n)(a_1, \ldots, a_n)$, and thus $(e_1, \ldots, e_n)$ is an identity for H. As well, since each G_i is a group, $a_i^{-1} \in G_i$ and hence $(a_1^{-1}, \ldots, a_n^{-1}) \in H$. Thus, since $(a_1, \ldots, a_n)(a_1^{-1}, \ldots, a_n^{-1}) = (e_1, \ldots, e_n) = (a_1^{-1}, \ldots, a_n^{-1})(a_1, \ldots, a_n)$, every element of H has an inverse, and therefore H is a group. ∎

DEFINITION

For groups $G_1, \ldots, G_n$, let

$$G_1 \times \cdots \times G_n = \{(g_1, \ldots, g_n) \mid g_i \in G_i\},$$

and for $(a_1, \ldots, a_n), (b_1, \ldots, b_n) \in G_1 \times \cdots \times G_n$, define

$$(a_1, \ldots, a_n)(b_1, \ldots, b_n) = (a_1 b_1, \ldots, a_n b_n).$$

The group $G_1 \times \cdots \times G_n$ is called the **(external) direct product** of the groups $G_1, \ldots, G_n$; the groups G_i are called **factors** of $G_1 \times \cdots \times G_n$. If the group operations on the factors are written additively, then the operation on the n-tuples is also written additively and the resulting group is sometimes called the **(external) direct sum;** in this case, the groups G_i are called **summands** of $G_1 \times \cdots \times G_n$.

The adjective "external" is used to differentiate the direct products defined here from their internal cousins described in the next chapter. When it is clear from the context which direct product is being discussed, the adjective is usually dropped.

For a general direct product $G_1 \times \cdots \times G_n$ of multiplicative groups, we abuse notation somewhat and use the same symbol e for the identity of each factor. In this way, we may write the identity of the product as $(e, \ldots, e)$. For direct products of specific groups, however, we adopt the notation appropriate to each factor. Consider, for instance, the following example.

Example 29.2   Consider $G = S_3 \times \mathbb{Z}_4$. The elements of G are of the form $(\sigma, [k]_4)$, where $\sigma \in S_3$ and $[k]_4 \in \mathbb{Z}_4$, e.g., $((12), [3]_4)$ and $((132), [2]_4)$. In particular, the identity of G is $(\iota, [0]_4)$. Note that the group operation in G is strictly coordinate by coordinate; multiplication is used for the first coordinate, while addition is used for the second coordinate. Since not both groups are additive, the operation in G is written multiplicatively:

$$((12), [3]_4)((132), [2]_4) = ((12)(132), [3]_4 + [2]_4) = ((13), [1]_4). \; ❖$$

We derive a few of the elementary properties of external direct products in the next four propositions (29.3, 29.5, 29.7, and 29.9). The first characterizes those external direct products that are Abelian. The second says the order of an external direct product is the product of the orders of its factors. The third says that the formation of direct products is commutative, and the fourth says that it is associative.

Proposition 29.3   *A finite external direct product of groups is Abelian if and only if each factor group is Abelian.*

Proof   Consider the groups $G_1, \ldots, G_n$. If each G_i is Abelian and if

$$(x_1, \ldots, x_n), (y_1, \ldots, y_n) \in G_1 \times \cdots \times G_n,$$

then

$$(x_1, \ldots, x_n)(y_1, \ldots, y_n) = (x_1 y_1, \ldots, x_n y_n)$$
$$= (y_1 x_1, \ldots, y_n x_n) = (y_1, \ldots, y_n)(x_1, \ldots, x_n),$$

and thus $G_1 \times \cdots \times G_n$ is Abelian. Conversely, suppose that $G_1 \times \cdots \times G_n$ is Abelian. Then for all $x_i, y_i \in G_i$,

$$(e, \ldots, x_i y_i, \ldots, e) = (e, \ldots, x_i, \ldots, e)(e, \ldots, y_i, \ldots, e)$$
$$= (e, \ldots, y_i, \ldots, e)(e, \ldots, x_i, \ldots, e) = (e, \ldots, y_i x_i, \ldots, e),$$

and hence $x_i y_i = y_i x_i$. So each factor group G_i is Abelian. ∎

Example 29.4   Consider the group $G = S_3 \times \mathbb{Z}_4$ of Example 29.2. Since S_3 is not Abelian, Proposition 29.3 implies that G is not Abelian. For instance,

$$((12), [0]_4)((13), [0]_4) = ((12)(13), [0]_4 + [0]_4) = ((123), [0]_4),$$

while

$$((13), [0]_4)((12), [0]_4) = ((12)(13), [0]_4 + [0]_4) = ((132), [0]_4). \; ❖$$

Recall that if G is a finite group, then $|G|$ denotes the order of G (i.e., the number of elements that G contains).

Proposition 29.5   *If $G_1, \ldots, G_n$ are finite groups, then*

$$|G_1 \times \cdots \times G_n| = |G_1| \cdots |G_n|.$$

Proof   Any element in $G_1 \times \cdots \times G_n$ has the form $(g_1, \ldots, g_n)$. Since there are $|G_i|$ choices for the ith coordinate, there are $|G_1| \cdots |G_n|$ elements altogether in $G_1 \times \cdots \times G_n$. ∎

Example 29.6   Consider the group $G = S \times \mathbb{Z}_4$ of Examples 29.2 and 29.4. Since S_3 has $3! = 6$ elements and $\mathbb{Z}_4$ has four elements, Proposition 2.5 implies that G has $6 \cdot 4 = 24$ elements. The elements of G are

$$(\iota, [0]_4), ((12), [0]_4), ((13), [0]_4), ((23), [0]_4), ((123), [0]_4), ((132), [0]_4)$$
$$(\iota, [1]_4), ((12), [1]_4), ((13), [1]_4), ((23), [1]_4), ((123), [1]_4), ((132), [1]_4)$$
$$(\iota, [2]_4), ((12), [2]_4), ((13), [2]_4), ((23), [2]_4), ((123), [2]_4), ((132), [2]_4)$$
$$(\iota, [3]_4), ((12), [3]_4), ((13), [3]_4), ((23), [3]_4), ((123), [3]_4), ((132), [3]_4). \; ❖$$

Our final results merely say that the formation of direct products is, up to isomorphism, commutative (Proposition 29.7) and associative (Proposition 29.9).

Proposition 29.7 *All finite external direct products involving the same collection of groups are isomorphic. Specifically, suppose that $G_1, \ldots, G_n$ are finite groups and that σ is a permutation of $\{1, \ldots, n\}$. Let*

$$\Phi : G_1 \times \cdots \times G_n \rightarrow G_{\sigma(1)} \times \cdots \times G_{\sigma(n)}$$

be the function

$$\Phi((g_1, \ldots, g_n)) = (g_{\sigma(1)}, \ldots, g_{\sigma(n)}).$$

Then Φ is an isomorphism.

Proof Suppose that $\Phi((g_1, \ldots, g_n)) = \Phi((h_1, \ldots, h_n))$. Then $(g_{\sigma(1)}, \ldots, g_{\sigma(n)}) = (h_{\sigma(1)}, \ldots, h_{\sigma(n)})$, and hence, for all i, $g_{\sigma(i)} = h_{\sigma(i)}$. Since σ is a permutation of $\{1, \ldots, n\}$, this implies that for all i, $g_i = h_i$. Then $(g_1, \ldots, g_n) = (h_1, \ldots, h_n)$ and hence Φ is one-to-one. If $(h_1, \ldots, h_n) \in G_{\sigma(1)} \times \cdots \times G_{\sigma(n)}$, then $(h_{\sigma^{-1}(1)}, \ldots, h_{\sigma^{-1}(n)}) \in G_1 \times \cdots \times G_n$, and $\Phi((h_{\sigma^{-1}(1)}, \ldots, h_{\sigma^{-1}(n)})) = (h_1, \ldots, h_n)$ so that Φ is onto. Finally, since the ith coordinate of $(g_1 h_1, \ldots, g_n h_n)$ is $g_i h_i$, the ith coordinate of $\Phi((g_1 h_1, \ldots, g_n h_n))$ is $g_{\sigma(i)} h_{\sigma(i)}$, and thus

$$\Phi((g_1, \ldots, g_n)(h_1, \ldots, h_n)) = \Phi((g_1 h_1, \ldots, g_n h_n)) = (g_{\sigma(1)} h_{\sigma(1)}, \ldots, g_{\sigma(n)} h_{\sigma(n)})$$
$$= (g_{\sigma(1)}, \ldots, g_{\sigma(n)})(h_{\sigma(1)}, \ldots, h_{\sigma(n)}) = \Phi((g_1, \ldots, g_n)) \Phi((h_1, \ldots, h_n)).$$

So Φ is an isomorphism. ∎

Example 29.8 Consider the groups $G_1 = \mathbb{Z}_4$, $G_2 = \mathbb{Z}_6$, and $G_3 = \mathbb{Z}_3$. Then Proposition 29.7 says that the function $\Phi : G_1 \times G_2 \times G_3 \rightarrow G_2 \times G_3 \times G_1$, defined by letting $\Phi((g_1, g_2, g_3)) = (g_2, g_3, g_1)$, is an isomorphism. In this case, the permutation σ is $\sigma = (123)$, and (for instance)

$$\Phi(([2]_4, [5]_6, [1]_3)) = ([5]_6, [1]_3, [2]_4). ❖$$

To prove in general that the formation of external direct products is associative requires some notational complication. So we will in fact prove a special case and leave the general proof to the reader.

Proposition 29.9 *A direct product of direct products is isomorphic to the direct product of the original factors. Specifically, suppose for each $i = 1, \ldots, m$, $G_i, G_{i,1}, \ldots, G_{i,n_i}$ are groups and $\Psi_i : G_{i,1} \times \cdots \times G_{i,n_i} \rightarrow G_i$ is an isomorphism. Let*

$$\Phi : G_{1,1} \times \cdots \times G_{1,n_1} \times \cdots \times G_{m,n_m} \rightarrow G_1 \times \cdots \times G_m$$

be the function

$$\Phi((g_{1,1}, \ldots, g_{1,n_1}, \ldots, g_{m,1}, \ldots, g_{m,n_m}))$$
$$= (\Psi_1(g_{1,1}, \ldots, g_{1,n_1}), \ldots, \Psi_m(g_{m,1}, \ldots, g_{m,n_m})).$$

Then Φ is an isomorphism.

Proof   To simplify notation, we will prove the result for $m = 2$ and $n_1 = 2 = n_2$. An analogous proof, whose formulation is left to the reader, may be used to prove the general result. Specifically, suppose that $A, B, C, D, X,$ and Y are groups and that $\Psi_1 : A \times B \to X$ and $\Psi_2 : C \times D \to Y$ are isomorphisms, and let $\Phi : A \times B \times C \times D \to X \times Y$ be the function $\Phi((a, b, c, d)) = (\Psi_1((a, b)), \Psi_2((c, d)))$. We must show that Φ is an isomorphism. Suppose that $\Phi((a, b, c, d)) = \Phi((r, s, t, u))$. Then $(\Psi_1((a, b)), \Psi_2((c, d))) = (\Psi_1((r, s)), \Psi_2((t, u)))$, and hence $\Psi_1((a, b)) = \Psi_1((r, s))$ and $\Psi_2((c, d))) = \Psi_2((t, u))$. Since Ψ_1 and Ψ_2 are both one-to-one, we have that $(a, b) = (r, s)$ and $(c, d) = (t, u)$ and hence that $a = r, b = s, c = t,$ and $d = u$. Then $(a, b, c, d) = (r, s, t, u)$, and thus Φ is one-to-one. Suppose next that $(x, y) \in X \times Y$. Since Ψ_1 is onto, there exists $(a, b) \in A \times B$ such that $\Psi_1((a, b)) = x$, and since Ψ_2 is onto, there exists $(c, d) \in C \times D$ such that $\Psi_2((c, d)) = y$. Then $(a, b, c, d) \in A \times B \times C \times D$ and $\Phi((a, b, c, d)) = (\Psi_1((a, b)), \Psi_2((c, d))) = (x, y)$ so that Φ is onto. Finally, since both Ψ_1 and Ψ_2 preserve multiplication,

$$\Phi((a, b, c, d)(r, s, t, u)) = \Phi((ar, bs, ct, du)) = (\Psi_1((ar, bs)), \Psi_2((ct, du)))$$
$$= (\Psi_1((a, b)(r, s)), \Psi_2((c, d)(t, u))) =$$
$$(\Psi_1((a, b))\Psi_1((r, s)), \Psi_2((c, d))\Psi_2((t, u)))$$
$$= (\Psi_1((a, b)), \Psi_2((c, d)))(\Psi_1((r, s)), \Psi_2((t, u))) =$$
$$\Phi((a, b, c, d))\Phi((r, s, t, u)),$$

and hence Φ is an isomorphism. ∎

Example 29.10   Proposition 29.9 says that $\mathbb{Z}_4 \times \mathbb{Z}_6 \times \mathbb{Z}_3 \times \mathbb{Z}_7$ is isomorphic to $(\mathbb{Z}_4 \times \mathbb{Z}_6) \times (\mathbb{Z}_3 \times \mathbb{Z}_7)$ via the natural function $\Phi((z_1, z_2, z_3, z_4)) = ((z_1, z_2), (z_3, z_4))$. ❖

Exercises

In Exercises 1–5, determine the number of elements in the group G.

1. $G = S_3 \times S_4$
2. $G = S_5 \times \mathbb{Z}_3$
3. $G = S_4 \times \mathbb{Z}_2 \times \mathbb{Z}_4$
4. $G = \mathbb{Z}_3 \times S_3 \times \mathbb{Z}_3$
5. $G = \mathbb{Z}_2 \times S_3 \times \mathbb{Z}_4 \times S_5$

In Exercises 6–10, list all the elements in the group G.

6. $G = \mathbb{Z}_2 \times \mathbb{Z}_3$ **7.** $G = \mathbb{Z}_2 \times \mathbb{Z}_4$

8. $G = S_3 \times S_3$ **9.** $G = \mathbb{Z}_3 \times S_3$

10. $G = \mathbb{Z}_3 \times S_3 \times \mathbb{Z}_2$

In Exercises 11–15, express the given element of $\mathbb{Z}_5 \times S_3$ in the form $([k]_5, \sigma)$, where $0 \leq k < 5$ and σ is written as a product of disjoint cycles.

11. $([4]_5, (23))\,([3]_5, (13))$

12. $([-2]_5, (123))\,([6]_5, (23))$

13. $([9]_5, (123))\,([-8]_5, (13))^{-1}\,([2]_5, (13))$

14. $([-7]_5, (13))^{-1}\,([13]_5, \iota)\,([0]_5, (23))$

15. $([22]_5, (132))^{-1}\,([-13]_5, (12))\,([64]_5, (13))^{-1}$

In Exercises 16–20, express the given element of $\mathbb{Z}_4 \times S_4 \times S_3$ in the form $([k]_4, \sigma, \tau)$, where $0 \leq k < 4$ and σ and τ are written as products of disjoint cycles.

16. $([1]_4, (14), (23))([3]_4, (143), (13))$

17. $([-5]_4, (12)(34), (23))([6]_4, (23), (132))$

18. $([5]_4, (1234), (23))\,([9]_4, (23), (132))^{-1}\,([2]_4, (13), (13))$

19. $([7]_4, (124), (13))^{-1}\,([9]_4, (34), \iota)\,([0]_4, (14), (23))$

20. $([18]_4, (134), (23))^{-1}\,([-13]_4, \iota, (23))\,([47]_4, (12), (123))^{-1}$

In Exercises 21–25, express the given element of $\mathbb{Z}_7 \times S_3 \times S_5 \times \mathbb{Z}_3$ in the form $([k]_7, \sigma, \tau, [j]_3)$, where $0 \leq k < 7$, $0 \leq j < 3$, and σ and τ are written as products of disjoint cycles.

21. $([1]_7, (14), (23), [1]_3)\,([3]_7, (143), (13), [2]_3)$

22. $([-5]_7, (12)(34), (23), [1]_3)\,([6]_7, (23), (132), [10]_3)$

23. $([5]_7, (1234), (23), [7]_3)\,([9]_7, (23), (132), [-8]_3)^{-1}\,([13]_7, (13), (13), [5]_3)$

24. $([7]_7, (124), (13), [-2]_3)^{-1}\,([9]_7, (34), \iota, [1]_3)\,([0]_7, (14), (23), [-23]_3)$

25. $([18]_7, (134), (23), [3]_3)^{-1}\,([-13]_7, \iota, (23), [7]_3)\,([47]_7, (12), (123), [47]_3)^{-1}$

26. Show that S_3 is not isomorphic to $\mathbb{Z}_2 \times \mathbb{Z}_3$.

27. Find an Abelian group with eight elements that is <u>not</u> isomorphic to $\mathbb{Z}_8$. Justify your answer.

28. Find an Abelian group with nine elements that is <u>not</u> isomorphic to $\mathbb{Z}_9$. Justify your answer.

29. Show that if A is a normal subgroup of a group G, and if B is a normal subgroup of a group H, then $A \times B$ is a normal subgroup of $G \times H$.

30. Find a group G of infinite order with a normal subgroup N such that G/N is isomorphic to $\mathbb{Z}_7 \times \mathbb{Z}_8$. Justify your answer.

31. Show that $\mathbb{Z}_{27}$ is <u>not</u> isomorphic to $\mathbb{Z}_3 \times \mathbb{Z}_3 \times \mathbb{Z}_3$.

32. Show that $\mathbb{Z}_{27}$ is <u>not</u> isomorphic to $\mathbb{Z}_3 \times \mathbb{Z}_9$.

33. Show directly (without using Proposition 29.7) that $\mathbb{Z}_5 \times S_3$ is isomorphic to $S_3 \times \mathbb{Z}_5$.

34. Show directly (without using Proposition 29.7) that $\mathbb{Z}_5 \times S_3 \times S_4 \times \mathbb{Z}_7$ is isomorphic to $S_4 \times \mathbb{Z}_7 \times \mathbb{Z}_5 \times S_3$.

Exercises 35–37 concern the groups $G = \mathbb{Z}_4 \times S_3$ and $H = \{[0]_4, [2]_4\} \times \{\iota, (123), (132)\}$.

35. Show that $H \triangleleft G$.

36. Find the number of elements in G/H. Justify your answer.

37. Show that G/H is Abelian.

38. For groups A, B, C, X, Y, and Z, show directly (without using Proposition 29.9) that $(A \times B \times C) \times (X \times Y \times Z)$ is isomorphic to $A \times B \times C \times X \times Y \times Z$.

39. For groups A, B, C, D, X, and Y, show directly (without using Proposition 29.9) that $(A \times B \times C \times D) \times (X \times Y)$ is isomorphic to $A \times B \times C \times D \times X \times Y$.

40. If p and q are distinct primes and P and Q are groups of orders p and q respectively, show that $P \times Q$ is isomorphic to $\mathbb{Z}_{pq}$.

Exercises 41 and 42 concern the groups $G = \mathbb{Z} \times \mathbb{Z}$ and $H = \{(n, n) \mid n \in \mathbb{Z}\}$.

41. Show that $H \triangleleft G$.

42. Show that H is not equal to $A \times B$ for any subgroups A and B of $\mathbb{Z}$.

43. Find groups $A \neq \{e\}$ and $B \neq \{e\}$ such that $\mathbb{Z}_{21}$ is isomorphic to $A \times B$. Justify your answer.

44. Find groups $A \neq \{e\}$, $B \neq \{e\}$, and $C \neq \{e\}$ such that $\mathbb{Z}_{70}$ is isomorphic to $A \times B \times C$. Justify your answer.

45. Suppose that m and n are positive integers with common prime factor p. Show that $\mathbb{Z}_{mn}$ is <u>not</u> isomorphic to $\mathbb{Z}_m \times \mathbb{Z}_n$.

46. Let G be a group and suppose that for each $i = 1, \ldots, n$, $H_i \triangleleft G_i$.

    (*a*)   Show that $H_1 \times \cdots \times H_n \triangleleft G_1 \times \cdots \times G_n$.

    (*b*)   Show that $(G_1 \times \cdots \times G_n)/(H_1 \times \cdots \times H_n)$ is isomorphic to $(G_1/H_1) \times \cdots \times (G_n/H_n)$.

47. Prove the following statement or give a counterexample. If M and N are normal subgroups of a group G such that G/M is isomorphic to G/N, then M is isomorphic to N.

48. If $m_1, \ldots, m_n$ are positive integers whose least common multiple is $m_1 \cdots m_n$, and if for each i, G_i is a cyclic group of order m_i, show that $G_1 \times \cdots \times G_n$ is isomorphic to $\mathbb{Z}_{m_1 \cdots m_n}$.

49. Prove Proposition 29.9 in full generality: If for each $i = 1, \ldots, m$, G_i, $G_{i,1}, \ldots, G_{i,n_i}$ are groups and $\Psi_i: G_{i,1} \times \cdots \times G_{i,n_i} \to G_i$ is an isomorphism, then the function

$$\Phi: G_{1,1} \times \cdots \times G_{1,n_1} \times \cdots \times G_{m,1} \times \cdots \times G_{m,n_m} \to G_1 \times \cdots \times G_m$$

defined by letting

$$\Phi((g_{1,1}, \ldots, g_{1,n_1}, \ldots, g_{m,1}, \ldots, g_{m,n_m}))$$
$$= (\Psi_1(g_{1,1}, \ldots, g_{1,n_1}), \ldots, \Psi_m(g_{m,1}, \ldots, g_{m,n_m})).$$

is an isomorphism.

Exercises 50–52 show that if G and H are groups, then the direct product $G \times H$ is the product of G and H in the sense of category theory (see Chapter 1 of Jacobson's *Basic Algebra II* or Chapter XV of MacLane and Birkhoff's *Algebra*).

50. Prove that the function $\pi_1: G \times H \to G$, $\pi_1(g, h) = g$, is an onto homomorphism.

51. Prove that the function $\pi_2: G \times H \to H$, $\pi_2(g, h) = h$, is an onto homomorphism.

52. Suppose that K is a group for which there exist onto homomorphisms $\alpha_1: K \to G$ and $\alpha_2: K \to H$. Show that there exists a unique homomorphism $f: K \to G \times H$ such that $\alpha_1 = \pi_1 \circ f$ and $\alpha_2 = \pi_2 \circ f$. This situation is illustrated in Figure 29.1.

FIGURE 29.1

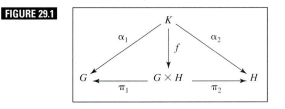

53. Suppose that $X = \{x_1, x_2\}$ is a two-element set and let $i: X \to \mathbb{Z} \times \mathbb{Z}$ be the function $i(x_1) = (1, 0)$ and $i(x_2) = (0, 1)$. Show that if A is an Abelian group and $s: X \to A$ is any function, then there exists a unique homomorphism

$f: \mathbb{Z} \times \mathbb{Z} \to A$ such that $f \circ i = s$. This says that $\mathbb{Z} \times \mathbb{Z}$ is the <u>free Abelian group on two elements</u> and is illustrated in Figure 29.2.

FIGURE 29.2

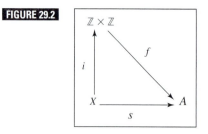

Exercises 54–57 show how to write the multiplicative group of nonzero complex numbers $\mathbb{C}^\times = \{z \in \mathbb{C} \mid z \neq 0\}$ as a direct product .

54. Show that the unit circle $U = \{z \in \mathbb{C} \mid |z| = 1\}$ is a group with respect to multiplication of complex numbers.

55. Show that $\mathbb{R}^+ = \{r \in \mathbb{R} \mid r > 0\}$ is a group with respect to multiplication of real numbers.

56. Show that $\mathbb{C}^\times$ is isomorphic to $\mathbb{R}^+ \times U$.

57. Show that $\mathbb{C}^\times$ is isomorphic to $\mathbb{R} \times U$.

HISTORICAL NOTE

J. H. M. Wedderburn

Joseph Henry Maclagen Wedderburn was born in Forfar, Scotland, on February 26, 1882, the tenth of fourteen children, and died in Princeton, New Jersey, on October 9, 1948. His father, Alexander Stormoth Maclagen Wedderburn, was a physician and the son of a minister; his mother, Anne Ogilvie, was the daughter of a lawyer from Dundee. Joseph attended Forfar Academy and then George Watson's College in Edinburgh. In 1898, he entered the University of Edinburgh, and in 1903, not only did he receive his M.A. with first class honors in mathematics but also his first paper, "On the isoclinal lines of a differential equation of the first order," appeared.

In the year after his graduation from Edinburgh, he attended the University of Leipzig and then the University of Berlin, and then, in 1904, he traveled to the United States to spend a year at the University of Chicago. This turned out to be a very fortuitous trip, resulting in two of his most important papers, "On the structure of hypercomplex number systems" and "A theorem on finite algebras."

He left Chicago to take up a position in Edinburgh, where he remained until 1909 when he returned to the United States to become one of Woodrow Wilson's Praeceptors at Princeton University. Wilson wanted "a large scale infusion of new blood, of scholars who would assume an intimate personal relation with small groups of undergraduates and impart to them something of their own enthusiasm for things of the mind."[1] And indeed the next five years were happy ones for Wedderburn.

However, in 1914, Germany's attack on France began World War I and ended Wedderburn's pleasant time at Princeton. Although he probably could have used his experience in the Territorial Forces before the war and his scientific prestige to obtain a commission in the British army, he instead volunteered as a private. In any case, he quickly rose through the ranks, becoming a lieutenant in the Seaforth Highlanders in November 1914. By the end of the war, he was a captain in Royal Engineers, where he used his scientific knowledge to help locate enemy batteries. He was the first resident of Princeton to leave to fight in the war and he was a soldier longer than any other resident of the town or member of the university.

At the end of the war, he returned to Princeton, where he remained for the rest of his life. He was very much an outdoorsman and, according to his friend Hugh Taylor, "found deep satisfaction in the wilderness, in the woods, canoeing along rivers and streams in the company of thoughtful men."[2]

At the end of the 1920s, Wedderburn suffered what seems to have been a nervous breakdown. Whatever the cause, he withdrew from his previous friendships and began to pursue a very solitary life. He dined alone, and his excursions to the mountains were solitary and silent. He complained to a friend who managed to break through the barriers that he "lived too much alone,"[3] But he was unable to help himself and eventually, three years after his retirement in 1945, he suffered a heart attack and died, alone in the house in Princeton in which he had lived for many years.

Mathematically, Wedderburn's main interest was not in theory of groups but in that of algebras, i.e., structures that are both rings and vector spaces. He showed that all finite division algebras (i.e., those in which each nonzero element has a multiplicative inverse) are commutative, and he was the first to investigate algebras over an arbitrary field. He characterized certain algebras, known as simple algebras, as algebras of matrices. And he showed that other algebras, known as semisimple algebras, were direct products of algebras, each of which could be written as the direct product of a division algebra and a matrix algebra over that division algebra. These results had far-reaching consequences in other branches of mathematics. They gave the complete structure of all projective geometries with a finite number of points and paved the way for Emmy Noether's great advances in the theory of ideals.

References

1. Quoted in Taylor, Hugh S. *Obituary Notices of Fellows,* Royal Society of London **6** (1949), p. 618 ff., reprinted in *The Nature and Growth of Modern Mathematics* by Edna E. Kramer, Princeton: Princeton University Press, 1981, p. 681.
2. *Ibid.,* p. 682.
3. *Ibid.,* p. 683.

30 Finite Internal Direct Products of Groups

Whenever it is possible, we would like to decompose a group into a direct product of simpler groups. That is, we would like to find a collection of simpler groups for which there exists an isomorphism from their direct product to the original group. The easiest place to look for candidates for the simpler groups is among the subgroups of the original group, and the easiest isomorphism to use is one that identifies the subgroups with themselves. For instance, if G is to be a direct product of its subgroups A and B, then there should exist an isomorphism $\Phi: A \times B \to G$ such that for all $a \in A$, $\Phi((a, e)) = a$, and for all $b \in B$, $\Phi((e, b)) = b$. Note that this means that Φ must take an element (a, b) to the product of its coordinates. For since Φ preserves multiplication,

$$\Phi((a, b)) = \Phi((a, e)(e, b)) = \Phi((a, e))\Phi((e, b)) = ab.$$

In view of this observation, we make the following general definition.

> **DEFINITION**
>
> A group G is the **(internal) direct product** of its subgroups $G_1, \ldots, G_n$ if the function $\Phi: G_1 \times \cdots \times G_n \to G$ defined by letting $\Phi(g_1, \ldots, g_n) = g_1 \cdots g_n$ is an isomorphism. We use the notation $G = G_1 \boxtimes \cdots \boxtimes G_n$ to mean that this function Φ is an isomorphism. If the group is written additively, then we say that G is the **(internal) direct sum** of $G_1, \ldots, g_n$ and we write $G = G_1 \oplus \cdots \oplus G_n$.

According to the definition, any internal direct product is isomorphic to an external direct product. It is shown in the exercises (Exercises 27) that any external direct product is, in a very natural way, an internal direct product.

Example 30.1 Consider the additive group $\mathbb{Z}_6$. It is easy to use Proposition 18.7 to check that $A = \{[0]_6, [3]_6\}$ and $B = \{[0]_6, [2]_6, [4]_6\}$ are subgroups of $\mathbb{Z}_6$. We will show that

$\mathbb{Z}_6$ is the internal direct product of A and B. Consider the function $\Phi:A \times B \to \mathbb{Z}_6$ defined by letting $\Phi(a, b) = a + b$. Since

$$\Phi((a, b) + (c, d)) = \Phi(a + c, b + d) = (a + c) + (b + d)$$
$$= (a + b) + (c + d) = \Phi(a, b) + \Phi(c, d),$$

Φ is a homomorphism, and it is clear from the following table that Φ is one-to-one and onto:

x	$([0]_6, [0]_6)$	$([0]_6, [2]_6)$	$([0]_6, [4]_6)$	$([3]_6, [0]_6)$	$([3]_6, [2]_6)$	$([3]_6, [4]_6)$
$\downarrow$	$\downarrow$	$\downarrow$	$\downarrow$	$\downarrow$	$\downarrow$	$\downarrow$
$\Phi(x)$	$[0]_6$	$[2]_6$	$[4]_6$	$[3]_6$	$[5]_6$	$[1]_6$

We conclude that Φ is an isomorphism and hence that $\mathbb{Z}_6$ is the internal direct product of A and B by definition. ❖

Checking that Φ is an isomorphism can be very tedious and cumbersome. We want to avoid this by finding conditions on the subgroups that are equivalent to it.

To state these conditions and for general use in the sequel, recall that if G is a group and if X is a subset of G, then (cf. Chapter 19) $S(X)$ denotes the subgroup of G generated by X. We will abbreviate this notation when it is convenient. For instance, if $x_1, \ldots, x_n \in G$, then $S(x_1, \ldots, x_n)$ will mean $S(\{x_1, \ldots, x_n\})$, if $x_1, \ldots, x_n \in G$ and $X \subseteq G$, then $S(x_1, \ldots, x_n, X)$ will mean $S(\{x_1, \ldots, x_n\} \cup X)$, and if $X_1, \ldots, X_n \subseteq G$, then $S(X_1, \ldots, X_n)$ will mean $S(X_1 \cup \cdots \cup X_n)$.

As well, since we will be interested only in internal direct sums of Abelian groups, we will restrict ourselves to Abelian groups for the remainder of this section and therefore, in keeping with standard practice, we will write the generic operation as addition rather than multiplication. Note that appropriately modified versions of the results that follow hold true for noncommutative groups as well; the statements for the general case appear in the exercises (Exercises 24–26).

We begin by showing that in the Abelian case, the function Φ is always a homomorphism.

Lemma 30.2   *If $X_1, \ldots, X_n$ are subgroups of an Abelian group G, then the function $\Phi:X_1 \times \cdots \times X_n \to G$, $\Phi(x_1, \ldots, x_n) = x_1 + \cdots + x_n$, is a homomorphism.*

Proof   If $(x_1, \ldots, x_n), (y_1, \ldots, y_n) \in X_1 \times \cdots \times X_n$, then

$$\Phi(x_1, \ldots, x_n) + \Phi(y_1, \ldots, y_n) = (x_1 + \cdots + x_n) + (y_1 + \cdots + y_n)$$
$$= (x_1 + y_1) + \cdots + (x_n + y_n)$$
$$= \Phi((x_1 + y_1, \ldots, x_n + y_n))$$
$$= \Phi((x_1, \ldots, x_n) + (y_1, \ldots, y_n)). \ \blacksquare$$

We next find conditions ensuring that Φ is onto.

Lemma 30.3   *Suppose that $X_1, \ldots, X_n$ are subgroups of an Abelian group G and that $x \in S(X_1, \ldots, X_n)$. Then $x = x_1 + \cdots + x_n$ for some $x_i \in X_i$.*

Proof   By definition of $S(X_1, \ldots, X_n)$, $x = s_1 + \cdots + s_m$, where $s_i \in X_1 \cup \cdots \cup X_n$ or $-s_i \in X_1 \cup \cdots \cup X_n$. Since the X_i are all groups, even in the latter case $s_i \in X_1 \cup \cdots \cup X_n$, i.e., $x = s_1 + \cdots + s_m$, where $s_i \in X_1 \cup \cdots \cup X_n$. But since G is commutative, we may collect all the s_i that are in X_1, and all s_i that are in X_2, etc. Adding each of the resulting sums gives us elements $x_i \in X_i$ such that $x = x_1 + \cdots + x_n$. $\blacksquare$

We may now characterize internal direct sums as follows. (Other characterizations are given in the exercises; the most commonly used ones are those given in Exercise 19 for Abelian groups and those given in Exercise 25 for arbitrary groups.)

Proposition 30.4   *Suppose that G is an Abelian group and that $X_1, \ldots, X_n$ are subgroups of G. Then $G = X_1 \oplus \cdots \oplus X_n$ if and only if*

(i) $G = S(X_1, \ldots, X_n)$;
(ii)   *for all $i = 2, \ldots, n$, $S(X_1, \ldots, X_{i-1}) \cap X_i = \{0\}$.*

Proof   Suppose that $G = X_1 \oplus \cdots \oplus X_n$, and let $\Phi: X_1 \times \cdots \times X_n \to G$ be the function $\Phi(x_1, \ldots, x_n) = x_1 + \cdots + x_n$. Since Φ is onto, each element of G may be written as the sum of elements taken from $X_1 \cup \cdots \cup X_n$, and thus (i) holds. For (ii), note that certainly $S(X_1, \ldots, X_{i-1}) \cap X_i \supseteq \{0\}$ and suppose conversely that $z \in S(X_1, \ldots, X_{i-1}) \cap X_i$. By Lemma 30.3, $z = z_1 + \cdots + z_{i-1}$ for $z_k \in X_k$, and hence $z = \Phi(z_1, \ldots, z_{i-1}, 0, \ldots, 0)$. But since $z \in X_i$, $z = \Phi(0, \ldots, z, \ldots, 0)$ as well, and therefore, since Φ is one-to-one, $(0, \ldots, z, \ldots, 0) = (z_1, \ldots, z_{i-1}, 0, \ldots, 0)$. It follows that $z = 0$ and hence that (ii) holds.

Conversely, suppose that (i) and (ii) hold, and note that by Lemma 30.2, Φ is a homomorphism. To see that Φ is onto, suppose that $x \in G$ and note that by Lemma 30.3, $x = x_1 + \cdots + x_n$ for $x_i \in X_i$. Then $(x_1, \ldots, x_n) \in X_1 \times \cdots \times X_n$ and $x = \Phi(x_1, \ldots, x_n)$ so that Φ is onto. To see that Φ is one-to-one, we will show that $\ker \Phi \subseteq \{0\}$ and then apply Lemma 24.4. So suppose that $(x_1, \ldots, x_n) \in \ker \Phi$, i.e., that $\Phi(x_1, \ldots, x_n) = 0$. Then $x_1 + \cdots + x_n = 0$ and hence $x_n = -(x_1 + \cdots + x_{n-1}) \in S(X_1, \ldots, X_{n-1})$. It then follows from (ii) that $x_n = 0$. A similar argument shows that each $x_i = 0$ and hence that $(x_1, \ldots, x_n) = (0, \ldots, 0)$. Thus, by Lemma 24.4, Φ is one-to-one. $\blacksquare$

Example 30.5 In Example 30.1, we showed directly that $\mathbb{Z}_6$ was the internal direct product of its subgroups $A = \{[0]_6, [3]_6\}$ and $B = \{[0]_6, [2]_6, [4]_6\}$. We can also deduce this from Proposition 30.4. We observed in Example 30.1 that A and B are subgroups of $\mathbb{Z}_6$. Since A is a subgroup of $S(A, B)$, 2 divides the order of $S(A, B)$ by Lagrange's theorem (Proposition 21.10), and similarly, since B is a subgroup of $S(A, B)$, 3 also divides the order of $S(A, B)$. Then $S(A, B)$ is a subgroup of $\mathbb{Z}_6$ with at least six elements and hence, since $\mathbb{Z}_6$ has exactly six elements, $\mathbb{Z}_6 = S(A, B)$. Since obviously $A \cap B = \{[0]_6\}$, Proposition 30.4 implies that $\mathbb{Z}_6$ is the internal direct sum of A and B. ❖

Example 30.6 It is easy to check that by Proposition 18.7, the following are subgroups of $\mathbb{Z}_{12}$.

$$X_1 = \{[0]_{12}, [3]_{12}, [6]_{12}, [9]_{12}\};$$
$$X_2 = \{[0]_{12}, [2]_{12}, [4]_{12}, [6]_{12}, [8]_{12}, [10]_{12}\};$$
$$X_3 = \{[0]_{12}, [6]_{12}\};$$
$$X_4 = \{[0]_{12}, [4]_{12}, [8]_{12}\}.$$

By Lagrange's theorem, both the order of X_1, which is 4, and the order of X_2, which is 6, divide the order of $S(X_1, X_2)$ and hence the order of $S(X_1, X_2)$ is at least 12. So $S(X_1, X_2) = \mathbb{Z}_{12}$. However, $X_1 \cap X_2 = \{[0]_{12}, [6]_{12}\} \neq \{[0]_{12}\}$ so that by Proposition 30.4, G is <u>not</u> the internal direct sum of X_1 and X_2. On the other hand, $X_3 \cap X_4 = \{[0]_{12}\}$ but $S(X_3, X_4) = X_2 \neq \mathbb{Z}_{12}$ so that again by Proposition 30.4, $\mathbb{Z}_{12}$ is <u>not</u> the internal direct sum of X_3 and X_4. However, since the order of X_1 is 4 and the order of X_4 is 3, the order of $S(X_1, X_3)$ is at least 12 and hence $S(X_1, X_3) = \mathbb{Z}_{12}$. But $X_1 \cap X_4 = \{[0]_{12}\}$, and thus Proposition 30.4 implies that $\mathbb{Z}_{12}$ is the internal sum of X_1 and X_4. ❖

We conclude this section by noting that, like external direct sums, the order of an internal direct sum is the product of the orders of its summands, and that the formation of internal direct sums is both associative and commutative.

Proposition 30.7 *Suppose that $X_1, \ldots, X_m$ are subgroups of a finite Abelian group G and that $G = X_1 \oplus \cdots \oplus X_m$.*

(i) *Then $|G| = |X_1| \cdots |X_m|$.*

(ii) *If $Y_1, \ldots, Y_m$ is a rearrangement of the X_i, then $G = Y_1 \oplus \cdots \oplus Y_m$.*

(iii) *If each X_i is an internal direct sum $X_{i,1} \oplus \cdots \oplus X_{i,n_i}$, then*

$$G = X_{1,1} \oplus \cdots \oplus X_{1,n_1} \oplus \cdots \oplus X_{m,1} \oplus \cdots \oplus X_{m,n_m}.$$

Proof Since G is isomorphic to $X_1 \times \cdots \times X_m$, (i) follows from Proposition 29.5.

For (ii), we know that there is a permutation σ of $\{1, \ldots, m\}$ such that for each $i, X_i = Y_{\sigma(i)}$. Then by Proposition 29.7, the function $\Phi((y_1, \ldots, y_m)) =$

$(y_{\sigma(1)}, \ldots, y_{\sigma(m)})$ is an isomorphism. But the function $\Psi((x_1, \ldots, x_m)) = x_1 + \cdots + x_m$ is also an isomorphism, and hence the composite function

$$\Psi(\Phi((y_1, \ldots, y_m))) = \Psi((y_{\sigma(1)}, \ldots, y_{\sigma(m)})) = y_{\sigma(1)} + \cdots + y_{\sigma(m)}$$

is an isomorphism (see Exercise 14). Since G is Abelian and σ is a permutation, $y_{\sigma(1)} + \cdots + y_{\sigma(m)} = y_1 + \cdots + y_m$, and hence the composite function is the function $\Psi(\Phi((y_1, \ldots, y_m))) = y_1 + \cdots + y_m$. It follows that $G = Y_1 \oplus \cdots \oplus Y_m$.

Statement (*iii*) can be deduced from Proposition 29.9 in much the same way statement (*ii*) was deduced from Proposition 29.7 (see Exercise 18). ∎

Example 30.8   It is easy to check that by proposition 18.7, the following are subgroups of $\mathbb{Z}_{30}$:

$$X_1 = \{[0]_{30}, [10]_{30}, [20]_{30}\};$$
$$X_2 = \{[0]_{30}, [6]_{30}, [12]_{30}, [18]_{30}, [24]_{30}\};$$
$$X_3 = \{[0]_{30}, [15]_{30}\};$$
$$X_4 = \{[0]_{30}, [5]_{30}, [10]_{30}, [15]_{30}, [20]_{30}, [25]_{30}\}.$$

By Lagrange's theorem (Proposition 21.10), both the order of X_2, which is 5, and the order of X_4, which is 6, divide the order of $S(X_2, X_4)$ and hence the order of $S(X_2, X_4)$ is at least 30. So $S(X_2, X_4) = \mathbb{Z}_{30}$, and since $X_2 \cap X_4 = \{[0]_{30}\}$,

$$\mathbb{Z}_{30} = X_2 \oplus X_4.$$

But also $S(X_4, X_2) = S(X_2, X_4) = \mathbb{Z}_{30}$, and thus, since $X_4 \cap X_2 = \{[0]_{30}\}$,

$$X_2 \oplus X_4 = \mathbb{Z}_{30} = X_4 \oplus X_2,$$

in agreement with Proposition 30.7 (*ii*). Furthermore, X_1 and X_3 are both contained in X_4 and by Lagrange's theorem, the order of $S(X_1, X_3)$ must be at least 6. So $S(X_1, X_3) = X_4$, and since $X_1 \cap X_3 = \{[0]_{30}\}$,

$$X_4 = X_1 \oplus X_3.$$

Similarly by Lagrange's theorem, $S(X_2, X_1, X_3) = \mathbb{Z}_{30}$. So since $S(X_1, X_3) \cap X_2 = X_4 \cap X_2 = \{[0]_{30}\}$,

$$\mathbb{Z}_{30} = X_2 \oplus X_1 \oplus X_3.$$

That is, in agreement with Proposition 30.7 (*iii*),

$$X_1 \oplus X_3 \oplus X_2 = \mathbb{Z}_{30} = X_2 \oplus X_4 = X_1 \oplus (X_3 \oplus X_2).$$

As well, in agreement with Proposition 30.7 (*i*),

$$|\mathbb{Z}_{30}| = 30 = 5 \cdot 6 = |X_2||X_4| \quad \text{and}$$
$$|\mathbb{Z}_{30}| = 30 = 5 \cdot 3 \cdot 2 = |X_2||X_1||X_3|. \quad ❖$$

Exercises

1. Show that $G_1 = \{[0]_{15}, [3]_{15}, [6]_{15}, [9]_{15}, [12]_{15}\}$ and $G_2 = \{[0]_{15}, [5]_{15}, [10]_{15}\}$ are subgroups of $\mathbb{Z}_{15}$ and that $\mathbb{Z}_{15} = G_1 \oplus G_2$.

2. Show that $G_1 = \{[0]_{20}, [4]_{20}, [8]_{20}, [12]_{20}, [16]_{20}\}$ and $G_2 = \{[0]_{20}, [5]_{20}, [10]_{20}, [15]_{20}\}$ are subgroups of $\mathbb{Z}_{20}$ and that $\mathbb{Z}_{20} = G_1 \oplus G_2$.

3. Show that $G_1 = \{[0]_{18}, [9]_{18}\}$ and $G_2 = \{[0]_{18}, [2]_{18}, [4]_{18}, [6]_{18}, [8]_{18}, [10]_{18}, [12]_{18}, [14]_{18}, [16]_{18}\}$ are subgroups of $\mathbb{Z}_{18}$ and that $\mathbb{Z}_{18} = G_1 \oplus G_2$.

4. For any Abelian group G, show that $G = G \oplus \{0\}$.

5. Let
    $$G_1 = \{[0]_{30}, [15]_{30}\},$$
    $$G_2 = \{[0]_{30}, [10]_{30}, [20]_{30}\},$$
    $$G_3 = \{[0]_{30}, [6]_{30}, [12]_{30}, [18]_{30}, [24]_{30}\}.$$
    Show that G_1, G_2, and G_3 are subgroups of $\mathbb{Z}_{30}$ and that $\mathbb{Z}_{30} = G_1 \oplus G_2 \oplus G_3$.

6. Let $G = \mathbb{Z}_3 \times \mathbb{Z}_{12}$ and let
    $$G_1 = \{([0]_3, [0]_{12}), ([1]_3, [0]_{12}), ([2]_3, [0]_{12})\},$$
    $$G_2 = \{([0]_3, [0]_{12}), ([1]_3, [4]_{12}), ([2]_3, [8]_{12})\},$$
    $$G_3 = \{([0]_3, [0]_{12}), ([0]_3, [3]_{12}), ([0]_3, [6]_{12}), ([0]_3, [9]_{12})\}.$$
    Show that G_1, G_2, and G_3 are subgroups of G and that $G = G_1 \oplus G_2 \oplus G_3$.

7. Suppose that G_1 and G_2 are subgroups of an Abelian group G. Show that $S(G_1, G_2) = G_1 \oplus G_2$ if and only if $G_1 \cap G_2 = \{0\}$.

In Exercises 8–12, let $G = \mathbb{Z}_3 \times \mathbb{Z}_3$ and let
$$G_1 = \{([0]_3, [0]_3), ([0]_3, [1]_3), ([0]_3, [2]_3)\},$$
$$G_2 = \{([0]_3, [0]_3), ([1]_3, [0]_3), ([2]_3, [0]_3)\},$$
$$G_3 = \{([0]_3, [0]_3), ([1]_3, [1]_3), ([2]_3, [2]_3)\}.$$

8. Show that G_1, G_2, and G_3 are all subgroups of G.

9. Show that $G = S(G_1, G_2, G_3)$.

10. Show that $G_i \cap G_j = \{([0]_3, [0]_3)\}$ whenever $i \neq j$.

11. Show that G is <u>not</u> the internal direct sum of G_1, G_2, and G_3.

12. Explain why Exercise 11 does <u>not</u> contradict Proposition 30.4.

13. Suppose that G, H, and K are groups and that $\Psi: G \to H$ and $\Theta: H \to K$ are homomorphisms. Show that $\Theta \circ \Psi: G \to K$ is also a homomorphism.

14. Suppose that G, H, and K are groups and that $\Psi: G \to H$ and $\Theta: H \to K$ are isomorphisms. Show that $\Theta \circ \Psi: G \to K$ is also an isomorphism.

15. Find an <u>infinite</u> Abelian group G with subgroups G_1, G_2, and G_3 such that $G = S(G_1, G_2, G_3)$ and $G_i \cap G_j = \{0\}$ whenever $i \neq j$ but $G \neq G_1 \oplus G_2 \oplus G_3$. Explain why the existence of such an example does not contradict Proposition 30.4.

16. Suppose that G is an Abelian group with subgroups A and B. If $G = A \oplus B$, show directly without using Proposition 30.7 that $G = B \oplus A$.

17. Suppose that G is an Abelian group with subgroups A, B, C, D, X, and Y. If $X = A \oplus B$, $Y = C \oplus D$, and $G = X \oplus Y$, show directly without using Proposition 30.7 that $G = A \oplus B \oplus C \oplus D$.

18. Prove part (*iii*) of Proposition 30.7 in detail.

19. Prove the following variation of Proposition 30.4. Suppose that G is an Abelian group and that $G_1, \ldots, G_n$ are subgroups of G. Then $G = G_1 \oplus \cdots \oplus G_n$ if and only if

    (*i*) $G = S(G_1, \ldots, G_n)$;
    (*ii*) for all $i = 1, \ldots, n$, $G_i \cap S(\cup_{k \neq i} G_k) = \{0\}$.

20. Suppose that G is an Abelian group and that $G_1, \ldots, G_n$ are subgroups of G. Prove that $G = G_1 \oplus \cdots \oplus G_n$ if and only if for all $g \in G$, there exist unique $g_1 \in G_1, \ldots, g_n \in G_n$ such that $g = g_1 + \cdots + g_n$.

21. Prove that if A and B are normal subgroups of a group G such that $A \cap B = \{e\}$, then for all $a \in A$ and $b \in B$, $ab = ba$.

22. Prove that if $G_1, \ldots, G_n$ are normal subgroups of a group G such that $G_i \cap G_j = \{e\}$ whenever $i \neq j$, then the function $\Phi: G_1 \times \cdots \times G_n \to G$, $\Phi(g_1, \ldots, g_n) = g_1 \cdots g_n$, is a homomorphism.

23. Prove that if $G_1, \ldots, G_n$ are normal subgroups of a group G and if $x \in S(G_1, \ldots, G_n)$, then $x = g_1 \cdots g_n$ for some $g_i \in G_i$.

24. Prove that if $G_1, \ldots, G_n$ are subgroups of a group G, then $G = G_1 \boxtimes \cdots \boxtimes G_n$ if and only if

    (*i*)   for all $i = 1, \ldots, n$, $G_i \lhd G$;
    (*ii*) $G = S(G_1, \ldots, G_n)$;
    (*iii*) for all $i = 2, \ldots, n$, $S(G_1, \ldots, G_{i-1}) \cap G_i = \{e\}$.

    (*Hint:* Use Exercises 21, 22, and 23.)

25. Suppose that G is a group and that $G_1, \ldots, G_n$ are subgroups of G. Prove that $G = G_1 \boxtimes \cdots \boxtimes G_n$ if and only if

    (*i*)   for all $i = 1, \ldots, n$, $G_i \lhd G$;
    (*ii*) $G = S(G_1, \ldots, G_n)$;
    (*iii*) for all $i = 1, \ldots, n$, $G_i \cap S(\cup_{k \neq i} G_k) = \{e\}$.

26. Suppose that G is a group and that $G_1, \ldots, G_n$ are subgroups of G. Prove that

$G = G_1 \boxtimes \cdots \boxtimes G_n$ if and only if each G_i is normal in G and for all $g \in G$, there exist unique $g_1 \in G_1, \ldots, g_n \in G_n$ such that $g = g_1 \cdots g_n$.

27. Show that any external direct product is an internal direct product. That is, suppose that $G, H_1, \ldots, H_n$ are groups. Show that if G is isomorphic to $H_1 \times \cdots \times H_n$, then there exist subgroups $G_1, \ldots, G_n$ of G such that (1) G_i is isomorphic to H_i for each i and (2) $G = G_1 \boxtimes \cdots \boxtimes G_n$.

Abelian Groups with Prime Power Order

Our goal in this chapter and the next is to write all finite Abelian groups as direct sums of cyclic groups of prime power order. In this chapter, we consider the special case in which the Abelian groups are of prime power order, and in the next chapter, we will deal with finite Abelian groups in general. So the objective of this chapter is to show that Abelian groups of prime power order can be written as direct sums of cyclic groups of prime power order.

It is easy to show that the resulting cyclic groups must themselves be of prime power order. So we begin by proving that.

Proposition 31.1 *Let p be prime and G be a finite group whose order is a power of p. If S is a subgroup of G, then the order of S is also a power of p.*

Proof By Lagrange's theorem (Proposition 21.10), the order of S divides the order of G, and hence p is the only prime divisor of the order of S. ∎

So it remains to show that an Abelian group of prime power order can be written as a direct sum of cyclic subgroups. Since for the remainder of this chapter, we will consider only <u>Abelian</u> groups, in keeping with standard practice, <u>we will write the generic operation as addition rather than as multiplication.</u>

Now suppose that p is a prime, that n is a positive integer, and that A is an Abelian group of order p^n. We wish to decompose A into a direct sum of cyclic subgroups. To do this, we will use the following subgroups of A. Note that if a is an element of an Abelian group A and if n is a positive integer, than na means $na + \cdots + a$ (n times). So if $a, b \in A$, then, since A is commutative, $na + nb = (a + \cdots + a) + (b + \cdots + b) = (a + b) + \cdots + (a + b) = n(a + b)$.

Proposition 31.2 *Let p be a prime and let A be a finite Abelian group and consider*

$$pA = \{pa \,|\, a \in A\} \quad and \quad S_p(A) = \{a \in A \,|\, pa = 0\}.$$

Then pA and $S_p(A)$ are both subgroups of A.

Proof Since $0 \in A$, $p0 \in pA$ and hence $pA \neq \emptyset$. And if $a, b \in A$, then $pa + pb = p(a + b) \in pA$. So by Proposition 18.7, pA is a subgroup of A. Similarly, since $p0 = 0$, $0 \in S_p(A)$ and hence $S_p(A) \neq \emptyset$. And if $a, b \in S_p(A)$, then $p(a + b) = pa + pb = 0$ and hence $a + b \in S_p(A)$. So $S_p(A)$ is a subgroup of A as well. ■

We will base the examples of this section on the groups $\mathbb{Z}_n$. To avoid cumbersome notation, we will abbreviate the sets $[k]_n$ by k, for $0 \leq k < n$. This amounts to creating new groups based on the sets $\{0, \ldots, n - 1\}$ with addition and multiplication modulo n. For instance, from this point of view, $\mathbb{Z}_2$ is the set $\{0, 1\}$ with addition given by the table:

+	0	1
0	0	1
1	1	0

And $\mathbb{Z}_4$ is the set $\{0, 1, 2, 3\}$ with addition given by the table:

+	0	1	2	3
0	0	1	2	3
1	1	2	3	0
2	2	3	0	1
3	3	0	1	2

Note that this simplification of notation introduces a substantial amount of ambiguity. For instance, the number 2 can be an element of any of the groups $\mathbb{Z}_n$ for any $n \geq 3$. So it is important to pay careful attention to precisely which group is the one under consideration.

Example 31.3 Obviously $\mathbb{Z}_2$ has order 2, and direct computation shows that $2\mathbb{Z}_2 = \{0 + 0, 1 + 1\} = \{0\}$. As well, since $0 + 0 = 0$ and $1 + 1 = 0$, $S_2(\mathbb{Z}_2) = \{0, 1\} = \mathbb{Z}_2$.

However, $\mathbb{Z}_4$ has order $4 = 2^2$, and

$$2\mathbb{Z}_4 = \{0 + 0, 1 + 1, 2 + 2, 3 + 3\} = \{0, 2, 0, 2\} = \{0, 2\}.$$

In this case, since $0 + 0 = 0$, $1 + 1 = 2 \neq 0$, $2 + 2 = 0$, $3 + 3 = 2 \neq 0$, $S_2(\mathbb{Z}_4) = \{0, 2\} \neq \mathbb{Z}_4$.

The group $\mathbb{Z}_8$ has order $8 = 2^3$, and we have $2\mathbb{Z}_8 = \{0, 2, 4, 6\}$ and $S_2(\mathbb{Z}_8) = \{0, 4\}$.

Finally, consider $\mathbb{Z}_2 \times \mathbb{Z}_4 = \{(0, 0), (0, 1), (0, 2), (0, 3), (1, 0), (1, 1), (1, 2), (1, 3)\}$. This is also a group of order $8 = 2^3$, and in this case, we have $2(\mathbb{Z}_2 \times \mathbb{Z}_4) = \{(0, 0), (0, 2)\}$ and $S_2(\mathbb{Z}_2 \times \mathbb{Z}_4) = \{(0, 0), (0, 2), (1, 0), (1, 2)\}$.

So these two subgroups can vary greatly from group to group. ❖

We want to decompose an Abelian group A of prime power order p^n into a direct sum of cyclic subgroups. Our strategy is first to write $S_p(A)$ as a direct sum of cyclic subgroups and then to incorporate this decomposition into a general decomposition of A.

We will need the following lemma which concerns the subgroup $S(a)$ of A generated by an element a of A. Recall that Proposition 21.14 describes $S(a)$ completely. In particular, if A is an additive Abelian group and if $a \in A$ has order m, then, since the operation on A is addition, $S(a) = \{0, a, 2a, \ldots, (m - 1)a\}$.

Lemma 31.4   *Suppose that p is a prime, that A is a finite Abelian group, that H is a subgroup of A, and that $a \in A$. If $pa = 0$ and $a \notin H$, then $S(a) \cap H = \{0\}$.*

Proof   Certainly $S(a) \cap H \supseteq \{0\}$, and hence it suffices to show that $S(a) \cap H \subseteq \{0\}$. To this end, let $z \in S(a) \cap H$ and suppose by way of contradiction that $0 \neq z$. Note that since $0 \neq z \in S(a)$ and $pa = 0$, Proposition 21.14 implies that for some $0 < k < p, z = ka$. But then since p is prime, the greatest common divisor of k and p is 1, and hence by Theorem 1.4, there exist integers m and n such that $1 = mk + np$. Then $a = 1 \cdot a = (mk + np)a = mka + npa = mka + 0 = mz \in H$, a contradiction. We conclude that $0 = z$ and hence that $S(a) \cap H \subseteq \{0\}$. ∎

Example 31.5   Consider the Abelian group $\mathbb{Z}_{12}$. It is easy to see that $H = \{0, 6\}$ is nonempty and closed and hence is a subgroup of $\mathbb{Z}_{12}$. Furthermore, 3 is prime and $4 \notin H$, and since $4 \neq 0, 4 + 4 = 8 \neq 0$, and $4 + 4 + 4 = 0$, 4 has order 3. Then $S(4) = \{0, 4, 8\}$, and as guaranteed by Lemma 31.4, $S(4) \cap H = \{0\}$.

It is easy to show in a similar fashion that 3 has nonprime order 4 and obviously $3 \notin H$. However, $S(3) = \{0, 3, 6, 9\}$ and so $S(3) \cap H = \{0, 6\} \neq \{0\}$. So Lemma 31.4 is not true without the assumption that p is prime. ❖

We now prove a result that will allow us first to decompose $S_p(A)$ into a sum of cyclic subgroups (Corollary 31.7) and later to decompose A itself (Theorem 31.12).

Proposition 31.6   *Suppose that p is a prime, that A is a finite Abelian group, and that H and B are subgroups of A. Then $H + B = \{h + b \mid h \in H, b \in B\}$ is a subgroup of A, and if $pB = \{0\}$, then there exist $x_1, \ldots, x_m \in B$ such that $H + B = H \oplus S(x_1) \oplus \cdots \oplus S(x_m)$.*

Proof Since H and B are subgroups, $0 \in H$ and $0 \in B$; then $0 = 0 + 0 \in H + B$ and hence $H + B \neq \varnothing$. Furthermore, if $x, y \in H + B$, then $x = h + b$ and $y = k + c$ for $h, k \in H$ and $b, c \in B$, and thus, since A is Abelian, $x + y = (h + b) + (k + c) = (h + k) + (b + c) \in H + K$. Since A is finite, it follows from Proposition 18.7 that $H + B$ is a subgroup of A.

Now suppose that $pB = \{0\}$. If $H + B = H$, we are done. Otherwise, there exists $x_1 \in B$ such that $x_1 \notin H$. By Lemma 31.4, $S(x_1) \cap H = \{0\}$ so that either $H + B = H \oplus S(x_1)$ or there exists $x_2 \in B$ such that $x_2 \notin H \oplus S(x_1)$. As before, Lemma 31.4 implies that $S(x_2) \cap [H \oplus S(x_1)] = \{0\}$ so that either $H + B = H \oplus S(x_1) \oplus S(x_2)$ or there exists $x_3 \in B$ such that $x_3 \notin H \oplus S(x_1) \oplus S(x_2)$. Since A is finite, this process must eventually stop, and when it does, we have a sequence $x_1, \ldots, x_m \in B$ such that $H + B = H \oplus S(x_1) \oplus \cdots \oplus S(x_m)$. ∎

Corollary 31.7 *Suppose that p is a prime and that A is a finite Abelian group. Then there exist $x_1, \ldots, x_m \in S_p(A)$ such that $S_p(A) = S(x_1) \oplus \cdots \oplus S(x_m)$.*

Proof Apply Proposition 31.6 with $H = \{0\}$ and $B = S_p(A)$. (This result may also be proved by using linear algebra—see Exercises 37 and 38.) ∎

Example 31.8 From Example 31.3, $S_2(\mathbb{Z}_2) = \{0, 1\} = S(1)$ and $S_2(\mathbb{Z}_8) = \{0, 4\} = S(4)$.

The group $\mathbb{Z}_2 \times \mathbb{Z}_4$ is more complicated because $S_2(\mathbb{Z}_2 \times \mathbb{Z}_4) = \{(0, 0), (0, 2), (1, 0), (1, 2)\}$ is not generated by any of its elements. However, since $(1, 2)$ has order 2, Proposition 21.14 implies that $S((1, 2)) = \{(0, 0), (1, 2)\}$ and similarly $S((1, 0)) = \{(0, 0), (1, 0)\}$. So $S((1, 2)) \cap S((1, 0)) = \{(0, 0)\}$. But $\{(0, 0), (0, 2), (1, 0), (1, 2)\}$ is a subgroup of $\mathbb{Z}_2 \times \mathbb{Z}_4$ and certainly each of its elements is a sum of $(1, 2)$ and $(0, 2)$. So $S((1, 2), (1, 0)) = \{(0, 0), (0, 2), (1, 0), (1, 2)\} = S_2(\mathbb{Z}_2 \times \mathbb{Z}_4)$ and hence $S_2(\mathbb{Z}_2 \times \mathbb{Z}_4) = S((1, 2)) \oplus S((1, 0))$. ❖

Corollary 31.7 says $S_p(A)$ can be decomposed into a direct sum of cyclic subgroups. We will decompose A by first decomposing pA as $pA = S(pz_1) \oplus \cdots \oplus S(pz_n)$ and then lifting that decomposition to A as $A = S(z_1) \oplus \cdots \oplus S(z_n) \oplus S(x_1) \oplus \cdots \oplus S(x_m)$ for $x_k \in S_p(A)$. The lifting of pA is accomplished as follows.

Proposition 31.9 *Suppose that p is a prime, that A is a finite Abelian group whose order is a power of p, and that $z_1, \ldots, z_n \in A$ are such that $pz_i \neq 0$. If $S(pz_1, \ldots, pz_n) = S(pz_1) \oplus \cdots \oplus S(pz_n)$, then $S(z_1, \ldots, z_n) = S(z_1) \oplus \cdots \oplus S(z_n)$.*

Proof Since certainly $S(S(z_1), \ldots, S(z_n)) = S(z_1, \ldots, z_n)$ (Exercise 16), it suffices by Proposition 30.4 to show that for all i, $S(z_1, \ldots, z_{i-1}) \cap S(z_i) = \{0\}$. To this end, let $w \in S(z_1, \ldots, z_{i-1}) \cap S(z_i)$. Since $w \in S(z_i)$, $w = kz_i$ for some integer k, and since $w \in S(z_1, \ldots, z_{i-1})$ and A is Abelian, $w = k_1 z_1 + \cdots + k_{i-1} z_{i-1}$ for integers $k_1, \ldots, k_{i-1}$. Then $pw = kpz_i \in S(pz_i)$ and $pw = p(k_1 z_1 + \cdots + k_{i-1} z_{i-1}) = k_1 pz_1 + \cdots + k_{i-1} pz_{i-1} \in S(pz_1, \ldots, pz_{i-1})$. So $pw = 0$. But then $k_{i-1} pz_{i-1} \in$

$S(pz_{i-1})$ and $k_{i-1}pz_{i-1} = -k_1pz_1 - \cdots - k_{i-2}pz_{i-2} \in S(pz_1, \ldots, pz_{i-2})$ so that $k_{i-1}pz_{i-1} = 0$. Continuing in the same fashion, we have that $pk_jz_j = k_jpz_j = 0$ for all $1 \le j \le i$, and hence by Corollary 21.6, every k_jz_j has order p or 1. So by Corollary 21.16 and Proposition 21.5, if k_jz_j has order 1, then $k_jz_j = 0$ and hence p divides k_j; and if k_jz_j has order p, then p also divides k_j. So for all $1 \le j \le i$, there exists d_j such that $k_j = pd_j$. Then $w = pd_iz_i = d_ipz_i \in S(pz_i)$ and $w = pd_1z_1 + \cdots + pd_{i-1}z_{i-1} = d_1pz_1 + \cdots + d_{i-1}pz_{i-1} \in S(pz_1, \ldots, pz_{i-1})$, and therefore $w = 0$. ∎

Example 31.10   The group $\mathbb{Z}_4 \times \mathbb{Z}_8$ has order $32 = 2^5$. Using the techniques of the previous examples, it can be shown that

$$S((2, 4)) = \{(0, 0), (2, 4)\},$$
$$S((0, 4)) = \{(0, 0), (0, 4)\}, \text{ and}$$
$$S((2, 4), (0, 4)) = S((2, 4)) \oplus S((0, 4)).$$

And since $(1, 2) + (1, 2) = (2, 4)$ and $(2, 2) + (2, 2) = (0, 4)$, we have, in agreement with Proposition 31.9, that

$$S((1, 2)) = \{(0, 0), (1, 2), (2, 4), (3, 6)\},$$
$$S((2, 2)) = \{(0, 0), (2, 2), (0, 4), (2, 6)\}, \text{ and}$$
$$S((1, 2), (2, 2)) = S((1, 2)) \oplus S((2, 2)).$$

Note that $(0, 1) \notin S((1, 2), (2, 2))$; so $S((1, 2), (2, 2)) \ne \mathbb{Z}_4 \times \mathbb{Z}_8$. We can enlarge the subgroup by using $(2, 2)$ in place of $(0, 4)$ in our first list of subgroups:

$$S((2, 4)) = \{(0, 0), (2, 4)\},$$
$$S((2, 2)) = \{(0, 0), (2, 2), (0, 4), (2, 6)\}, \text{ and}$$
$$S((2, 4), (2, 2)) = S((2, 4)) \oplus S((2, 2)).$$

And, in agreement with Proposition 31.9,

$$S((1, 2)) = \{(0, 0), (1, 2), (2, 4), (3, 6)\},$$
$$S((1, 1)) = \{(0, 0), (1, 1), (2, 2), (3, 3), (0, 4), (1, 5), (2, 6), (3, 7)\}, \text{ and}$$
$$S((1, 2), (1, 1)) = S((1, 2)) \oplus S((1, 1)).$$

Note that now $S((1, 2), (1, 1)) = \mathbb{Z}_4 \times \mathbb{Z}_8$. So we have decomposed $\mathbb{Z}_4 \times \mathbb{Z}_8$ into a direct sum of cyclic subgroups by using Proposition 31.9. Note also that this is not the only way of decomposing $\mathbb{Z}_4 \times \mathbb{Z}_8$ into a direct sum of cyclic subgroups. Using the subgroups corresponding to $\mathbb{Z}_4$ and $\mathbb{Z}_8$, we have $\mathbb{Z}_4 \times \mathbb{Z}_8 = S((1, 0)) \oplus S((0, 1))$.

On the other hand, the group $\mathbb{Z}_2 \times \mathbb{Z}_4 \times \mathbb{Z}_8$ also has order a power of 2; its order is $64 = 2^6$. As before, we can show that

$$S((0, 2, 4)) = \{(0, 0, 0), (0, 2, 4)\},$$
$$S((0, 2, 2)) = \{(0, 0, 0), (0, 2, 2), (0, 0, 4), (0, 2, 6)\}, \text{ and}$$
$$S((0, 2, 4), (0, 2, 2)) = S((0, 2, 4)) \oplus S((0, 2, 2)),$$

and, in agreement with Proposition 31.9,

$$S((0, 1, 2)) = \{(0, 0, 0), (0, 1, 2), (0, 2, 4), (0, 3, 6)\},$$
$$S((0, 1, 1)) = \{(0, 0, 0), (0, 1, 1), (0, 2, 2), (0, 3, 3), (0, 0, 4), (0, 1, 5),$$
$$(0, 2, 6), (0, 3, 7)\}, \text{ and}$$
$$S((0, 1, 2), (0, 1, 1)) = S((0, 1, 2)) \oplus S((0, 1, 1)).$$

However, $(1, 0, 0) \notin S((0, 1, 2), (0, 1, 1))$ so that in this case as well, $S((0, 1, 0), (0, 1, 1)) \notin \mathbb{Z}_2 \times \mathbb{Z}_4 \times \mathbb{Z}_8$.

Furthermore, this is the best we can do. For if $z \in \mathbb{Z}_2 \times \mathbb{Z}_4 \times \mathbb{Z}_8$, then $2z = (0, 2m, 2n)$ and hence $2z \in S((0, 2, 0)) \oplus S((0, 0, 2))$. In fact, $2(\mathbb{Z}_2 \times \mathbb{Z}_4 \times \mathbb{Z}_8) = S((0, 2, 0)) \oplus S((0, 0, 2))$. So $S((0, 2, 0)) \oplus S((0, 0, 2))$ is the largest subgroup that can be generated by elements of $\mathbb{Z}_2 \times \mathbb{Z}_4 \times \mathbb{Z}_8$ of the form $2z$. This means that it is impossible to decompose $\mathbb{Z}_2 \times \mathbb{Z}_4 \times \mathbb{Z}_8$ by using Proposition 31.9. The difficulty is of course that $S((0, 1, 2), (0, 1, 1))$ is missing elements like $(1, 0, 0)$ in $S_2(\mathbb{Z}_2 \times \mathbb{Z}_4 \times \mathbb{Z}_8)$. If we adjoin these elements, we recover all of $\mathbb{Z}_2 \times \mathbb{Z}_4 \times \mathbb{Z}_8$: $S_2(\mathbb{Z}_2 \times \mathbb{Z}_4 \times \mathbb{Z}_8) + S((0, 1, 2), (0, 1, 1)) = \mathbb{Z}_2 \times \mathbb{Z}_4 \times \mathbb{Z}_8$. ❖

The example shows that the subgroup generated by the set of elements z such that $pz \neq 0$ may exclude elements in $S_p(A)$ and that for the group $\mathbb{Z}_2 \times \mathbb{Z}_4 \times \mathbb{Z}_8$, it suffices to adjoin this subgroup. The following result shows that adding these elements always generates all of A.

Proposition 31.11   *Suppose that p is a prime, that A is a finite Abelian group whose order is a power of p, and that $z_1, \ldots, z_n \in A$ are such that $pA = S(pz_1, \ldots, pz_n)$. Then $A = S(z_1, \ldots, z_n) + S_p(A)$.*

Proof   Certainly $A \supseteq S(z_1, \ldots, z_n) + S_p(A)$. Conversely, suppose that $a \in A$. Since A is Abelian, $pa = k_1pz_1 + \cdots + k_npz_n$ for integers $k_1, \ldots, k_n$, and hence $p(-k_1z_1 - \cdots - k_nz_n + a) = 0$. Then $-k_1z_1 - \cdots - k_nz_n + a \in S_p(A)$ and thus, since $k_1z_1 + \cdots + k_nz_n \in S(z_1, \ldots, z_n)$, $a = (k_1z_1 + \cdots + k_nz_n) + (-k_1z_1 - \cdots - k_nz_n + a) \in S(z_1, \ldots, z_n) + S_p(A)$. We conclude that $A \subseteq S(z_1, \ldots, z_n) + S_p(A)$ and hence that $A = S(z_1, \ldots, z_n) + S_p(A)$. ∎

Proposition 31.11 allows us to combine the decomposition of $S_p(A)$ given in Corollary 31.7 with the decomposition described in Proposition 31.9 in the following way.

Theorem 31.12   *Every finite Abelian group whose order is the power of a prime is a direct sum of cyclic subgroups.*

Proof   Suppose that p is a prime, that m is a positive integer, and that $A \neq \{0\}$ is a finite Abelian group of order p^m. We will show that A is the direct sum of cyclic groups by using the second principle of mathematical induction (Theorem 1.16).

(i)   If $m = 1$, then $A = S_p(A)$ and thus by Corollary 31.7, there exist $x_1, \ldots, x_m \in A$ such that $A = S(x_1) \oplus \cdots \oplus S(x_m)$.

(ii) Suppose that $m > 1$, that A has order p^m, and that if H is an Abelian group of order less than p^m, then H is the direct sum of cyclic subgroups. If $A = S_p(A)$, then A is the direct sum of cyclic subgroups by Corollary 31.7. So suppose that $A \neq S_p(A)$ and let $a \notin S_p(A)$. Since the order of A is a power of p, Corollary 21.16 implies that the order of a is also a power of p, say p^k, and since $a \notin S_p(A)$, $1 < k$. Then $p^{k-1}a \neq 0$ while $p(p^{k-1}a) = p^k a = 0$, and hence the function $x \to px$ is an onto function from A to pA which is not one-to-one. So pA has fewer elements than A, and thus by the induction hypothesis, there exist $z_1, \ldots, z_n \in A$ such that $pA = S(pz_1) \oplus \cdots \oplus S(pz_n)$. Then $S(z_1, \ldots, z_n) = S(z_1) \oplus \cdots \oplus S(z_n)$ by Proposition 31.9 and $A = S(z_1, \ldots, z_n) + S_p(A)$ by Proposition 31.11. Thus by Proposition 31.6, there exist $x_1, \ldots, x_m \in S_p(A)$ such that $A = S(z_1, \ldots, z_n) \oplus S(x_1) \oplus \cdots \oplus S(x_m)$ and hence by Proposition 30.7, $A = S(z_1) \oplus \cdots \oplus S(z_n) \oplus S(x_1) \oplus \cdots \oplus S(x_m)$. ∎

Example 31.13   We may apply Theorem 31.12 to $A = \mathbb{Z}_2 \times \mathbb{Z}_4 \times \mathbb{Z}_8$ in the following way. We saw in Example 31.10 that $S((0, 2, 4), (0, 2, 2)) = S((0, 2, 4)) \oplus S((0, 2, 2))$ and $S((0, 1, 2), (0, 1, 1)) = S((0, 1, 2)) \oplus S((0, 1, 1))$. Furthermore, $2A = S((0, 2, 0), (0, 0, 2)) = S((0, 2, 4), (0, 2, 2))$, and hence by Proposition 31.11, $A = S_2(A) + S((0, 1, 2), (0, 1, 1))$. Direct calculation shows that $S_2(A) = \{(0, 0, 0), (1, 0, 0), (0, 2, 0), (1, 2, 0), (0, 0, 4), (1, 0, 4), (0, 2, 4), (1, 2, 4)\}$. Since $(1, 2, 4)$ has order 2, $S((1, 2, 4)) = \{(0, 0, 0), (1, 2, 4)\}$ and since no elements of $S((0, 1, 2), (0, 1, 1))$ can have first coordinate 1, $S((1, 2, 4)) \cap S((0, 1, 2), (0, 1, 1)) = \{(0, 0, 0)\}$. So $S((1, 2, 4)) + S((0, 1, 2), (0, 1, 1)) = S((1, 2, 4)) \oplus S((0, 1, 2), (0, 1, 1)) = S((1, 2, 4)) \oplus S((0, 1, 2)) \oplus S((0, 1, 1))$. The order of this subgroup of A is $2 \cdot 4 \cdot 8 = 64$, and hence since A has order 64 as well, in fact $\mathbb{Z}_2 \times \mathbb{Z}_4 \times \mathbb{Z}_8 = A = S((1, 2, 4)) \oplus S((0, 1, 2)) \oplus S((0, 1, 1))$.

Note that we may also decompose $\mathbb{Z}_2 \times \mathbb{Z}_4 \times \mathbb{Z}_8$ by using subgroups corresponding to the factors of the direct product: $\mathbb{Z}_2 \times \mathbb{Z}_4 \times \mathbb{Z}_{8A} = A = S((1, 0, 0)) \oplus S((0, 1, 0)) \oplus S((0, 0, 1))$. ❖

Exercises

In Exercises 1–11, for the given prime p and the given Abelian group A, (a) list all the elements in pA, (b) list all the elements in $S_p(A)$, and (c) find $x_1, \ldots, x_m \in S_p(A)$ such that $S_p(A) = S(x_1) \oplus \cdots \oplus S(x_m)$.

1. $p = 2, A = \mathbb{Z}_{10}$ **2.** $p = 5, A = \mathbb{Z}_{10}$

3. $p = 2, A = \mathbb{Z}_8$ **4.** $p = 3, A = \mathbb{Z}_{24}$

5. $p = 2, A = \mathbb{Z}_{24}$

6. $p = 2, A = \mathbb{Z}_2 \times \mathbb{Z}_4$

7. $p = 3, A = \mathbb{Z}_3 \times \mathbb{Z}_{12}$

8. $p = 5, A = \mathbb{Z}_2 \times \mathbb{Z}_{10}$

9. $p = 2, A = \mathbb{Z}_2 \times \mathbb{Z}_3 \times \mathbb{Z}_8$

10. $p = 2, A = \mathbb{Z}_2 \times \mathbb{Z}_3 \times \mathbb{Z}_6 \times \mathbb{Z}_8$

11. $p = 2, A = \mathbb{Z}_2 \times \mathbb{Z}_2 \times \mathbb{Z}_8 \times \mathbb{Z}_8$

12. Suppose that G is a group whose order is a power of a prime p. Show that if H is a subgroup of G, then the order of H is also a power of p.

13. Suppose that G is a group whose order is a power of a prime p. Show that if H is a normal subgroup of G, then the order of G/H is also a power of p.

14. Suppose that p is a prime and that H is a normal subgroup of a group G. Show that if both H and G/H have order a power of p, then G also has order a power of p.

15. Suppose that p is a prime and that A is a finite Abelian group. Show that if pA has order a power of p, then so does A.

16. Suppose that G is an Abelian group and that $z_1, \ldots, z_n \in G$. Show that $S(S(z_1), \ldots, S(z_n)) = S(z_1, \ldots, z_n)$.

Exercises 17–22 concern an Abelian group A, a prime p, and a subset H of A. In each exercise, show first that H is a subgroup of A and that $A = S(S_p(A), H)$. Then find $x_1, \ldots, x_m \in S_p(A)$ such that $A = H \oplus S(x_1) \oplus \cdots \oplus S(x_m)$. Remember to justify your answer.

17. $A = \mathbb{Z}_{10}, p = 5, H = \{[0]_{10}, [5]_{10}\}$

18. $A = \mathbb{Z}_2 \times \mathbb{Z}_4, p = 2, H = \{[0]_2\} \times \mathbb{Z}_4$

19. $A = \mathbb{Z}_2 \times \mathbb{Z}_6, p = 2, H = \{[0]_2\} \times \{[0]_6, [2]_6, [4]_6\}$

20. $A = \mathbb{Z}_5 \times \mathbb{Z}_5 \times \mathbb{Z}_{10}, p = 5, H = \{[0]_5\} \times \{[0]_5\} \times \{[0]_{10}, [5]_{10}\}$

21. $A = \mathbb{Z}_2 \times \mathbb{Z}_2 \times \mathbb{Z}_4 \times \mathbb{Z}_8, p = 2, H = \{[0]_2\} \times \{[0]_2\} \times \mathbb{Z}_4 \times \mathbb{Z}_8$

22. $A = \mathbb{Z}_3 \times \mathbb{Z}_3 \times \mathbb{Z}_6 \times \mathbb{Z}_{12}, p = 3$,
$H = \{[0]_3\} \times \{[0]_3\} \times \{[0]_6, [3]_6\} \times \{[0]_{12}, [3]_{12}, [6]_{12}, [9]_{12}\}$

Exercises 23–28 concern an Abelian group A and a prime p. In each exercise, show first that A is a group whose order is a power of p, and then find $x_1, \ldots, x_m \in S_p(A)$ and $z_1, \ldots, z_n \in A$ such that $pA = S(pz_1) \oplus \cdots \oplus S(pz_n)$ and $A = S(z_1) \oplus \cdots \oplus S(z_n) \oplus S(x_1) \oplus \cdots \oplus S(x_m)$.

23. $A = \mathbb{Z}_8, p = 2$

24. $A = \mathbb{Z}_3 \times \mathbb{Z}_3 \times \mathbb{Z}_3, p = 3$

25. $A = \mathbb{Z}_3 \times \mathbb{Z}_9, p = 3$

26. $A = \mathbb{Z}_2 \times \mathbb{Z}_2 \times \mathbb{Z}_8, p = 2$

27. $A = \mathbb{Z}_2 \times \mathbb{Z}_2 \times \mathbb{Z}_{16}, p = 2$

28. $A = \mathbb{Z}_2 \times \mathbb{Z}_2 \times \mathbb{Z}_8 \times \mathbb{Z}_{16}, p = 2$

29. Find a finite Abelian group G whose order is not a power of p for any prime p but which has the property such that for any subgroup $S \neq G$ there is a prime p such that the order of S is a power of p.

30. Let p be a prime and let A be a cyclic group of order p. Show that $S_p(A) \supset pA$. (Recall that $[X \supset Y] \Leftrightarrow [X \supseteq Y$ and $X \neq Y]$.)

31. Let p be a prime and let A be a cyclic group of order p^2. Show that $S_p(A) = pA$.

32. Let p be a prime and let A be a cyclic group of order p^k for $k \geq 3$. Show that $S_p(A) \subset pA$. (Recall that $[X \subset Y] \Leftrightarrow [X \subseteq Y$ and $X \neq Y]$.)

33. Let p be a prime and let A be a finite Abelian group of order p. Show that $A = S(S_p(A), pA)$.

34. Find a prime p and a finite Abelian group A such that $A \neq S(S_p(A), pA)$.

35. Let p be a prime, let A be a finite Abelian group, and let B be a subgroup of A. Let $B^c = \{x \in S_p(A) \mid x \notin B\} \cup \{0\}$ be the complement of B in $S_p(A)$. Show that B^c is the union of cyclic groups.

36. Suppose that p is a prime, that n is a positive integer, and that A is an Abelian group of order p^n. If $x_1, \ldots, x_m \in A$ are such that $A = S(x_1) \oplus \cdots \oplus S(x_m)$, show that the maximum of the orders of the x_i is the maximum order of any element of A.

Exercises 37 and 38 use linear algebra to prove Corollary 31.7. Let A be a finite Abelian group and let p be a prime. Recall that $\mathbb{Z}_p$ is a field by Proposition 4.3.

37. Show that $S_p(A)$ is a vector space over $\mathbb{Z}_p$.

38. Since A is finite, $S_p(A)$ is finite and hence has a finite basis $\{x_1, \ldots, x_m\}$ as a vector space over $\mathbb{Z}_p$. Show that $S_p(A) = S(x_1) \oplus \cdots \oplus S(x_m)$.

39. Suppose that p is a prime, that n is a positive integer, and that G is a group that contains an element of order p^n. Show that for all $0 \leq k \leq n$, G contains an element of order p^k.

32 The Fundamental Theorem of Finite Abelian Groups

Our objective in this chapter is to show that any finite Abelian group can be expressed as a direct sum of cyclic subgroups of prime power order in only one way (this is the "fundamental theorem" of the title). We already know that a finite Abelian group of prime power order can be written as a direct sum of cyclic subgroups (Theorem 31.12). We will first generalize this to all finite Abelian groups and we will then show how to distinguish between the resulting direct sums.

In this chapter, as in Chapter 31, we will consider only <u>Abelian</u> groups, and hence, in keeping with standard practice, <u>we will write the generic operation as addition rather than multiplication</u>.

Furthermore, to avoid cumbersome notation, we will abbreviate, as we did in Chapter 31, the sets $[k]_n$ by k, for $0 \leq k < n$. So we will, for example, write $\mathbb{Z}_6$ as $\{0, 1, 2, 3, 4, 5\}$ instead of $\{[0]_6, [1]_6, [2]_6, [3]_6, [4]_6, [5]_6\}$.

Note that it is <u>not</u> the case that a finite Abelian group can be written as a direct sum of cyclic subgroups in only one way. The restriction to cyclic subgroups of prime power order is essential. For instance, it is easy to see that $\mathbb{Z}_6 = S(1)$ and that $\mathbb{Z}_6 = S(2) \oplus S(3)$; in fact, for any positive integers m and n with no common prime factor, $\mathbb{Z}_{mn} = S(1)$ and $\mathbb{Z}_{mn} = S(m) \oplus S(n)$.

For our first result, that every finite Abelian group can be written as a direct sum of cyclic subgroups, we need first to write such a group as a direct sum of subgroups of prime power order. The subgroups we use are the following.

DEFINITION

For a prime p and an Abelian group A, let

$$A_p = \{a \in A \mid \text{the order of } a \text{ is a power of } p\}.$$

Proposition 32.1   *If p is a prime and A is a finite Abelian group, then A_p is a subgroup of A.*

Proof   To see that A_p is a subgroup of A, observe that since 0 has order $1 = p^0$, $0 \in A_p$ and hence $A_p \neq \varnothing$. Furthermore, if $a, b \in A_p$, then $p^\alpha a = 0 = p^\beta b$ for nonnegative integers α and β, and hence, since A is Abelian, $p^{\alpha+b}(a + b) = p^\beta(p^\alpha a) + p^\alpha(p^\beta b) = 0$, i.e., $a + b \in A_p$. Therefore, by Proposition 18.7, A_p is a subgroup of A. ∎

For instance, direct calculation shows that $(\mathbb{Z}_6)_2 = \{0, 3\}$ and $(\mathbb{Z}_6)_3 = \{0, 2, 4\}$. Note that 0 has order 1 and $1 = p^0$ for any prime p. So $(\mathbb{Z}_6)_p = \{0\}$ for any prime p other than 2 or 3.

To apply Theorem 31.12 to A_p, we need to know that the order of A_p is a power of p. In fact, as we will show, any Abelian group whose elements all have order a power of p must itself have order a power of p. In Chapter 34, we will use Cauchy's theorem to show that this result holds even for noncommutative groups.

Proposition 32.2   *Let p be a prime and suppose that G is a finite Abelian group whose every element has order a power of p. Then the order of G is also a power of p.*

Proof   We will prove the result by using the second principle of mathematical induction (Theorem 1.16) applied to the order of G.

(i)   If $|G| = 1$, then $|G| = p^0$.

(ii)   Suppose that $|G| > 1$ and that if B is an Abelian group whose order is less than that of G and all of whose elements have order a power of p, then B has order a power of p. Since $|G| > 1$, there exists $e \neq g \in G$, and since G is Abelian, $S(g)$ is a normal subgroup of G by Proposition 23.4. So we may consider the group $B = G/S(g)$, whose order according to Proposition 23.8 is $|B| = |G|/|S(g)|$. But since $e \neq g$, $|S(g)| > 1$, and thus $|B| < |G|$. Furthermore, if $X \in B$, then $X = x + S(g)$ for some $x \in G$, and hence for some p^k, $p^k X = p^k(x + S(g)) = p^k x + S(g) = 0 + S(g)$. So X has order a power of p by Proposition 21.5, and therefore by the induction hypothesis, the order of B is a power of p. But by Proposition 21.14, $S(g)$ has order a power of p, and thus since $|G| = |B||S(g)|$, G must also have order a power of p. ∎

Corollary 32.3   *If p is a prime and A is a finite Abelian group, then A_p is a subgroup of A whose order is a power of p.*

Proof   The result follows from Propositions 32.1 and 32.2. ∎

The primary decomposition theorem, which we prove next, says that every Abelian group can be written as a direct sum of its subgroups A_p of prime power order. That a finite Abelian group can be written as a direct sum of cyclic subgroups of prime power order is the basis theorem (Corollary 32.5).

Proposition 32.4   **Primary Decomposition Theorem**

Let A be a finite Abelian group of order m and suppose that $m = p_1^{r_1} \cdots p_n^{r_n}$ is the prime decomposition of m. Then $A = A_{p_1} \oplus \cdots \oplus A_{p_n}$.

Proof   By Propositon 32.1, A_{p_i} is a subgroup of A for all i, and thus by Proposition 30.4, it suffices to show (i) that $A = S(A_{p_1}, \ldots, A_{p_n})$ and (ii) that for all i, $S(A_{p_1}, \ldots, A_{p_{i-1}}) \cap A_{p_i} = \{0\}$.

(i)   Certainly $A \supseteq S(A_{p_1}, \ldots, A_{p_n})$. To prove the reverse containment, let $0 \neq x \in A$ and note that by Corollary 21.16, x has finite order r dividing m. If $r = q_1^{s_1} \cdots q_k^{s_k}$ is the prime decomposition of r, then for each $i = 1, \ldots, k$, $r_i = \dfrac{r}{q_i^{s_i}}$ is a positive integer. Clearly the greatest common divisor of the r_i is 1 and hence by Theorem 1.4 applied k times (cf. Exercise 47 in Chapter 1), there exist integers $\mu_1, \ldots, \mu_k$ such that $1 = \mu_1 r_1 + \cdots + \mu_k r_k$. Then $x = (\mu_1 r_1)x + \cdots + (\mu_k r_k)x$. However, r divides m and hence by the fundamental theorem of arithmetic (Proposition 1.17), for each i, there exists j such that $p_j = q_i$. Then $p_j^{s_i}(\mu_i r_i)x = (q_i^{s_i}\mu_i r_i)x = (\mu_i r)x = 0$, and therefore, by Proposition 21.5, $(\mu_i r_i)x \in A_{p_j}$, and it follows that $x \in S(A_{p_1}, \ldots, A_{p_n})$.

(ii)   Since A_{p_i} and $S(A_{p_1}, \ldots, A_{p_{i-1}})$ are subgroups, we must have $S(A_{p_1}, \ldots, A_{p_{i-1}}) \cap A_{p_i} \supseteq \{0\}$. Conversely, suppose that $z \in S(A_{p_1}, \ldots, A_{p_{i-1}}) \cap A_{p_i}$. Since $z \in A_{p_i}$, $(p_i^{t_i})z = 0$ for some t_i. Since A is Abelian and each A_{p_k} is a subgroup and since $z \in S(A_{p_1}, \ldots, A_{p_{i-1}})$, we may also write $z = a_1 + \cdots + a_{i-1}$ for $a_k \in A_{p_k}$. But $(p_k^{t_k})a_k = 0$ for some t_k, and thus, if $T = p_1^{t_1} \cdots p_{i-1}^{t_{i-1}}$, $Ta_k = 0$ for all k and hence $Tz = 0$. But the greatest common divisor of $p_i^{t_i}$ and T is certainly 1, and hence by Theorem 1.4, there exist μ and ν such that $1 = \mu p_i^{t_i} + \nu T$. Then $z = (\mu p_i^{t_i})z + (\nu T)z$, and since, as we previously showed, $p_i^{t_i}z = 0$ and $Tz = 0$, it follows that $z = 0$ and hence that $S(A_{p_1}, \ldots, A_{p_{i-1}}) \cap A_{p_i} = \{0\}$. ∎

For instance, the decomposition given for $\mathbb{Z}_6$, $\mathbb{Z}_6 = S(2) \oplus S(3)$, is the decomposition determined by Proposition 32.4. For $6 = 2 \cdot 3$ is the prime decomposition of 6, and thus, since $(\mathbb{Z}_6)_2 = \{0, 3\} = S(3)$ and $(\mathbb{Z}_6)_3 = \{0, 2, 4\} = S(2)$, this decomposition may be written $\mathbb{Z}_6 = S(2) \oplus S(3) = S(3) \oplus S(2) = (\mathbb{Z}_6)_2 \oplus (\mathbb{Z}_6)_3$.

Corollary 32.5   **Basis Theorem**

Every finite Abelian group is the direct sum of cyclic subgroups of prime power order.

Proof   If A is a finite Abelian group, then $A = A_{p_1} \oplus \cdots \oplus A_{p_n}$ by Proposition 32.4, and by Theorem 31.12, each A_{p_i} is a direct sum of cyclic subgroups of prime power order. But then by Proposition 30.7, A is the direct sum of cyclic subgroups of prime power order. ∎

Example 32.6   Consider $A = \mathbb{Z}_4 \times \mathbb{Z}_6$. Then the order of A is $4 \cdot 6 = 24 = 2^3 \cdot 3$, and direct calculation shows that $A_2 = \{(0, 0), (1, 0), (2, 0), (3, 0), (0, 3), (1, 3), (2, 3), (3, 3)\}$ and $A_3 = \{(0, 0), (0, 2), (0, 4)\}$. It is easy to see that

$$A_2 = \{(0, 0), (1, 0), (2, 0), (3, 0)\} \oplus \{(0, 0), (2, 3)\} = S((1, 0)) \oplus S((2, 3)).$$

And since $A_3 = S((0, 2))$, the primary decomposition theorem ensures that

$$A = A_2 \oplus A_3 = S((1, 0)) \oplus S((2, 3)) \oplus S((0, 2)).$$

But $S((1, 0))$ has order $4 = 2^2$, $S(((2, 3))$ has order 2, and $S((0, 2))$ has order 3. So this is a decomposition of the type described by the basis theorem. ❖

Since by Proposition 21.18 all cyclic groups of order n are isomorphic to $\mathbb{Z}_n$, and since by definition each internal direct sum is isomorphic to an external direct product, the basis theorem says that all finite Abelian groups can be constructed by forming external direct products of groups $\mathbb{Z}_{p^n}$ for primes p. We want to know more than this. We want to know when these products are distinct. We will determine this by showing that any decomposition of a finite Abelian group into a direct sum of cyclic subgroups of prime power order has the same number of summands of each order. This is the so-called fundamental theorem of finite Abelian groups. To prove this, we will need to isolate the special case in which the nonzero elements all have order p.

Proposition 32.7   *Suppose that p is prime and that H and K are Abelian groups of the same order such that $pH = \{0\}$ and $pK = \{0\}$. If $H = S(x_1) \oplus \cdots \oplus S(x_n)$, where $x_i \neq 0$, and $K = S(y_1) \oplus \cdots \oplus S(y_m)$, where $y_j \neq 0$, then $n = m$ and each cyclic subgroup has the same order, viz., p.*

Proof   Since $pH = \{0\}$, Corollary 21.6 implies that each element of H has order 1 or p. Thus, since $x_i \neq 0$ for all i, each x_i has order p, and hence by Proposition 21.14, each $S(x_i)$ has order p. For the same reasons, each $S(y_j)$ also has order p. But then by Proposition 30.7, H has order p^n and K has order p^m. So by hypothesis, $p^n = p^m$ and hence $m = n$. ∎

Example 32.8   Certainly the group $A = \mathbb{Z}_3 \times \mathbb{Z}_3$ is such that $3A = \{(0, 0)\}$ and by using the subgroups corresponding to the copies of $\mathbb{Z}_3$ in each coordinate, we have $A = S((1, 0)) \oplus S((0, 1))$. Note that each cyclic subgroup has order 3 and that since A has no element of order 9, A itself is not cyclic.

Another Abelian group of order 9 is the subgroup $B = \{\iota, (123), (132), (456), (123)(456), (132)(456), (465), (465)(123), (465)(132)\}$ of S_6. (It is easy to see both that B is a subgroup of S_6 and that B is Abelian by constructing its multiplication table—we leave this to the reader.) In this case, we have $B = S((123)) \oplus S((456))$. As with A, each cyclic subgroup has order 3 and since B has no element of order 9, B is also not cyclic.

Both A and B are Abelian groups of order 9, and $3A = \{0\}$ and $3B = \{0\}$, and we have seen that, in agreement with Proposition 32.7, A and B have similar decompositions.

On the other hand, the Abelian group $\mathbb{Z}_9$, which is also of order 9, can be written as a direct sum of cyclic groups with one rather than two summands. However, Proposition 32.7 does not apply to this group because $3\mathbb{Z}_9 = \{0, 3, 6\} \neq \{0\}$. ❖

We are now in a position to prove the fundamental theorem. (Note that in the statement of the theorem, "nontrivial" means that the group is not $\{0\}$.)

Theorem 32.9   **Fundamental Theorem of Finite Abelian Groups**

Any two decompositions of a nontrivial finite Abelian group into direct sums of nontrivial cyclic subgroups of prime power orders have the same number of summands of each order.

Proof   The primary decomposition theorem tells us that any nontrivial finite Abelian group A may be written as a direct sum of its subgroups A_{p_i}. By Corollary 21.16, the elements of any nontrivial cyclic subgroup of prime power order must all have order a power of the same prime, and hence the cyclic subgroup must be contained in one of these A_{p_i}. So we need only prove the result for Abelian groups of prime power orders. That is, we need only consider nontrivial finite Abelian groups A such that $p^n A = \{0\}$ for some prime p and positive integer n. We will do this by induction on n. Note that in this case, all subgroups are of prime power order by Lagrange's theorem (Proposition 21.10).

(i)   If $pA = \{0\}$, then any two decompositions of A into a direct sum of nontrivial cyclic subgroups have the same number of summands and each summand has order p by Proposition 32.7.

(ii)   Suppose that $n > 0$ and that for any nontrivial Abelian group B such that $p^n B = \{0\}$, any decomposition of B into a direct sum of nontrivial cyclic subgroups has the same number of summands of each order. Suppose further that A is a nontrivial finite Abelian group such that $p^{n+1} A = \{0\}$ and that A has two decompositions as the direct sum of cyclic subgroups. By Proposition 30.7, we can write the decompositions as follows:

$$A = S(x_1) \oplus \cdots \oplus S(x_r) \oplus S(x_{r+1}) \oplus \cdots \oplus S(x_{r+s}), \text{ and}$$
$$A = S(y_1) \oplus \cdots \oplus S(y_u) \oplus S(y_{u+1}) \oplus \cdots \oplus S(y_{u+v}),$$

where

$$px_i \neq 0 \text{ for } 1 \leq i \leq r, px_{r+i} = 0 \text{ but } x_{r+i} \neq 0 \text{ for } 1 \leq i \leq s, \text{ and}$$
$$py_j \neq 0 \text{ for } 1 \leq j \leq u, px_{s+j} = 0 \text{ but } y_{s+j} \neq 0 \text{ for } 1 \leq j \leq v.$$

Then $p^n(pA) = p^{n+1}A = \{0\}$ and

$$pA = S(px_1) \oplus \cdots \oplus S(px_r), px_i \neq 0, \text{ and}$$
$$pA = S(py_1) \oplus \cdots \oplus S(py_u), py_j \neq 0.$$

So by the induction hypothesis, $r = u$ and the two direct sums have the same number of summands of each order. Certainly the order of x_i is p times the order of px_i, and hence by Proposition 21.14, the order of $S(x_i)$ is p times the order of $S(px_i)$. So the two direct sums $S(x_1) \oplus \cdots \oplus S(x_r)$ and $S(y_1) \oplus \cdots \oplus S(y_u)$ also have the same number of summands of each order. But then by Proposition 30.7, $S(x_1) \oplus \cdots \oplus S(x_r)$ and $S(y_1) \oplus \cdots \oplus S(y_u)$ have the same orders so that by division, $S(x_{r+1}) \oplus \cdots \oplus S(x_{r+s})$ and $S(y_{u+1}) \oplus \cdots \oplus S(y_{u+v})$ have the same orders. And hence we may apply Proposition 32.7 and conclude that $s = v$ and each of the cyclic subgroups in these direct sums has order p. It then follows that the two decompositions of A given above have the same number of cyclic subgroups of each order. ∎

The fundamental theorem, in conjunction with Proposition 29.7 and Proposition 32.10 and Corollary 32.11 below, yields the following list on the next page of Abelian groups of order less than 16.

Proposition 32.10 *Suppose that m and n are positive integers whose greatest common divisor is 1. Then $\mathbb{Z}_m \times \mathbb{Z}_n$ is isomorphic to $\mathbb{Z}_{mn}$.*

Proof Suppose that k is a positive integer and that $k([1]_m, [1]_n) = ([0]_m, [0]_n)$. Then $[k]_m = [0]_m$ and $[k]_n = [0]_n$, and hence both m and n divide k. Since m and n have no common prime factors, this implies that mn divides k. So the order of $([1]_m, [1]_n)$ is at least mn. But since the order of $\mathbb{Z}_m \times \mathbb{Z}_n$ is mn (by Proposition 29.5), the order of $([1]_m, [1]_n)$ divides mn. It follows that the order of $([1]_m, [1]_n)$ is mn and hence that $\mathbb{Z}_m \times \mathbb{Z}_n$ is cyclic. Then by Proposition 21.18, $\mathbb{Z}_m \times \mathbb{Z}_n$ is isomorphic to $\mathbb{Z}_{mn}$. ∎

Corollary 32.11 *Suppose that p and q are distinct primes. If A is an Abelian group of order pq, then A is isomorphic to $\mathbb{Z}_{pq}$.*

Proof By the primary decomposition theorem (Proposition 32.4), $A = A_p \oplus A_q$, and by Corollary 32.3, the order of A_p is a power of p and the order of A_q is a power of q. But by Proposition 30.7, $pq = |A| = |A_p||A_q|$, and therefore the order of A_p must be p and the order of A_q must be q. So by Corollary 21.19, A_p is isomorphic to $\mathbb{Z}_p$ and A_q is isomorphic to $\mathbb{Z}_q$. So by Proposition 32.10 (and Exercise 22), A is isomorphic to $\mathbb{Z}_{pq}$. ∎

Abelian Groups of Order Less Than 16		
Order	Number of Groups	Distinct Groups
1	1	$\mathbb{Z}_1$
2	1	$\mathbb{Z}_2$
3	1	$\mathbb{Z}_3$
4	2	$\mathbb{Z}_4, \mathbb{Z}_2 \times \mathbb{Z}_2$
5	1	$\mathbb{Z}_5$
6	1	$\mathbb{Z}_6$
7	1	$\mathbb{Z}_7$
8	3	$\mathbb{Z}_8, \mathbb{Z}_4 \times \mathbb{Z}_2, \mathbb{Z}_2 \times \mathbb{Z}_2 \times \mathbb{Z}_2$
9	2	$\mathbb{Z}_9, \mathbb{Z}_3 \times \mathbb{Z}_3$
10	1	$\mathbb{Z}_{10}$
11	1	$\mathbb{Z}_{11}$
12	2	$\mathbb{Z}_{12}, \mathbb{Z}_3 \times \mathbb{Z}_2 \times \mathbb{Z}_2$
13	1	$\mathbb{Z}_{13}$
14	1	$\mathbb{Z}_{14}$
15	1	$\mathbb{Z}_{15}$

Exercises

In Exercises 1–6, for the group A and the prime p, find A_p.

1. $A = \mathbb{Z}_{10}, p = 5$

2. $A = \mathbb{Z}_{12}, p = 2$

3. $A = \mathbb{Z}_6 \times \mathbb{Z}_6, p = 2$

4. $A = \mathbb{Z}_6 \times \mathbb{Z}_{18}, p = 3$

5. $A = \mathbb{Z}_{12} \times \mathbb{Z}_{18}, p = 2$

6. $A = \mathbb{Z}_{12} \times \mathbb{Z}_{18}, p = 3$

In Exercises 7–13, find all Abelian groups of order n.

7. $n = 16$

8. $n = 17$

9. $n = 18$

10. $n = 20$

11. $n = 22$

12. $n = 27$

13. $n = 36$

In Exercises 14–17, for the group G, verify the primary decomposition theorem (Proposition 32.4).

14. $G = \mathbb{Z}_{10}$

15. $G = \mathbb{Z}_{12}$

16. $G = \mathbb{Z}_6 \times \mathbb{Z}_6$

17. $G = \mathbb{Z}_6 \times \mathbb{Z}_{18}$

In Exercises 18–21, for the group G, verify the basis theorem (Corollary 32.5).

18. $G = \mathbb{Z}_{10}$

19. $G = \mathbb{Z}_{12}$

20. $G = \mathbb{Z}_6 \times \mathbb{Z}_6$

21. $G = \mathbb{Z}_6 \times \mathbb{Z}_{18}$

22. Fill in the details of the proof of Corollary 32.11 by showing that if $G = A \oplus B$, if A is isomorphic to X and B is isomorphic to Y, and if $X \times Y$ is isomorphic to H, then G is isomorphic to H.

23. For an Abelian group A and a prime p, show that A/A_p has no elements of order p.

24. Let A be an Abelian group and let p be a prime such that A has order $p^a k$, where p does not divide k. Show that the order of A_p is p^a.

25. Let p be a prime and let G be a cyclic group whose order is a power of p. Show that if p^a divides the order of G, then G has a subgroup of order p^a.

26. Let p be a prime. Show that every cyclic group G whose order is a power of p has a normal subgroup N such that G/N has order p.

In Exercises 27–29, show that the group G is not isomorphic to the group H.

27. $G = \mathbb{Z}_2 \times \mathbb{Z}_2 \times \mathbb{Z}_4$, $H = \mathbb{Z}_4 \times \mathbb{Z}_4$

28. $G = \mathbb{Z}_2 \times \mathbb{Z}_3 \times \mathbb{Z}_{12}$, $H = \mathbb{Z}_4 \times \mathbb{Z}_9 \times \mathbb{Z}_2$

29. $G = \mathbb{Z}_{25} \times \mathbb{Z}_3 \times \mathbb{Z}_8$, $H = \mathbb{Z}_{10} \times \mathbb{Z}_{15} \times \mathbb{Z}_4$

30. Show that $\mathbb{Z}_8 \times \mathbb{Z}_5 \times \mathbb{Z}_9$, $\mathbb{Z}_{40} \times \mathbb{Z}_9$, and $\mathbb{Z}_8 \times \mathbb{Z}_{45}$ are all isomorphic.

31. Show that $\mathbb{Z}_2 \times \mathbb{Z}_4 \times \mathbb{Z}_3$, $\mathbb{Z}_6 \times \mathbb{Z}_4$, and $\mathbb{Z}_2 \times \mathbb{Z}_{12}$ are all isomorphic.

32. Show that if G is a finite Abelian group of order m and that k divides m, then G has a subgroup of order k.

33. Show that for finite Abelian groups A and B, A is isomorphic to B if and only if A_p is isomorphic to B_p for all primes p.

34. Find an Abelian group A and a subgroup B of A such that A/B is not isomorphic to any subgroup of A.

35. Suppose that A and B are finite Abelian groups and that for each positive integer n, A and B have the same number of elements of order n. Show that A is isomorphic to B.

36. Suppose that A, B, and C are all finite Abelian groups. Show that if $A \times B$ is isomorphic to $C \times A$, then B is isomorphic to C.

HISTORICAL NOTE

Leopold Kronecker

Leopold Kronecker was born in Liegnitz, Germany (now Legnica, Poland), on December 7, 1823, and died in Berlin, Germany, on December 29, 1891. His father, Isidor Kronecker, and his mother, Johanna Prausnitzer, were wealthy and provided private tutoring at home for Leopold until he entered Liegnitz Gymnasium. At the Gymnasium, he became a student of Ernst Kummer who recognized his ability and encouraged him in mathematics; they remained friends for the remainder of Kronecker's life. Even though Kronecker was Jewish, he received Evangelical religious instruction at the Gymnasium and eventually, in the last year of his life, he formally converted to Christianity. He matriculated at the University of Berlin in 1841 but followed Kummer to Breslau (now Wroclaw, Poland) in 1844, where he remained for two semesters before returning to Berlin to take his doctorate in 1845.

Rather than continue his life in academia, he then returned to Liegnitz to work in the family business. One of his duties was to dissolve the banking business of an uncle, and in 1848, he married the daughter of that uncle, Fanny Prausnitzer, with whom he had six children. While he was in Liegnitz, he did not ignore mathematics but continued his research and carried on a lively correspondence with Kummer. In fact, in 1853, he announced that he had proved that the roots of every integral polynomial equation with Abelian Galois group over the rationals live in an extension field of the rationals generated by ζ_n's.

By 1855, he was independently wealthy and returned to Berlin to pursue mathematics. Simultaneously, Kummer came to Berlin as the successor of Dirichlet who in his turn was appointed the successor of Gauss at Göttingen. Kronecker and Kummer became friends with Weierstrass, and with his help, Kronecker became a member of the Berlin Academy in 1861. As a member of the Academy, he had the right to deliver a series of lectures at the University of Berlin and he gave such a series of lectures in the winter of 1862. He tended to ramble, however, and by the end of the semester only a few of the students remained.

He enjoyed his life in Berlin, so much so that when in 1868 he was offered the chair at Göttingen that had been held first by Gauss and then by Dirichlet and Riemann, he refused it. He traveled a great deal in Europe and visited many of his foreign colleagues. Because of his hard work for the Berlin Academy, he became very influential in its affairs and was instrumental in recruiting many of its foreign members.

In spite of all the travel and administrative work, Kronecker continued his mathematical research and in 1870 gave the defining axioms for Abelian groups and proved the fundamental theorem of finite Abelian groups. His formulation asserted that every finite Abelian group has a fundamental system of generators $\Theta_1, \Theta_2, \ldots, \Theta_k$ of orders $n_1, n_2, \ldots, n_k$ such that each element of the group can be written as a unique product of the Θ_k's, each n_{i+1} divides each n_i, and the order of the group is $n_1 n_2 \cdots n_k$.

His friendship with Weierstrass deteriorated and over time they became enemies. Certainly their temperaments were different, but more importantly they had diametrically opposed views of mathematics, Weierstrass championing Cantor's belief in degrees of infinity and Kronecker disdaining them. The basis of Kronecker's position can be summarized by quoting his well-known dictum: "God made the integers and all the rest is the work of man."[1] From Weierstrass's point of view, such a position undercut all of analysis, in particular all of his own mathematics. So it is not surprising that he took pointed exception to Kronecker's views. On his side, Kronecker prefigured Brouwer whose school of intuitionism continued to attract adherents through much of the twentieth century.

The enmity between Kronecker and Weierstrass was personal as well as mathematical. For New Year's, 1885, H. A. Schwarz, Weierstrass's student and Kummer's son-in-law, sent Kronecker a greeting which included the words: "He who does not honor the Smaller, is not worthy of the Greater."[2] Kronecker was a short man and had become increasingly sensitive about his height as he had grown older. He read this allusion to his stature as both an intellectual comment and a physical one and would have nothing more to do with Schwarz.

In 1891, Weierstrass and Kronecker were at odds over the qualifications of Kronecker's student Hensel and Weierstrass's student Knoblach when Kronecker's wife died. The argument became moot when Kronecker himself died a few months later.

References

1. Kline, Morris. *Mathematics: The Loss of Certainty.* New York: Oxford University Press, 1980, p. 232.
2. Bierman, Kurt-R. "Kronecker, Leopold," *Dictionary of Scientific Biography,* vol. VII, Gillispie, Charles Coulston, ed. New York: Charles Scribner's Sons, 1970, p. 507.

Dihedral Groups

Certain small groups arise as transformation groups of plane figures. While the geometric point of view is intuitively appealing and provides strong motivation to study these groups, it is more efficient from the point of view of abstract group theory to view them in terms of generators and relations. We want to use the dihedral groups to help classify the small finite groups. So in what follows, while we will define the dihedral groups in terms of plane transformations, it is the more abstract characterization that we will use for the classification. We will also apply this point of view to the subgroup $Q = \{\pm 1, \pm i, \pm j, \pm k\}$ of the multiplicative quaternions in order to classify all groups of order 8. We begin by defining the dihedral groups geometrically.

In general, for any $n \geq 3$, consider a regular plane polygon P_n. That is, P_n has n edges all of the same length and n equal angles. Consecutively label the vertices $1, \ldots, n$, and draw a line L from vertex 1 through the centre of the polygon. The polygons P_3, P_4, and P_5 are given below.

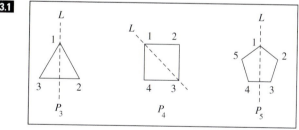

FIGURE 33.1

Now let X be the clockwise rotation sending each vertex into the adjacent one, let Y be the reflection of P_n about L, and let D_n denote the set of all possible compositions of the transformations X and Y. (To be consistent with functional notation, we will agree that the first transformation to be applied in a given composition will be the one on the right.)

As an example, we will explicitly construct D_4.

Example 33.1   To construct D_4, we first construct all powers of X:

FIGURE 33.2

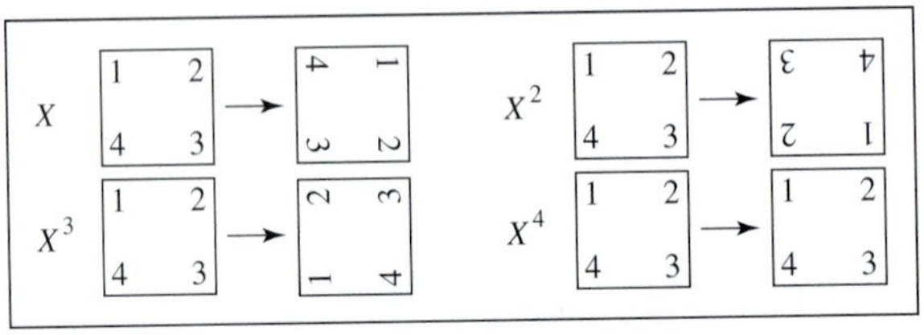

Note that X^4 leaves the square unchanged and hence if I denotes the identity function, then $X^4 = I$, and $X^5 = X$, $X^6 = X^2$, ... The powers of Y are

FIGURE 33.3

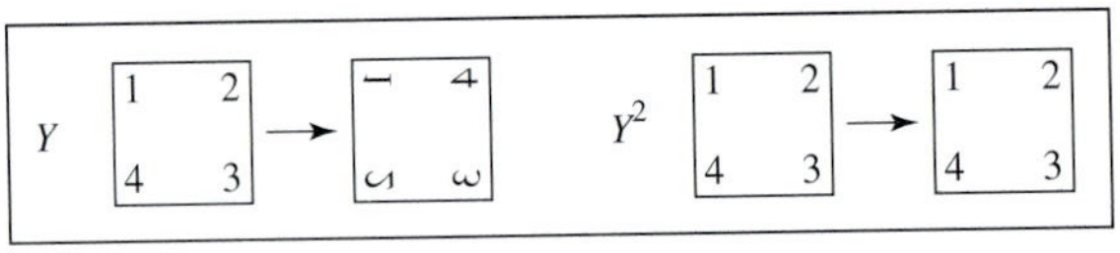

And in this case, $Y^2 = I$ so that $Y^3 = Y$, $Y^4 = I$, ... Next note that $XY = YX^3$:

FIGURE 33.4

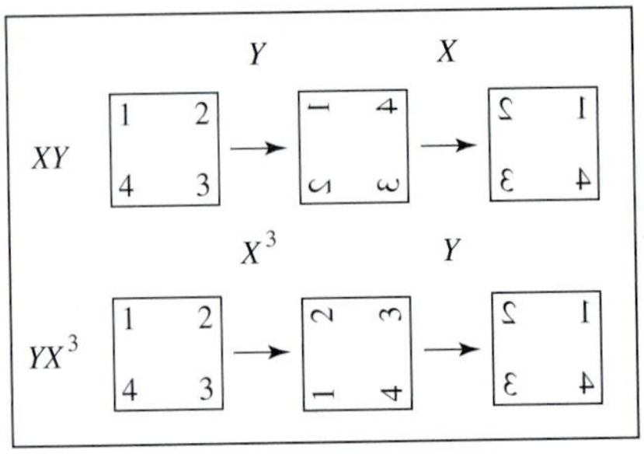

Then any product of X's and Y's can be transformed into a product of the form $Y^m X^n$ for positive integers m and n, and each such product must equal one of I, X, X^2, X^3, Y, YX, YX^2, or YX^3. For instance, $X^2 I Y X Y^4 X^5 = X^2 I Y X I X = X^2 Y X^2 = XYX^3 X^2 = YX^3 X^3 X^2 = Y$. Thus D_4 consists of the transformations in Figures 33.5a and b:

FIGURE 33.5a

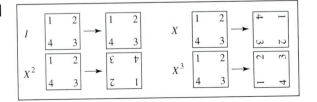

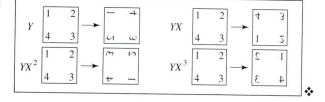

FIGURE 33.5b

In general, for any positive n, $X^n = I$, $Y^2 = I$, and $XY = YX^{n-1}$ so that $D_n = \{I, X, \ldots, X^{n-1}, Y, YX, \ldots, YX^{n-1}\}$. Composition of transformations is certainly associative. As well, since $Y^2 = I$, Y has an inverse Y with respect to composition; since $X^k X^{n-k} = I = X^{n-k} X^k$, X^k has inverse X^{n-k}; and since $(YX^k)(YX^k) = YYX^{(n-1)k}X^k = I$, YX^k is its own inverse. It follows that D_n is a group with respect to composition of transformations.

DEFINITION

The group D_n is called the ***nth dihedral group.***

As previously indicated, the geometric definition of D_n is not the most efficient characterization from the point of view of abstract group theory. From that point of view, it is better to think of D_n in terms of generators and relations. Indeed, all groups may be written in this manner, some more fruitfully than others. But we will not be concerned with the general construction here; rather, we will content ourselves with taking an ad hoc approach and merely deriving the specific presentations of the groups D_n and the subgroup Q of the nonzero quaternions in this chapter. In the sequel we will do the same for other small groups.

How can D_n be characterized abstractly? Certainly it has two generators, x and y, x being of order n and y being of order 2. The key relationship between x and y is that $xy = yx^{n-1}$ because it allows every element to be written in the form $y^i x^j$ for $i = 0,1$ and $j = 0, \ldots, n - 1$. It turns out that these three conditions are sufficient to characterize D_n.

Proposition 33.2   *Suppose that $n \geq 3$ and that G is a group with generators x and y such that x has order n, y has order 2, and $xy = yx^{n-1}$. Then $G = \{e, x, \ldots, x^{n-1}, y, yx, \ldots, yx^{n-1}\}$ and since all these elements are distinct, G has order $2n$.*

Proof   If $z \in G$, then z is a product of powers of x and y. But according to the division algorithm (Theorem 1.3), for any positive integer k, $k = nq + r$, $0 \leq r < n$, and hence $x^k = x^{nq+r} = x^r$, i.e., any power of x can be replaced by a power x^r, where $0 \leq r < n$. Similarly, any power of y can be replaced by e or y. Since any occurrence of xy can be replaced by yx^{n-1}, z may thus be written $z = y^i x^k$ for $i \in \{0, 1\}$

and $k \in \{0, \ldots, n - 1\}$. It follows that $G = \{e, x, \ldots, yx^{n-1}, y, yx, \ldots, yx^{n-1}\}$. But by Proposition 21.14, the elements $e, x, \ldots x^{n-1}$ of G are all distinct, and thus so are the elements $y, yx, \ldots yx^{n-1}$. If $x^i = yx^k$ for some i and k, then y is a power of x and hence commutes with x so that $yx = xy = yx^{n-1}$ and hence $e = x^{n-2}$. This contradicts our assumption that x has order n. It follows that the elements of $\{e, x, \ldots, x^{n-1}, y, yx, \ldots, yx^{n-1}\}$ are all distinct and hence that G has order $2n$. ∎

Corollary 33.3   *Suppose that $n \geq 3$ and that G is a group with generators x and y such that x has order n, y has order 2, and $xy = yx^{n-1}$. Then G is isomorphic to D_n.*

Proof   Since $D_n = \{I, X, \ldots, X^{n-1}, Y, YX, \ldots, YX^{n-1}\}$, the function $f: G \to D_n$ which identifies x with X and y with Y is both one-to-one and onto. For $z, w \in G$, there exist $a, r \in \{0,1\}$ and $b, s \in \{0, \ldots, n - 1\}$ such that $z = y^a x^b$ and $w = y^r x^s$. If $r = 0$, let A be a and let B be the remainder of $b + s$ divided by n; if $r = 1$, let A be the remainder of $a + b$ divided by 2 and let B be the remainder of $(n - 1)b + s$ divided by n. Then $zw = y^a x^b y^r x^s = y^A x^B$, and $Y^A X^B = Y^a X^b Y^r X^s$ so that

$$f(zw) = f(y^A x^B) = Y^A X^B = Y^a X^b Y^r X^s = f(z)f(w). \quad ∎$$

Example 33.4   By Proposition 19.8, the group $S_3 = \{\iota, (12), (13), (23), (123), (132)\}$ is generated by $x = (123)$ and $y = (12)$. And since $(123)^3 = \iota$ and $(12)^2 = \iota$, it follows from Corollary 21.6 that x has order 3 and y has order 2. Furthermore, $xy = (123)(12) = (13) = (12)(132) = (12)(123)^2 = yx^{3-1}$. So by Corollary 33.3, S_3 is isomorphic to D_3. ❖

Proposition 34.15, proved in the next chapter, will show that there are no other noncommutative groups of order 6. On the other hand, there are at least two noncommutative groups of order 8. For certainly D_4 is a noncommutative group of order 8, and so is the subgroup $Q = \{\pm 1, \pm i, \pm j, \pm k\}$ of the multiplicative quaternions. And by Proposition 21.9, Q and D_4 cannot be isomorphic because -1 is the only element in Q of order 2 while x^2 and y are different elements of order 2 in D_4. We will show in Proposition 33.8 that these are the only noncommutative groups of order 8 by writing Q in terms of generators and relations.

Proposition 33.5   *Suppose that G is a group with generators x and y such that x has order 4, $y^2 = x^2$, and $xy = yx^{-1}$. Then G is isomorphic to Q.*

Proof   By a proof similar to that of Proposition 33.2, $G = \{e, x, x^2, x^3, y, yx, yx^2, yx^3\}$ and all these elements are distinct. Observe that i and $-j$ generate Q and that i has order 4, that $(-j)^2 = -1 = i^2$, and that $i(-j) = -k = (-j)(-i) = (-j)i^{-1}$. A proof similar to that of Corollary 33.3 then shows that G is isomorphic to Q (Exercise 24). ∎

We will also need the following two results, each of which is interesting in its own right.

Proposition 33.6   *If G is a group in which the square of every element is the identity, then G is Abelian.*

Proof   If $x, y \in G$, then $xy = x^{-1}y^{-1} = x^{-1}(xy)^2 y^{-1} = x^{-1}xyxyy^{-1} = yx$. ∎

Proposition 33.7   *If S is a subgroup of index 2 in a finite group G, then S is a normal subgroup of G.*

Proof   If $y \in S$, then certainly $yS = Sy$. Note that a proof similar to that showing that each left coset of S has the same number of elements as S (cf. Proposition 20.4) shows that each right coset also has the same number of elements as S (Exercise 20 in Chapter 20). But then since G has only two left cosets, it has only enough elements for two right cosets, and thus if $y \notin S$, then both the left coset yS and the right coset Sy must consist of all those elements of G that are not in S, i.e., $yS = \{g \in G \mid g \notin S\} = Sy$. It follows from Proposition 23.1 that S is a normal subgroup of G. ∎

We can now show that Q and D_4 are the only noncommutative groups of order 8. Since the fundamental theorem of finite Abelian groups determines all Abelian groups of order 8, we have thus determined all groups of order 8.

Proposition 33.8   *The only noncommutative groups of order 8 are Q and D_4.*

Proof   Let G be a noncommutative group of order 8. Since G is not Abelian, Proposition 33.6 implies that G has at least one element whose order is neither 1 nor 2. By Corollary 21.16, the order of this element can only be 4 or 8. If its order is 8, then G is cyclic and hence Abelian by Corollary 21.17. Thus we may assume that G contains an element x of order 4. By Lagrange's theorem, the index of $S(x)$ in G is 2, and hence by Proposition 33.7, $S(x) \lhd G$. Let $y \notin S(x)$. Since $4 < |S(x,y)| \le 8$, Lagrange's theorem implies that $|S(x,y)| = 8$, and thus that $G = S(x,y)$. Since $G/S(x)$ has order 2, $yS(x)$ can only have order 1 or 2. In either case, $y^2 S(x) = (yS(x))^2 = S(x)$ so that $y^2 \in S(x)$. But Corollary 21.16 implies that y has order 1, 2, 4, or 8; and since $y \ne e$, y cannot have order 1, and since G is not cyclic, y cannot have order 8. So y has order 2 or 4 and hence y^2 has order 1 or 2. Therefore, since x^2 is the only element in $S(x)$ of order 2, either $y^2 = e$ or $y^2 = x^2$. However, since $S(x) \lhd G$, $y^{-1}xy \in S(x)$, and $y^{-1}xy$ has the same order as x (by Exercise 18). Then $y^{-1}xy$ has order 4, and hence either $y^{-1}xy = x$ or $y^{-1}xy = x^3$. If $y^{-1}xy = x$, then y commutes with x and hence G is Abelian, a contradiction. Thus, either $y^2 = e$ and $xy = yx^3$ or (since $x^3 = x^{-1}$) $y^2 = x^2$ and $xy = yx^{-1}$. In the first case, G is isomorphic to D_4 by Corollary 33.3; in the second case, G is isomorphic to Q by Proposition 33.5. ∎

Exercises

In Exercises 1–3, specifically construct all the transformations in the given dihedral group.

1. D_3 **2.** D_5 **3.** D_6

In Exercises 4–6, construct the subgroup lattice of the given dihedral group.

4. D_3 **5.** D_4 **6.** D_5

In Exercises 7–11, find all the subgroups of the given dihedral group D_n whose orders are some power of the given prime p.

7. $D_3, p = 2$ **8.** $D_3, p = 3$ **9.** $D_4, p = 2$

10. $D_5, p = 2$ **11.** $D_5, p = 5$

In Exercises 12–16, find all the subgroups of the given dihedral group D_n which have order a power of the given prime p and which are not properly contained in any other subgroup whose order is a power of p.

12. $D_3, p = 2$ **13.** $D_3, p = 3$ **14.** $D_4, p = 2$

15. $D_5, p = 2$ **16.** $D_5, p = 5$

17. Find all groups of order 8.

18. Suppose that G is a group. Show that if $x \in G$ has order n and $g \in G$, then $g^{-1}xg$ also has order n.

Exercises 19 and 20 concern the subgroup G of S_4 generated by (12) and (1324).

19. Find all the elements in G.

20. Show that G is isomorphic to D_4.

21. Show that any subgroup of D_n that has exactly two cosets is normal.

22. Show that if S is a subgroup of a finite group G that has exactly two cosets, then S is a normal subgroup of G.

23. Show that the subset

$$H = \{i, (1324)(5768), (1526)(3847), (1728)(3546), (12)(34)(56)(78),$$
$$(1827)(3645), (1625)(3748), (1423)(5867)\}$$

of S_8 is a subgroup of S_8 which is isomorphic to Q.

24. Suppose that G is a group with generators x and y such that x has order 4, $y^2 = x^2$, and $xy = yx^{-1}$. Complete the proof of Proposition 33.5 by showing in detail that $G = \{e, x, x^2, x^3, y, yx, yx^2, yx^3\}$, that all these elements are distinct, and that therefore G is isomorphic to Q.

25. Suppose that G is a finite group. Show that every element in G has order at most 2 if and only if G is isomorphic to a direct product of the form $\mathbb{Z}_2 \times \cdots \times \mathbb{Z}_2$.

26. Prove the following generalization of Proposition 33.7: If S is a subgroup of index 2 in a group G, then S is a normal subgroup of G.

HISTORICAL NOTE

Felix Klein

Felix Klein was born in Düsseldorf, Germany, on April 25, 1849, and died in Göttingen, Germany, on June 22, 1925. April 25 saw not only the birth of Klein but also the crushing of the Revolution of 1848. The Germans were unable to throw off the yoke of their Prussian rulers, and Klein's father, the secretary to the Regierungspräsident, was doubly happy, with the birth of his son to complement the victory on the battlefield.

When Felix was seventeen, he began his university studies at the University of Bonn. He studied physics and mathematics and was soon chosen as a laboratory assistant by Julius Plücker, who had recently returned to geometry after many years of studying physics. When Plücker died in 1868, he bequeathed to Klein the task of finishing his book on geometry.

Klein spent little time working on the book in Bonn, for he graduated within the year and, after a short visit to Göttingen, he traveled to Berlin and Paris. According to Klein, "The most important event of my time in Berlin was certainly that, toward the end of October, at a meeting of the Berlin Mathematical Society, I made the acquaintance of the Norwegian Sophus Lie. We had in our work been led from different points of view to the same questions, or, at least, to kindred ones. Thus it came about that we met every day and kept up an animated exchange of ideas. Our intimacy was all the closer, because at first we found very little interest for our geometrical interests in our immediate neighbourhood."[1] Klein and Lie went to Paris and made the acquaintance of Camille Jordan and Gaston Darboux. They had intended to continue on to England in the winter but had to change their plans when the beginning of the Franco-Prussian War forced Klein to return to Germany.

In Lie's case, the friendship with Darboux turned out to be very fortunate. He decided to walk to Italy and was arrested at Fountainbleu on the suspicion of being a spy. The French authorities held him for four weeks and released him only after Darboux convinced them that the illegible German papers that Lie was carrying were in fact Klein's mathematical scribblings.

Klein's return to Germany saw him join the ambulance corps, but he soon contracted typhoid fever and was forced to leave active duty. In 1871, he became a lecturer at Göttingen and the next year he was made a full professor of mathematics at the small Bavarian university of Erlangen.

It was here that he conceived his famous "Erlanger Programm." During the nineteenth century, a number of geometries (e.g., projective geometry and hyperbolic geometry) had been discovered and studied independently. Klein wanted to unify geometry by viewing each new geometry as a set of invariants of some appropriate group. Klein's resulting "Programm" guided work in geometry for many years, and Klein returned to it several times during his career.

It was also at Erlangen that, in the words of his friend William Henry Young, "he met and married the beautiful, cultured Anna Hegel."[2] She was the daughter of one of the professors at the university and also the granddaughter of the philosopher Hegel. Klein and his new wife did not stay at Erlangen long, for in 1875, he was offered and accepted a position at the Technische Hochschule in Munich. At that time, many Germans would have regarded this as a backward step because he was leaving a university and going to a technical institute, but Klein, having seen the École Polytechnique in Paris, viewed things differently. Indeed, his move to Munich did not hurt his career; five years later, he moved to the University of Leipzig, and six years after that, he returned to Göttingen.

While at Leipzig, he had turned down an offer to fill Sylvester's chair at Johns Hopkins in Baltimore, and his happiness at his return to Göttingen more than justified this decision. He built a house there, on the one hand close to his work at the university and on the other close to the Göttingerwald, a forest that stretched far into the mountains. It was his habit to walk in the forest with colleagues after lunch and then retire to his office, where he would continue his discussions or prepare his lectures.

Lie became Klein's successor at Leipzig but their friendship was strained when Lie took umbrage at Klein's suggestion that he was Klein's pupil. "Rather," he said, "the contrary was the case."[3] In his turn, Klein resented this comment. According to Klein's wife, Lie was ill when he made his comments, and he and Klein were reconciled in the following fashion. "One summer evening, as we came home from an excursion, there, in front of our door, sat a sick man. 'Lie!' we cried, in joyful surprise. The two friends shook hands, looked into one another's eyes, all that had passed since their last meeting was forgotten. Lie stayed with us one day, the dear old friend, and yet changed. I cannot think of him and his tragic fate without emotion. Soon after, he died, but not before the great mathematician had been received in Norway like a king."[4]

Although at the time, Klein was by far the more famous of the two men, only a cursory look at the mathematical literature is needed to see that it is Lie's work rather than Klein's that has had the more lasting impact on contemporary mathematics.

Klein was not only a famous mathematician but also a popular teacher. He supervised forty-eight Ph.D. dissertations, including that of Grace Chisholm. In 1893, no woman had yet received a Ph.D. from a Prussian university. Nonetheless, in that year, three women arrived at Göttingen to study with Klein—two Americans who had met him when he visited Chicago the previous year, and Grace Chisholm from Britain. Eighteen months later, in 1895, with Klein's encouragement, Chisholm, whom Klein referred to as his "favorite pupil,"[5] became the first woman to receive a German Ph.D. based on regular examinations.

Klein's mathematical work was not restricted to geometry. He worked in function theory as well and was also interested in the solutions of fifth degree equations. He used geometric arguments and the dihedral group to determine the complete theory of this equation. He also found all finite groups of fractional linear transformations by showing that such groups were precisely the groups of Euclidean rotations, and then noting that, according to Jordan, these groups were the cyclic groups and the dihedral, tetrahedral, octahedral, and icosahedral groups.

References

1. Quoted in Kramer, Edna E. *The Nature and Growth of Modern Mathematics.* Princeton: Princeton University Press, 1981, p. 431.
2. *Ibid.,* p. 402.
3. *Ibid.,* p. 432.
4. *Ibid.,* p. 432.
5. *Ibid.,* p. 402.

34

Cauchy's Theorem

The group axioms assert that every group contains an identity and that each element of a group has an inverse. Are there any other elements that have special properties and that a group must contain? We will show in this chapter that a finite group whose order is divisible by a prime p must always contain an element of order p. This assertion is Cauchy's theorem, and we will need it to determine all noncommutative groups of orders less than 16.

As indicated above, Cauchy's theorem says that if a prime divides the order of a group, then the group has an element whose order is that prime. Proposition 22.9 is a special case of this result (for the prime 5) and the proof given there can be generalized to a proof of the theorem for any prime (Exercise 43 in Chapter 22). However, the result is important enough for there to be several other proofs, each of which has its own value. We give one of these proofs in this chapter. It uses a counting device called the class equation and derives the general result from the special case for Abelian groups.

We will conclude this chapter by using Cauchy's theorem to show that for any odd prime p, a group of order $2p$ must be cyclic or dihedral (Proposition 34.15), and by using the class equation to show that for any prime p, all groups of order p^2 are Abelian (Proposition 34.17).

Cauchy's Theorem for Abelian Groups

The proof of Cauchy's theorem for Abelian groups relies on an induction argument that utilizes some of the methods of the following example.

Example 34.1 Consider the group $\mathbb{Z}_{10}$. We note that $\mathbb{Z}_{10}$ has order 10 and that the prime 2 divides 10. We want to find a process that we can apply to any nonzero element $[k]_{10}$ in $\mathbb{Z}_{10}$ and that will yield an element in $\mathbb{Z}_{10}$ of order 2. We break the argument into two parts.

(1)   If $[k]_{10}$ has order n divisible by 2, then $\frac{n}{2}[k]_{10}$ has order 2. For certainly $\frac{n}{2}[k]_{10} \neq [0]_{10}$ and $2\left(\frac{n}{2}[k]_{10}\right) = n[k]_{10} = [0]_{10}$.

(2)   If the order n of $[k]_{10}$ is not divisible by 2, then $n = 5$ by Corollary 21.16. In this case, let $N = S([k]_{10})$. By Proposition 21.14, the order of N must then also be 5 and hence by Lagrange's theorem (Proposition 21.10), the order of $\mathbb{Z}_{10}/N$ must be 2. Hence if $[j]_{10} \notin N$, $[j]_{10} + N$ has order 2 in $\mathbb{Z}_{10}/N$ and thus $2[j]_{10} \in N$. But by Corollary 21.16, the order of $2[j]_{10}$ divides the order of N and N has order 5. So $2(5[j]_{10}) = 5(2[j]_{10}) = [0]_{10}$, and if $5[j]_{10} = [0]_{10}$, then by Corollary 21.6, $[j]_{10} + N$ has order 5, a contradiction. So $5[j]_{10} \neq [0]_{10}$, and thus Corollary 21.6, $5[j]_{10}$ has order 2. ❖

The first part of the argument given in Example 34.1 generalizes readily to an arbitrary finite Abelian group. We isolate this in the following result. (Since A is Abelian, we will follow the usual convention and write its operation as addition rather than multiplication.)

Lemma 34.2   *Let p be a prime, let A be an Abelian group, and let $x \in A$ have finite order n. If p divides n, then $\frac{n}{p}x$ is an element of A of order p.*

Proof   Since p divides n, $\frac{n}{p}$ is an integer and $\frac{n}{p}x \in A$. As well, since p divides n, $n > 1$ and hence $0 \neq x$. And, since $\frac{n}{p} < n$, $\frac{n}{p}x \neq 0$. But $p\left(\frac{n}{p}x\right) = nx = 0$ and hence by Corollary 21.6, $\frac{n}{p}x$ has order p. ∎

The second part of the argument given in Example 34.1 depends on the order of $\mathbb{Z}_{10}/S([k]_{10})$. In general, while we would not know exactly what this order is, we would know that it is less than the order of the original group, and hence an induction argument would allow us to conclude that the quotient group contains an element of order the given prime. The details of this induction argument are as follows. (As in the preceding lemma, we will use addition as the generic operation.)

Theorem 34.3   **Cauchy's Theorem for Abelian Groups**

If p is a prime and A is a finite Abelian group whose order is divisible by p, then A has an element of order p.

Proof   The proof is by induction on m, the order of A. Note that since p is prime and divides m, $m \geq 2$.

(i)   If $m = 2$, then $p = 2$ and A has exactly one element $a \neq 0$. But then a must have order $2 = p$.

(ii)   We suppose that m, the order of A, is divisible by the prime p. Our induction hypothesis is that if H is a group whose order is both less than m and divisible by p, then H has an element of order p. To see that A has an element of order p, begin by observing that since p divides the order of A, there exists

$0 \neq a \in A$, and that by Corollary 21.16, a has finite order, say n. If p divides n, then A has an element of order p by Lemma 34.2. Suppose therefore that p does not divide n and observe that by Lagrange's theorem (Proposition 21.10), the order of $A/S(a)$ must be $\frac{m}{n}$. But p, being a prime that divides m but not n, must divide $\frac{m}{n}$, i.e., p must divide the order of $A/S(a)$. Furthermore, since $a \neq 0$, $n > 1$, and hence the order of $A/S(a)$ is less than m. So by the induction hypothesis, $A/S(a)$ has an element $y + S(a)$ of order p. Let r denote the order of y. We have $py + S(a) = p(y + S(a)) = 0 + S(a)$ and hence $py \in S(a)$. But by Proposition 21.14, $S(a)$ has order n and hence by Lagrange's theorem, the order of py divides n. Then $(np)y = n(py) = 0$ and thus by Proposition 21.5, r divides np. Suppose that p does not divide r. Then p does not appear in the prime decomposition of r and thus since r divides np, r must divide n. Then by Proposition 21.5, $ny = 0 \in S(a)$ and hence $n(y + S(a)) = 0 + S(a)$. But then, since the order of $y + S(a)$ is p, Proposition 21.5 implies that p must divide n, a contradiction. We conclude that p does divide r and hence that A has an element of order p by Lemma 34.2. ∎

The Class Equation and Cauchy's Theorem for Arbitrary Groups

We want to prove Cauchy's theorem for noncommutative groups as well as for commutative groups. So suppose we have a finite group G and a prime p that divides the order of G. We will argue by induction on the order of G. The smallest possible order for G is p and in this case G is Abelian by Corollary 21.17. And if G is Abelian, then G has an element of order p by Theorem 34.3. For the second step of the induction, we assume that any group having a smaller order divisible by p must have an element of order p. And we can certainly assume that G is not Abelian, since otherwise we may apply Theorem 34.3. We want a proper subgroup of G whose order is divisible by p. How can we find such a subgroup? Since G is not commutative, one place to look is among the following subgroups.

Proposition 34.4   *For any group G,*

$$Z(G) = \{x \in G \,|\, gx = xg \text{ for all } g \in G\}$$

is an Abelian group which is a normal subgroup of G, and for any $x \in G$,

$$C(x) = \{g \in G \,|\, gx = xg\}$$

is a subgroup of G.

Proof   It is not difficult to check that $Z(G)$ and all the $C(x)$ satisfy the conditions of Proposition 18.2 and hence are subgroups of G (see Exercises 41 and 42). It is equally easy to check that $Z(G)$ also satisfies Proposition 23.1 (v) and hence is normal in G (see Exercise 41). If $x, y \in Z(G)$, then $y \in G$ and hence $xy = yx$ by definition of $Z(G)$. So $Z(G)$ is Abelian. ■

Note that $Z(G)$ is the set of elements in G that commute with everything in G and that $C(x)$ is the set of elements in G that commute with x. It follows immediately from these observations that if G is Abelian, then $Z(G) = G$ and that for any group G, $C(x) = G$ if and only if $x \in Z(G)$.

Example 34.5 Consider the quaternion group $Q = \{\pm 1, \pm i, \pm j, \pm k\}$. Since ± 1 certainly commute with all elements in Q, $Z(Q) \supseteq \{\pm 1\}$, and since none of i, j, k commute with each other, $Z(Q) \subseteq \{\pm 1\}$. It follows that $Z(Q) = \{\pm 1\}$ and hence that $C(1) = Q = C(-1)$. Furthermore, ± 1 and $\pm i$ certainly commute with i, while $\pm j$ and $\pm k$ do not, so that $C(i) = \{\pm 1, \pm i\}$. Similarly, $C(-i) = \{\pm 1, \pm i\}$, $C(j) = \{\pm 1, \pm j\} = C(-j)$, and $C(k) = \{\pm 1, \pm k\} = C(-k)$. ❖

Example 34.6 Consider $(12) \in S_4$. We want to find all functions $\sigma \in C(12)$. Since $\sigma(12) = (12)\sigma$, $\sigma(12)\sigma^{-1} = (12)$, and applying $\sigma(12)\sigma^{-1}$ to $\sigma(1)$, we have $[\sigma(12)\sigma^{-1}](\sigma(1)) = [\sigma(12)](1) = \sigma(2)$. Since σ is one-to-one, $\sigma(2) \neq \sigma(1)$ and thus $\sigma(12)\sigma^{-1}$ moves $\sigma(1)$. Thus, if $\sigma(12)\sigma^{-1} = (12)$, it must be that $\sigma(1) \in \{1, 2\}$. A similar argument shows that $\sigma(2) \in \{1, 2\}$ as well. The only functions $\sigma \in S_4$ satisfying both these conditions are $\iota, (12), (34)$, and $(12)(34)$. (For a complete list of the functions in S_4, see Example 16.3.) And it is easy to check that indeed $\sigma(12)\sigma^{-1} = (12)$ for each of these functions. So $C(12) = \{\iota, (12), (34), (12)(34)\}$. A similar argument shows that $C(13) = \{\iota, (13), (24), (13)(24)\}$, and by definition, $Z(S_4)$ is contained in both $C(12)$ and $C(13)$. But certainly $Z(S_4)$ contains ι and hence $\{\iota\} \subseteq Z(S_4) \subseteq C(12) \cap C(13) = \{\iota\}$, i.e., $Z(S_4) = \{\iota\}$. ❖

Now if G is a group and if p divides the order of $Z(G)$, then $Z(G)$, and hence G, has an element of order p by Proposition 34.4 and Theorem 34.3. Furthermore, if $x \notin Z(G)$, then $C(x) \neq G$ and hence if p divides the order of any such $C(x)$, then by our induction hypothesis, $C(x)$, and hence G, contains an element of order p. Thus we need to show that p either divides the order $Z(G)$ or divides the order $C(x)$ for some $x \notin Z(G)$. Since the only assumption we are making about p is that it divides the order of G, we therefore need to relate the order of G to the orders of $Z(G)$ and the subgroups $C(x)$ for $x \notin Z(G)$.

By Lagrange's theorem (Proposition 21.10), the order of $C(x)$ is the order of G divided by the index of $C(x)$ in G, and hence we may indirectly relate the order of $C(x)$ to the order of G by relating the index of $C(x)$ in G to the order of G. And the following result gives a way of determining this index.

Proposition 34.7   *If x is an element of a finite group G, then the index of C(x) in G is exactly the number of distinct elements of the set $\{gxg^{-1} \mid g \in G\}$.*

Proof   Since the index of $C(x)$ in G is the number of left cosets of $C(x)$, we can establish the proposition by constructing a one-to-one and onto function from the set of left cosets to the set $\{gxg^{-1} \mid g \in G\}$. So define a function f from the set of left cosets of $C(x)$ to the set $\{gxg^{-1} \mid g \in G\}$ by letting $f(gC(x)) = gxg^{-1}$. Since the same coset may be generated by different elements, we first need to show that f is well defined. To this end, suppose that $gC(x) = hC(x)$. Then $h^{-1}g \in C(x)$ by Proposition 20.4, and hence $h^{-1}gx = xh^{-1}g$. It follows that $gxg^{-1} = hxh^{-1}$ and hence that $f(gC(x)) = f(hC(x))$. So f is well defined. To see that f is a one-to-one, suppose that $gxg^{-1} = hxh^{-1}$. Then $h^{-1}gx = xh^{-1}g$ and hence $h^{-1}g \in C(x)$. It follows that $gC(x) = hC(x)$ and hence that f is one-to-one. Finally let $X \in \{gxg^{-1} \mid g \in G\}$. Then $X = gxg^{-1}$ for some $g \in G$, $gC(x)$ is a left coset of $C(x)$, and $f(gC(x)) = gxg^{-1} = x$. So f is also onto. ∎

Example 34.8   Consider the quaternion group $Q = \{\pm 1, \pm i, \pm j, \pm k\}$. We saw in Example 34.5 that $C(i) = \{\pm 1, \pm i\}$, and thus by Proposition 23.8, $Q/C(i)$ has $\frac{8}{4} = 2$ elements. Since $(\pm 1)i(\pm 1)^{-1} = i = (\pm i)i(\pm i)^{-1}$ and $(\pm j)i(\pm j)^{-1} = -i = (\pm k)i(\pm k)^{-1}$, $\{qiq^{-1} \mid q \in Q\} = \{\pm i\}$, which also has two elements, in agreement with Propositon 34.7. ❖

Example 34.9   We saw in Example 34.6 that in S_4, $C((12)) = \{\iota, (12), (34), (12)(34)\}$. Since the order of $C((12))$ is 4, Lagrange's theorem (Proposition 21.10), together with Proposition 16.6, implies that the index of $C((12))$ in S_4 is 6.

We can calculate the cosets of $C((12))$ as follows. One coset is certainly

$$\iota\, C((12)) = C((12)) = \{\iota, (12), (34), (12)(34)\}.$$

Since $(13) \notin C(12)$, we can form a new coset

$$(13)\, C((12)) = \{(13), (123), (134), (1234)\}.$$

Since $(14) \notin C((12)) \cup (13)\, C((12))$, we can form another coset

$$(14)\, C((12)) = \{(14), (124), (143), (1243)\}.$$

Continuing in the same fashion, we can construct the other three cosets

$$(23)\, C((12)) = \{(23), (132), (234), (1342)\},$$
$$(24)\, C((12)) = \{(24), (142), (243), (1432)\},$$
$$(13)(24)\, C((12)) = \{(13)(24), (1423), (1324), (14)(23)\}.$$

According to Proposition 34.7, there are six permutations in $\{\sigma(12)\sigma^{-1} \mid \sigma \in S_4\}$, and according to its proof, we can find these functions by choosing one σ from each coset and calculating $\sigma(12)\sigma^{-1}$:

$$\text{for any } \sigma \in C((12)), \sigma(12)\sigma^{-1} = (12),$$
$$\text{for any } \sigma \in (13)C((12)), \sigma(12)\sigma^{-1} = (23),$$
$$\text{for any } \sigma \in (14)C((12)), \sigma(12)\sigma^{-1} = (24),$$
$$\text{for any } \sigma \in (23)C((12)), \sigma(12)\sigma^{-1} = (13),$$
$$\text{for any } \sigma \in (24)C((12)), \sigma(12)\sigma^{-1} = (14),$$
$$\text{for any } \sigma \in (13)(24)C((12)), \sigma(12)\sigma^{-1} = (34).$$

And thus we have that $\{\sigma(12)\sigma^{-1} \mid \sigma \in S_4\} = \{(12), (23), (24), (13), (14), (34)\}$, which, in agreement with Proposition 34.7, has six elements. ❖

According to Proposition 34.7, the number of elements in the set $\{gxg^{-1} \mid g \in G\}$ can be used as a surrogate for the index of $C(x)$. This will be very useful, and so for notational convenience we introduce the following abbreviation for this set.

DEFINITION

Suppose that x is an element of a group G. A **conjugate** of x is an element of the form gxg^{-1} for some $g \in G$. The set of all conjugates of x is denoted by x^G:$x^G = \{gxg^{-1} \mid g \in G\}$.

So to relate the index of $C(x)$ in G, and hence the order of $C(x)$, to the order of G, we need to relate the number of elements in x^G to the order of G. We will do this by observing that these sets, together with $Z(G)$, partition G into disjoint subsets and hence that the order of G must be the order of $Z(G)$ plus the number of elements in each of the distinct sets x^G. (The astute reader will realize that we are really just observing that these sets are equivalence classes of the equivalence relation on G determined by conjugation—see Exercises 43 and 44.)

Lemma 34.10   *Suppose that x and y are elements of a group G. Then either $x^G = y^G$ or $x^G \cap y^G = \emptyset$.*

Proof   Suppose that $x^G \cap y^G \neq \emptyset$. Then there exists $z \in x^G \cap y^G$ so that $z = axa^{-1}$ and $z = byb^{-1}$ for some $a, b \in G$. Then $x = a^{-1}za$ and hence for any $\alpha \in x^G$,

$$\alpha = gxg^{-1} = ga^{-1}zag^{-1} = ga^{-1}byb^{-1}ag^{-1} = (ga^{-1}b)y(ga^{-1}b)^{-1} \in y^G.$$

So $x^G \subseteq y^G$. A similar argument shows that $x^G \supseteq y^G$, and hence $x^G = y^G$. ∎

Recall that for any group G, $|G|$ denotes the order of G, and for any subgroup of G, $[G:S]$ denotes the index of S in G, i.e., the number of distinct left cosets of S.

Proposition 34.11   **The Class Equation**

Let G be a finite group. Then there exists a subset N of G such that the subsets
$x^G, x \in N$, *partition the complement of Z(G) into disjoint subsets, and for any such*
set N,

$$|G| = |Z(G)| + \sum_{x \in N}[G:C(x)].$$

Proof   If $x \in Z(G)$, then $gxg^{-1} = xgg^{-1} = e$ for any $g \in G$ so that $x^G = \{x\}$, and hence
by Lemma 34.10, the sets $x^G, x \notin Z(G)$, partition the complement of $Z(G)$ into a
collection of subsets that are either equal or disjoint. Pick N so that each element of
N contributes exactly one such subset to this collection. Then the number of elements in the complement of $Z(G)$ is the sum of the number of elements in each of
these disjoint subsets. But by Proposition 34.7, the number of elements in the subset
x^G is just $[G:C(x)]$, and hence the number of elements in the complement of $Z(G)$ is
just $\sum_{x \in N}[G:C(x)]$. Since the number of elements in $Z(G)$ is $|Z(G)|$, the number
of elements in all of G must be $|Z(G)| + \sum_{x \in N}[G:C(x)]$. ∎

Example 34.12   As in Examples 34.5 and 34.8, consider the quaternion group $Q =$
$\{\pm 1, \pm i, \pm j, \pm k\}$. According to Example 34.5, $Z(Q) = \{\pm 1\}$, and hence
$\{\pm i, \pm j, \pm k\}$ is the set of elements of Q that are not in $Z(Q)$. And according to Example 34.8, $i^Q = \{\pm i\}$. Similar arguments show that $j^Q = \{\pm j\}$ and $k^Q = \{\pm k\}$.
So the set $N = \{i, j, k\}$ satisfies the condition given in Proposition 34.11. According
to Example 34.5, $C(i)$, $C(j)$, and $C(k)$ each have four elements and hence by Proposition 23.8, each quotient group has two elements, i.e., $[Q:C(i)] = 2 = [Q:C(j)] =$
$[Q:C(k)]$. The class equation for Q then reads:

$$|Q| = |Z(Q)| + ([Q:C(i)] + [Q:C(j)] + [Q:C(k)]) = 2 + (2 + 2 + 2) = 8. \; ❖$$

The class equation relates the order of G to the orders of its subgroups $Z(G)$ and
$C(x)$ for $x \notin Z(G)$. As previously noted, for the general case of Cauchy's theorem,
we need such a relationship because we want to show that p either divides the order of
$Z(G)$ or divides the order of $C(x)$ for some $x \notin Z(G)$. The class equation enters the
argument as follows. If p does <u>not</u> divide the order of $C(x)$ for any $x \notin Z(G)$, then (as
we will show in the proof) p must divide every $[G:C(x)]$ and hence by the class equation, p must divide the order of $Z(G)$. The details of the proof are as follows.

Theorem 34.13   **Cauchy's Theorem**

If G is a finite group and if p is a prime dividing the order of G, then G contains an
element of order p.

Proof   The proof is by induction on the order of G.

(*i*)   If the order of G is 2, then 2 is the only prime dividing the order of G and
the element of G that is not the identity has order 2.

(*ii*)   We suppose that G is a finite group whose order is divisible by the prime p. Our induction hypothesis is that if H is a group such that $|H| < |G|$ and if p divides $|H|$, then H has an element of order p. Suppose then that p divides $|G|$. If $x \notin Z(G)$, then $|C(x)| < |G|$, and hence if p divides $|C(x)|$, then by the induction hypothesis, $C(x)$, and hence G, has an element of order p. So suppose that p does <u>not</u> divide $|C(x)|$ for any $x \notin Z(G)$, and let N be a subset of G such that the sets x^G, $x \in N$, partition the complement of $Z(G)$ into disjoint subsets. Then for any $x \in N$, since p divides $|G|$ and $|G| = |C(x)|[G:C(x)]$ by Lagrange's theorem (Proposition 21.10), p must divide $[G:C(x)]$. But by the class equation (Proposition 34.11), $|Z(G)| = |G| - \sum_{x \in N}[G:C(x)]$, and hence p must also divide $|Z(G)|$. But then by Cauchy's theorem for Abelian groups (Theorem 34.3), $Z(G)$, and hence G, contains an element of order p. ∎

We record the following corollary for use in the sequel.

Corollary 34.14   *Let p be prime and G be a finite group. If every element of G has order a power of p, then G has order a power of p.*

Proof   Suppose that the order of G is not a power of p. Then there exists a prime $q \neq p$ such that q divides the order of G. Then by Cauchy's theorem, G has an element of order q, in contradiction of our hypothesis. ∎

Groups of Orders 2*p* and *p²*

We will conclude this chapter by using its previous results to classify all groups of order $2p$ for any odd prime p (Proposition 34.15) and all groups of order p^2 for any prime p (Proposition 34.17).

For an odd prime p, we can use Cauchy's theorem to show that any group of order $2p$ is either cyclic or dihedral.

Proposition 34.15   *If p is an odd prime and G is a group of order $2p$, then G is either cyclic or dihedral.*

Proof   By Cauchy's theorem (Theorem 34.13), G contains an element x of order p and an element y of order 2. By Proposition 21.14, $S(x)$ has order p and hence by Lagrange's theorem (Proposition 21.10), $[G:S(x)] = 2$. Thus by Proposition 33.7, $S(x)$ is a normal subgroup of G, and therefore $yxy^{-1} = x^k$ for some $0 \leq k < p$. But

$$x = y^{-1}(yxy^{-1})y = y^{-1}x^k y = y^{-1}\underbrace{xx\cdots x}y$$
$$= y^{-1}xy^{-1}yxy^{-1}\cdots y^{-1}xy = (y^{-1}xy)^k = (x^k)^k = x^{k^2}$$

and hence $x^{k^2 - 1} = e$. Then by Proposition 21.5, p divides $k^2 - 1 = (k-1)(k+1)$ and hence, since p is prime, p divides either $k - 1$ or $k + 1$. Since $0 \leq k < p$, this means that either $k = 1$ or $k = p - 1$. In the first case, $yx = xy^{-1} = xy$ and hence

G is Abelian. So by Theorem 32.9, in this case, G is isomorphic to $\mathbb{Z}_{2p}$ and hence is cyclic. In the second case, $xy = y^{-1}x^{p-1} = yx^{p-1}$ so that by Corollary 33.3, G is isomorphic to D_p. ∎

We have already seen (Corollary 21.19) that if p is a prime, than all groups of order p are isomorphic to $\mathbb{Z}_p$. The same may not be true for p^2. For example, while $\mathbb{Z}_2 \times \mathbb{Z}_2$ has order 4, it has no elements of order 4 and hence by Proposition 21.9 it cannot be isomorphic to $\mathbb{Z}_4$. However, we can use the class equation to show in general that all groups of order p^2 are Abelian. For suppose that G has order p^2. Then Lagrange's theorem implies that $Z(G)$ has order 1, p, or p^2, and therefore, since G is Abelian when $G = Z(G)$, it suffices to show that $Z(G)$ cannot have order 1 or p. We can use the class equation to eliminate 1 whenever the order of G is a power of p (Proposition 34.16) and we can use a proof by contradiction to eliminate the case $|Z(G)| = p$ whenever $|G| = p^2$ (Proposition 34.17).

Proposition 34.16 *Let p be prime and G be a finite group whose order is a power of p. If G has more than one element, then so does $Z(G)$.*

Proof Since G has more than one element, $|G| = p^n$ for $n > 0$. If G is Abelian, then $Z(G) = G$ and hence $Z(G)$ has more than one element. So suppose that G is not Abelian, and let x be an element of G which is not in $Z(G)$. Since $C(x)$ is a subgroup of G, $|C(x)|[G:C(x)] = p^n$ by Lagrange's theorem (Proposition 21.10), and hence $[G:C(x)] = p^m$ for $m \geq 0$. But since $x \notin Z(G)$, $[G:C(x)] > 1$ so that $m > 0$. Then in the class equation (Proposition 34.13), the term $\sum_{x \in N}[G:C(x)]$ is divisible by p, and thus, since $|G|$ is divisible by p and $|Z(G)| = |G| - \sum_{x \in N}[G:C(x)]$, $|Z(G)|$ must also be divisible by p. We conclude that $|Z(G)| \geq p$ and hence that $Z(G)$ must contain more than one element. ∎

Proposition 34.17 *If p is prime and if G is a group of order p^2, then G is Abelian.*

Proof Suppose by way of contradiction that G is not Abelian. By Lagrange's theorem (Propositon 21.10), $|Z(G)|$ is 1, p, or p^2. Since G is not Abelian, $|Z(G)| \neq p^2$, and by Proposition 34.16, $|Z(G)| \neq 1$. So $|Z(G)| = p$, and thus by Lagrange's theorem, $|G/Z(G)| = p$ so that by Corollary 21.17, $G/Z(G)$ is cyclic. Suppose that $gZ(G)$ generates $G/Z(G)$, and let $x, y \in G$. Then $x = g^n a$ and $y = g^m b$ for $a, b \in Z(G)$, and hence $xy = g^n a g^m b = g^{n+m}ab = g^m b g^n a = yx$. This contradicts our assumption that G was not Abelian. ∎

Exercises

In Exercises 1–6, for the given element $[i]_n \in \mathbb{Z}_n$, find a multiple $k[i]_n$ of $[i]_n$ of order p.

1. $[2]_{10} \in \mathbb{Z}_{10}, p = 5$ **2.** $[3]_{10} \in \mathbb{Z}_{10}, p = 2$

3. $[3]_{10} \in \mathbb{Z}_{10}, p = 5$ **4.** $[1]_{12} \in \mathbb{Z}_{12}, p = 3$

5. $[5]_{12} \in \mathbb{Z}_{12}, p = 3$ **6.** $[5]_{12} \in \mathbb{Z}_{12}, p = 2$

In Exercises 7–13, for the element g of the given group, find a power g^n of order k; Q is the subgroup $Q = \{\pm1, \pm i, \pm j, \pm k\}$ of the multiplicative group of nonzero quaternions.

7. $g = i \in Q, k = 2$ **8.** $g = -j \in Q, k = 2$

9. $g = (1234) \in S_4, k = 2$ **10.** $g = (132465) \in S_6, k = 3$

11. $g = (132)(46) \in S_6, k = 3$ **12.** $g = ((13246), [7]_{12}) \in S_6 \times \mathbb{Z}_{12}, k = 6$

13. $g = ((12654), [5]_{12}) \in S_6 \times \mathbb{Z}_{12}, k = 15$

In Exercises 14–18, find all the elements in the given groups; Q is the subgroup $Q = \{\pm1, \pm i, \pm j, \pm k\}$ of the multiplicative group of nonzero quaternions.

14. $Z(\mathbb{Z}_{10})$ **15.** $Z(S_3)$ **16.** $Z(S_3 \times \mathbb{Z}_{10})$

17. $Z(S_3 \times Q)$ **18.** $Z(S_4 \times Q)$

In Exercises 19–28, for the element x in the group G find all the elements in $C(x)$; Q is the subgroup $Q = \{\pm1, \pm i, \pm j, \pm k\}$ of the multiplicative group of nonzero quaternions.

19. $G = \mathbb{Z}_{10}, x = [7]_{10}$ **20.** $G = S_3, x = \iota$

21. $G = S_3, x = (12)$ **22.** $G = S_3, x = (132)$

23. $G = S_3 \times \mathbb{Z}_{10}, x = ((12), [3]_{10})$ **24.** $G = S_3 \times Q, x = ((123), -1)$

25. $G = S_3 \times Q, x = ((132), -i)$ **26.** $G = S_4, x = (23)$

27. $G = S_4, x = (12)(34)$ **28.** $G = S_4, x = (1234)$

In Exercises 29–34, find all the conjugates of the element x in the group G.

29. $G = \mathbb{Z}_{10}, x = [3]_{10}$ **30.** $G = S_3, x = \iota$

31. $G = S_3, x = (12)$.   **32.** $G = S_3, x = (132)$.

33. $G = S_4, x = (12)(34)$.   **34.** $G = S_4, x = (1234)$.

In Exercises 35–40, find a subset N of the group G such that the subsets $x^G, x \in N$, partition the complement of $Z(G)$ into disjoint subsets; Q is the subgroup $Q = \{\pm1, \pm i, \pm j, \pm k\}$ of the multiplicative group of nonzero quaternions.

35. $G = \mathbb{Z}_{12}$ **36.** $G = \mathbb{Z}_9$ **37.** $G = S_3$

38. $G = S_3 \times \mathbb{Z}_{10}$ **39.** $G = S_3 \times Q$ **40.** $G = S_4$

41. Show in detail that for any group G, $Z(G)$ is a normal subgroup of G.

42. Show in detail that for any group G and any element $x \in G$, $C(x)$ is a subgroup of G.

43. Let G be a group and define a relation $\approx$ on G as follows:

$$x \approx y \text{ if and only if there exists } g \in G \text{ such that } y = gxg^{-1}.$$

Show that $\approx$ is an equivalence relation on G.

44. Let G be a group. For the equivalence relation $\approx$ defined in Exercise 43, show that for all $x \in G$, the equivalence class $[x]$ generated by x is just the set x^G of all conjugates of x.

45. Show that $Q/Z(Q)$ is Abelian.

46. Show that Q contains no subgroup that is isomorphic to $Q/Z(Q)$.

47. Suppose that G is the only subgroup of itself that has order divisible by 7. Show that G is isomorphic to $\mathbb{Z}_7$.

48. Suppose that G is a finite group and p is a prime that divides the order of G. Show that there exists $x \in G$ such that p divides $|C(x)|$.

49. If p and q are distinct primes, show that any group G of order pq must contain subgroups A and B such that $|A| = p$, $|B| = q$, $A \cap B = \{e\}$, and $G = S(A, B)$.

50. Suppose that G is a finite group and p is a prime that divides the order of G. If p does not divide $|Z(G)|$, show that there exists $x \in G$ such that $x \notin Z(G)$ and p does not divide $[G:C(x)]$.

51. Show that if G is a finite group and if $e \neq x \in G$, then $1 < |C(x)|$.

52. Show that if G is a finite group and if x is an element of G that is not in $Z(G)$, then $1 < [G:C(x)] < |G|$.

53. Suppose that p and q are distinct primes and that G is a group of order pq. If $Z(G)$ has order 1, show that $C(x)$ has order p for some $x \in G$. (*Hint:* See Exercise 52.)

54. Show that $Z(S_5)$ has order 1.

55. Suppose that G is a group. Show that if $G/Z(G)$ is cyclic, then G is Abelian.

56. Suppose that p is prime and that G is a noncommutative group of order p^3. Show that $Z(G)$ has order p.

57. If p is prime, show that any group of order p^3 is solvable.

58. Show that any subset N of S_4 such that the subsets x^G, $x \in N$, partition the complement of $Z(S_4)$ into disjoint subsets must consist of four elements w, x, y, z such that $|C(w)| = 3$, $|C(x)| = 4$, $|C(y)| = 4$, and $|C(z)| = 8$.

59. Find a subset N of S_5 such that the subsets x^G, $x \in N$, partition the complement of $Z(S_5)$ into disjoint subsets. Remember to justify your answer.

35

The Sylow Theorems

Cauchy's theorem says that if a prime p divides the order of a finite group, then the group has an element of order p. Since the subgroup generated by an element has the same order as the element, it follows that if a prime divides the order of a finite group, then the group also has a subgroup of order p. Will the same result be true for <u>any</u> divisor of the order of the group? The answer to this question is in general no. In the next section, using the results of this section, we will construct a group of order 12 that has no subgroup of order 6. However, if the same question is asked about prime power divisors, the answer is yes, a result that is usually called the first Sylow theorem. The remaining Sylow theorems deal with the number of, and relationship between, subgroups of maximal prime power order. As we will see, these theorems are extremely useful for counting elements of prime power order and, from this information, determining the structure of the group from its order. We begin by proving the first Sylow theorem.

Proposition 35.1   **First Sylow Theorem**

Suppose that p is a prime, k is a nonnegative integer, and G is a finite group whose order is divisible by p^k. Then G contains a subgroup of order p^k.

Proof   The proof is by induction on $|G|$. Note that if $k = 0$, then G has a subgroup of order p^k, viz., $\{e\}$.

(i)   If $|G| = 1$, then $k = 0$ and, as we have observed, G has the desired subgroup.

(ii)   Suppose that p^k divides G for some $k \geq 0$. Our induction hypothesis is that any group H, whose order is both less than that of G and divisible by p^r for some $r \geq 0$, has a subgroup of order p^r, and we want to show that G has a subgroup of order p^k. We have observed that the desired subgroup exists when $k = 0$; so we will assume that $k > 0$. Note that either p divides $|Z(G)|$ or it does not.

If p divides $|Z(G)|$, then Cauchy's theorem for Abelian groups (Theorem 34.3) implies that $Z(G)$ has an element z of order p. Since $z \in Z(G)$, $S(z) \triangleleft G$; since $|S(z)| = p$ and p^k divides $|G|, p^{k-1}$ divides $|G/S(z)|$ by Lagrange's theorem (Proposition 21.10); and since $e \neq z, |G/S(z)| < |G|$. Thus by the induction hypothesis, $G/S(z)$ has a subgroup K of order p^{k-1}. Let $H = \{x \in G \mid xS(z) \in K\}$. Then Proposition 18.7 implies that H is a subgroup of G (by Exercise 48). But since $z \in Z(G)$, $S(z) \triangleleft H$, and hence by Lagrange's theorem, $|H| = |H/S(z)||S(z)| = p^{k-1}p = p^k$.

On the other hand, if p does not divide $|Z(G)|$, then the class equation (Theorem 34.11) implies that p does not divide $[G:C(x)]$ for some $x \notin Z(G)$. But by Lagrange's theorem $|G| = [G:C(x)]|C(x)|$ and thus, since p^k divides $|G|, p^k$ must divide $|C(x)|$. If $|C(x)| = |G|$, then $G = C(x)$ and hence $x \in Z(G)$, a contradiction of our choice of x. So $|C(x)| < |G|$ and hence by the induction hypothesis, $C(x)$—and thus G—contains a subgroup of order p^k. ∎

Proposition 35.1 says nothing about normality and indeed none of the subgroups whose existence the proposition guarantees may be normal. For instance, the subgroups of S_3 of order 2 are $\{\iota, (12)\}$, $\{\iota, (13)\}$, and $\{\iota, (23)\}$, not one of which is normal.

Sometimes the name "first Sylow theorem" is given to a special case of Proposition 35.1, viz., that if a prime power p^k divides the order of a group and p^{k+1} does not, then the group must have a subgroup of order p^k. Such a subgroup is of course a maximal subgroup of order a power of p and because of their use in the Sylow theorems, such maximal subgroups are called Sylow p-subgroups.

DEFINITION

Let p be a prime and let G be a finite group of order $p^k m$, where p does not divide m. A **p-subgroup** of G is a subgroup of order p^l, where $l \leq k$; a **Sylow p-subgroup** of G is a subgroup of order p^k.

Since the quaternion group $Q = \{\pm 1, \pm i, \pm j, \pm k\}$ has order $8 = 2^3$, all its subgroups are 2-subgroups, and Q itself is the only Sylow 2-subgroup. Analogous statements will be true for any group of prime power order and hence p-subgroups and Sylow p-subgroups of such groups are not very useful. However, the Sylow p-subgroups of a group whose order is divisible by at least two primes will be strictly contained in the group and thus might be useful for analyzing it.

Example 35.2   By Corollary 33.3, the dihedral group D_6 is generated by x and y, where x has order 6, y has order 2, and $xy = yx^5$, and it has twelve elements: $e, x, x^2, x^3, x^4, x^5, y, yx,$ yx^2, yx^3, yx^4, yx^5. It is easy to check that $\{e, x^3, y, yx^3\}$, $\{e, x^3, yx, yx^4\}$, and

$\{e, x^3, yx^2, yx^5\}$ are subgroups of order 4 and hence are Sylow 2-subgroups. It will follow from the Sylow theorems (Theorem 35.13) that D_6 can have at most three Sylow 2-subgroups and hence that these are the only Sylow 2-subgroups. Furthermore, it is easy to check that the elements of order 2 are $x^3, y, yx, yx^2, yx^3, yx^4, yx^5$ and that D_6 has no elements of order 2^k for any $k \geq 2$. It follows that the 2-subgroups of D_6 are

$$\{e\}, \{ex^3\}, \{e, y\}, \{e, yx\}, \{e, yx^2\}, \{e, yx^3\}, \{e, yx^4\}, \{e, yx^5\}$$
$$\{e, x^3, y, yx^3\}, \{e, x^3, yx, yx^4\}, \{e, x^3, yx^2, yx^5\}.$$

It is easy to check that $Z(D_6) = \{e, x^3\}$ and hence that $\{e\}$ and $\{e, x^3\}$ are both normal. However, none of the other 2-subgroups is normal. For example, $x(yx^3)x^{-1} = yx$ and hence neither $\{e, yx^3\}$ nor $\{e, x^3, y, yx^3\}$ is normal, with similar computations showing that none of the remaining subgroups is normal.

It is also easy to check that $\{e, x^2, x^4\}$ is a subgroup and hence a Sylow 3-subgroup of D_6. By Corollary 21.16, a 3-subgroup cannot contain any elements of orders 2 or 6, and hence this subgroup is the only Sylow 3-subgroup. It follows that the only 3-subgroups of D_6 are $\{e\}$ and $\{e, x^2, x^4\}$. For any k, $(yx^k)(x^2)(yx^k)^{-1} = yx^kx^2yx^k = x^{5(k+2)+k} = x^{6k+10} = x^4$ and similarly $yx^kx^4(yx^k)^{-1} = x^2$. Since certainly $x^kx^ax^{-k} = x^a$ for any a, it follows that $\{e, x^2, x^4\}$ is normal in D_6. ❖

While, as we have seen, a p-subgroup may not be a normal subgroup of its group, it is certainly a normal subgroup of itself. An important tool in the study of finite groups in general and of Sylow p-subgroups in particular is the largest subgroup containing a given subgroup as a normal subgroup. In the Abelian case, this will of course exist and be the group itself. Such a subgroup exists even in the noncommutative case and is called the normalizer.

DEFINITION

For any subgroup S of a group G, the **normalizer** of S in G is the set $N_G(S) = \{g \in G \mid gSg^{-1} = S\}$. If no ambiguity results, we write $N(S)$ for $N_G(S)$.

Proposition 35.3 *For any subgroup S of a finite group G, $N(S)$ is the largest subgroup of G that contains S as a normal subgroup.*

Proof Since $eSe^{-1} = S, e \in N(S)$, and hence $N(S) \neq \emptyset$. If $g, h \in N(S)$, then $(gh^{-1})S(gh^{-1})^{-1} = g(h^{-1}Sh)g^{-1} = g^{-1}Sg = S$. It follows that $gh \in N(S)$, and thus by Proposition 18.7, that $N(S)$ is a subgroup of G. By definition of $N(S)$, $S \triangleleft N(S)$ and $N(S)$ contains any subgroup that has S as a normal subgroup. ∎

Example 35.4 As in Example 35.2, consider $D_6 = \{e, x, x^2, x^3, x^4, x^5, y, yx, yx^2, yx^3, yx^4, yx^5\}$. We saw in the previous example that $\{e, x^3\} \triangleleft D_6$, and hence $N(\{e, x^3\}) = D_6$. On the other hand, we also saw that $\{e, y\}$ is not normal in D_6 so that $N(\{e, y\}) \neq D_6$. Now since $\{e, y\}$ is a subgroup of index 2 in $\{e, x^3, y, yx^3\}$, Proposition 33.7 implies that $\{e, y\} \lhd \{e, x^3, y, yx^3\}$, and hence $N(\{e, y\}) \supseteq \{e, x^3, y, yx^3\}$. But then $12 > |N(\{e, y\})| \geq 4$, and hence by Lagrange's theorem (Proposition 21.10), $N(\{e, y\})$ must have order 4. It follows that $N(\{e, y\}) = \{e, x^3, y, yx^3\}$. ❖

We will use the normalizer in several ways. One way will be to use the following striking relationship between a Sylow p-subgroup and its normalizer, viz., that a Sylow p-subgroup contains any p-subgroup of its normalizer.

Proposition 35.5   *Suppose that p is a prime, that G is a finite group whose order is divisible by p, and that S is a Sylow p-subgroup of G. If T is a p-subgroup of $N(S)$, then $T \subseteq S$.*

Proof   Since S is a Sylow p-subgroup of G, S has order p^k and G has order $p^k m$, where p does not divide m. Then by Lagrange's theorem (Proposition 21.10), $N(S)$ has order $p^k n$, where p does not divide n, and the quotient group $N(S)/S$ has order n. Now let $t \in T$. Since T is a p-subgroup, t has order p^a for some a. Then $(tS)^{p^a} = t^{p^a}S = eS = S$, and hence by Proposition 21.5, tS has order p^b for some b. But by Corollary 21.16, $(tS)^n = S$, and hence by Proposition 21.5, p^b divides n. So since p does not divide n, b must be 0, and therefore $tS = (tS)^{p^0} = S$. Then $t \in S$ by Proposition 20.4, and hence $T \subseteq S$. ∎

Example 35.6   Certainly $S = \{e, y\}$ is a 2-subgroup of D_{30}, and since $30 = 2 \cdot 15$, it is a Sylow 2-subgroup. Suppose that $yx^k \in N(S)$. Then, as noted in Chapter 33, $(yx^k)^2 = e$, and hence $\{e, yx^k\}$ is also a 2-subgroup of $N(S)$. So by Proposition 35.5, $\{e, yx^k\} \subseteq S$. Then, $yx^k = y$ and thus $k = 0$. Suppose next that $x^k \in N(S)$. Then, since $y \in S \subseteq N(S)$, $yx^k \in N(S)$ and hence, as we showed previously, $k = 0$ in this case as well. We conclude that $N(S) \subseteq \{e, y\}$, and thus that $N(S) = S$. ❖

Corollary 35.7   *Suppose that p is a prime and that G is a finite group whose order is divisible by p. If S is a Sylow p-subgroup of G, then S is the only Sylow p-subgroup of $N(S)$.*

Proof   Certainly S is a Sylow p-subgroup of $N(S)$, and by Proposition 35.5 any other Sylow p-subgroup, T, of $N(S)$ is contained in S, and hence, since $|T| = |S|$, $T = S$. ∎

If S is a Sylow p-subgroup of a group G, then $gSg^{-1} = S$ for $g \in N(S)$. What can be said about gSg^{-1} for $g \notin N(S)$? Although it will not be equal to S, it will, as we show next, be another Sylow p-subgroup.

Proposition 35.8   *Let p be a prime, G a finite group, and S a subgroup of G. For all $g \in G$, gSg^{-1} is also subgroup of G; if S is a Sylow p-subgroup, then gSg^{-1} is also a Sylow p-subgroup.*

Proof   Since $e \in S$, $e = geg^{-1} \in gSg^{-1}$, and thus $gSg^{-1} \neq \varnothing$. Furthermore, if $x, y \in S$, then $(gxg^{-1})(gyg^{-1}) = gxyg^{-1} \in gSg^{-1}$, and therefore by Proposition 18.7, gSg^{-1} is a subgroup of G.

Now suppose that the order of G is $p^k m$, where p does not divide m, and that S is a Sylow p-subgroup of G. It is easy to see that the function $f(x) = gxg^{-1}$ establishes a one-to-one and onto correspondence between S and gSg^{-1} (Exercise 45) and thus that S and gSg^{-1} are subgroups of the same order. But then since S has order p^k, gSg^{-1} also has order p^k and hence is a Sylow p-subgroup of G. ∎

If x is an element of a group G, then in Chapter 34 we called elements of the form gxg^{-1} for $g \in G$ "conjugates" of x and used x^G to denote the set of all conjugates of x by elements of G. We now adopt similar terminology and notation for subgroups. Note that in this terminology, Proposition 35.8 says that conjugates of Sylow p-subgroups are Sylow p-subgroups.

DEFINITION

A subgroup A of a group G is a **conjugate** of a subgroup S of G if there exists $g \in G$ such that $A = gSg^{-1}$. If H is a subgroup of G, then the set of all conjugates of S by elements of H is denoted by S^H: $S^H = \{hSh^{-1} \mid h \in H\}$.

Example 35.9   By Example 35.2, the Sylow 2-subgroups of $D_6 = \{e, x, x^2, x^3, x^4, x^5, y, yx, yx^2, yx^3, yx^4, yx^5\}$ are $S_1 = \{e, x^3, y, yx^3\}$, $S_2 = \{e, x^3, yx, yx^4\}$, and $S_3 = \{e, x^3, yx^2, yx^5\}$. Thus, according to Proposition 35.8, the only possible conjugates of S_1 are S_1 itself or S_2 or S_3. Indeed, computing all the conjugates of S_1, we have

$$S_1 = e\, S_1\, e^{-1} = x^3\, S_1\, x^{-3} = y\, S_1\, y^{-1} = yx^3\, S_1\, (yx^3)^{-1},$$
$$S_2 = x\, S_1 x^{-1} = x^4\, S_1\, x^{-4} = yx^2\, S_1\, (yx^2)^{-1} = yx^5\, S_1\, (yx^5)^{-1},$$
$$S_3 = x^2\, S_1\, x^{-2} = x^5\, S_1\, x^{-5} = yx\, S_1\, (yx)^{-1} = yx^4\, S_1\, (yx^4)^{-1}.$$

Note that if $gS_ig^{-1} = S_1 = hS_kh^{-1}$, then $S_i = (g^{-1}h)S_k(g^{-1}h)^{-1}$ and hence the calculations show that any S_i is a conjugate of any S_k.

We have shown that $\{S_1, S_2, S_3\}$ is the set of all conjugates of S_1, i.e., $S_1^{D_6} = \{S_1, S_2, S_3\}$. Of course, the list may be shorter if we restrict the elements of D_6 we use to form the conjugates. For example, $S_1^{\{e,y\}} = \{gS_1g^{-1} \mid g \in \{e, y\}\} = \{S_1\}$ and $S_1^{S_2} = \{gS_1g^{-1} \mid g \in S_2\} = \{S_1, S_3\}$, while $S_1^{\{e,x,x^2,x^3,x^4,x^5\}} = \{gS_1g^{-1} \mid g \in \{e, x, x^2, x^3, x^4, x^5\}\} = \{S_1, S_2, S_3\}$.

Now, by definition, $S_1 \subseteq N(S_1) \subseteq D_6$. Furthermore, since $xyx^{-1} = yx^5x^{-1} = yx^4 \notin S_1$, $xS_1x^{-1} \neq S_1$, and hence $x \notin N(S_1)$. So, $N(S_1) \neq D_6$, and thus $4 \leq |N(S_1)| < 12$. Then by Lagrange's theorem (Proposition 21.10), $N(S_1)$ has order 4, and hence $N(S_1) = S_1$. So in each of these cases, the number of conjugates of S_1 is the index of $N(S_1) \cap H$ in H, where H is the subgroup that is used to generate the conjugates. That is, the index of $N(S_1) \cap D_6 = S_1$ in D_6 is 3 and $S_1^{D_6}$ has three elements; the index of $N(S_1) \cap \{e, y\} = \{e, y\}$ in $[e, y]$ is 1 and $S_1^{\{e,y\}}$ has one element; the index of $N(S_1) \cap S_2 = \{e, x^3\}$ in S_2 is 2 and $S_1^{S_2}$ has two elements; and the index of $N(S_1) \cap \{e, x, x^2, x^3, x^4, x^5\} = \{e, x^3\}$ in $\{e, x, x^2, x^3, x^4, x^5\}$ is 3 and $S_1^{\{e,x,x^2,x^3,x^4,x^5\}}$ has three elements. ❖

We show next that we can always count conjugates by using the normalizer the way we did in Example 35.9.

Proposition 35.10   *Let G be a finite group and let S and H be subgroups of G. Then the number of distinct elements in S^H is the index of $N_G(S) \cap H$ in H.*

Proof   Let $X = N_G(S) \cap H$. We need to show that the number of distinct elements in S^H is the same as the number of cosets in H/X. To this end, define $f: S^H \to H/X$ by letting $f(hSh^{-1}) = hX$. We need to show that f is a function that is one-to-one and onto. Note that since it is possible for hSh^{-1} to equal kSk^{-1} for $h \neq k$ in H, we must first show that f is well defined. So suppose that $hSh^{-1} = kSk^{-1}$ for $h, k \in H$. Then $k^{-1}hS(k^{-1}h)^{-1} = S$, hence $k^{-1}h \in X$, and thus $kX = hX$ by Proposition 20.4. So f is well defined. But if $f(hSh^{-1}) = f(kSk^{-1})$, then $kX = hX$, so that $k^{-1}h \in X$ and hence $k^{-1}hS(k^{-1}h)^{-1} = S$. Then $hSh^{-1} = kSk^{-1}$. So f is one-to-one. To see that f is onto, consider $\chi \in H/X$. We have that $\chi = hX$ for some $h \in H$ and $hSh^{-1} \in S^H$. Then $\chi = f(hSh^{-1})$ and hence f is onto. We conclude that the number of elements in S^H is the index of $X = N_G(S) \cap H$ in H. ∎

Now suppose that p is a prime, G is a finite group, and S is a Sylow p-subgroup of G; then $S^G = \{gSg^{-1} \mid g \in G\}$ is a complete set of conjugates of S. Since G is finite, S^G is finite, say $S^G = \{S_1, \ldots, S_m\}$, and by Proposition 35.8, each S_i is a Sylow p-subgroup of G. The remaining Sylow theorems assert that S^G contains all Sylow p-subgroups of G, that p divides $m - 1$, and that any subgroup whose order is a power of p is contained in one of the S_i's. The proof relies on the following lemma.

Lemma 35.11   *Let p be a prime, G a group, T a p-subgroup of G, S a Sylow p-subgroup of G, and $S^G = \{S_1, \ldots, S_n\}$ the complete set of conjugates of S in G. Define a binary relation $\approx_T$ on S^G by letting $S_i \approx_T S_k$ if and only if $S_i = zS_kz^{-1}$ for some $z \in T$. Then $\approx_T$ is an equivalence relation and the number of elements in each equivalence class of $\approx_T$ is a power of p. Furthermore, an equivalence class $[S_i]_T$ determined by this relation has one element if and only if $T \subseteq S_i$.*

Proof   Since $e \in T$ and $eS_ie^{-1} = S_i$, $\approx_T$ is reflexive. If $S_i \approx_T S_k$, then $S_i = zS_kz^{-1}$ for some $z \in T$. So $S_k = z^{-1}S_i(z^{-1})^{-1}$ and hence $S_k \approx_T S_i$ so that $\approx_T$ is symmetric. Finally, if $S_i \approx_T S_k$ and $S_k \approx_T S_j$, then $S_i = zS_kz^{-1}$ and $S_k = wS_jw^{-1}$ for some $z, w \in T$. Then $S_i = (zw)S_j(zw)^{-1}$ and hence $S_i \approx_T S_j$ so that $\approx_T$ is transitive. So $\approx_T$ is an equivalence relation. But for any S_i,

$$[S_i]_T = \{S_k \mid S_k \approx_T S_i\} = \{S_k \mid S_k = zS_iz^{-1} \text{ for some } z \in T\} = \{zS_iz^{-1} \mid z \in T\},$$

and hence by Proposition 35.10, the number of elements in $[S_i]_T$ is the index of $N(S_i) \cap T$ in T. By Lagrange's theorem, this index, and hence the number of elements in the equivalence class, divides the order of T. Since T is a p-subgroup, its order must be a power of p. So the number of elements in $[S_i]_T$ must also be a power of p. If $T \subseteq S_i$, then $zS_iz^{-1} = S_i$ for all $z \in T$ and hence S_i is equivalent only to itself, i.e., $[S_i]_T = \{S_i\}$. Conversely, if $[S_i]_T$ has only one element, then $[S_i]_T = \{S_i\}$ and hence $zS_iz^{-1} = S_i$ for all $z \in T$. Then $T \subseteq N(S_i)$ and since T is a p-subgroup, $T \subseteq S_i$ by Proposition 35.5. ∎

Example 35.12   Let $G = D_6$. As in Example 35.9, let $S_1 = \{e, x^3, y, yx^3\}$, $S_2 = \{e, x^3, yx, yx^4\}$, and $S_3 = \{e, x^3, yx^2, yx^5\}$ denote the Sylow 2-subgroups of G. Then, as we saw in that example, $S_1^G = \{S_1, S_2, S_3\}$.

According to Example 35.2, $T = \{e, yx^3\}$ is a 2-subgroup of G. Consider the relation $\approx_T$ of Lemma 35.11 on S_1^G: $S_i \approx_T S_k$ if an only if $S_i = zS_kz^{-1}$ for some $z \in T$. Since $eS_1e^{-1} = S_1 = (yx^3)S_1(yx^3)^{-1}$, the equivalence class $[S_1]_T$ generated by S_1 is just $[S_1]_T = \{S_1\}$. And since $S_3 = (yx^3)S_2(yx^3)^{-1}$, the equivalence class $[S_2]_T$ generated by S_2 is $[S_2]_T = \{S_2, S_3\}$. Note that, in agreement with Lemma 35.11, $[S_1]_T$ has $1 = 2^0$ element and $[S_2]_T$ has $2 = 2^1$ elements. Furthermore, $[S_1]_T$ has one element and S_1 contains T and $[S_2]_T$ has more than one element and S_2 does not contain T.

Similarly, $H = \{e, x^3\}$ is a 2-subgroup of G, and hence we may also use it to define the relation $\approx_H$ on S_1^G. But in this case, $S_i = (x^3)S_i(x^3)^{-1}$ for any i and hence $[S_i]_H = \{S_i\}$ for any i. This also agrees with Lemma 35.11 because $1 = 2^0$ and $H \subseteq S_i$ for any i. ❖

Theorem 35.13   **Sylow Theorems**

Let p be a prime, let G be a finite group of order $p^k m$ where p does not divide m, and let σ_p be the number of Sylow p-subgroups of G. Then

(i)   any two Sylow p-subgroups of G are conjugate in G:
(ii) $\sigma_p = rp + 1$ for some nonnegative integer r;
(iii) σ_p divides m;
(iv)   any p-subgroup is contained in a Sylow p-subgroup.

Proof.   Let S be a Sylow p-subgroup of G and let $S^G = \{gSg^{-1} \mid g \in G\} = \{S_1, \ldots, S_n\}$ be the complete set of conjugates of S in G. Using $T = S$, define the equivalence relation $\approx_S$ of Lemma 35.11 on S^G. By Proposition 22.7, S^G is the disjoint union of its equivalence classes and hence the number of elements it contains is the sum of the number of elements of each of its distinct classes. But $S \in S^G$, and by Lemma 35.11, $[S]_S$ contains the single element S. As well, Lemma 35.11 implies that the number of elements in each other equivalence class is divisible by p, and therefore

that the number of equivalence classes is of the form $rp + 1$ for some nonnegative integer r. Suppose that some Sylow p-subgroup W is not in S^G. Using $T = W$, define the equivalence relation $\approx_W$ of Lemma 35.11 on S^G. As shown earlier, the number of elements in S^G must be the sum of the number of elements in each of the equivalence classes of this relation. But since $W \notin S^G$, Lemma 35.11 implies that the number of elements in each of the equivalence classes of this relation must be divisible by p, and hence that the number of elements of S^G must be divisible by p. This is a contradiction and thus $W \in S^G$, i.e., S^G must contain all the Sylow p-subgroups of G. We have thus proved (*i*) and (*ii*). For (*iii*), note that by Proposition 35.10 with $H = G$, n, the number of elements in S^G, is $[G:N(S)]$, and that by Lagrange's theorem $|N(S)||[G:N(S)]| = |G| = |S||[G:S]$. Since S is a subgroup of $N(S)$, $|S|$ divides $|N(S)|$ and hence $[G:N(S)]$ divides $[G:S]$. But by definition, $|S| = p^k$ and hence $[G:S] = m$. We have thus shown that n divides m. However, by (*i*), $n = \sigma_p$, and hence (*iii*) must hold. For (*iv*), suppose that V is a p-subgroup of G and using $T = V$, define the equivalence relation $\approx_V$ of Lemma 35.11 on S^G. By the lemma, the number of elements in each equivalence class $[S_i]_V$ is a power of p, and since by (*ii*) the total number of elements in S^G is of the form $rp + 1$, it follows that one of the equivalence classes $[S_i]_V$ must contain only one element. By Lemma 35.11, $V \subseteq S_i$. ∎

Corollary 35.14   *Let p be a prime, let G be a finite group, and let P be a Sylow p-subgroup of G. Then P is the only Sylow p-subgroup of G if and only if $P \lhd G$.*

Proof   The corollary follows from part (*i*) of Theorem 35.13. ∎

We have observed that the group D_6 has $12 = 2^2 \cdot 3$ elements (Proposition 33.2 and Example 35.2). Thus, by Theorem 35.13 (*iii*), σ_2 divides 3. But by Theorem 35.13 (*ii*), $\sigma_2 = 2r + 1$ for some r, and hence σ_2 must be 1 or 3. In Example 35.2, we found three Sylow 2-subgroups of D_6, and thus Proposition 35.13 implies that there are no more Sylow 2-subgroups. We will exploit the Sylow theorems in similar ways in the next section to determine all groups of orders 12 and 15.

Exercises

In Exercises 1–6, show that the group G contains a subgroup of order n.

1. $G = D_{10}$, $n = 4$ **2.** $G = S_5$, $n = 8$ **3.** $G = D_{20}$, $n = 8$

4. $G = D_{18}$, $n = 4$ **5.** $G = S_6$, $n = 9$ **6.** $H = S_8$, $n = 64$

Exercises 7–17 all concern the group S_3.

7. Find all the 2-subgroups of S_3.

8. Find all the 3-subgroups of S_3.

9. Find all the Sylow 2-subgroups of S_3 and determine which ones are normal.

10. Find all the Sylow 3-subgroups of S_3 and determine which ones are normal.

11. Show that $S = \{\iota, (12)\}$ is a subgroup of S_3 and find the normalizer of S in S_3.

12. Show that $S = \{\iota, (123), (132)\}$ is a subgroup of S_3 and find the normalizer of S in S_3.

13. Show that $S = \{\iota, (12)\}$ is a subgroup of S_3 and find all the conjugates of S in S_3.

14. Show that $S = \{\iota, (123), (132)\}$ is a subgroup of S_3 and find all the conjugates of S in S_3.

15. Show that $S = \{\iota, (123), (132)\}$ and $H = \{\iota, (12)\}$ are subgroups of S_3. Find all the conjugates in S^H. Verify that the number of conjugates in S^H is the index of $N(S) \cap H$ in H.

16. Show that $S = \{\iota, (12)\}$ and $H = \{\iota, (123), (132)\}$ are subgroups of S_3. Find all the conjugates in S^H. Verify that the number of conjugates in S^H is the index of $N(S) \cap H$ in H.

17. Show that $T = \{\iota, (23)\}$ is a 2-subgroup of S_3 and that $S = \{\iota, (12)\}$ is a Sylow 2-subgroup of S_3. Then determine S^{S_3} and verify Lemma 35.11 by calculating the equivalence classes of S^{S_3} for the equivalence relation $\approx_T$ determined by T.

Exercises 18–23 all concern the group $Q = \{\pm 1, \pm i, \pm j, \pm k\}$.

18. Find all the 2-subgroups of Q.

19. Find all the Sylow 2-subgroups of Q and determine which ones are normal.

20. Show that $S = \{\pm 1, \pm i\}$ is a subgroup of Q and find the normalizer of S in Q.

21. Show that $S = \{\pm 1, \pm k\}$ is a subgroup of Q and find all the conjugates of S in Q.

22. Show that $S = \{\pm 1, \pm i\}$ and $H = \{\pm 1, \pm j\}$, are subgroups of Q. Find all the conjugates in S^H. Verify that the number of conjugates in S^H is the index of $N(S) \cap H$ in H.

23. Show that $T = \{\pm 1, \pm i\}$ is a 2-subgroup of Q and that Q is a Sylow 2-subgroup of itself. Then determine Q^Q and verify Lemma 35.11 by calculating the equivalence classes of Q^Q for the equivalence relation $\approx_T$ determined by T.

Exercises 24–33 all concern the group D_5.

24. Find all the 2-subgroups of D_5.

25. Find all the 5-subgroups of D_5.

26. Find all the Sylow 2-subgroups of D_5 and determine which ones are normal.

27. Find all the Sylow 5-subgroups of D_5 and determine which ones are normal.

28. Show that $S = \{e, yx^2\}$ is a subgroup of D_5 and find the normalizer of S in D_5.

29. Show that $S = \{e, x, x^2, x^3, x^4\}$ is a subgroup of D_5 and find the normalizer of S in D_5.

30. Show that $S = \{e, yx\}$ is a subgroup of D_5 and find all the conjugates of S in D_5.

31. Show that $S = \{e, x, x^2, x^3, x^4\}$ is a subgroup of D_5 and find all the conjugates of S in D_5.

32. Show that $S = \{e, yx\}$ and $H = \{e, yx^4\}$ are subgroups of D_5. Find all the conjugates in S^H. Verify that the number of conjugates in S^H is the index of $N(S) \cap H$ in H.

33. Show that $T = \{e, yx^4\}$ is a 2-subgroup of D_5 and that $S = \{e, yx\}$ is a Sylow 2-subgroup of D_5. Then determine S^{D_5} and verify Lemma 35.11 by calculating the equivalence classes of S^{D_5} for the equivalence relation $\approx_T$ determined by T.

Exercises 34–43 all concern the group S_4.

34. Find all the 2-subgroups of S_4.

35. Find all the 3-subgroups of S_4.

36. Find all the Sylow 2-subgroups of S_4 and determine which ones are normal.

37. Find all the Sylow 3-subgroups of S_4 and determine which ones are normal.

38. Show that $S = \{\iota, (12), (34), (12)(34)\}$ is a subgroup of S_4 and find the normalizer of S in S_4.

39. Show that $S = \{\iota, (134), (143)\}$ is a subgroup of S_4 and find the normalizer of S in S_4.

40. Show that $S = \{\iota, (24)\}$ is a subgroup of S_4 and find all the conjugates of S in S_4.

41. Show that $S = \{\iota, (1234), (13)(24), (1432)\}$ is a subgroup of S_4 and find all the conjugates of S in S_4.

42. Show that $S = \{\iota, (12), (34), (12)(34)\}$ and $H = \{\iota, (1234), (13)(24), (1432)\}$ are subgroups of S_4. Find all the conjugates in S^H. Verify that the number of conjugates in S^H is the index of $N(S) \cap H$ in H.

43. Show that $S = \{\iota, (1234), (13)(24), (1432)\}$ and $H = \{\iota, (123), (132)\}$ are subgroups of S_4. Find all the conjugates in S^H. Verify that the number of conjugates in S^H is the index of $N(S) \cap H$ in H.

44. Let G be a group, let S be a subgroup of G, and let $g \in G$. Define $f: S \to gSg^{-1}$ by letting $f(s) = gsg^{-1}$. Show that f is one-to-one and onto.

45. Show that $S = \{(e, [0]_3), (x^3, [0]_3), (yx, [0]_3), (yx^4, [0]_3)\}$ is a Sylow 2-subgroup of $D_6 \times \mathbb{Z}_3$. Find $N(S)$.

46. Suppose that G and H are groups, that $f: G \to H$ is a homomorphism, and that S is a subgroup of H. Show that $M = \{x \in G \mid f(x) \in S\}$ is a subgroup of G.

47. Suppose that G and H are groups, that $f: G \to H$ is a homomorphism, and that S is a normal subgroup of H. Show that $M = \{x \in G \mid f(x) \in S\}$ is a normal subgroup of G.

48. Suppose that N is a normal subgroup of a group G and that S is a subgroup of G/N. Show that $\{g \in G \mid gN \in S\}$ is a subgroup of G.

49. Suppose that N is a normal subgroup of a group G and that S is a normal subgroup of G/N. Show that $\{g \in G \mid gN \in S\}$ is a normal subgroup of G.

50. Suppose that p is a prime and that G is a finite group of order $p^k m$ for some positive integer m. Show that G has a subgroup H such that $[G:H] = m$.

51. Suppose that G and H are groups and that $f: G \to H$ is a homomorphism. Show that $N(\ker f) = G$.

52. Prove or give a counterexample. If G is a group such that $N(S) = G$ for all subgroups S, then G is Abelian.

53. Show that a group G of order 285 contains a normal subgroup N such that $\{e\} \neq N \neq G$.

54. Suppose that p is a prime, that $0 < m < p$, and that G is a group of order pm. Show that if S is a subgroup of G of order p, then $S \triangleleft G$.

55. Suppose that p is a prime and S is a Sylow p-subgroup of a group G. Show that if $gSg^{-1} \neq S$ for $g \in G$, then $N(S)$ does not contain gSg^{-1}.

56. Suppose that G is a group with subgroups A and B such that $A \subseteq B$. Show that $N(A) \cap B$ is the largest subgroup of B in which A is normal.

57. Suppose that p is a prime and that S is a Sylow p-subgroup of a finite group G. Show that if $gSg^{-1} \subseteq S$, then $g \in N(S)$.

58. Suppose that p is a prime and that S is a Sylow p-subgroup of a group G. Show that for any $g \in G$, $gN(S)g^{-1} = N(gSg^{-1})$.

59. If A and B are subgroups of a group G, show that $N(A) \cap N(B) \subseteq N(A \cap B)$.

60. Find a group G with subgroups A and B such that $N(A) \cap N(B) \neq N(A \cap B)$.

61. Suppose that p is a prime, that G is a finite group, that H is a normal subgroup of G, and that S is a Sylow p-subgroup of H. Show that $G = HN(S)$.

62. Suppose that p is a prime and show that a finite group, all of whose Sylow p-subgroups are normal, is the internal direct product of its Sylow p-subgroups.

63. Find a prime p and a group G with Sylow p-subgroups X, Y, and Z such that $X \cap Z = \{e\}$ but $X \cap Y \neq \{e\}$.

64. Suppose that p and q are primes, that $p > q$, and that G is a group of order pq. If g is an element of G of order p, show that $S(g) \triangleleft G$.

65. Suppose that G is a finite group and that H is a subgroup of G. If $H \neq G$, show that G has an element which is not contained in any conjugate of H.

66. Suppose that G is a finite group in which all maximal proper subgroups are conjugate. Show that G is cyclic.

Peter Ludvig Sylow

Peter Ludvig Mejdell Sylow was born in Kristiana (now Oslo), Norway, on December 12, 1832, and died in Kristiana on September 7, 1918. His father was a captain in the cavalry and later became a government minister. Upon graduation from the Kristiana Cathedral School in 1850, Peter entered Kristiana University where he won a mathematics prize in 1853. He graduated in 1855 and, unable to obtain a position at a university, took the examination to become a school teacher. Eventually, in 1858, he found a position as a high school teacher in the town of Frederikshald (now Halden).

In 1860, he traveled to Copenhagen to present his reconstruction of an unfinished paper of Abel; he argued that Abel knew much more than had previously been suspected about the solutions of polynomial equations. The next year, he won a travel grant to study in Berlin and Paris, where he attended lectures on geometry, mechanics, limits, and the theory of equations, and studied number theory and the theory of equations on his own. Weierstrass, whom Sylow wanted to meet, was ill, but he was able to make the acquaintance of Carl Borchardt, the editor of *Crelle's Journal*, and met several times with Kronecker.

When Sylow returned to Norway, he taught at Kristiana University for a year (1862–1863) on a temporary appointment and then returned to the high school in Frederikshald. During his time at the university, he gave a series of lectures whose intent was to explain the approach of Abel and Galois to the theory of equations. Cauchy's theorem was included and following its proof, Sylow asked, "What if m is divisible by v^δ? Can the above be extended?"[1] So it would appear that at this time, he was already considering the questions that he was later so famously to answer.

He applied for a position at the university in 1869 but was not appointed. However, in 1873, he was granted a leave of absence from his high school teaching duties so that he could help Sophus Lie edit a new edition of Abel's collected works. He worked full time at this for four years and then had to return to Frederikshald. It took another four years for the two of them to complete the book, which was published in 1881.

Finally, in 1898, at the urging of Lie, when Sylow was 65 years old, a special chair was created for him at Kristiana University and he left Frederikshald. He continued lecturing at the university until his death twenty years later. In 1902, with Elling Holst, he published Abel's correspondence.

Sylow's fame rests on his eponymous theorems, which he proved in a ten-page paper entitled *"Théorè mes sur les groupes de substitutions,"* published in *Mathematische Annalen* in 1872. These results have remained fundamental to the study of finite groups ever since.

References

1. Quoted in Birkeland, Bent. "Ludwig Sylow's Lectures on Algebraic Equations and Substitutions, Christiana (Oslo), 1862: An Introduction and a Summary," *Historia Mathematics* 23 (1996): 185.

Groups of Order Less Than 16

In the preceding chapters, we have determined all groups of orders p, $2p$, or p^2 for any prime p and all groups of order 8. Thus, in particular, we have determined all groups of orders 2, 3, 4, 5, 6, 7, 8, 9, 10, 11, 13, and 14. The only orders less than 16 that remain to be considered are thus 12 and 15. The objective of this chapter is to use the Sylow theorems to determine all groups of these orders. In fact, in the case of 15, we will do much better; we will characterize groups of order pq, where p and q are prime, $p > q$, and q does not divide $p - 1$. (The remaining groups of order pq are characterized in Exercise 30.) Specifically, we have the following.

Proposition 36.1   *Let p and q be primes such that $p > q$ and suppose that G is a group of order pq. If q does not divide $p - 1$, then G is cyclic.*

Proof   By Cauchy's theorem (Theorem 34.13), G contains an element x of order p. Since $S(x)$ has order p (Proposition 21.14), $S(x)$ is a Sylow p-subgroup of G, and thus by the Sylow theorems (Theorem 35.13), the number of conjugates of $S(x)$ is $kp + 1$ and divides q. So since $p > q$, $k = 0$, and hence $S(x) \lhd G$. Similarly, G contains an element y of order q and the number of conjugates of $S(y)$ is $nq + 1$ and divides p. So since p is prime, $nq + 1 = 1$ or $nq + 1 = p$. The latter case contradicts our hypothesis that q does not divide $p - 1$, and thus $nq + 1 = 1$. We conclude that $n = 0$ and hence that $S(y) \lhd G$. As well, since both p and q divide $|S(x, y)|$, $pq \leq |S(x, y)| \leq |G| = pq$. Therefore, $|S(x, y)| = pq$ and hence $S(x, y) = G$. Note next that if $z \in S(x) \cap S(y)$, then the order of z, being a divisor of both p and q, must be 1 and thus $z = e$; so $S(x) \cap S(y) = \{e\}$. But since $S(x)$ and $S(y)$ are both normal, $xyx^{-1}y^{-1} = x(yx^{-1}y^{-1}) \in S(x)$ and $xyx^{-1}y^{-1} = (xyx^{-1})y^{-1} \in S(y)$ so that $xyx^{-1}y^{-1} = S(x) \cap S(y)$. So $xyx^{-1}y^{-1} = e$, and thus x and y commute. Then, since $G = S(x, y)$, G is Abelian (Exercise 8), and thus, since $S(x) \cap S(y) = \{e\}$, Proposition 30.4 implies that G is isomorphic to $S(x) \times S(y)$. But by Corollary 21.19, $S(x)$ is isomorphic to $\mathbb{Z}_p$ and $S(y)$ is isomorphic to $\mathbb{Z}_q$. So G is isomorphic to $\mathbb{Z}_{pq}$ (by Exercise 48 in Chapter 29) and hence must be cyclic by Proposition 21.9. ∎

Proposition 36.1, in conjunctions with Proposition 21.18, implies that $\mathbb{Z}_{15}$ is the only group of order 15, and thus, to complete the classification of groups of orders less than 16, it only remains to consider groups of order 12. In contrast to the case for order 15, there certainly exist noncommutative groups of order 12, e.g., D_6 and the following subgroup of S_4.

Example 36.2   Let A_4 be the subset of S_4:

$$A_4 = \{i, (123), (132), (124), (142), (134), (143),$$
$$(234), (243), (12)(34), (13)(24), (14)(23)\}.$$

It is straightforward to check that for all $\sigma\, \tau \in A_4$, $\sigma\, \tau \in A_4$, and hence that A_4 is a subgroup of S_4 by Proposition 18.7. Clearly A_4 has order 12 and since $(123)(124) \neq (124)(123)$, A_4 is not Abelian. Note that no element of A_4 has order 6. However, D_6 does have an element of order 6, and thus, by Proposition 21.9, any group that is isomorphic to D_6 must also have an element of order 6. We conclude that A_4 is not isomorphic to D_6.

Subgroups analogous to A_4 may in fact be defined in any S_n; we will consider the general construction in Chapter 37. ❖

Are there any other noncommutative groups of order 12? The following result says that we will not find any more in S_4.

Proposition 36.3   *The only subgroup of S_4 of order 12 is A_4.*

Proof   Suppose that H is a subgroup of S_4 of order 12. Note that by Lagrange's theorem, H has index 2 and hence by Proposition 33.7, $H \lhd S_4$. Since H has order 12, Cauchy's theorem (Theorem 34.13) implies that H has an element of order 3, and according to Example 16.3, the only elements in S_4 of order 3 are the 3-cycles. So some 3-cycle (abc) must be in H. But since $H \lhd S_4$, $(bcd) = (ad)(abc)(ad)^{-1} \in H$ as well. So by Proposition 21.14, H contains at least two different subgroups of order 3, viz., $S((abc)) = \{\iota, (abc), (acb)\}$ and $S((bcd)) = \{\iota, (bcd), (bdc)\}$. Then by the Sylow theorems (Theorem 35.13 (*ii*) and (*iii*)), H must contain exactly four subgroups of order 3. Since distinct subgroups of order 3 can have only the identity in common, this implies that H must have exactly eight elements of order 3. But A_4 contains all the 3-cycles in S_4, and hence $H \cap A_4$ is a subgroup of A_4 which has at least eight elements. By Lagrange's theorem (Proposition 21.10), the only such subgroup is A_4 itself, and hence $A_4 \subseteq H$. So since A_4 and H both have twelve elements, $A_4 = H$. ∎

Are there any more groups of order 12? There is at least one more; it is described in the following example.

Example 36.4   Let $C_3 = \{e, a, a^2\}$ be a cyclic group of order 3 and let $C_4 = \{e, b, b^2, b^3\}$ be a cyclic group of order 4. Let $T = C_3 \times C_4 = \{(x, y) \,|\, x \in C_3, y \in C_4\}$ and let $*$ be the following binary operation on T:

$$(x, y) * (z, w) = \begin{cases} (xz, yw) & \text{if } y = e \text{ or } y = b^2 \\ (xz^2, yw) & \text{if } y = b \text{ or } y = b^3. \end{cases}$$

By definition, T is closed with respect to $*$, and it is straightforward (but tedious) to show that $*$ is associative. Direct calculations show that the element (e, e) is an identity for $*$, and for $(x, y) \in T$,

$$(r, s) = \begin{cases} (x^2, y^3) & \text{if } y = e \text{ or } y = b^2 \\ (x, y^3) & \text{if } y = b \text{ or } y = b^3 \end{cases}$$

is an inverse of (x, y) with respect to $*$. It follows that $(T, *)$ is a group which clearly has order 12. It is easy to check that (a, b^2) has order 6 and hence that, as in Example 36.2, A_4 cannot be isomorphic to T. Similarly, $(e, b) \in T$ has order 4, while direct calculation (Exercise 5) shows that D_6 has no elements of order 4. So by Proposition 21.9, T cannot be isomorphic to D_6.

In fact, the elements (a, b^2) and (e, b) generate T. For since (a, b^2) has order 6 and (e, b) has order 4, both 6 and 4 divide $|S((a, b^2), (e, b))|$ and hence the order of this group must be 12. It follows that $T = S((a, b^2), (e, b))$. These elements are related to each other as follows:

$$(a, b^2)^3 = (e, b)^2 = ((a, b^2) * (e, b))^2,$$

and, as we show next, any group of order 12 with generators satisfying these relations must be isomorphic to T.

The group T is an example of a semidirect product. We will consider the general construction in Chapter 38. ❖

Proposition 36.5   *Suppose that G is a group with generators x of order 6 and y of order 4 such that $x^3 = y^2 = (xy)^2$. Then G is isomorphic to T.*

Proof   Since $x^6 = e$, $y^4 = e$, and $xy = (xy)^2 y^{-1} x^{-1} = y^2 y^3 x^5 = yx^5$, it is possible to write any product of x and y in the form $y^i x^k$ for $0 \le i \le 3$ and $0 \le k \le 5$. Since $y^2 = x^3$, i can be further restricted to $0 \le i \le 1$, and hence $G = \{e, x, x^2, x^3, x^4, x^5, y, yx, yx^2, yx^3, yx^4, yx^5\}$. We will show that all these elements are distinct. Since x has order 6, the x^i are all distinct and the yx^i are all distinct. Suppose that $yx^i = x^k$ for $0 \le i < k \le 5$ or $0 \le k < i \le 5$. Then $y = x^{k-i}$ for $0 < k - i \le 5$ or $y = x^{i-k}$ for $0 < i - k \le 5$ so that y commutes with x, and hence $x^3 = (xy)^2 = x^2 y^2 = x^5$, which is a contradiction. It follows that the twelve elements are all distinct. A proof similar to that of Corollary 33.3 then shows that G is isomorphic to T. ∎

We have now found three distinct, noncommutative groups of order 12, viz., D_6, A_4, and T. Are there any more? The answer is no but the proof is more complicated than those of similar assertions given above. The first step is the following result, which is interesting in its own right.

Proposition 36.6 *If B is a subgroup of index n in a group G, then there is a homomorphism $F: G \to S_n$ whose kernel is contained in B.*

Proof For $g \in G$, define $f_g: G/B \to G/B$ by letting $f_g(xB) = gxB$. Note that if $xB = yB$, then $gxB = gyB$ and hence f_g is well defined. Furthermore, if $gxB = gyB$, then $xB = yB$ and hence f_g is one-to-one. And if $\chi \in G/B$, then $\chi = xB$ for some $x \in G$ and $g^{-1}xB \in G/B$; so since $f_g(g^{-1}xB) = xB, f$ is onto. We conclude that f_g is a permutation of G/B and hence, since B has n cosets, that the equation $F(g) = f_g$ defines a function $F: G \to S_n$. Since $f_g \circ f_h(xB) = ghxB = f_{gh}(xB)$, $F(gh) = f_{gh} = fg \circ f_h = F(g) \circ F(h)$ and hence F is a homomorphism. If $g \in \ker(F)$, then $f_g = F(g) = \iota$, hence $gB = f_g(B) = \iota(B) = B$, and thus $g \in B$. ■

Example 36.7 We observed in Example 35.2 that $B = \{e, x^3, y, yx^3\}$ is a subgroup of D_6. Since D_6 has twelve elements, Lagrange's theorem (Proposition 21.10) implies that the index of B in D_6 is 3. Direct calculation shows that $xB = \{x, x^4, yx^5, yx^2\}$, $x^2B = \{x^2, x^5, yx^4, yx\}$, and $D_6/B = \{B, xB, x^2B\}$. The function $F: D_6 \to S_3$ defined in the proof of Proposition 36.6 takes $z \in D_6$ to the permutation f_z of D_6/B (where f_z is the function $f_z(wB) = zwB$). We can describe F in cycle notation by associating 1 with B, 2 with xB, and 3 with x^2B, and then computing each f_z and translating it into cycle notation. For instance, f_x takes B to xB $(1 \to 2)$, xB to x^2B $(2 \to 3)$, and x^2B to $x^3B = B$ $(3 \to 1)$. So f_x is the permutation (123). And f_y takes B to B $(1 \to 1)$, xB to x^2B $(2 \to 3)$, and x^2B to xB $(3 \to 2)$. So f_y is the permutation (23). Similarly,

w	$F(w) = f_w$	w	$F(w) = f_w$
e	ι	y	(23)
x	(123)	yx	(13)
x^2	(132)	yx^2	(12)
x^3	ι	yx^3	(23)
x^4	(123)	yx^4	(13)
x^5	(132)	yx^5	(12)

Note that, in agreement with Proposition 36.6, $\ker F = \{g \in D_6 \mid F(g) = \iota\} = \{e, x^3\} \subseteq B$. ❖

To prove that D_6, A_4, and T are the only noncommutative groups of order 12 we will show that any group of order 12 that is not isomorphic to A_4 must contain an element x of order 6 (Lemma 36.8). Then, picking an element y not in $S(x)$, we will use a case-by-case analysis to show that if G is not Abelian, then x and y must satisfy either the equations for the generators of D_6 or the equations for the generators of T. We begin by proving the lemma.

Lemma 36.8 _____ *Every group of order 12 that is not isomorphic to A_4 contains an element of order 6.*

Proof   Let G be a group of order 12 that is not isomorphic to A_4. By Cauchy's theorem (Theorem 34.13), G has an element b of order 3. By Proposition 21.14, $S(b)$ has order 3 and hence by Lagrange's theorem (Proposition 21.10) $S(b)$ has index 4. Thus by Proposition 36.6, there exists a homomorphism $F : G \rightarrow S_4$ whose kernel is contained in $S(b)$. Since $S(b)$ has order 3, Lagrange's theorem implies that the order of ker F is either 1 or 3. But if this order is 1, then F is a one-to-one homomorphism by Lemma 24.4 and hence G is isomorphic to a subgroup of S_4 by Proposition 18.13. Proposition 36.3 then implies that G is isomorphic to A_4, a contradiction. We conclude that $|\ker F| = 3$ and hence that ker $F = S(b)$. It follows from Proposition 23.5 that $S(b) \lhd G$ and hence from Corollary 35.14 that $S(b)$ is the only Sylow 3-subgroup of G. Any element of order 3 must be contained in the subgroup it generates and hence by Proposition 35.13 must be contained in a Sylow 3-subgroup. Therefore, b and b^2 are the only elements of G of order 3. It is easy to see that a conjugate of an element of order 3 must also have order 3 (Exercise 7) and hence b has at most two conjugates. Thus by Proposition 34.7, the index of $C(b)$ in G is either 1 or 2, and hence by Lagrange's theorem, the order of $C(b)$ is either 12 or 6. In either case, Cauchy's theorem implies that $C(b)$ has an element a of order 2. Since $a \in C(b)$, a commutes with b and therefore ab has order 6 (see Exercise 10). ∎

Proposition 36.9 _____ *The only noncommutative groups of order 12 are D_6, A_4, and T.*

Proof   Let G be a noncommutative group of order 12 and suppose that G is not isomorphic to A_4. By Lemma 36.8, G contains an element x of order 6. Then $|S(x)| = 6$ and $[G:S(x)] = 2$. Pick $y \in G$ such that $y \notin S(x)$. Then $|S(x, y)| > 6$ and thus by Lagrange's theorem (Proposition 21.10), $|S(x, y)| = 12$ so that $G = S(x, y)$. By Proposition 33.7, $S(x) \lhd G$. Since $|G/S(x)| = [G:S(x)] = 2$, $y^2 S(x) = (yS(x))^2 = S(x)$ and hence, by Proposition 20.4, $y^2 \in S(x)$. We will show that if $y^2 \notin \{e, x^3\}$, then G is Abelian (in contradiction of our choice of G), that if $y^2 = e$, then G is dihedral, and that if $y^2 = x^3$, then G is isomorphic to T.

(1) If $y^2 = x$ or if $y^2 = x^5$, then y has order 12 so that G is cyclic and hence Abelian (Exercise 9).

(2) Suppose that $y^2 = x^2$ and note that $yx^3y^{-1} \in S(x)$ because $S(x) \lhd G$. It is easy to see that yx^3y^{-1} has order 2 (Exercise 7) and since x^3 is the only element in $S(x)$ of order 2, $yx^3y^{-1} = x^3$, i.e., $yx^3 = x^3y$. Then $yx^2x = x^2xy$ and hence

$$y^2yx = yy^2x = yx^2x = x^2xy = y^2xy$$

so that $yx = xy$. Since $G = S(x, y)$, it follows that G is Abelian (Exercise 8).

(3) Suppose that $y^2 = x^4$ and note that, as in (2), $yx^3y^{-1} = x^3$ and hence $yx^9y^{-1} = x^9$. Then $yx^9 = x^9y$ and hence

$$y^4yx = yy^2y^2x = yx^4x^4x = yx^9 = x^9y = x^4x^4xy = y^2y^2xy = y^4xy$$

so that $yx = xy$. As in (2), it follows that G is Abelian.

(4) Suppose that $y^2 = e$ and consider $a = xy$. Since $a \notin S(x)$, $|S(a, x)| > 6$ and hence $G = S(a, x)$. Then since $|G/S(x)| = 2$, $a^2S(x) = S(x)$ and hence $a^2 \in S(x)$. So by (1), (2), and (3), $a^2 \in \{e, x^3\}$. Suppose that $a^2 = x^3$. Then

$$y^{-1}xy = yxy = x^{-1}xyxy = x^{-1}a^2 = x^{-1}x^3 = x^2$$

and hence $y^{-1}xy$ has order 3. Since $y^{-1}xy$ has order 6, this is a contradiction and thus $a^2 \neq x^3$. It follows that $a^2 = e$, hence that $xy = y^{-1}x^{-1} = yx^5$, and thus that G is isomorphic to D_6 by Corollary 33.3.

(5) Suppose finally that $y^2 = x^3$. Observe that y must have order 4 and again consider $a = xy$. As in (4), $G = S(a, x)$ and $a^2 \in \{e, x^3\}$. Suppose that $a^2 = e$. Then

$$y^{-1}xy = y^{-1}y^{-1}x^{-1}xyxy = y^{-2}x^5a^2 = y^{-2}x^5 = y^{-2}x^3x^2 = y^{-2}y^2x^2 = x^2$$

and hence $y^{-1}xy$ has order 3. As in (4), this is impossible and thus $a^2 = x^3$. We conclude that G is isomorphic to T by Proposition 36.5. ∎

We can now expand the table presented at the end of Chapter 32 to include the noncommutative groups.

Groups of Order Less Than 16		
Order	Number of Groups	Distinct Groups
1	1	$\mathbb{Z}_1$
2	1	$\mathbb{Z}_2$
3	1	$\mathbb{Z}_3$
4	2	$\mathbb{Z}_4, \mathbb{Z}_2 \times \mathbb{Z}_2$
5	1	$\mathbb{Z}_5$
6	2	$\mathbb{Z}_6, D_3 \approx S_3$
7	1	$\mathbb{Z}_7$
8	5	$\mathbb{Z}_8, \mathbb{Z}_4 \times \mathbb{Z}_2, \mathbb{Z}_2 \times \mathbb{Z}_2 \times \mathbb{Z}_2, D_4, Q$
9	2	$\mathbb{Z}_9, \mathbb{Z}_3 \times \mathbb{Z}_3$

(continued)

Groups of Order Less Than 16 (continued)		
Order	Number of Groups	Distinct Groups
10	2	$\mathbb{Z}_{10}, D_5$
11	1	$\mathbb{Z}_{11}$
12	5	$\mathbb{Z}_{12}, \mathbb{Z}_3 \times \mathbb{Z}_2 \times \mathbb{Z}_2, D_6, A_4, T$
13	1	$\mathbb{Z}_{13}$
14	2	$\mathbb{Z}_{14}, D_7$
15	1	$\mathbb{Z}_{15}$

Exercises

In Exercises 1–4, find all groups of order n.

1. $n = 77$ **2.** $n = 143$ **3.** $n = 169$ **4.** $n = 65$

5. Show that D_6 has no elements of order 4.

6. Suppose that G is a group of order 12 which has no subgroups of order 6. Show that G is isomorphic to A_4.

7. Suppose that G is a finite group and that $x, g \in G$. Show that x and gxg^{-1} both have the same order.

8. Show that x and y are elements of a group G such that $xy = yx$. Show that $S(x, y)$ is Abelian.

9. Suppose that G is a group and that $x \in G$ has order 6. If $y \in G$ is such that $y^2 = x$ or if $y^2 = x^5$, show that y has order 12.

10. Suppose that p and q are distinct primes and that G is a finite group with elements x of order p and y of order q such that x and y commute. Show that xy has order pq.

11. Show that $Z(A_4) = \{\iota\}$.

12. Find the subgroup lattice of A_4.

13. Show that even though 6 divides the order of A_4, A_4 has no subgroup of order 6.

14. Show that $Z(T) = S((a, b^2)^3)$.

15. Show that $T/Z(T)$ is isomorphic to D_3.

16. Find the subgroup lattice of S_4.

17. For the group $Q = \{\pm 1, \pm i, \pm j, \pm k\}$ and the subgroup $S(i)$, use the method of Example 36.7 to describe a homomorphism $F : Q \to S_n$, where n is the index of $S(i)$ in G.

18. For the group $Q = \{\pm 1, \pm i, \pm j, \pm k\}$ and the subgroup $S(-1)$, use the method of Example 36.7 to describe a homomorphism $F : Q \rightarrow S_n$, where n is the index of $S(-1)$ in G.

19. For the group T and the subgroup $S((e, b))$ of Example 36.4, use the method of Example 36.7 to describe a homomorphism $F : Q \rightarrow S_n$, where n is the index of $S((e, b))$ in G.

20. For the group A_4 and the subgroup $S((123))$, use the method of Example 36.7 to describe a homomorphism $F : Q \rightarrow S_n$, where n is the index of $S((123))$ in G.

In Exercises 21–26, suppose that G is a group of order 12, that x is an element of G of order 6, and that $y \notin S(x)$.

21. Show that $G = S(x, y)$.

22. Show that $y^2 \in S(x)$.

23. Show that if $y^2 = x$, then y has order 12.

24. Show that if $y^2 = x^5$, then y has order 12.

25. Show that if $y^2 = x^2$, then G is Abelian.

26. Show that if $y^2 = x^4$, then G is Abelian.

27. Suppose that G is a group of order 12 which is not Abelian and which has a unique element of order 2. Show that G is isomorphic to T.

28. Suppose that G is a group of order 12 which has no normal subgroups of order 3. Show that G is isomorphic to A_4.

29. Suppose that G is a group of order 12 with generators x and y, each of order 3, such that $(xy)^2 = e$. Show that G is isomorphic to A_4.

30. Suppose that p and q are primes, that $p > q$, and that q divides $p - 1$. If G is a noncyclic group of order pq, show that there exist $x, y \in G$ such that $x^p = e = y^q$ and $xy = yx^s$, where p does not divide $s - 1$ but p does divide $s^q - 1$.

37

Groups of Even Permutations

In Chapter 36, we defined the group A_4 of order 12. In this chapter, we want to describe the general construction that A_4 exemplifies. That is, for each n, we will define a subgroup A_n in such a way that the subgroup A_4 given by this construction is the same as that defined in Example 36.2. We will show that the only normal subgroups of A_5, the analogous subgroup of S_5, are $\{\iota\}$, and A_5 itself.

In fact, A_5 is the smallest noncommutative group with this property. We will prove part of this result by using the Sylow theorems to show that all noncommutative groups of order less than 31 have at least three normal subgroups. We will leave the remaining cases for the exercises (Exercises 5–10).

In general, A_n is the following subgroup of S_n. (Recall that a transposition is a permutation (ab), where $a \neq b$.)

DEFINITION

A permutation is **even** if it can be written as a product of an even number of transpositions; it is **odd** if it can be written as a product of an odd number of transpositions. The set of even permutations in S_n is denoted by A_n.

Since $\iota = (ab)(bc)$, ι is even and hence $A_n \neq \varnothing$, and since the sum of two even numbers is even, the product of two even permutations must be even. It follows from Proposition 18.7 that A_n is a subgroup of S_n.

DEFINITION

The group A_n is called the **alternating group** of degree n.

In Example 36.2, we showed that $\{i, (123), (132), (124), (142), (134), (143), (234), (243), (12)(34), (13)(24), (14)(23)\}$ is a subgroup of S_4 and named it A_4. In view of the preceding definition, this requires some justification.

Example 37.1   Let S be the following subset of S_4:

$$S = \{\iota, (123), (132), (124), (142), (134), (143),$$
$$(234), (243), (12)(34), (13)(24), (14)(23)\}.$$

We will show that S is the group A_4 of our definition. Any 3-cycle can be written as a product of an even number of transpositions: $(abc) = (ab)(bc)$. So every element of S can be written as a product of an even number of transpositions, and hence $S \subseteq A_4$. We will show later (Proposition 37.4) that A_4 has $4!/2 = 12$ elements. Thus, since S is a 12-element subset of A_4, $S = A_4$. ❖

We need to prove that A_n has $n!/2$ elements. We will do this by showing that no permutation can be both even and odd (Lemma 37.3). First we isolate part of this proof in the following result.

Lemma 37.2 — *Suppose that α and β are transpositions in S_n and that $x \in \{1, \ldots, n\}$. If $\alpha\beta(x) \neq x$, then there exists γ and δ such that $\alpha\beta = \gamma\delta$ and $\gamma(x) \neq x$ and $\delta(x) = x$.*

Proof   If $\beta(x) = x$, we can choose $\gamma = \alpha$ and $\delta = \beta$. Otherwise, $\beta = (xy)$ for some $x \neq y \in \{1, \ldots, n\}$. If α is disjoint from β, then $\alpha(x) = x$ and $\alpha\beta = \beta\alpha$ so that we may choose $\gamma = \beta$ and $\delta = \alpha$. The other cases are: $\alpha = (xy)$, $\alpha = (xz)$, or $\alpha = (yz)$ for some $z \in \{1, \ldots, n\}$ such that $z \neq x$ and $z \neq y$. If $\alpha = (xy)$, then $\alpha\beta = \iota$ and hence $\alpha\beta(x) = x$, a contradiction. If $\alpha = (xz)$, then $\alpha\beta = (xz)(xy) = (xyz) = (xy)(yz)$ so that we may choose $\gamma = (xy)$ and $\delta = (yx)$. Finally, if $\alpha = (yz)$, then $\alpha\beta = (yz)(xy) = (xyz) = (xz)(yz)$ so that we may choose $\gamma = (xz)$ and $\delta = (yz)$. ∎

Lemma 37.3 — *For $n \geq 2$, no permutation in S_n can be both even and odd.*

Proof   If a permutation σ can be written as a product of an even number of transpositions $\sigma = \tau_1 \cdots \tau_e$ and also as a product of an odd number of transpositions $\sigma = \rho_1 \cdots \rho_d$, then the identity can be written as a product of an odd number of transpositions: $\iota = \rho_d^{-1} \cdots \rho_1^{-1} \tau_1 \cdots \tau_e$. Thus it suffices to show that the identity cannot be written as a product of an odd number of transpositions. We will show that $\iota \neq \tau_1 \cdots \tau_{2n+1}$ by induction on n.

(i)   If $n = 0$, then $\tau_1 \cdots \tau_{2n+1} = \tau_1$ and certainly $\iota \neq (ab)$ for any transposition (ab).

(ii)   Suppose that no product of $2n + 1$ transpositions can equal ι, and consider a product of transpositions of length $2(n + 1) + 1 = 2n + 3$: $\tau_1 \cdots \tau_{2n+3}$. Let $\tau_{2n+3} = (xy)$. If $\tau_1 \cdots \tau_{2n+3}$ can be written as a product of $2n + 1$ transpositions, then the induction hypothesis implies that $\tau_1 \cdots \tau_{2n+3} \neq \iota$. So suppose that $\tau_1 \cdots \tau_{2n+3}$ cannot be written as a product of $2n + 1$ transpositions. If $\tau_{2n+2}\tau_{2n+3} = \iota$, then $\tau_1 \cdots \tau_{2n+3} = \tau_1 \cdots \tau_{2n+1}$ and hence $\tau_1 \cdots \tau_{2n+3}$ can be written as a product of $2n + 1$ transpositions.

So we must have $\tau_{2n+2}\tau_{2n+3} \neq \iota$. Suppose that $\tau_{2n+2}\tau_{2n+3}(x) = x$. Then, since $\tau_{2n+3} = (xy)$, $\tau_{2n+2}(y) = x$, and thus, since τ_{2n+2} is a transposition, $\tau_{2n+2} = (xy)$, in which case $\tau_{2n+2}\tau_{2n+3} = \iota$, a contradiction. So $\tau_{2n+2}\tau_{2n+3}(x) \neq x$. Then by Lemma 37.2, $\tau_1 \cdots \tau_{2n+3} = \tau_1 \cdots \tau_{2n+1}\gamma_{2n+2}T_{2n+3}$, where $\gamma_{2n+2}(x) \neq x$ and $T_{2n+3}(x) = x$. If $\tau_{2n+1}\gamma_{2n+2} = \iota$, then $\tau_1 \cdots \tau_{2n+3}$ can be written as a product of $2n + 1$ transpositions. So suppose that $\tau_{2n+1}\gamma_{2n+2} \neq \iota$. Then an argument analogous to the preceding one shows that $\tau_{2n+1}\gamma_{2n+2}(x) \neq x$ so that, again by Lemma 37.2, $\tau_1 \cdots \tau_{2n+3} = \tau_1 \cdots \tau_{2n}\gamma_{2n+1}T_{2n+2}T_{2n+3}$, where $\gamma_{2n+1}(x) \neq x$ and $T_{2n+2}(x) = x$ and $T_{2n+3}(x) = x$. Continuing in the same fashion, we can eventually write $\tau_1 \cdots \tau_{2n+3} = \gamma_1 T_2 \cdots T_{2n+3}$ for transpositions $\gamma_1, T_2, \ldots, T_{2n+3}$ such that $T_i(x) = x$ for all $i > 1$ and $\gamma_1(x) \neq x$. But then $\tau_1 \cdots \tau_{2n+3} \neq \iota$ because $\tau_1 \cdots \tau_{2n+3}(x) = \gamma_1(x) \neq x$. ∎

Proposition 37.4   *For $n \geq 2$, the subgroup A_n of S_n has order $n!/2$.*

Proof   Let O_n denote the set of odd permutations in S_n, pick a transposition τ, and define $f : A_n \rightarrow O_n$ by letting $f(\sigma) = \sigma\tau$. Note that since τ is a transposition, $\tau\tau = \iota$. If $f(\sigma) = f(\rho)$, then $\sigma\tau = \rho\tau$ and thus $\sigma = \sigma\tau\tau = \rho\tau\tau = \rho$ so that f is one-to-one. Furthermore, if ω is odd, then $\omega\tau$ is even and $f(\omega\tau) = \omega\tau\tau = \omega$ so that f is onto. We conclude that A_n and O_n have the same number of elements. But by Proposition 19.6, every permutation can be written as a product of transpositions and hence is either even or odd. So $S_n = A_n \cup O_n$, and by Lemma 37.3, $A_n \cap O_n = \emptyset$. So A_n contains exactly half the elements of S_n and thus by Proposition 16.6, A_n has $n!/2$ elements. ∎

We saw in Chapter 36 that A_4 is one of the noncommutative groups of order 12. The next alternating group, A_5, is even more noteworthy. It is the smallest noncommutative group that is simple in the following sense.

DEFINITION

A group G is **simple** if G and $\{e\}$ are its only normal subgroups.

By Lagrange's theorem (Proposition 21.10), a group G of prime order has only two subgroups, G and $\{e\}$, and hence must be simple. On the other hand, we have the following negative results.

Example 37.5   We will show that $A_4 = \{\iota, (123), (132), (124), (142), (134), (143), (234), (243), (12)(34), (13)(24), (14)(23)\}$ is not simple. Since $(12)(34)$, $(13)(24)$, and $(14)(23)$ are the only elements in A_4 of order 2, $\{\iota, (12)(34), (13)(24), (14)(23)\}$ is the only Sylow 2-subgroup of A_4. It follows from Corollary 35.14 that $\{\iota, (12)(34), (13)(24), (14)(23)\}$ is normal in A_4 and hence that A_4 is not simple. ❖

Example 37.6   We will show that the group $(T, *)$ of Example 36.4 is not simple. We have already observed that (a, b^2) is an element of order 6 in T, and therefore by Proposition 21.14, that $S((a, b^2))$ is a subgroup of T of order 6. So by Proposition 33.7, $S((a, b^2)) \lhd T$. But since T is not Abelian, $S((a, b^2)) \neq T$, and since $(a, b^2) \neq (e, e)$, $S((a, b^2)) \neq \{(e, e)\}$. It follows that T is not simple. ❖

Proposition 37.7   *If $n > 1$, then D_n is not simple.*

Proof   By definition, D_n contains an element x of order n. So by Proposition 21.14, the subgroup, $S(x)$, which x generates, is a subgroup of D_n of order n. But by Lagrange's theorem (Proposition 21.10), the index of $S(x)$ in D_n is $2n/n = 2$, and hence by Proposition 33.7, $S(x) \lhd D_n$. ∎

Proposition 37.8   *If G is a noncommutative group whose order is the power of a prime p, then G is not simple.*

Proof   By Proposition 34.4, $Z(G) \lhd G$, and by Proposition 34.16, $Z(G) \neq \{e\}$. Since G is not commutative, $Z(G) \neq G$. ∎

These results show that the only simple groups in the table at the end of Chapter 36 are those of prime order and hence are Abelian. Furthermore, since $16 = 2^4$, Proposition 37.8 implies that no group of order 16 is simple. The next composite integer is 18, and the following general result shows that we will find no group of order 18 that is simple.

Proposition 37.9   *Let p be prime and suppose that G is a group of order $p^n m$, where p does not divide m and $n > 0$. If $kp + 1$ fails to divide m for all $k > 0$, then G has a normal subgroup of order p^n.*

Proof   By the first Sylow theorem (Proposition 35.1), G has a subgroup H of order p^n, and by the other Sylow theorems (Proposition 35.13), the number of conjugates of H is equal to $kp + 1$ and divides m. Thus, by hypothesis, $k = 0$ and hence H has only one conjugate. Then H is normal by Corollary 35.14. ∎

Since $18 = 2 \cdot 3^2$, taking $p = 3$ in Proposition 37.9 shows that no group of order 18 is simple. Similarly, choosing $p = 5$ for $20 = 2^2 \cdot 5$, $p = 7$ for $21 = 3 \cdot 7$, and $p = 11$ for $22 = 2 \cdot 11$ in Proposition 37.9 shows that no group of order 20, 21, or 22 is simple. And since 23 is prime, any group of order 23 is Abelian. Since $2 + 1 = 3$ divides 3 and $3 + 1 = 4$ divides $8 = 2^3$, Proposition 37.9 does not apply to groups of order $24 = 2^3 \cdot 3$. However, as the following argument shows, no group of order 24 is simple.

Example 37.10   Suppose that G is a group of order $24 = 2^3 \cdot 3$. By the Sylow theorems (Theorem 35.13), G contains a subgroup W of order $2^3 = 8$, which by Lagrange's theorem

(Proposition 21.10) has index 3 in G. Then by Proposition 36.6, there is a homomorphism $F: G \to S_3$ whose kernel is contained in W. Since $|W| = 8$, $W \neq G$ and hence $\ker(F) \neq G$. Furthermore, by Proposition 18.13, $F(G)$ is isomorphic to a subgroup of S_3, and by the homomorphism theorem for groups (Theorem 24.6), $G/\ker(F)$ is isomorphic to $F(G)$. So $G/\ker(F)$ contains at most six cosets. Since, by Lagrange's theorem, $24 = |G| = |\ker F| \, |G/\ker F|$, $\ker F$ has at least four elements. So $\ker(F) \neq \{e\}$. But by Proposition 23.5, $\ker(F) \lhd G$, and hence G is not simple. ❖

Proposition 34.17 implies that all groups of order $25 = 5^2$ are Abelian. Proposition 37.9 with $p = 13$ implies that no group of order $26 = 2 \cdot 13$ is simple. Proposition 37.8 implies that no group of order $27 = 3^3$ is simple. Proposition 37.9 with $p = 7$ implies that no group of order $28 = 2^2 \cdot 7$ is simple. And since 29 is prime, all groups of order 29 are Abelian. The following argument shows that no group of order $30 = 2 \cdot 3 \cdot 5$ is simple.

Example 37.11   Suppose that G is a group of order $30 = 2 \cdot 3 \cdot 5$. By the Sylow theorems (Theorem 35.13), G contains either one or six subgroups of order 5 and either one or ten subgroups of order 3. If G contains only one subgroup of order 5 or one subgroup of order 3, then that subgroup will be normal. So suppose that G contains six subgroups of order 5 and ten subgroups of order 3. Then G contains at least twenty-four elements of order 5 and twenty elements of order 3. Since G has only thirty elements, this is impossible. We conclude that G cannot be simple. ❖

We previously claimed that in contrast to these examples, A_5 is simple. (In fact, an analogous but more complicated argument shows that A_n is simple for any $n \geq 5$.)

Example 37.12   We will show that A_5 is simple. That is, we will show that if $N \neq \{\iota\}$ is a normal subgroup of A_5, then $N = A_5$.

(1) We first show that A_5 is generated by the 3-cycles. Since $(abc) = (ab)(bc)$, A_5 contains all the 3-cycles. Conversely, by definition, every element in A_5 can be written as a product of pairs of transpositions and any product of pairs of transpositions can be written as a product of 3-cycles: $(ab)(ab) = \iota = (123)(132)$, $(ab)(ac) = (acb)$, $(ab)(cd) = (abc)(bcd)$.

(2) We next show that if the normal subgroup N contains one 3-cycle, it contains all 3-cycles. Suppose that $(abc) \in N$. First note that all 3-cycles of the form (abx) are in N because $(abx) = [(ab)(cx)](abc)^2[(ab)(cx)]^{-1}$ and $N \lhd A_5$. Next note that any 3-cycle of the form (axb) is in N because $(axb) = (abx)^2$ and thus any 3-cycle of the form (axy) or (bxy) is in N because $(axy) = (aby)(axb)$ and $(bxy) = (ayb)(abx)$. Finally, for distinct $x, y, z \notin \{a, b\}$, any 3-cycle of the form (xyz) is in N because $(xyz) = (azx)(axy)$.

(3) We conclude by showing that N contains a 3-cycle. Let $\iota \neq \nu \in N$. If ν is not a 3-cycle, then $\nu = (ab)(cd)$ or $\nu = (abcde) = (ab)(bc)(cd)(de)$ (Exercise 16). If $\nu = (ab)(cd)$, then

$$(abe) = [(aeb)(ab)(cd)(aeb)^{-1}](ab)(cd) \in N$$

because $N \lhd A_5$. Similarly, if $\nu = (abcde)$, then

$$(ace) = (abcde)^{-1}[(abc)(abcde)(abc)^{-1}] \in N$$

because $N \lhd A_5$. Thus in all cases, N contains a 3-cycle.

In summary, (3) implies that N contains at least one 3-cycle, then (2) implies that N contains all 3-cycles, and then (1) implies that $N = A_5$. We conclude that A_5 is simple. ❖

Our work so far shows that A_5 is simple and that no noncommutative group of order less than 31 is simple. It is easy to see that A_5 is not commutative and Proposition 37.4 implies that A_5 has order 60. So it remains to show that no noncommutative group whose order is strictly between 30 and 60 is simple. This is left to the exercises.

Exercises

1. List the even permutations in S_3.

2. List the odd permutations in S_4.

3. Determine those n for which all n-cycles are even.

4. Determine those n for which all n-cycles are odd.

In Exercises 5–11, assume that G is a simple noncommutative group of order n.

5. Show that the order of G is not the power of a prime.

6. Use Exercise 5 and Proposition 37.9 to show that the only possible values for n that are less than 60 are 12, 24, 30, 36, 48, and 56.

7. Use the method of Example 37.10 to show that n cannot be 36.

8. Use the method of Example 37.10 to show that n cannot be 48.

9. Show that n cannot be 56.

10. Show that $n \geq 60$.

11. Find as many integers $60 < k < 100$ as you can such that $n \neq k$. Remember to justify your answer.

12. If $n \geq 3$, show that every even permutation in S_n can be written as a product of 3-cycles.

13. Show that for $n \geq 3$, A_n is generated by the 3-cycles of S_n.

14. If $N \lhd A_n$ and if N contains a 3-cycle, show that $N = A_n$.

Exercises 15–18 show that proper subgroups of A_5 have order no greater than 12.

15. Show that A_5 has no subgroup of order 15.

16. Show that A_5 has no subgroup of order 20.

17. Show that A_5 has no subgroup of order 30.

18. Show that if S is a subgroup of A_5 such that $S \neq A_5$, then $|S| \leq 12$.

19. Show that if $A_5 \lhd S_5$ and that if $S \lhd S_5$ such that $\{\iota\} \neq S \neq S_5$, then $S = A_5$.

20. Show that if $\iota \neq \sigma \in A_5$ is not a 3-cycle, then σ is either a 5-cycle or a product of two disjoint 2-cycles.

21. Suppose that G is a simple group with a subgroup H of index n. Show that if $n > 1$, then there exists a one-to-one homomorphism $F : G \rightarrow S_n$.

22. Assume that A_6 is simple and show that it contains no subgroup of prime index.

In Exercises 23–27, suppose that p, q, and r are primes such that $p < q < r$ and G is a group of order pqr; let σ_p denote the number of Sylow p-subgroups of G, σ_q denote the number of Sylow q-subgroups, and σ_r denote the number of Sylow r-subgroups.

23. If no Sylow r-subgroup of G is normal, show that $\sigma_r = pq$.

24. If no Sylow q-subgroup of G is normal, show that $\sigma_q \geq r$.

25. If no Sylow p-subgroup of G is normal, show that $\sigma_p \geq q$.

26. If G is simple, show that it must have at least $pqr + (q - 1)^2$ elements.

27. Show that G is not simple.

HISTORICAL NOTE

Camille Jordan

Marie Ennemond Camille Jordan was born in Lyons, France, on January 5, 1838, and died in Milan, Italy, on January 22, 1922. The family was well-do-do, his father (Alexander Jordan) being a civil engineer and his mother (Joséphine Puvis de Chavannes) the sister of a well-known painter. His granduncle, also named Camille Jordan, was active in French politics from the Revolution in 1789 to the Bourbon restoration. A cousin, Alexis Jordan, was a botanist for whom the species "jordanons" is named. Camille was a brilliant student, and in 1855, following his father's lead, he matriculated at the École Polytechnique.

Upon graduation, he began work as a mining engineer, first at Privas and then at Paris, and in 1862, he married Isabelle Munet. They had six sons, two daughters, and many grandchildren. Most of his mathematical work, including 120 mathematical papers, was done before he retired from engineering in 1885. From 1873 to 1921, he taught at the École Polytchnique and the Collège de France.

His generally pleasant life took a turn for the worse with the loss of one of his daughters in 1912 and then was overwhelmed with tragedy with the onset of World War I. Two of his sons died in battle in 1914 and a third in 1915. Then his grandson, a professor at the Sorbonne, was killed in 1916. These tragedies were compounded in 1918 by the death of his wife.

Jordan contributed to many branches of mathematics and is well known for his work in linear algebra and topology. His *Cours d'Analyse* (1882–1887) was the definitive text in analysis for many years, and his *Traité des Substitutions* (1870) became the standard reference for group theorists and introduced the term "group." In the short note "*Sur les groupes de mouvements*" (1867), he announced a complete determination of all possible groups of displacements of rigid bodies in Euclidean 3-space. His list missed the crystallographic groups, but was otherwise complete. The discrete groups of rotations that he found were the finite cyclic groups, the dihedral groups, the tetrahedral group, the octahedral group, and the icosahedral group. In "*Sur la limite de transitivité des groupes non alternés*" (1875), he showed that the alternating group A_n is the only nontrivial proper normal subgroup of S_n if $n \neq 4$ and is simple if $n \geq 5$.

38

Semidirect Products

For groups G and H, we have seen in Chapter 29 that the set $G \times H = \{(g, h) \mid g \in G, h \in H\}$ is a group with respect to coordinatewise multiplication: $(x, y)(z, w) = (xz, yw)$. We have seen in Chapter 36 that this is not the only way to define a multiplication on $G \times H$.

Specifically, the group T of Example 36.4 was constructed by placing an unusual multiplication on the set $C_3 \times C_4$, where $C_3 = \{e, a, a^2\}$ was a cyclic group of order 3 and $C_4 = \{e, b, b^2, b^3\}$ was a cyclic group of order 4. The usual multiplication on $C_3 \times C_4$ is of course the one just defined: $(x, y)(z, w) = (xz, yw)$; the new multiplication was defined:

$$(x, y) * (z, w) = \begin{cases} (xz, yw) & \text{if } y = e \text{ or } y = b^2 \\ (xz^2, yw) & \text{if } y = b \text{ or } y = b^3 \end{cases}$$

In each case, the second coordinate is the same, viz., yw; the difference between the two multiplications lies in the first coordinate. But the first coordinate depends on the second coordinate, and we can describe this dependence by using isomorphisms from C_3 to itself. Specifically, let $g_1 : C_3 \to C_3$ be the identity function $g_1(x) = x$, and let $g_2 : C_4 \to C_4$ be the square function $g_2(x) = x^2$. Then g_2 is the function that interchanges a and a^2:

x:	e	a	a^2
$\downarrow$	$\downarrow$	$\downarrow$	$\downarrow$
$g_2(x)$:	e	a^2	a

Obviously, g_1 is an isomorphism, and the table shows that g_2 is one-to-one and onto. And since C_4 is Abelian, g_2 is also a homomorphism: $g_2(xy) = (xy)^2 = xyxy =$

$x^2y^2 = g_2(x)g_2(y)$. Then the new multiplication $*$ may be described in terms of g_1 and g_2 as follows:

$$(x, y) * (z, w) = \begin{cases} (xg_1(z), yw) & \text{if } y = e \text{ or } y = b^2 \\ (xg_2(z), yw) & \text{if } y = b \text{ or } y = b^3 \end{cases}$$

So if, with each $y \in C_4$, we associate a function $\Theta_y \in \text{Aut}(C_3)$ in the following way:

y:	e	b	b^2	b^3
↓	↓	↓	↓	↓
Θ_y:	g_1	g_2	g_1	g_2

then the multiplication $*$ may be succinctly written: $(x, y) * (z, w) = (x\Theta_y(z), yw)$.

We will show how to generalize this construction so that it always produces a group and hence we will be able to use many of the sets $G \times H$ to produce more than one group. To be sure that the operation $*$ satisfies the group axioms, we will need to know that the set of isomorphisms is a group with respect to composition of functions and that associating y with Θ_y is a homomorphism of groups.

We begin by showing that the set of isomorphisms forms a group.

Proposition 38.1 *For any group G, let $\text{Aut}(G)$ denote the set of all automorphisms of G, i.e., the set of all homomorphisms $f: G \to G$ which are both one-to-one and onto. Then $\text{Aut}(G)$ is a group with respect to composition of functions.*

Proof Let $f, g \in \text{Aut}(G)$. Clearly, $f \circ g: G \to G$, and for all $a, b \in G$, $(f \circ g)(ab) = f(g(ab)) = f(g(a)g(b)) = f(g(a))f(g(b)) = (f \circ g)(a)(f \circ g)(b)$. So $f \circ g$ is a homomorphism. As well, if $(f \circ g)(a) = (f \circ g)(b)$, then $f(g(a)) = f(g(b))$ and hence $g(a) = g(b)$ because f is one-to-one and $a = b$ because g is one-to-one. So $f \circ g$ is one-to-one. And since f is onto, for any $a \in G$, there exists $r \in G$ such that $f(r) = a$ and since g is onto, there exists $s \in K$ such that $g(s) = r$. Then $(f \circ g)(s) = a$ and hence $f \circ g$ is onto. So $f \circ g \in \text{Aut}(G)$ and $\text{Aut}(G)$ is closed. For associativity, note that for any $a \in G$.

$$(f \circ (g \circ h))(a) = f(g \circ h(a)) = f(g(h(a))) = (f \circ g)(h(a)) = ((f \circ g) \circ h)(a),$$

and hence $f \circ (g \circ h) = (f \circ g) \circ h$. The identity of $\text{Aut}(G)$ is of course the identity function on G: $\iota(g) = g$. It is easy to check that ι is a homomorphism from G to G which is one-to-one and onto so that $\iota \in \text{Aut}(G)$, and since $\iota(f(a)) = f(a) = f(\iota(a))$ for any $f \in \text{Aut}(G)$ and any $a \in G$, $f \circ \iota = f = \iota \circ f$ and hence ι is the identity with respect to composition. For inverses, note that if $f \in \text{Aut}(G)$, then, since f is one-to-one and onto, it has an inverse function t, i.e., there exists a function $t: G \to G$

such that $f \circ t = \iota = t \circ f$ (t is the function defined by letting $t(x)$ be the unique element y of G such that $f(y) = x$). We need to show that $t \in \text{Aut}(G)$. If $t(a) = t(b)$ for $a, b \in G$, then $a = f(t(a)) = f(t(b)) = b$ and hence t is one-to-one. If $c \in G$, then $f(c) \in G$ and $t(f(c)) = c$ so that t is onto. If $a, b \in G$, then $t(ab) = t(f(t(a))f(t(b))) = t(f(t(a)t(b))) = t(a)t(b)$, and therefore t is a homomorphism. We conclude that $t \in \text{Aut}(G)$. ∎

Example 38.2   Let $C_4 = \{e, b, b^2, b^3\}$ be a cyclic group of order 4. By Lemma 18.12, any isomorphism $f: C_4 \to C_4$ must satisfy $f(e) = e$. Since b^2 has order 2, $f(b^2)$ also has order 2 by Proposition 21.9, and hence $f(b^2) = b^2$. So the only possibilities for f are:

f_1		
e	$\to$	e
b	$\to$	b
b^2	$\to$	b^2
b^3	$\to$	b^3

f_2		
e	$\to$	e
b	$\to$	b^3
b^2	$\to$	b^2
b^3	$\to$	b

Each of these functions is obviously one-to-one and onto. As well, f_1 is the identity function and hence obviously a homomorphism. And f_2 is the cube, $f_2(x) = x^3$, and thus, since C_4 is Abelian, it too is a homomorphism: $f_2(xy) = (xy)^3 = x^3 y^3 = f_2(x)f_2(y)$. So $\text{Aut}(C_4) = \{f_1, f_2\}$. ❖

Example 38.3   Let $C_3 = \{e, a, a^2\}$ be a cyclic group of order 3. By Lemma 18.12, any isomorphism $g: C_3 \to C_3$ must satisfy $g(e) = e$. So the only possibilities for g are:

g_1		
e	$\to$	e
a	$\to$	a
a^2	$\to$	a^2

g_2		
e	$\to$	e
a	$\to$	a^2
a^2	$\to$	a

We previously observed that each of these functions is an isomorphism. So $\text{Aut}(C_3) = \{g_1, g_2\}$. ❖

Note that making the association we used to define T amounts to defining a function $\Theta: C_4 \to \text{Aut}(C_3)$, $y \mapsto \Theta_y$, and that this function Θ is a homomorphism in the sense that $\Theta_{xy} = \Theta_x \circ \Theta_y$ for all $x, y \in C_4$. For by definition,

$$\Theta_{b^k} = \begin{cases} g_1 & \text{if } k \text{ is even} \\ g_2 & \text{if } k \text{ is odd} \end{cases}$$

and hence

$$\Theta_{b^k} \circ \Theta_{b^i} = \begin{cases} g_1 \circ g_1 & \text{if } k \text{ is even and } i \text{ is even} \\ g_1 \circ g_2 & \text{if } k \text{ is even and } i \text{ is odd} \\ g_2 \circ g_1 & \text{if } k \text{ is odd and } i \text{ is even} \\ g_2 \circ g_2 & \text{if } k \text{ is odd and } i \text{ is odd} \end{cases}$$

$$= \begin{cases} g_1 & \text{if } k + i \text{ is even} \\ g_2 & \text{if } k + i \text{ is odd} \end{cases}$$

$$= \Theta_{b^{k+i}} = \Theta_{b^k b^i}.$$

We generalize this construction of T by using an arbitrary homomorphism Θ. Specifically, suppose that G and H are groups and consider the set $G \times H = \{(g, h) \mid g \in G, h \in H\}$. Suppose further that $\Theta : H \to \mathrm{Aut}(G), h \mapsto \Theta_h$, is a homomorphism: $\Theta_{xy} = \Theta_x \circ \Theta_y$, and define a multiplication $*_\Theta$ on $G \times H$ by letting $(x, y) *_\Theta (z, w) = (x\Theta_y(z), yw)$.

We will show that $G \times H$ is a group with respect to the multiplication $*_\Theta$. For any $(x, y), (z, w) \in G \times H$, since $\Theta_y \in \mathrm{Aut}(G)$, $\Theta_y(z) \in G$ and hence $(x, y) *_\Theta (z, w) = (x\Theta_y(z), yw) \in G \times H$; so the operation $*_\Theta$ is closed. The following calculations show that, since Θ is a homomorphism and each Θ_y is a homomorphism, $*_\Theta$ is associative:

$$(a, b) *_\Theta [(x, y) *_\Theta (z, w)] = (a, b) *_\Theta (x\Theta_y(z), yw) = (a\Theta_b(x\Theta_y(z)), b(yw))$$
$$= (a\Theta_b(x)\Theta_b(\Theta_y(z)), b(yw)) = (a\Theta_b(x)\Theta_{by}(z)), (by)w)$$
$$= (a\Theta_b(x), by) *_\Theta (z, w) = [(a, b) *_\Theta (x, y)] *_\Theta (z, w).$$

Also, since each Θ_y is a homomorphism, $\Theta_y(e) = e$ for all y, and, since Θ is a homomorphism, Θ_e must be the identity function in $\mathrm{Aut}(G)$: $\Theta_e(x) = x$. Therefore, (e, e) is the identity with respect to $*_\Theta$:

$$(x, y) *_\Theta (e, e) = (x\Theta_y(e), ye) = (xe, ye)$$
$$= (x, y) = (ex, ey) = (e\Theta_e(x), ey) = (e, e) *_\Theta (x, y).$$

Furthermore, for any $(x, y) \in G \times H$, $(\Theta_{y^{-1}}(x^{-1}), y^{-1}) \in G \times H$, and since

$$(x, y) *_\Theta (\Theta_{y^{-1}}(x^{-1}), y^{-1}) = (x\Theta_y(\Theta_{y^{-1}}(x^{-1})), yy^{-1}) = (x\Theta_{yy^{-1}}(x^{-1}), yy^{-1})$$
$$= (x\Theta_e(x^{-1}), yy^{-1}) = (xx^{-1}, yy^{-1})$$
$$= (e, e) = (\Theta_{y^{-1}}(e), e)$$
$$= (\Theta_{y^{-1}}(x^{-1}x), y^{-1}y = (\Theta_{y^{-1}}(x^{-1})\Theta_{y^{-1}}(x), y^{-1}y)$$
$$= (\Theta_{y^{-1}}(x^{-1}), y^{-1}) *_\Theta (x, y),$$

$(\Theta_{y^{-1}}(x^{-1}), y^{-1})$ is the inverse of (x, y) with respect to $*_\Theta$.

> **DEFINITION**
>
> Let G and H be groups, let $\Theta: H \to \text{Aut}(G)$, $y \mapsto \Theta_y$, be a homomorphism, and define $*_\Theta$ on $G \times H$ by letting $(x, y) *_\Theta (z, w) = (x\Theta_y(z), yw)$. Then the group $(G \times H, *_\Theta)$ is called the **semidirect product of G and H with respect to Θ** and is denoted by $G \times_\Theta H$.

Example 38.4 We will first show that S_3 cannot be written as a direct product of proper subgroups. For if A and B are subgroups of S_3 such that $A \neq S_3$ and $B \neq S_3$, then by Propositions 21.10 and 21.17, both A and B are commutative so that $A \times B$ is commutative. But then $A \times B$ cannot be isomorphic to S_3.

We will next find subgroups N and H such that S_3 is a semidirect product of N and H. Specifically, recall that $N = \{\iota, (123), (132)\}$ is a normal subgroup of S_3 and that $H = \{\iota, (12)\}$ is a subgroup of S_3 (Examples 18.11 and 21.11). Since N is a cyclic group of order 3, Example 38.3 implies that $\text{Aut}(N) = \{g_1, g_2\}$, where g_1 and g_2 are the functions $g_1(x) = x$ and $g_2(x) = x^2$. Let $\Theta: H \to \text{Aut}(N)$ be the function

$$\Theta_y = \begin{cases} g_1 & \text{if } y = \iota \\ g_2 & \text{if } y = (12) \end{cases}$$

It is easy to check that Θ is a homomorphism (Exercise 8) and hence that we can form the semidirect product $N \times_\Theta H$ of N and H with respect to Θ. Since

$$((123), (12)) *_\Theta ((123), \iota) = ((123)(132), (12)\iota) = (\iota, (12))$$
$$\neq ((132), (12)) = ((123)(123), \iota(12))$$
$$= ((123), \iota) *_\Theta ((123), (12)),$$

$N \times_\Theta H$ is a noncommutative group of order 6 and by Proposition 34.15 and Example 33.4, all such groups are isomorphic. So $N \times_\Theta H$ is isomorphic to S_3 (Exercise 9). ❖

Example 38.5 We will show that the quaternion group $Q = \{\pm 1, \pm i, \pm j, \pm k\}$ is not isomorphic to any semidirect product $A \times_\Theta B$ of proper subgroups A and B of Q. For $A \times_\Theta B$ contains subgroups $H_1 = \{(a, 1) \mid a \in A\}$ and $H_2 = \{(1, b) \mid b \in B\}$ such that $H_1 \cap H_2 = \{(1, 1)\}$. However, the subgroup $\{\pm 1\}$ is contained in every subgroup of Q except $\{1\}$ (see Exercise 21.40), and hence Q can contain no subgroups corresponding to H_1 and H_2. So Q cannot be isomorphic to $A \times_\Theta B$ (cf. Exercise 26). ❖

We were able to characterize direct products of Abelian groups internally (Proposition 30.4). We conclude this chapter by showing that an analogue of this result is true for semidirect products of arbitrary groups. We first show that a semidirect product $G \times_\Theta H$ always contains a normal subgroup $\overline{G}$ and a subgroup $\overline{H}$ such that $G \times_\Theta H = \overline{G}\,\overline{H}$. and $\overline{G} \cap \overline{H} = \{(e, e)\}$. Then we show that conversely whenever a group contains such subgroups, it may be written as a semidirect product of those subgroups.

Proposition 38.6   *Let G and H be groups and suppose that $\Theta : G \to \operatorname{Aut}(H)$ is a homomorphism. Consider the following subsets of $G \times_\Theta H$: $\overline{G} = \{(g, e) \mid g \in G\}$ and $\overline{H} = \{(e, h) \mid h \in H\}$. Then $\overline{G}$ is a normal subgroup of $G \times_\Theta H$, $\overline{H}$ is a subgroup of $G \times_\Theta H$, $G \times_\Theta H = \overline{G}\,\overline{H}$, and $\overline{G} \cap \overline{H} = \{(e, e)\}$.*

Proof   Since $e \in G$, $(e, e) \in \overline{G}$ and hence $\overline{G} \neq \varnothing$. And for (x, e), $(y, e) \in \overline{G}$, $(x, e) *_\Theta (y, e) = (x\Theta_e(z), ee) = (xz, e) \in \overline{G}$, and $(x, e)^{-1} = (\Theta_{e^{-1}}(x^{-1}), e^{-1}) = (x^{-1}, e) \in \overline{G}$. So by Proposition 18.2, $\overline{G}$ is a subgroup of $G \times_\Theta H$. Furthermore, for $(g, h) \in G \times_\Theta H$,

$$
\begin{aligned}
(g, h)^{-1} *_\Theta (x, e) *_\Theta (g, h) &= (\Theta_{h^{-1}}(g^{-1}), h^{-1}) *_\Theta (x, e) *_\Theta (g, h) \\
&= (\Theta_{h^{-1}}(g^{-1}), h^{-1}) *_\Theta (x\Theta_e(g), eh) \\
&= (\Theta_{h^{-1}}(g^{-1})\Theta_{h^{-1}}(x\Theta_e(g)), h^{-1}eh) \\
&= (\Theta_{h^{-1}}(g^{-1}xg), e) \in \overline{G}
\end{aligned}
$$

so that $\overline{G}$ is normal in $G \times_\Theta H$. Similarly, since $e \in H$, $(e, e) \in \overline{H}$ and hence $\overline{H} \neq \varnothing$. And for (e, x), $(e, y) \in \overline{H}$, $(e, x) *_\Theta (e, y) = (e\Theta_x(e), xy) = (ee, xy) = (e, xy) \in \overline{H}$, and $(e, x)^{-1} = (\Theta_{x^{-1}}(e^{-1}), x^{-1}) = (\Theta_{x^{-1}}(e), x^{-1}) = (e, x^{-1}) \in \overline{H}$. So $\overline{H}$ is a subgroup of $G \times_\Theta H$. Then since $\overline{G}$ and $\overline{H}$ are subgroups of $G \times_\Theta H$, certainly $G \times_\Theta H \supseteq \overline{G}\,\overline{H}$. On the other hand, if $(x, y) \in G \times_\Theta H$, then $(x, y) = (xe, ey) = (x\Theta_e(e), ey) = (x, e) *_\Theta (e, y) \in \overline{G}\,\overline{H}$, and hence $G \times_\Theta H \subseteq \overline{G}\,\overline{H}$. Finally note that by definition of $\overline{G}$ and $\overline{H}$, $\overline{G} \cap \overline{H} = \{(e, e)\}$. ∎

Proposition 38.7   *Suppose that G is a group, that N is a normal subgroup of G, that H is a subgroup of G, that $G = NH$, and that $N \cap H = \{e\}$. For each $q \in H$, let $\Theta_q : N \to N$ be the function $\Theta_q(n) = qnq^{-1}$. Then $q \mapsto \Theta_p$ defines a homomorphism $\Theta : H \to \operatorname{Aut}(N)$, and letting $T(n, q) = nq$ defines an isomorphism $T : N \times_\Theta H \to G$.*

Proof   Note first that for $q \in H$ and $n \in N$, $\Theta_q(n) = qnq^{-1} \in N$ because $N \lhd G$, and hence Θ_q is indeed a function from N to N. Furthermore, $\Theta_q(nm) = qnmq^{-1} = qnq^{-1}qmq^{-1} = \Theta_q(n)\Theta_q(m)$ so that Θ_q is a homomorphism. If $\Theta_q(n) = \Theta_q(m)$, then $qnq^{-1} = qmq^{-1}$ and hence $n = m$ so that Θ_q is one-to-one. If $n \in N$, then $q^{-1}nq \in N$ because $N \lhd G$, and $\Theta_q(q^{-1}nq) = qq^{-1}nqq^{-1} = n$ so that Θ_q is onto. It follows that $\Theta_q \in \operatorname{Aut}(N)$. And if $p, q \in H$, then for any $n \in N$,

$$
\Theta_p \circ \Theta_q(n) = \Theta_p(\Theta_q(n)) = \Theta_p(qnq^{-1}) = pqnq^{-1}p^{-1} = (pq)n(pq)^{-1} = \Theta_{pq}(n),
$$

and hence $\Theta_p \circ \Theta_q = \Theta_{pq}$ so that $\Theta : H \to \operatorname{Aut}(N)$ is a homomorphism. It remains to show that T is an isomorphism. For (n, p), $(m, q) \in N \times_\Theta H$,

$$T((n, p) *_\Theta (m, q)) = T((n\Theta_p(m), pq))$$
$$= n\Theta_p(m)pq = npmp^{-1}pq = npmq = T((n, p))T((m, q)),$$

and hence T is a homomorphism. If $T((n, p)) = T((m, q))$, then $np = mp$, and hence $m^{-1}n = qp^{-1}$. But $m^{-1}n \in N$ and $qp^{-1} \in H$; so $m^{-1}n = e = qp^{-1}$ and thus $m = n$ and $p = q$, i.e., $(n, p) = (m, q)$. So T is one-to-one. Finally, if $g \in G$, then, since $G = NH$, there exist $n \in N$ and $q \in H$ such that $g = nq$. Then $(n, q) \in N \times_\Theta H$ and $T((n, q)) = nq = g$ so that T is onto. ∎

Example 38.8   We can use Proposition 38.7 to generate the representation of S_3 given in Example 38.4 as follows. As before, observe that $N = \{\iota, (123), (132)\}$ is a normal subgroup of S_3 and that $H = \{\iota, (12)\}$ is a subgroup of S_3. Since $(123)(12) = (13)$ and $(132)(12) = (23)$, $S_3 = NH$, and clearly $N \cap H = \{\iota\}$. So by Proposition 38.7, S_3 is isomorphic to $N \times_\Theta H$, where $\Theta: H \to \text{Aut}(N)$ is the function which takes $q \in H$ to $\Theta_q \in \text{Aut}(N)$ defined by $\Theta_q(n) = qnq^{-1}$. We've seen that $\text{Aut}(N) = \{g_1, g_2\}$, where g_1 and g_2 are the functions $g_1(x) = x$ and $g_2(x) = x^2$. Now $\Theta_\iota(n) = \iota n \iota^{-1} = n$ so that Θ_ι is the identity function, i.e., $\Theta_\iota = g_1$. And $\Theta_{(12)}$ is the following function:

n:	ι	(123)	(132)
↓	↓	↓	↓
$\Theta_{(12)}(n)$:	$(12)\iota(12)^{-1} = \iota$	$(12)(123)(12)^{-1} = (132)$	$(12)(132)(12)^{-1} = (123)$

That is, $\Theta_{(12)} = g_2$. Then $N \times_\Theta H$ is exactly the same semidirect product as that found in Example 38.4. There we argued that Proposition 34.15 is forced $N \times_\Theta H$ to be isomorphic to S_3; Proposition 38.7 provides the specific isomorphism. ❖

Exercises

In Exercises 1 and 2, show that if G is a group of order m and H is a group of order n, then there is only one semidirect product $G \times_\Theta H$.

1. $m = 1, n = 1$
2. $m = 1, n = 2$

In Exercises 3–7, suppose that G is a group of order m and H is a group of order n and find all semidirect products $G \times_\Theta H$.

3. $m = 2, n = 1$
4. $m = 2, n = 2$
5. $m = 2, n = 3$
6. $m = 3, n = 2$
7. $m = 3, n = 3$

8. Show that the function $\Theta:Q \to \text{Aut}(N)$ defined in Example 38.4 is a homomorphism.

9. Show that if G is a noncommutative group of order 6, then G is isomorphic to S_3.

10. Suppose that G and H are isomorphic groups and that G has subgroups $A \neq \{e\}$ and $B \neq \{e\}$ such that $A \cap B = \{e\}$. Show that H has subgroups $X \neq \{e\}$ and $Y \neq \{e\}$ such that $X \cap Y = \{e\}$.

In Exercises 11–14, use a semidirect product to find a noncommutative group of order n.

11. $n = 16$
12. $n = 24$
13. $n = 40$
14. $n = 100$

In Exercises 15–18, find k nonisomorphic noncommutative groups of order n. Remember to show that the groups you find are nonisomorphic.

15. $k = 2, n = 16$
16. $k = 3, n = 16$
17. $k = 2, n = 40$
18. $k = 3, n = 40$

Exercises 19–23 concern extensions of groups defined as follows. For groups G and H, an extension of G by H is a group E which has a normal subgroup N such that G is isomorphic to N and H is isomorphic to E/N.

19. Show that S_3 is an extension of $N = \{\iota, (123), (132)\}$ by $H = \{\iota, (12)\}$.

20. Show that $Q = \{\pm1, \pm i, \pm j, \pm k\}$ is an extension of $N = \{\pm1, \pm\iota\}$ by $H = \{\pm1\}$.

21. Show that $Q = \{\pm1, \pm i, \pm j, \pm k\}$ is an extension of $N = \{\pm1\}$ by $H = \mathbb{Z}_2 \times \mathbb{Z}_2$

22. Suppose that G and H are groups and that $\Theta:G \to \text{Aut}(H)$, $y \mapsto \Theta_y$, is a homomorphism. Show that the semidirect product $G \times_\Theta H$ is an extension of G by H.

23. Prove the following statement or give a counterexample. If a group E is an extension of a group G by a group H, then E is isomorphic to a semidirect product of G and H.

24. Show that $\mathbb{Z}_6$ is a semidirect product of $\mathbb{Z}_2$ and $\mathbb{Z}_3$.

25. Show that $\mathbb{Z}_8$ is not a semidirect product of groups with orders strictly less than 8.

26. Show that if p is a prime and n is a positive integer, then $\mathbb{Z}_{p^n}$ is not a semidirect product of groups with orders strictly less than p^n.

Appendix A
The Greek Alphabet

Greek Letter		Greek Name	English Equivalent
Small	Capital		
α	A	alpha	a
β	B	beta	b
γ	Γ	gamma	g
δ	Δ	delta	d
ϵ	E	epsilon	e (short)
ζ	Z	zeta	z
η	H	eta	e (long)
θ	Θ	theta	th
ι	I	iota	i
κ	K	kappa	k
λ	Λ	lambda	l
μ	M	mu	m
ν	N	nu	n
ξ	Ξ	xi	x
o	O	omicron	o (short)
π	Π	pi	p
ρ	P	rho	r
σ	Σ	sigma	s
τ	T	tau	t
υ	Y	upsilon	u
ϕ	Φ	phi	f (or ph)
χ	X	chi	ch
ψ	Ψ	psi	ps
ω	Ω	omega	o (long)

Appendix B
Proving Theorems

The purpose of this appendix is to review some important techniques of proof that are commonly used in mathematics. Since most theorems are stated in the form "if . . . then," we will deal almost exclusively with implications.

We first present the basic logical connectives and discuss their truth or falsity by means of truth tables, and then we introduce quantifiers. This puts us in a position where we can discuss various fundamental techniques of proof. Specifically, we discuss direct proof, disproof by counterexample, proving two statements equivalent, using the contrapositive, proof by contradiction, proving more than two statements equivalent, and proving a conclusion involving "or." At the end, we mention three common errors in the hope that pointing them out specifically will eradicate them. Note that proof by induction is not so much a technique of proof as a property of the integers. For that reason, it is not treated here but rather in Chapter 1, in which we derive various results about the integers from the well-ordering principle. There are of course techniques of proof and subtleties of argument that we do not consider. The reader who wants to investigate this material in more detail should consult a book on mathematical logic.

Logical Connectives

George Boole began his treatise *The Mathematical Analysis of Logic* in the following way: ". . . the validity of the processes of analysis does not depend upon the interpretation of the symbols which are employed, but solely upon the laws of their combination. Every system of interpretation which does not affect the truth of the relations supposed, is equally admissible, and it is thus that the same process may, under one scheme of interpretation, represent the solution of a question on the properties of numbers, under another, that of a geometrical problem, and under a third, that of dynamics or optics."[1] For instance, if a statement X and a statement Y are both true, then the statement "X and Y" is also true. Thus "$2 + 2 = 4$ and $1 + 3 = 4$" is a true statement because both the statements "$2 + 2 = 4$" and "$1 + 3 = 4$" are true statements. Similarly, for lines in the plane, the statement "lines parallel to parallel lines are parallel and lines perpendicular to parallel lines are parallel" is true because it is indeed the case that, in the plane, "lines parallel to parallel lines are parallel" and "lines perpendicular to parallel lines are parallel."

From this point of view, the following elements of a sentence become its crucial building blocks: conjunction ("and"), disjunction ("or"), negation ("not"), implication ("implies"), and equivalence ("if and only if"). However, since equivalence may be expressed in terms of the other relations—"X if and only if Y" is the same as "(X implies Y) and (Y implies X)"—we may restrict our attention to the other four connectives.

Boole's idea was to determine the truth of a sentence's constituent parts and then let rules governing the connectives determine the truth of the sentence. For instance, "$0 = 1$ or $2 + 2 = 4$" is true because the connective "or" requires only that at least one of the two statements be true, and $2 + 2$ does indeed equal 4.

What rules govern the connectives? Certainly "X and Y" is true if and only if both X and Y are true, "X or Y" is true if and only if X is true or Y is true or both are true, and "not X" is true if and only if X is not true.

Implication is somewhat more complicated. The first thing to note is that we are requiring that the truth of "X implies Y" depend only on the truth of X and the truth of Y. With this in mind, it is not difficult to analyze implications with true hypotheses. For suppose X is true. Since "X implies X" is certainly true, it must be the case that "X implies Y" is true whenever X is true and Y is true. On the other hand, "X implies (not X)" is certainly false and hence "X implies Y" is false whenever X is true and Y is false.

Common usage like this says little about the truth of statements with false hypotheses. However, common usage does require that implication be transitive, i.e., that the statement

$$((X \text{ implies } Y) \text{ and } (Y \text{ implies } Z)) \text{ implies } (X \text{ implies } Z) \ (*)$$

be true. From this observation, it follows that a false hypothesis always yields a true implication. For if X and Y are true and Z is false, then "Y implies Z" is false and hence the hypothesis "(X implies Y) and (Y implies Z)" is false. Since X is true and Z is false, the conclusion "X implies Z" is false, and thus the implication $(*)$ is a true implication with both a false hypothesis and a false conclusion. Since we are requiring that the truth of an implication depend only on the truth of its constituents, it must be that an implication with a false hypothesis and a false conclusion is always true. On the other hand, if X and Z are true and Y is false, then "X implies Y" is false and hence again the hypothesis "(*X* implies *Y*) and (*Y* implies *Z*)" is false. Since in this case *X* and *Z* are both true, the conclusion "*X* implies *Z*" is true, and thus the implication (*) is a true implication with a false hypothesis but a true conclusion. From this, it follows that an implication with a false hypothesis and a true conclusion is also always true. In summary, the only implication that is not true is one with a true hypothesis and a false conclusion.

Truth Tables

A very clear and concise way of describing the rules just presented is to use truth tables. In this formalism, false statements are given a value *F* and true statements are given a value *T*. Then he truth of "*X* implies *Y*" depends on the truth of *X* and the truth of *Y* in the following way:

X	*Y*	*X* implies *Y*
T	*T*	*T*
T	*F*	*F*
F	*T*	*T*
F	*F*	*T*

Similar tables may be constructed for "not," "or," or "and":

X	not *X*
T	*F*
F	*T*

X	*Y*	*X* or *Y*
T	*T*	*T*
T	*F*	*T*
F	*T*	*T*
F	*F*	*F*

X	*Y*	*X* and *Y*
T	*T*	*T*
T	*F*	*F*
F	*T*	*F*
F	*F*	*F*

We can construct truth tables for more complicated statements by combining these tables. For instance, the truth table for "(*X* and *Y*) implies (*X* or (not *Y*))" can be constructed as follows:

X	Y	X and Y	not Y	X or (not Y)	(X or (not Y)) implies (X and Y)
T	T	T	F	T	T
T	F	F	T	T	F
F	T	F	F	F	T
F	F	F	T	T	F

Note that different sentences may generate identical truth tables. For instance, "(not X) or (not Y)" has the same truth value as "not (X and Y)" and "(not X) and (not Y)" has the same truth value as "not (X or Y)":

X	Y	X and Y	not (X and Y)	not X	not Y	(not X) or (not Y)
T	T	T	F	F	F	F
T	F	F	T	F	T	T
F	T	F	T	T	F	T
F	F	F	T	T	T	T

X	Y	X or Y	not (X or Y)	not X	not Y	(not X) and (not Y)
T	T	T	F	F	F	F
T	F	T	F	F	T	F
F	T	T	F	T	F	F
F	F	F	T	T	T	T

And "(not X) or Y" has the same truth table as "X implies Y":

X	Y	not X	(not X) or Y
T	T	F	T
T	F	F	F
F	T	T	T
F	F	T	T

Statements with the same truth tables are **logically equivalent** in the sense that they are true or false in exactly the same situations, and hence we can deduce the truth of one from the truth of the other. This observation has several important consequences.

First, we can see immediately from the tables how to negate "and" and "or" because "not (X and Y)" is logically equivalent to "(not X) or (not Y)" and "not (X or Y)" is logically equivalent to "(not X) and (not Y)."

Second, as we have observed, "X implies Y" is logically equivalent to "(not X) or Y." So "implies" is definable in terms of "not" and "or," and hence any statement, even one involving implication, may be written by using only "not," "or," and "and." Furthermore, we can negate "X implies Y" by negating "(not X) or Y." That is, "not (X implies Y)" is logically equivalent to "X and (not Y)."

Mathematical theorems usually boil down to an implication "X implies Y." An inevitable question following the proof of "X implies Y" is whether "Y implies X" is also true. For this reason, it is useful to be able to refer to implications that are derived in particularly simple ways from a given implication. Specifically,

"Y implies X" is the **converse** of "X implies Y,"

"(not Y) implies (not X)" is the **contrapositive** of "X implies Y."

The contrapositive is particularly important because it is logically equivalent to the original implication:

X	Y	not X	not Y	(not Y) implies (not X)
T	T	F	F	T
T	F	F	T	F
F	T	T	F	T
F	F	T	T	T

Quantifiers

Of course, most interesting mathematical statements are not of such a simple form as "$0 = 1$ or $2 + 2 = 4$." They are more like the assertion "if n is positive, then n is odd." In particular, they are almost certain to involve variables. Now the truth of "if n is positive, then n is odd" depends on what n is. We can infer from the words "positive" and "odd" that n is meant to be an integer and hence what is really being asserted is "for all integers n, if n is positive, then n is odd." Unlike the initial assertion, this one is either true or false. (In fact, of course, it is false.) The difference between the two assertions is that in the second one, the variable is quantified by the phrase "for all." Mathematical assertions typically involve variables and those variables are typically restricted by using "for all" or "there exists."

Note that the two quantifiers "for all" and "there exists" frequently appear together, as in the statement "for all integers n, if n is positive, then there exists an integer k such that $k^2 = n$." The result is again a statement that is either true or false.

Note also that "there exists" can also appear as "for some." For instance, the statement "there exists an integer n such that $n^2 - 2n + 1 = 0$" is the same as the statement "$n^2 - 2n + 1 = 0$ for some integer n."

It is, of course, important to be able to negate statements involving quantifiers. This is easy to do, but care must be taken to ensure that the resulting sentence is syntactically correct.

First suppose that for each n, a particular statement involving n is either true or false. Then if it is not the case that the statement is true for every n, there must be some n for which it is false. So the negation of "for all x, $P(x)$" is "there exists x, not $P(x)$." Conversely, if it is not the case that there exists some n for which the statement is true, then it must be false for every n. That is, the negation of "there exists x, $P(x)$" is "for all x, not $P(x)$."

Finally note that blind application of these principles may lead to statements that are grammatically incorrect. For instance, consider the statement

for all integers n, either n is negative or there exists an integer k such that $k^2 = n$.

To negate this, recall that (1) the quantifier "for all" switches to "there exists," (2) the negation of "X or Y" is "(not X) and (not Y)," and (3) the quantifier "there exists" switches to "for all." The negation of the displayed statement is then

there exists an integer n such that n is nonegative and for all integers k, $k^2 \neq n$.

Note that the resulting sentence must be phrased in such a way that it is grammatically correct. For instance, the following negation of the original statement, while it is formed according to rules (1), (2), and (3), is not grammatically correct:

there exists an integer n, n is nonnegative and for all integers k such that $k^2 \neq n$.

We now turn our attention to fundamental techniques of proof. In what follows, recall that in an implication "X implies Y," X is called the **hypothesis** and Y is called the **conclusion** of the implication.

Direct Proof

A direct proof proves an assertion of the form "X implies Y" by proceeding directly from X to Y. Each step is an axiom or a previously proved result or follows logically from what precedes it. For instance, consider the following.

Proposition B.1   *If a nonempty subset of the integers is closed with respect to addition, then it is closed with respect to subtraction.*

Proof   Suppose that S is a nonempty subset of the integers which is closed with respect to subtraction. Since S is nonempty, it contains an element s, and since it is closed with respect to subtraction, $0 = s - s \in S$. So for any element $y \in S$, $-y = 0 - y \in S$, and hence for any elements $x, y \in S$, since S is closed with respect to subtraction, $x + y = x - (-y) \in S$. So S is closed with respect to addition. ∎

Note the steps used in the proof:

(1)   The set S is defined and it is noted that
   (*a*) S is nonempty, and
   (*b*) S is closed with respect to subtraction.
(2)   By (*a*), there exists $s \in S$.
(3)   By (*b*) and step (2), $0 \in S$.
(4)   By (*b*) and step (3), $y \in S$ implies $-y \in S$.
(5)   By (*b*) and step (4), S is additively closed.

Each step is one of the hypotheses or follows from a previous step. This is typical of a direct proof.

Another example of a direct proof is the following.

Proposition B.2   *If n is an odd integer, then $n^2 - 1$ is divisible by 8.*

Proof   Since n is odd, $n = 2k + 1$ for some integer k. Then

$$n^2 - 1 = 4k^2 + 4k + 1 - 1 = 4k(k + 1).$$

Then since either k or $k + 1$ is divisible by 2, $n^2 - 1$ is divisible by $4 \cdot 2 = 8$. ∎

Counterexamples

As previously indicated, the proof of an implication inevitably raises the question of whether the converse is true. The converse of Proposition B.1 is:

If a nonempty subset of the integers is closed with respect to addition, then it is closed with respect to subtraction.

We will show that this statement is false.

To do this, we of course need to show that its negation is true. So, as a first step, we need to negate the statement, and to do this, it is important to understand any hidden quantifiers. Phrased carefully, with quantifiers specified, this statement is:

For all nonempty subsets S of the integers,

if S is closed with respect to addition,

then S is closed with respect to subtraction.

So to negate the statement, we must change the quantifier from "for all" to "there exists" and then negate the implication. To negate the implication, recall that the negation of "X implies Y" is "X and (not Y)." So in this case, the negated statement is:

There exists a nonempty subset S of the integers such that

S is closed with respect to addition but

S is not closed with respect to subtraction.

Thus, to show that the converse of Proposition B.1 is false, we must show that there is at least one subset of the integers with the given properties. That is, we must produce an example, called a **counterexample,** of a set that is closed with respect to addition but not subtraction. Note that a counterexample is just that, an example. We are not required to prove, nor is it true, that every subset of the integers that is closed with respect to addition is not closed with respect to subtraction. Rather, we want to exhibit a single specific subset of the integers with the desired properties. It is very important to keep this in mind: To show that a statement introduced by the quantifier "for all" is false requires exhibiting a specific example for which the statement is false.

In the case of the converse of Proposition B.1, it is easy to find the desired subset.

Example B.3   Let S be the set of positive integers. We will show that S is closed with respect to addition but not subtraction. Certainly the sum of two positive integers is positive, and hence S is closed with respect to addition. However, 2 and 3 are in S but $2 - 3 = -1$ is not in S, so S is not closed with respect to subtraction. ❖

We thus conclude that the converse of Proposition B.1 is false.

It is quite common that, as in this case, the statement to be proved false begins with the quantifier "for all." Then the negated statement begins with the quantifier "there exists" and thus requires a counterexample. On the other hand, if the original statement begins with the quantifier "there exists," then the negated statement will begin with the quantifier "for all" and hence will require a general proof.

Another assertion that can be disproved by means of a counterexample is the statement "If $2^n - 1$ is prime, then n is even."

Example B.4   Since 31 is prime and $2^5 - 1 = 31$, n need not be even for $2^n - 1$ to be prime. Note that 3 is prime and $2^2 - 1 = 3$. So it is also not the case that if $2^n - 1$ is prime, then n is odd. ❖

Equivalent Statements

The other possibility is that the converse of a true implication is also true. In that case, the two statements are equivalent in the sense that each implies the other. As previously mentioned, the reverse implication "*X* implies *Y*" is usually phrased "*X* only if *Y*" so that an equivalence is typically phrased "*X* if and only if *Y*." To prove such an assertion requires two proofs, one for each implication.

Proposition B.5   *A nonempty subset of the integers is closed with respect to subtraction if and only if it is closed with respect to addition and contains the additive inverse of each of its elements.*

Proof   ($\Rightarrow$) By Proposition B.1, if S is a nonempty subset of the integers which is closed with respect to subtraction, then S is closed with respect to addition. We also showed in the proof of Proposition B.1 that if y is an element of such a set S, then $-y$ is also in S. So S contains the additive inverse of each of its elements.

($\Leftarrow$) Suppose that S is a nonempty subset of the integers which is closed with respect to addition and which contains the additive inverse of each of its elements. Then, for $x, y \in S$, $-y \in S$ and hence $x - y = x + (-y) \in S$. So S is closed with respect to subtraction. ■

Note that since two proofs are required, it is important to indicate clearly which implication is being proved. Parenthetical guides can help but of much more practical use is the clear stating of the hypotheses that are being assumed. They supply not only a reminder of what assumptions are being made, but also a guide to the flow of the proof.

Contrapositives

Of course, not every true statement is amenable to a direct proof of the given implication. One very useful alternative method is to give a direct proof of the contrapositive. For as we noted before, the contrapositive is logically equivalent to the original implication and hence proving the contrapositive proves the original implication.

Proposition B.6   *If n is a positive integer whose square is odd, then n is also odd.*

Proof   The contrapositive of the theorem is "If n is an even positive integer, then so is its square." So suppose that $n = 2k$ for a positive integer k. Then $n^2 = 4k^2$, and hence n^2 is even. So the contrapositive of the theorem is true and hence so is the theorem. ■

The contrapositive can also be useful when applying a previously proved implication.

Proposition B.7 *If S is a nonempty set of the integers such that $\{2s \mid s \in S\}$ is not contained in S, then there exist a, b ∈ S such that a − b ∉ S.*

Proof By hypothesis, there exists $s \in S$ such that $s + s = 2s \notin S$. So S is not closed with respect to addition. So by the contrapositive of proposition B.1, S is not closed with respect to subtraction, i.e., there exist $a, b \in S$ such that $a - b \notin S$. ■

Proof by Contradiction

Proving the contrapositive of an implication can be viewed in the following way. To prove an implication, one assumes that the hypothesis is true and proves that the conclusion is true. Proving the contrapositive assumes that the conclusion is false and then shows that the hypothesis is also false.

But the latter argument amounts to first assuming that the hypothesis is true and then further that the conclusion is false, and then proving that the hypothesis is both true and false, which is of course impossible. So one of the two assumptions must be wrong. This wrong assumption can only be that the conclusion is false and therefore the conclusion must be true.

This line of argument can be used in more general circumstances and is called a proof by contradiction. The negation of the conclusion is assumed, and then an obviously false statement is derived. It is a very good idea to warn the reader when embarking on such a proof.

Proposition B.8 *If n is an integer greater than one, then the smallest divisor of n that is greater than one is prime.*

Proof Let d be the smallest divisor of n that is greater than one, and suppose by way of contradiction that d is not prime. Then $d = ab$ for integers a and b such that $1 < a < d$ and $1 < b < d$. But since d divides n, $n = kd$ for some positive integer k. So $n = kab$ and hence b is a divisor of n that is greater than one but smaller than d. This contradicts our choice of d and hence we conclude that d must be prime. ■

Although the proof of Proposition B.8 is phrased as a proof by contradiction, it can be viewed as a proof of a contrapositive. For the proposition may be restated as follows:

For all integers n that are greater than one,

for all divisors d of n that are greater than one,

if d is the smallest such divisor,

then d is prime.

The proof of Proposition B.8 shows that if such a divisor d is not prime, then d is not the smallest such divisor of n, and this is precisely the contrapositive of the restated implication.

A proof by contradiction that cannot be reduced to a proof of the contrapositive is the following.

Proposition B.9   *There are an infinite number of prime numbers.*

Proof   Suppose by way of contradiction that there are only a finite number of prime numbers $p_1, \ldots, p_k$, and consider the integer is $n = p_1 \cdots p_k + 1$. Let d be the smallest divisor of n that is greater than one. By Proposition B.8, d is prime, and thus, since $p_1, \ldots, p_k$ is a complete list of the primes, $d = p_i$ for some i. Then d divides $n - p_1 \cdots p_k = 1$, and this is impossible because $d > 1$. This is a contradiction, and hence the number of primes must be infinite. ∎

The contradiction that is found in the proof of Proposition B.9 is that d is simultaneously greater than one and less than or equal to one. That is not a contradiction of an hypothesis; rather it is a statement of the form "X and (not X)," which is always false. This type of contradiction occurs frequently.

Lists of Equivalent Statements

Sometimes we need to show that more than two statements are equivalent. To do this, it is usually most efficient to prove a circle of implications rather than to prove that each pair is equivalent. For instance, to prove that X, Y, and Z are all equivalent requires showing the six implications

$$X \Rightarrow Y, \ Y \Rightarrow X,$$
$$X \Rightarrow Z, \ Z \Rightarrow X, \text{ and}$$
$$Y \Rightarrow Z, \ Z \Rightarrow Y.$$

It is quicker to observe that all six implications follow from the three implications:

$$X \Rightarrow Y, \ Y \Rightarrow Z, \text{ and } Z \Rightarrow X.$$

For the first two show that $X \Rightarrow Z$, the second two show that $Y \Rightarrow X$, and the first and third show that $Z \Rightarrow Y$.

As an example of this technique, consider the following.

Proposition B.10   *If n is a positive integer, then the following statements are equivalent:*

   (*i*)   *n is odd,*
   (*ii*)   *$n + 1$ is even,*
   (*iii*)   *n^2 is odd.*

Proof   That (*i*) implies (*ii*) is obvious.
   To show that (*ii*) implies (*iii*), suppose that $n + 1$ is even. That is, $n + 1 = 2k$ for some positive integer k and hence $4k^2 = (n + 1)^2 = n^2 + 2n + 1$. Then

$n^2 = 4k^2 - 2n - 1$, and since $4k^2 - 2n = 2(2k^2 - n)$, $4k^2 - 2n$ is even. It follows that n^2, being an even number minus one, is odd.

That (*iii*) implies (*i*) is Proposition B.6. ∎

Conclusions Involving "Or"

Efficiently proving an implication of the form "X implies (Y or Z)" requires a special technique.

We can analyze the situation as follows. To prove the implication "X implies (Y or Z)," we assume X and want to show "Y or Z." There are two possibilities: Either Y holds or it does not. If Y holds, then certainly "Y or Z" holds and we are done. On the other hand, if Y does not hold, then "not Y" does hold. So if we can show that "(X and (not Y)) implies Z," then in the latter case, we will have shown that Z holds, and hence that "Y or Z" holds.

So to prove "X implies (Y or Z)," it suffices to prove "(X and (not Y)) implies Z." Since certainly "Y or Z" is logically equivalent to "Z or Y," the implication "X implies (Y or Z)" will also follow if we can prove "(X and (not Z)) implies Y." Note that only <u>one</u> of these alternate implications must be proved; the original implication follows from either of the alternate implications.

Of course, these observations also follow from what we already know about logically equivalent statements. Specifically, we know the following:

"X implies (Y or Z)" is logically equivalent to

"(not X) or (Y or Z)," which is the same as

"(not X or Y) or Z," which is logically equivalent to

"(not (X and (not Y))) or Z," which is logically equivalent to

"(X and (not Y)) implies Z."

Proposition B.11   *If k and n are integers and $|k + n| = |k|$, then $kn < 0$ or $n = 0$.*

Proof   Suppose that $|k + n| = |k|$ and that $kn \not< 0$. Then $kn \geq 0$ and as well

$$k^2 = |k|^2 = |k + n|^2 = (k + n)^2 = k^2 + 2kn + n^2.$$

So $2kn = -n^2 \leq 0$, and thus $kn \leq 0$ so that in fact $kn = 0$. From this, it follows that if $k \neq 0$, $n = 0$, and if $k = 0$, then $|n| = |0 + n| = |0|$ so that $n = 0$ in this case as well. We conclude that if $|k + n| = |k|$, then either $kn < 0$ or $n = 0$. ∎

Note that we could also prove Proposition B.11 by supposing that $|k + n| = |k|$ and that $n \neq 0$ and then proving that $kn < 0$.

Another example is the following.

Proposition B.12   *If n is an integer, then either $n = 3k$ for some integer k or $n^2 = 3k + 1$ for some integer k.*

Proof   Suppose that $n \neq 3k$ for any integer k. Then there exists an integer q such that $n = 3q + 1$ or $n = 3q + 1$. In the first case, we have

$$n^2 = (3q + 1)^2 = 9q^2 + 6q + 1 = 3(3q^2 + 2q) + 1,$$

and in the second case, we have

$$n^2 = (3q + 2)^2 = 9q^2 + 12q + 4 = 3(3q^2 + 4q + 1) + 1.$$

So in both cases, there exists an integer k such that $n^2 = 3k + 1$. ∎

Three Important Considerations

(1)   <u>Always use complete, grammatically correct sentences.</u> The following scratch work should never form part of a formal proof:

V vector space. *B* basis. *B* linearly independent.

Rather, express the same argument in grammatically correct sentences:

Let *V* be a vector space and let *B* be a basis of *V*. Then *B* is linearly independent.

(2)   <u>Never start a sentence with a symbol.</u> Starting sentences with symbols can lead to unfortunate situations. The following example is not proper English usage because the second sentence does not start with a capital letter.

Let $f(x) = x^2$. f is a function.

And the following correction is not mathematically correct because the name of the function is f rather than F.

Let $f(x) = x^2$. F is a function.

Such problems can be avoided by adopting the general practice of never starting a sentence with a symbol.

(3)   <u>Do not assume what you want to prove.</u> This probably seems obvious. But it is surprisingly easy to fall into the following trap.

Proposition B.13   *For $0 < x < \frac{\pi}{2}$, $\cot(2x) = \frac{1}{2}\cot(x) - \frac{1}{2}\tan(x)$.*

Incorrect Proof
$$\cot(2x) = \frac{1}{2}\cot(x) - \frac{1}{2}\tan(x)$$

$$\frac{\cos(2x)}{\sin(2x)} = \frac{\cos(x)}{2\sin(x)} - \frac{\sin(x)}{2\cos(x)}$$

$$\frac{\cos^2(x) - \sin^2(x)}{2\sin(x)\cos(x)} = \frac{\cos^2(x) - \sin^2(x)}{2\sin(x)\cos(x)} \quad ✔$$

The first step of this alleged proof is $\cot(2x) = \frac{1}{2}\cot(x) - \frac{1}{2}\tan(x)$, which is of course what is to be proved. So the technique of proof amounts to starting with what is to be proved, deducing a true statement, and then concluding that the original statement is true. This method can be used to prove that $1 = -1$:

$$1 = -1$$
$$(1)^2 = (-1)^2$$
$$1 = 1 \quad \checkmark$$

This method should never be used. If it hides a correct proof, then it is easy to find it. The expressions on the left-hand side should all be equal, as should all the expressions on the right-hand side. So a correct proof can be found by equating expressions down the left-hand side and then up the right-hand side. In this case, we can construct the following correct proof. ∎

Proposition B.13   *For $0 < x < \frac{\pi}{2} < \cot(x) = \frac{1}{2}\cot(x) - \frac{1}{2}\tan(x)$.*

Correct Proof   Let $0 < x < \frac{\pi}{2}$. Then

$$\cot(2x) = \frac{\cos(2x)}{\sin(2x)} = \frac{\cos^2(x) - \sin^2(x)}{2\sin(x)\cos(x)}$$
$$= \frac{\cos(x)}{2\sin(x)} - \frac{\sin(x)}{2\cos(x)} = \frac{1}{2}\cot(x) - \frac{1}{2}\tan(x). \quad ∎$$

HISTORICAL NOTE

George Boole

George Boole was born in Lincoln, England, on November 2, 1815, and died in Ballintemple, Cork, Ireland, on December 8, 1864. His father, John Boole, was a cobbler with a keen interest in mathematics and the making of optical instruments, and his mother, Mary Ann Joyce, had been employed as a lady's maid before her marriage. Although his parents married when John was twenty-nine and Mary Ann was twenty-six, they had to wait ten years for George, the first of their four children, to arrive. John's many outside activities and his poor business sense meant that George had to educate himself for the most part, albeit with great encouragement from both his parents. When George was fourteen, John published his son's translation of Meleager's "Ode to Spring" in the local newspaper. It was a sufficiently good rendering to draw unfounded accusations of plagiarism.

When George was sixteen, his father's business collapsed. George had been considering a career in the church but, assailed by religious doubts and with his family to support, he became instead a teacher at a school in Doncaster, some forty miles away. At this time, in 1831, he discovered a calculus text by Laroix, and upon reading it, he began his career in mathematics. While he was walking across a field in 1833, he was struck by the thought that logical relations could be expressed in algebraic form, an idea that eventually flowered into his major contribution to mathematics.

Lonely and unhappy in Doncaster, he moved to Liverpool and then back to Lincoln, and in 1834, at the age of nineteen, he opened his own school. At the same time, he became involved with the newly opened Mechanics Institute, and his address on Newton became his first scientific publication when it appeared in 1835. He contributed regularly to the recently founded *Cambridge Mathematical Journal* and won a medal for his work on operators in 1844. In 1847, stimulated by a controversy raging between Augustus De Morgan and the Scottish philosopher William Hamilton (not the Irish mathematician William Rowan Hamilton), Boole returned to the study of logic, and hurriedly wrote *The Mathematical Analysis of Logic,* which appeared in 1847. Then, in 1849, he moved to Ireland to become professor of mathematics in the newly established Queen's College in Cork. Although he worked very hard at his teaching and other university duties, he found time to continue his mathematical work and was able to rethink and polish the ideas presented in 1847. The result was *An Investigation of the Laws of Thought,* which appeared in 1854.

Boole was apparently a man of romantic disposition who regularly fell in and out of love. However, he had never married, partly perhaps because of his precarious financial situation. In 1850, eighteen-year-old Mary Everest arrived in Cork on a visit to her uncle, John Ryall, Vice-President and Professor of Greek at Queen's College. (Another uncle was George Everest, after whom Mount Everest was named.) At the suggestion of a friend of Mary's, Boole became her tutor, and when her father died in 1855 leaving her ill and destitute, they were married in spite of the differences in their ages, Mary's health, and Boole's small income. They were proved right in the end; by Mary's account, the marriage was a "sunny dream."[2] Nine months and one week after they were married, the first of their five daughters was born. According to one acquaintance who was in an alley behind the College looking for a chimney sweep, Boole was so delighted at becoming a father that he was "passionately shaking hands with a ragged and barefoot man, and saying 'I had to come and tell you dear friends; I've got a baby and she *is* such a beauty."[3]

On November 24, 1864, Boole walked three miles in the pouring rain in order not to miss a lecture. He lectured in his wet clothes, came down with a fever, and died on December 8.

Boole's family had a difficult life after he died and his daughters had few educational opportunities. In spite of this, his third daughter, Alicia, managed to become a mathematician, introducing the now common term "polytope" to describe convex regular solids and showing that there were exactly six polytopes in four dimensions.

She published geometrical results sporadically throughout her life. His fourth daughter became a chemist and his youngest daughter, Ethel Lilian, became a writer. Her greatest success was a novel called *The Gadfly,* which in the hundred years since it was published has sold 5 million copies in the former Soviet Union.

Boole's mathematical work marks the divide between metaphysical logic and mathematical logic. Before Boole, logic was viewed as derived from ordinary language; after him, it was viewed as a construct which can be interpreted in everyday language. The consequences of this change of viewpoint have been enormous. Boole's fundamental insight, Boolean algebra, is the basis of much of modern computer science.

References

1. Quoted in Boole, George. *The Mathematical Analysis of Logic.* Oxford: Basil Blackwell, 1965; a reprint of the edition published in 1847 by MacMillan, Barclay & MacMillan, Cambridge, p. 3.
2. MacHale, Desmond. *George Boole.* Dublin: Boole Press, 1985, p. 111.
3. *Ibid.,* p. 157.

Appendix C
Vector Spaces Over Fields

This appendix ties up some loose ends in the preceding sections by recalling some of the proofs from elementary linear algebra in the general setting of vector spaces over fields. We begin with the definition of such a vector space.

DEFINITION

A set V is a **vector space** over a field F if V has defined on it a closed, associative, and commutative binary operation $+$ with respect to which it has an identity and all of its elements have inverses (cf. Chapters 3 and 4), and if, for every $\delta \in F$ and $\mathbf{v} \in V$, there exists an element $\delta\mathbf{v} \in V$ for which the following conditions hold for all $\delta, \gamma \in F$ and all $\mathbf{v}, \mathbf{w} \in V$:

(i) $\delta(\mathbf{v} + \mathbf{w}) = \delta\mathbf{v} + \delta\mathbf{w}$

(ii) $(\delta + \gamma)\mathbf{v} = \delta\mathbf{v} + \gamma\mathbf{v}$

(iii) $\delta(\gamma\mathbf{v}) = (\delta\gamma)\mathbf{v}$

(iv) $1\mathbf{v} = \mathbf{v}.$

The additive identity of V is denoted by $\mathbf{0}$. A vector space is **nontrivial** if it has at least two elements. The elements of V are called **vectors** and the elements of F are called **scalars.** A subset of V which is a vector space with respect to the inherited operations is called a **subspace** of V over F.

For the proofs in the sequel, we will need the following computational tools.

Lemma C.1   *Let F be a field and suppose that V is a vector space over F. Then*

(i) $0\mathbf{v} = \mathbf{0}$ *for all* $\mathbf{v} \in V$;

(ii) $\delta\mathbf{0} = \mathbf{0}$ *for all* $\delta \in F$.

Proof   (i)   For all $\mathbf{v} \in V$, $0\mathbf{v} = 0\mathbf{v} + 0\mathbf{v} - 0\mathbf{v} = (0 + 0)\mathbf{v} - 0\mathbf{v} = 0\mathbf{v} - 0\mathbf{v} = \mathbf{0}.$

(ii) Similarly, for all $\delta \in F$, $\delta\mathbf{0} = \delta\mathbf{0} + \delta\mathbf{0} - \delta\mathbf{0} = \delta(0 + 0) - \delta\mathbf{0} = \delta\mathbf{0} - \delta\mathbf{0} = \mathbf{0}.$ ∎

Next we recall the definition of a basis of a vector space and collect some facts about linear dependence and independence.

DEFINITION

Let V be a vector space over a field F. The subset $\{\mathbf{v}_1, \ldots, \mathbf{v}_n\}$ of V is **linearly independent** if

for all $\delta_1, \ldots, \delta_n \in F$, $\delta_1\mathbf{v}_1 + \cdots + \delta_n\mathbf{v}_n = \mathbf{0}$ implies that
$$\delta_1 = \delta_2 = \cdots = \delta_n = 0;$$

$\{\mathbf{v}_1, \ldots, \mathbf{v}_n\}$ is **linearly dependent** if it is not linearly independent, i.e., if

there exist $\delta_1, \ldots, \delta_n \in F$, not all 0, such that $\delta_1\mathbf{v}_1 + \cdots + \delta_n\mathbf{v}_n = \mathbf{0}$;

$\{\mathbf{v}_1, \ldots, \mathbf{v}_n\}$ **spans** V if

for all $\mathbf{u} \in V$, there exist $\gamma_1, \ldots, \gamma_n \in F$ such that $\mathbf{u} = \gamma_1\mathbf{v}_1 + \cdots + \gamma_n\mathbf{v}_n$.

A **basis** of V is a linearly independent subset of V which spans V. An expression of the form $\gamma_1\mathbf{v}_1 + \cdots + \gamma_n\mathbf{v}_n$, where $\gamma_1, \ldots, \gamma_n \in F$, is called a **linear combination** of the vectors $\mathbf{v}_1, \ldots, \mathbf{v}_n$.

Lemma C.2   *Suppose that V is a vector space over the field F and that S and T are finite subsets of V. Then*

(i)   *if $S \subseteq T$ and S is linearly dependent, T is linearly dependent;*

(ii)   *if $S \subseteq T$ and T is linearly independent, S is linearly independent;*

(iii)   *if $\mathbf{0} \in S$, S is linearly dependent;*

(iv)   *if $\mathbf{0} \neq \mathbf{v} \in V$, $\{\mathbf{v}\}$ is linearly independent;*

(v)   *if $\{\mathbf{w}_1, \ldots, \mathbf{w}_m\}$ is linear independent and does not span V, then there exists $\mathbf{w}_{m+1} \in V$ such that $\{\mathbf{w}_1, \ldots, \mathbf{w}_{m+1}\}$ is linearly independent.*

Proof   Let $S = \{\mathbf{s}_1, \ldots, \mathbf{s}_k\}$ and $T = \{\mathbf{t}_1, \ldots, \mathbf{t}_m\}$.

(i) Suppose that $\sigma_1\mathbf{s}_1 + \cdots + \sigma_k\mathbf{s}_k = \mathbf{0}$ for $\sigma_i \in F$, not all 0, and let $\tau_i = \sigma_j$ if $\mathbf{t}_i = \mathbf{s}_j$ and $\tau_i = 0$ if $\mathbf{t}_i \notin S$. Then not all the τ_i are 0 and by Lemma C.1 (i), $\tau_1\mathbf{t}_1 + \cdots + \tau_m\mathbf{t}_m = \sigma_1\mathbf{s}_1 + \cdots + \sigma_k\mathbf{s}_k = \mathbf{0}$. Therefore T is linearly dependent.

(ii) Statement (ii) is the contrapositive of statement (i).

(iii) Suppose that $\mathbf{0} = \mathbf{s}_i$ and let $\sigma_j = 1$ if $j = i$ and $\sigma_j = 0$ if $j \neq i$. By Proposition 5.1 (vii), $1 \neq 0$, and hence not all the σ_i are 0. But by Lemma C.1 (i) and (ii), $\sigma_1\mathbf{s}_1 + \cdots + \sigma_k\mathbf{s}_k = 1 \cdot \mathbf{0} = \mathbf{0}$ and hence S is linearly dependent.

(iv) If $\sigma\mathbf{v} = 0$ for some $0 \neq \sigma \in F$, then $\mathbf{v} = 1\mathbf{v} = \sigma^{-1}(\sigma\mathbf{v}) = \sigma^{-1}\mathbf{0} = \mathbf{0}$ by Lemma C.1 (ii), a contradiction of our choice of $\mathbf{v}$. So $\sigma\mathbf{v} = \mathbf{0}$ implies $\sigma = 0$, and hence $\{\mathbf{v}\}$ is linearly independent.

(*v*) Since $\{\mathbf{w}_1, \ldots, \mathbf{w}_m\}$ does not span V, there exists an element $\mathbf{w}_{m+1} \in W$ which cannot be written as a linear combination of $\mathbf{w}_1, \ldots, \mathbf{w}_m$. If $\{\mathbf{w}_1, \ldots, \mathbf{w}_{m+1}\}$ is not linearly independent, then there exist $\alpha_1, \ldots, \alpha_{m+1} \in F$, not all 0, such that $\alpha_1 \mathbf{w}_1 + \cdots + \alpha_{m+1} \mathbf{w}_{m+1} = \mathbf{0}$. If $\alpha_{m+1} = 0$, then we have $\alpha_1, \ldots, \alpha_m \in F$, not all 0, such that $\alpha_1 \mathbf{w}_1 + \cdots + \alpha_m \mathbf{w}_m = \mathbf{0}$, a contradiction of our assumption that $\{\mathbf{w}_1, \ldots, \mathbf{w}_m\}$ is linearly independent; thus $\alpha_{m+1} \neq 0$. But then $\mathbf{w}_{m+1} = (-\alpha_{m+1}^{-1}\alpha_1)\mathbf{w}_1 + \cdots + (-\alpha_{m+1}^{-1}\alpha_m)\mathbf{w}_m$, a contradiction of our assumption that $\mathbf{w}_{m+1}$ cannot be written as a linear combination of $\mathbf{w}_1, \ldots, \mathbf{w}_m$. We conclude that $\{\mathbf{w}_1, \ldots, \mathbf{w}_{m+1}\}$ must be linearly independent. ∎

We now have the tools necessary to prove the elementary properties of vector spaces that we have used in the previous sections. We begin by proving Proposition 12.1 (*i*).

Proposition C.3 *Suppose that V is a nontrivial vector space over the field F which is spanned by the finite set $\{\mathbf{w}_1, \ldots, \mathbf{w}_k\}$. Then there exists a subset $\{\mathbf{b}_1, \ldots, \mathbf{b}_n\}$ of $\{\mathbf{w}_1, \ldots, \mathbf{w}_k\}$ which is a basis of V over F.*

Proof Since $\{\mathbf{w}_1, \ldots, \mathbf{w}_k\}$ spans V and V has at least two elements, $\{\mathbf{w}_1, \ldots, \mathbf{w}_k\}$ contains a nonzero element, and hence by Lemma C.2 (*iv*), $\{\mathbf{w}_1, \ldots, \mathbf{w}_k\}$ has at least one linearly independent subset. Let $\mathbb{L}$ denote the set of all linearly independent subsets of $\{\mathbf{w}_1, \ldots, \mathbf{w}_k\}$ and pick a subset $\{\mathbf{b}_1, \ldots, \mathbf{b}_n\}$ in $\mathbb{L}$ whose number of elements is maximal among all the subsets in $\mathbb{L}$. That is, $\{\mathbf{b}_1, \ldots, \mathbf{b}_n\}$ is linearly independent and $\{\mathbf{b}_1, \ldots, \mathbf{b}_n, \mathbf{w}_i\}$ is linearly dependent for any $\mathbf{w}_i \notin \{\mathbf{b}_1, \ldots, \mathbf{b}_n\}$. But then there exist $\delta_1, \ldots, \delta_n, \gamma_i \in F$, not all zero, such that $\delta_1 \mathbf{b}_1 + \cdots + \delta_n \mathbf{b}_n + \gamma_i \mathbf{w}_i = \mathbf{0}$. If $\gamma_i = 0$, then $\delta_1 \mathbf{b}_1 + \cdots + \delta_n \mathbf{b}_n = \mathbf{0}$ and not all the δ_j are 0. Since this contradicts our assumption that $\{\mathbf{b}_1, \ldots, \mathbf{b}_n\}$ is linearly independent, we conclude that $\gamma_i \neq 0$, and hence that $\mathbf{w}_i = (-\gamma_i^{-1}\delta_1)\mathbf{b}_1 + \cdots + (-\gamma_i^{-1}\delta_n)\mathbf{b}_n$. Since such an equation holds for every $\mathbf{w}_i \notin \{\mathbf{b}_1, \ldots, \mathbf{b}_n\}$, every linear combination of the elements of $\{\mathbf{w}_1, \ldots, \mathbf{w}_k\}$ may be written as a linear combination of the elements $\{\mathbf{b}_1, \ldots, \mathbf{b}_n\}$, i.e., $\{\mathbf{b}_1, \ldots, \mathbf{b}_n\}$ spans V. Since $\{\mathbf{b}_1, \ldots, \mathbf{b}_n\}$ was chosen to be linearly independent, $\{\mathbf{b}_1, \ldots, \mathbf{b}_n\}$ is a basis of V. ∎

The following result has Proposition 12.1 (*ii*) as an immediate corollary.

Proposition C.4 *Suppose that V is a vector space over a field F with a finite basis $\{\mathbf{b}_1, \ldots, \mathbf{b}_n\}$. Then*

(*i*) *any subset of V with $n + 1$ distinct elements is linearly dependent;*

(*ii*) *any subset with less than n elements cannot span V.*

Proof (*i*) Suppose that $\mathbf{v}_1, \ldots, \mathbf{v}_{n+1}$ are distinct vectors in V, and note that if $\mathbf{v}_i = \mathbf{0}$ for any i, then $\{\mathbf{v}_1, \ldots, \mathbf{v}_{n+1}\}$ is linearly dependent by Lemma C.2 (*iii*). So suppose that none of the $\mathbf{v}_i$ are $\mathbf{0}$ and observe that since $\{\mathbf{b}_1, \ldots, \mathbf{b}_n\}$ spans V, we have

$$\mathbf{v}_1 = \alpha_{1,1}\mathbf{b}_1 + \cdots + \alpha_{1,n}\mathbf{b}_n \tag{1}$$

$$\mathbf{v}_2 = \alpha_{2,1}\mathbf{b}_1 + \cdots + \alpha_{2,n}\mathbf{b}_n \tag{2}$$

$$\vdots \qquad\qquad\qquad \vdots$$

$$\mathbf{v}_{n+1} = \alpha_{n+1,1}\mathbf{b}_1 + \cdots + \alpha_{n+1,n}\mathbf{b}_n. \tag{$n+1$}$$

Since $\mathbf{v}_1 \neq \mathbf{0}$, some $\alpha_{1,k} \neq 0$ and hence we may solve equation (1) for $\mathbf{b}_k$ in terms of $\mathbf{v}_1$ and the other $\mathbf{b}_j$. Using this solution, we may substitute for $\mathbf{b}_k$ in each of the remaining equations. Now consider the revised equation (2). The right-hand side involves $\mathbf{v}_1$ and all the $\mathbf{b}_j$ except $\mathbf{b}_k$, and by Proposition 5.1 (*vii*), the coefficient of $\mathbf{v}_2$ on the left-hand side is not 0. Thus, if the coefficients of all the $\mathbf{b}_j$ are all 0, then $\{\mathbf{v}_1, \mathbf{v}_2\}$ is linearly dependent and hence $\{\mathbf{v}_1, \ldots, \mathbf{v}_{n+1}\}$ is linearly dependent by Lemma C.2 (*i*). On the other hand. if the coefficient of some $\mathbf{b}_l$ is not 0, then we may solve equation (2) for $\mathbf{b}_l$ and substitute the result into the remaining equations. Continuing in this fashion, either we find that $\{\mathbf{v}_1, \ldots, \mathbf{v}_{n+1}\}$ is linearly dependent before considering equation ($n+1$) or, upon completing the substitution into equation ($n+1$), we have an equation involving $\mathbf{v}_1, \ldots, \mathbf{v}_n$ on the right-hand side and $\mathbf{v}_{n+1}$ on the left-hand side, not all of whose coefficients are 0. Thus, in the later case as well, $\{\mathbf{v}_1, \ldots, \mathbf{v}_{n+1}\}$ is linearly dependent.

(*ii*) Suppose that $m < n$ and that $\{\mathbf{v}_1, \ldots, \mathbf{v}_m\}$ spans V. Then

$$\mathbf{b}_1 = \delta_{1,1}\mathbf{v}_1 + \cdots + \delta_{1,m}\mathbf{v}_m \tag{1}$$

$$\mathbf{b}_2 = \delta_{2,1}\mathbf{v}_1 + \cdots + \delta_{2,m}\mathbf{v}_m \tag{2}$$

$$\vdots \qquad\qquad\qquad \vdots$$

$$\mathbf{b}_n = \delta_{n,1}\mathbf{v}_1 + \cdots + \delta_{n,m}\mathbf{v}_m. \tag{n}$$

As in the proof of (*i*), some $\delta_{1,k} \neq 0$ so that we can solve equation (1) for $\mathbf{v}_k$ and substitute the result into the remaining equations. If the coefficients of all the $\mathbf{v}_i$'s remaining in the new equation (2) are 0, then the equation only involves $\mathbf{b}_1$ on the right-hand side and $\mathbf{b}_2$ on the left-hand side, and the coefficient of $\mathbf{b}_2$ is not 0. But then $\{\mathbf{b}_1, \mathbf{b}_2\}$, and hence $\{\mathbf{b}_1, \ldots, \mathbf{b}_n\}$ must be linearly dependent, a contradiction. So the coefficient of some $\mathbf{v}_l$ is not 0 and hence we may solve equation (2) for $\mathbf{v}_l$ and substitute the result into the remaining equations. Since $m < n$, we can continue this process until we reach equation (n), which will then have $\mathbf{b}_1, \ldots, \mathbf{b}_{n-1}$ on the right-hand side and $\mathbf{b}_n$ on the left-hand side and not all of whose coefficients will be 0. Then $\{\mathbf{b}_1, \ldots, \mathbf{b}_n\}$ will be linearly dependent, a contradiction. ■

Corollary C.5   *Suppose that V is a vector space over a field F with a finite basis $\{\mathbf{b}_1, \ldots, \mathbf{b}_n\}$. Then every basis of V over F has n elements.*

Proof If $\{\mathbf{v}_1, \ldots, \mathbf{v}_m\}$ is also a basis of V, then Proposition C.4 (*i*) implies that $m \leq n$ because $\{\mathbf{v}_1, \ldots, \mathbf{v}_m\}$ is linearly independent, and Proposition C.4 (*ii*) implies that $m \geq n$ because $\{\mathbf{v}_1, \ldots, \mathbf{v}_m\}$ spans V. ■

We conclude by proving Proposition 12.1 (*iii*) and (*iv*).

Proposition C.6 *Suppose that V is a vector space over a field F with a finite basis $\{\mathbf{b}_1, \ldots, \mathbf{b}_n\}$ and that W is a nontrivial subspace of V. Then W has a basis with no more than n elements. If $W \neq V$, W has a basis with less than n elements.*

Proof Note first that if W has a finite basis $\{\mathbf{w}_1, \ldots, \mathbf{w}_m\}$, then $m \leq n$. For since $\{\mathbf{w}_1, \ldots, \mathbf{w}_m\}$ is a linearly independent subset of W, then it is also a linearly independent subset of V, and hence if $m > n$, Lemma C.2 (*ii*) implies that $\{\mathbf{w}_1, \ldots, \mathbf{w}_{n+1}\}$ is linearly independent. Since this contradicts Proposition C.4 (*i*), we conclude that $m \leq n$.

Since W is a nontrivial vector space, we may pick $\mathbf{0} \neq \mathbf{u}_1 \in W$. Note that by Lemma C.2 (*iv*), $\{\mathbf{u}_1\}$ is linearly independent. If $\{\mathbf{u}_1\}$ spans W, then $\{\mathbf{u}_1\}$ is a basis of W and as noted before, $1 \leq n$. So suppose that $\{\mathbf{u}_1\}$ does not span W. Then by Lemma C.2 (*v*), there exists $\mathbf{u}_2 \in W$ such that $\{\mathbf{u}_1, \mathbf{u}_2\}$ is linearly independent. If $\{\mathbf{u}_1, \mathbf{u}_2\}$ spans W, then $\{\mathbf{u}_1, \mathbf{u}_2\}$ is a basis of W and as noted before, $2 \leq n$. This process can only continue for n steps because by Proposition C.4 (*i*), V, and hence W, has no linearly independent subsets with $n + 1$ elements.

Now suppose that $W \neq V$. We have shown that W has a finite basis $\{\mathbf{w}_1, \ldots, \mathbf{w}_m\}$ such that $m \leq n$. Since $W \neq V$, $\{\mathbf{w}_1, \ldots, \mathbf{w}_m\}$ does not span V, and hence by Lemma C.2 (*v*), there exists $\mathbf{w}_{m+1} \in V$ such that $\{\mathbf{w}_1, \ldots, \mathbf{w}_{m+1}\}$ is linearly independent. If $m = n$, then $\{\mathbf{w}_1, \ldots, \mathbf{w}_{m+1}\}$ is a linearly independent subset of V with more than n elements in contradiction of Proposition C.4 (*i*). We conclude that $m < n$. ■

Appendix D
Constructions with Straightedge and Compass

In this appendix, we will solve a problem even older than that of solving polynomial equations by radicals. The ancient Greeks were very sophisticated geometers, but they were unable to find methods of trisecting angles or duplicating cubes by means of straightedge and compass alone. We will use the results of Chapters 1–12 to show that it is not possible to trisect every angle, nor is it possible to duplicate every cube. As with the case of solvability by radicals, our first job is to phrase the questions precisely.

What does it mean to construct a figure by straightedge and compass? We start with a finite set of points, $P_1, \ldots, P_n$, in Euclidean two-dimensional space and then find all lines joining these points and all circles having centers at these points and radii equal to line segments determined by these points. We then form a new set of points consisting of all points of intersection of these lines and circles. This gives us a bigger set of points to which we can apply the same construction. Any point that can be obtained in a finite number of steps in this fashion is said to be **constructed by straightedge and compass from** $P_1, \ldots, P_n$. We let $C(P_1, \ldots, P_n)$ denote the set of all points that can be so constructed.

If $n = 1$, then $C(P_1) = \{P_1\}$. If $n \geq 2$, then we may certainly choose a Cartesian coordinate system such that $P_1 = (0, 0)$ and $P_2 = (1, 0)$. An angle Θ can then be constructed if and only if the point $(\cos \Theta, \sin \Theta)$ on the unit circle can be constructed, and therefore an angle Θ can be trisected if and only if $P_4 \in C(P_1, P_2, P_3)$, where

$$P_1 = (0, 0), \quad P_2 = (1, 0), \quad P_3 = (\cos \Theta, \sin \Theta), \quad \text{and} \quad P_4 = \left(\cos \frac{\Theta}{3}, \sin \frac{\Theta}{3} \right).$$

To duplicate a cube means that another cube can be constructed whose volume is twice that of the given cube. Since a cube is completely determined by one side, a cube of side s can be duplicated if and only if $P_4 \in C(P_1, P_2, P_3)$, where

$$P_1 = (0, 0), \quad P_2 = (1, 0), \quad P_3 = (s, 0), \quad \text{and} \quad P_4 = (s\sqrt[3]{2}, 0).$$

To show that neither of these constructions is always possible, we first want to describe $C(P_1, \ldots, P_n)$ algebraically. We do this by identifying the complex numbers $\mathbb{C}$ with Euclidean two-dimensional space in the usual fashion (cf. Chapter 3). We then have the following algebraic description of $C(P_1, \ldots, P_n)$.

Theorem D.1   *If $n \geq 2$ and $P_1, \ldots, P_n$ are any points on the Euclidean plane such that $P_1 = (0, 0)$ and $P_2 = (1, 0)$, then $C(P_1, \ldots, P_n)$ is the smallest subfield F of $\mathbb{C}$ that satisfies the following conditions:*

    (i)   F contains $P_1, \ldots, P_n$
    (ii) for all $v \in F$, $\bar{v} \in F$,
    (iii) for all $v \in F$, if $u \in \mathbb{C}$ is such that $u^2 = v$, then $u \in F$.

Proof   By hypothesis, $C(P_1, \ldots, P_n)$ contains at least two elements. Thus, to show that $C(P_1, \ldots, P_n)$ is a field we must show that $C(P_1, \ldots, P_n)$ is closed with respect to addition, has additive inverses, is closed with respect to multiplication, and has multiplicative inverses. In addition, we must show that $C(P_1, \ldots, P_n)$ is closed with respect to the formation of complex conjugates and the extraction of square roots. Once we have done this, we must show that any subfield that contains $P_1, \ldots, P_n$ and that is closed with respect to the formation of complex conjugates and the extraction of square roots must contain $C(P_1, \ldots, P_n)$. We begin by showing that $C(P_1, \ldots, P_n)$ is additively closed.

For any complex numbers v and w, we can construct the complex number $v + w$ as indicated in Figure D.1. Since $P_0 = (0, 0)$, the modulus $|v|$ may serve as the radius of a circle with center w and similarly we may draw a circle of radius $|w|$ around v. Then $v + w$ is a point of intersection of these two circles.

For any complex number v, $-v$ may be constructed by drawing a circle through v with center $P_0 = (0, 0)$. As indicated in Figure D.2, $-v$ is a point of intersection of this circle with the line through $(0, 0)$ and v.

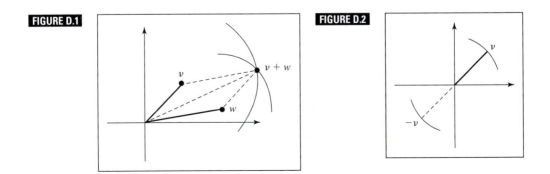

FIGURE D.1

FIGURE D.2

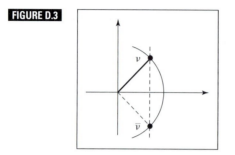

FIGURE D.3

Figure D.3 illustrates how to construct $\bar{v}$. Draw a circle through v with center $P_0 = (0, 0)$ and a line through v perpendicular to the real axis. The other point of intersection of this circle with the constructed perpendicular will be $\bar{v}$.

We still have to show that $C(P_1, \ldots, P_n)$ is multiplicatively closed, has multiplicative inverses, and has square roots. For these proofs, we will write complex numbers in their polar form: $z = r(\cos\Theta + i\sin\Theta)$. Note that if we can construct the positive real number r and a line at an angle Θ to the real axis, then we can construct z. For z will be the point of intersection of the given line with the circle of radius r and center at $P_0 = (0, 0)$.

Now consider the complex numbers v and w; write them in polar form: $v = r(\cos\Theta + i\sin\Theta)$ and $w = s(\cos\Psi + i\sin\Psi)$. Then $vw = rs(\cos(\Theta + \Psi) + i\sin(\Theta + \Psi))$, and thus it suffices to construct the point rs on the real axis and a line with angle $\Theta + \Psi$ to the real axis.

The construction of rs is illustrated in Figure D.4. The line through $P_0 = (0, 0)$ and $P_1 = (1, 0)$ is just the x-axis. The y-axis is just a line perpendicular to the x-axis going through $(0, 0)$ and hence we can construct it in accordance with the preceding rules. The unit circle is the circle with center $P_0 = (0, 0)$ which goes through $P_1 = (1, 0)$ and hence we can also construct it. The point $(0, 1)$ is just the intersection of this line and the y-axis. The point $(r, 0)$ is a point of intersection of the x-axis and the circle through v with center $(0, 0)$. Draw the line L through $(0, 1)$ and $(r, 0)$ and then construct the point $(0, s)$ in a manner similar to that used to construct $(r, 0)$.

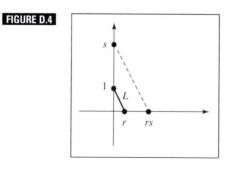

FIGURE D.4

Since parallel lines can be drawn in accordance with the preceding rules, we may draw the line K through $(0, s)$ parallel to L. This line K intersects the x-axis at a point $(x, 0)$. Since K is parallel to L, they have the same slope and hence $\dfrac{1 - 0}{0 - r} = \dfrac{s - 0}{0 - x}$. It follows that $x = rs$, and thus we have constructed rs on the x-axis.

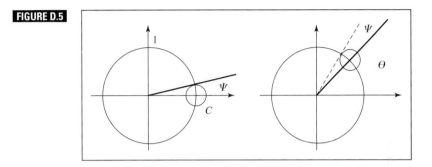

FIGURE D.5

The construction of $\Theta + \Psi$ is illustrated in Figure D.5. Draw the unit circle and the line from $(0, 0)$ through w. Using $(1, 0)$ as center, draw a circle C through the point of intersection. Draw the line from $(0, 0)$ through v. Using the point of intersection of this line and the unit circle, draw a circle whose radius is the same as that of C. Finally draw a line from $(0, 0)$ through the point of intersection of this last circle and the unit circle. This line will have angle $\Theta + \Psi$ with the x-axis.

Now let $0 \neq v \in \mathbb{C}$ and recall that the argument of v^{-1} is $-\Theta$. Observe that the argument of $\bar{v}$ is also $-\Theta$, and hence, since we can construct $\bar{v}$, we can construct a line at an angle $-\Theta$ to the real axis. Thus, to construct v^{-1} it suffices to construct r^{-1} on the real axis. This construction is similar to the construction of rs shown before and illustrated in Figure D.6.

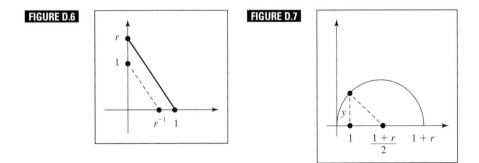

FIGURE D.6　　　　**FIGURE D.7**

The square roots of v will be $\pm z$, where $|z| = \sqrt{r}$ and $\arg(z) = \dfrac{\Theta}{2}$. Since angle bisectors can be constructed in accordance with the preceding rules, it suffices to construct $\sqrt{r}$ on the real axis. Figure D.7 illustrates this construction. Construct $1 + r$ on the x-axis and then construct its midpoint $\dfrac{1 + r}{2}$. Draw the circle of radius

$\frac{1+r}{2}$ with center at $\left(\frac{1+r}{2}, 0\right)$. Construct a perpendicular to the x-axis through $(1, 0)$. The line from $(1, 0)$ to the point of intersection of the circle and the perpendicular will have length $\sqrt{r}$. For, referring to the illustration, we have $\left(\frac{1+r}{2} - 1\right)^2 + y^2 = \left(\frac{1+r}{2}\right)^2$ and hence

$$y^2 = \frac{1}{4}(r^2 + 2r + 1 - r^2 + 2r - 1) = r.$$

We have now shown that $C(P_1, \ldots, P_n)$ is a subfield of $\mathbb{C}$ which satisfies the given condition. It remains to show that if $\mathbb{F}$ is any subfield of $\mathbb{C}$ which satisfies conditions (*i*), (*ii*), and (*iii*), then $\mathbb{F} \supseteq C(P_1, \ldots, P_n)$. For this, by the definition of $C(P_1, \ldots, P_n)$, it suffices to show that any points of intersection of lines or circles determined by points $\mathbb{F}$ are themselves in $\mathbb{F}$.

We first observe that if $a + bi \in \mathbb{F}$, then a, b, and $\sqrt{a^2 + b^2}$ are all in $\mathbb{F}$. For $a - bi = \overline{a + bi} \in \mathbb{F}$ and hence $a = \frac{1}{2}(a + bi + a - bi) \in \mathbb{F}$. Furthermore, $i = \sqrt{-1} \in \mathbb{F}$ and hence $b = \frac{1}{i}(a + bi - a) \in \mathbb{F}$. Since $\mathbb{F}$ contains square roots, $\sqrt{a^2 + b^2} \in \mathbb{F}$. Now if $a + bi$ and $c + di$ are distinct points in $\mathbb{F}$, then the line joining them has equation $y - b = \frac{d - b}{c - a}(x - a)$ (if $a \neq c$) or $x = a$ (if $a = c$), both of which are equivalent to

$$(c - a)y - (c - a)b = (d - b)x - (d - b)a.$$

Since $a, b, c, d \in \mathbb{F}$, this line has equation $ry + sx + t = 0$ for $r, s, t \in \mathbb{F}$. As well, since the distance between $a + bi$ and $c + di$ is $\sqrt{(a - c)^2 + (b - d)^2}$, it is in $\mathbb{F}$, and then an argument similar to the one for straight lines shows that a circle with center in $\mathbb{F}$ and radius equal to the distance between points in $\mathbb{F}$ has an equation of the form

$$x^2 + y^2 + rx + sy + t = 0$$

for $r, s, t \in \mathbb{F}$.

So consider lines $ry + sx + t = 0$ and $uy + vx + w = 0$ determined by points in $\mathbb{F}$. If they intersect, then they have different slopes and hence $su \neq rv$. So $\det\begin{pmatrix} r & s \\ u & v \end{pmatrix} = rv - su \neq 0$, and by Cramer's rule, the coordinates of the point of intersection of the two lines are

$$x_0 = \frac{\det\begin{pmatrix} r & -t \\ u & -w \end{pmatrix}}{\det\begin{pmatrix} r & s \\ u & v \end{pmatrix}} \quad \text{and} \quad y_0 = \frac{\det\begin{pmatrix} -t & s \\ -w & v \end{pmatrix}}{\det\begin{pmatrix} r & s \\ u & v \end{pmatrix}}.$$

Since each coordinate is in $\mathbb{F}$, the point of intersection itself, $x_0 + y_0 i$, is also in $\mathbb{F}$.

To find the x-coordinates of the points of intersection on the line $y = mx + b$ an the circle $x^2 + y^2 + rx + sy + t = 0$, it suffices to solve the quadratic

$$x^2 + (mx + b)^2 + rx + s(mx + b) + t = 0.$$

Since $\mathbb{F}$ has square roots, the solutions to this equation are in $\mathbb{F}$ and hence the points of intersection are in $\mathbb{F}$. Similarly, the points of intersection of the line $x = a$ and the circle $x^2 + y^2 + rx + sy + t = 0$ are also in $\mathbb{F}$.

Finally, any coordinates that lie simultaneously on the circles $x^2 + y^2 + rx + sy + t = 0$ and $x^2 + y^2 + ux + vy + w = 0$ must also satisfy the equation

$$\begin{aligned} 0 &= (x^2 + y^2 + rx + sy + t) - (x^2 + y^2 + ux + vy + w) \\ &= (r - u)x + (s - v)y + (t - w). \end{aligned}$$

Since this is an equation of a line whose coefficients are in $\mathbb{F}$, such coordinates must be in $\mathbb{F}$ and hence the points of intersection must be in $\mathbb{F}$ as well. ∎

We want to use the characterization of constructible complex numbers described in Theorem D.1 to derive a criterion for nonconstructibility. This criterion is given in Proposition D.3. The crux of the proof is contained in the next lemma.

Lemma D.2   *Let $v_1, \ldots, v_n \in \mathbb{C}$ and let $F = \mathbb{Q}(v_1, \ldots, v_n, \overline{v_1}, \ldots, \overline{v_n})$. Suppose that C_2 is the set of $z \in \mathbb{C}$ for which there exist $r_1, \ldots, r_m \in \mathbb{C}$ such that $r_1^2 \in F, r_j^2 \in F(r_1, \ldots, r_{j-1})$ for $1 < j \le m$, and $z \in F(r_1, \ldots, r_m)$. Then C_2 is a subfield of $\mathbb{C}$ containing $C(v_1, \ldots, v_n)$.*

Proof   Let $0 \ne z \in C_2$ and $w \in C_2$. Then $z \in F(r_1, \ldots, r_m)$ and $w \in F(s_1, \ldots, s_k)$, where r_j and s_l satisfy the given conditions. Then $z + w, -z, zw$, and z^{-1} are all in $F(r_1, \ldots, r_m, s_1, \ldots, s_k)$ and $F(r_1, \ldots, r_m, s_1, \ldots, s_k)$ is an extension field satisfying the given conditions. So C_2 is a subfield of $\mathbb{C}$. By definition, $v_1, \ldots, v_n \in C_2$. If $z \in C_2$, say $z \in F(r_1, \ldots, r_m)$ where $r_1, \ldots, r_m$ satisfy the given conditions, and $w^2 = z$, then $r_1, \ldots, r_m, w$ satisfy the given conditions and hence $w \in C_2$. By Proposition 3.3, F is closed with respect to conjugation and hence if $r_j^2 \in F(r_1, \ldots, r_{j-1})$, then $\overline{r_j}^2 \in F(\overline{r_1}, \ldots, \overline{r_{j-1}})$, and it follows that if $z \in C_2$, then $\overline{z} \in C_2$. By Theorem D.1, C_2 contains $C(v_1, \ldots, v_n)$. ∎

Proposition D.3   *Let $v_1, \ldots, v_n \in \mathbb{C}$ and let $F = \mathbb{Q}(v_1, \ldots, v_n, \overline{v_1}, \ldots, \overline{v_n})$. If z is constructible from $v_1, \ldots, v_n$, then z is algebraic over F and $[z:F]$ is a power of 2.*

Proof   By Lemma D.2, there exists an extension field L of F such that $z \in L$ and $L = F(r_1, \ldots, r_m)$, where $r_1^2 \in F$ and $r_j^2 \in F(r_1, \ldots, r_{j-1})$ for $1 < j \le m$. By Proposition 12.5 applied m times,

$$[L:F] = [F(r_1, \ldots, r_m):F(r_1, \ldots, r_{j-1})] \cdots [F(r_1):F].$$

By the conditions on the r_j, each of the dimensions $[F(r_1, \ldots, r_j) : F(r_1, \ldots, r_{j-1})]$ is either 1 or 2, and hence $[L:F]$ is a power of 2. By Proposition 12.10, z is algebraic over F, and by Proposition 12.8, $[L:F(z)]$ is finite. Therefore, $[L:F] = [L:F(z)][F(z):F]$ by Proposition 12.5 and it follows that $[F(z):F]$ must be a power of 2. But by Proposition 12.4, $[z:F] = [F(z):F]$. ∎

We are now ready to find an angle that cannot be trisected and a cube that cannot be duplicated. Note that the angle $\frac{\pi}{2}$ can easily be constructed from the angle $\frac{3\pi}{2}$. As well, by the proof of Theorem D.1, the product rs can be constructed from the real numbers r and s. It follows that the cube of side $\sqrt[3]{4}$ can be constructed from the cube of side $\sqrt[3]{2}$. Thus there are certainly angles that can be trisected and cubes that can be duplicated. What we want to show is that not every angle can be trisected and that not every cube can be duplicated.

Proposition D.4   *Not every angle can be trisected.*

Proof   We will show that the angle $\frac{\pi}{3}$ cannot be trisected. Suppose by way of contradiction that the angle $\frac{\pi}{9}$ can be constructed from $\frac{\pi}{3}$. If $P_1 = (0,0)$, $P_2 = (1,0)$, and $P_3 = \left(\cos\left(\frac{\pi}{3}\right), \sin\left(\frac{\pi}{3}\right)\right) = \left(\frac{1}{2}, \frac{1}{2}\sqrt{3}\right)$, then our assumption says that $\left(\cos\left(\frac{\pi}{9}\right), \sin\left(\frac{\pi}{9}\right)\right) \in C\left(P_1, P_2, P_3\right)$. Note that the line perpendicular to the x-axis going through $\left(\cos\left(\frac{\pi}{9}\right), \sin\left(\frac{\pi}{9}\right)\right)$ intersects the x-axis at the point $Q = \left(\cos\left(\frac{\pi}{9}\right), 0\right)$ and hence that also $Q \in C(P_1, P_2, P_3)$. We will show that this last statement leads to a contradiction.

Consider the complex numbers $v_1 = 0$, $v_2 = 1$, and $v_3 = \frac{1}{2} + \frac{1}{2}i\sqrt{3}$ and let

$$F = \mathbb{Q}(v_1, v_2, v_3, \overline{v_1}, \overline{v_2}, \overline{v_3}) = \mathbb{Q}(\sqrt{-3}).$$

If $r = \cos\left(\frac{\pi}{9}\right)$, then we are assuming that r is constructible from v_1, v_2, and v_3, and thus by Proposition D.3, that $[r:F]$ is a power of 2. We will arrive at a contradiction by showing that 3 divides $[r:F]$. To see this, we begin by observing that

$$[\mathbb{Q}(r, \sqrt{-3}) : \mathbb{Q}(r)][r:\mathbb{Q}] = [\mathbb{Q}(r, \sqrt{-3}):\mathbb{Q}] = [r:F][F:\mathbb{Q}]$$
$$= [r:F][\mathbb{Q}(\sqrt{-3}):\mathbb{Q}] = [r:F] \cdot 2$$

and hence that $[r:\mathbb{Q}]$ divides $[r:F] \cdot 2$. It thus suffices to show that $[r:\mathbb{Q}] = 3$; for then since 3 does not divide 2, it must then divide $[r:F]$ by Proposition 1.10.

To see that $[r:\mathbb{Q}] = 3$, first note that for all Θ,

$$\cos 3\Theta = \cos \Theta \cos 2\Theta - \sin \Theta \sin 2\Theta$$
$$= \cos \Theta(2\cos^2\Theta - 1) - \sin \Theta(2\sin \Theta \cos \Theta)$$
$$= 2\cos^3\Theta - \cos \Theta - 2(1 - \cos^2\Theta)\cos \Theta$$
$$= 4\cos^3\Theta - 3\cos \Theta$$

so that in particular for $\Theta = \frac{\pi}{9}, \frac{1}{2} = 4 \cos^3\left(\frac{\pi}{9}\right) - 3 \cos\left(\frac{\pi}{9}\right)$, i.e., r solves the equation $4x^3 - 3x - \frac{1}{2} = 0$. Let $m(x) = x^3 - \frac{3}{4}x - \frac{1}{8}$. Then $m(r) = 0$ and clearly $m(x)$ is monic. It thus suffices to show that $m(x)$ is irreducible over $\mathbb{Q}$; for then $m(x)$ will be the minimum polynomial of r over $\mathbb{Q}$ and hence $[r:\mathbb{Q}] = 3$. But if $m(x)$ is reducible, then it may be written as a product of polynomials of lower degree and hence $p(x) = 8\, m\left(\frac{1}{2}x\right) = x^3 - 3x - 1$ may also be written as a product of polynomials of lower degree: $p(x) = (a_1x + a_0)(b_2x^2 + b_1x + b_0)$. Proposition 11.3 allows us to assume that $a_1, a_0, b_2, b_1,$ and b_0 are all integers, and thus since $1 = a_1b_2$ and $-1 = a_0b_0$, $a_1 = \pm 1$ and $a_0 = \pm 1$. But then $p(x)$ must have 1 or -1 as a root, which it obviously does not. We conclude that $p(x)$, and hence $m(x)$, is irreducible and hence that $[r:\mathbb{Q}] = 3$. ∎

Proposition D.5   *Not every cube can be duplicated.*

Proof   Consider the cube with side of length 1; it has volume 1. Suppose by way of contradiction that we can construct a cube of volume 2. This means that we can construct the point $P_3 = (\sqrt[3]{2}, 0)$ from the points $P_1 = (0, 0)$ and $P_2 = (1, 0)$. Then by Proposition D.3, $[\sqrt[3]{2}:\mathbb{Q}]$ is a power of 2. But clearly $x^3 - 2$ is the minimum polynomial of $\sqrt[3]{2}$ over $\mathbb{Q}$ so that in fact $[\sqrt[3]{2}:\mathbb{Q}] = 3$. This is a contradiction. ∎

Answers to Odd-Numbered Computational Exercises

Chapter 1

3. $5 = (-2)15 + (1)35$

5. $3 = (35)105 + (17)216$

7. $13 = (-2)65 + (1)143$

Chapter 2

1. $x = \sqrt[3]{\dfrac{-1 + \sqrt{5}}{2}} + \sqrt[3]{\dfrac{-1 - \sqrt{5}}{2}},$

$\left(\dfrac{-1 + \sqrt{-3}}{2}\right)\sqrt[3]{\dfrac{-1 + \sqrt{5}}{2}} + \left(\dfrac{-1 - \sqrt{-3}}{2}\right)\sqrt[3]{\dfrac{-1 - \sqrt{5}}{2}},$

$\left(\dfrac{-1 - \sqrt{-3}}{2}\right)\sqrt[3]{\dfrac{-1 + \sqrt{5}}{2}} + \left(\dfrac{-1 + \sqrt{-3}}{2}\right)\sqrt[3]{\dfrac{-1 - \sqrt{5}}{2}};$

one real root

3. $y = \dfrac{\sqrt{2}}{2} - 1, \dfrac{\sqrt{2}}{2} - 1, -\dfrac{\sqrt{2}}{2} - 1, -\dfrac{\sqrt{2}}{2} - 1;$ four real roots

5. $x = \dfrac{\sqrt{5} \pm \sqrt{5 + 4\sqrt{5}}}{2} + 1, \dfrac{-\sqrt{5} \pm \sqrt{5 - 4\sqrt{5}}}{2} + 1;$ two real roots

7. $x = 1,$

$\sqrt[3]{\dfrac{5}{2} + \sqrt{\dfrac{25}{4} - \dfrac{125}{27}}} + \sqrt[3]{\dfrac{5}{2} - \sqrt{\dfrac{25}{4} - \dfrac{125}{27}}} + 1,$

$\left(\dfrac{-1 + \sqrt{-3}}{2}\right)\sqrt[3]{\dfrac{5}{2} + \sqrt{\dfrac{25}{4} - \dfrac{125}{27}}} + \left(\dfrac{-1 - \sqrt{-3}}{2}\right)\sqrt[3]{\dfrac{5}{2} - \sqrt{\dfrac{25}{4} - \dfrac{125}{27}}} + 1,$

$\left(\dfrac{-1 - \sqrt{-3}}{2}\right)\sqrt[3]{\dfrac{5}{2} + \sqrt{\dfrac{25}{4} - \dfrac{125}{27}}} + \left(\dfrac{-1 + \sqrt{-3}}{2}\right)\sqrt[3]{\dfrac{5}{2} - \sqrt{\dfrac{25}{4} - \dfrac{125}{27}}} + 1;$

two real roots

9. $x = -4, -1, -1$; three real roots

11. $x = \dfrac{\sqrt{5}}{4} + \dfrac{\sqrt{-10 + 2\sqrt{5}}}{4} - \dfrac{1}{4}$, $x = \dfrac{\sqrt{5}}{4} - \dfrac{\sqrt{-10 + 2\sqrt{5}}}{4} - \dfrac{1}{4}$,

$x = \dfrac{\sqrt{5}}{4} + \dfrac{\sqrt{-10 - 2\sqrt{5}}}{4} - \dfrac{1}{4}$, $x = \dfrac{\sqrt{5}}{4} - \dfrac{\sqrt{-10 - 2\sqrt{5}}}{4} - \dfrac{1}{4}$;

no real roots

13. $x = \pm\sqrt{\alpha + \beta - \dfrac{p}{3}}$

$x = \pm\sqrt{\left(-\dfrac{1}{2} + \dfrac{\sqrt{-3}}{2}\right)\alpha + \left(-\dfrac{1}{2} - \dfrac{\sqrt{-3}}{2}\right)\beta - \dfrac{p}{3}}$

$x = \pm\sqrt{\left(-\dfrac{1}{2} - \dfrac{\sqrt{-3}}{2}\right)\alpha + \left(-\dfrac{1}{2} + \dfrac{\sqrt{-3}}{2}\right)\beta - \dfrac{p}{3}}$

where $\alpha = \sqrt[3]{-\dfrac{b}{2} + \sqrt{\dfrac{b^2}{4} + \dfrac{a^3}{27}}}$ and $\beta = \sqrt[3]{-\dfrac{b}{2} - \sqrt{\dfrac{b^2}{4} + \dfrac{a^3}{27}}}$

Chapter 3

1. $\dfrac{16}{113} + \dfrac{14}{113}i$

3. $-\dfrac{46}{13} + \dfrac{30}{13}i$

5. $\sqrt{2} + \sqrt{2}i$

7. $\dfrac{1}{2} - \dfrac{\sqrt{3}}{2}i$

9. $|-i| = 1$, $\arg(-i) = \dfrac{3\pi}{2}$

11. $|-1 - i| = \sqrt{2}$, $\arg(-1 - i) = \dfrac{5\pi}{4}$

13. $|-2 + 2i| = 2\sqrt{2}$, $\arg(-2 + 2i) = \dfrac{3\pi}{4}$

15. $|-\zeta_4| = 1$, $\arg(-\zeta_4) = \dfrac{3\pi}{2}$

17. $|\zeta_3(\zeta_5)^4| = 1$, $\arg(\zeta_3(\zeta_5)^4) = \dfrac{4\pi}{15}$

19. $|(\zeta_3)(-1 + i\sqrt{3})(4\zeta_6)^3| = 128$, $\arg((\zeta_3)(-1 + i\sqrt{3})(4\zeta_6)^3) = \dfrac{\pi}{3}$

21. $|(-\sqrt{3} - i)(2\zeta_5)^3(1 + i)(3\zeta_{10})| = 48\sqrt{2}$,

$\arg((-\sqrt{3} - i)(2\zeta_5)^3(1 + i)(3\zeta_{10})) = \dfrac{49\pi}{60}$

23. $-3i = 3\left(\cos\left(\dfrac{3\pi}{2}\right) + i\sin\left(\dfrac{3\pi}{2}\right)\right)$

25. $-3 + 3i = 3\sqrt{2}\left(\cos\left(\dfrac{3\pi}{4}\right) + i\sin\left(\dfrac{3\pi}{4}\right)\right)$

Chapter 3 (continued)

27. $7(\zeta_{11})^9 = 7\left(\cos\left(\dfrac{18\pi}{11}\right) + i\sin\left(\dfrac{18\pi}{11}\right)\right)$

29. $(\sqrt{3} - i)(2\zeta_6)^4 = 32\left(\cos\left(\dfrac{7\pi}{6}\right) + i\sin\left(\dfrac{7\pi}{6}\right)\right)$

31. $\left[-\dfrac{1}{2} + \dfrac{\sqrt{3}}{2}i\right]^2 = -\dfrac{1}{2} - \dfrac{\sqrt{3}}{2}i$, $\quad \left[-\dfrac{1}{2} + \dfrac{\sqrt{3}}{2}i\right]^3 = 1$

35. $\sqrt[10]{2}\left(\cos\left(\dfrac{\pi}{4}\right) + i\sin\left(\dfrac{\pi}{4}\right)\right)$, $\quad \sqrt[10]{2}\left(\cos\left(\dfrac{13\pi}{20}\right) + i\sin\left(\dfrac{13\pi}{20}\right)\right)$,

$\sqrt[10]{2}\left(\cos\left(\dfrac{21\pi}{20}\right) + i\sin\left(\dfrac{21\pi}{20}\right)\right)$, $\quad \sqrt[10]{2}\left(\cos\left(\dfrac{29\pi}{20}\right) + i\sin\left(\dfrac{29\pi}{20}\right)\right)$,

$\sqrt[10]{2}\left(\cos\left(\dfrac{37\pi}{20}\right) + i\sin\left(\dfrac{37\pi}{20}\right)\right)$

37. $\sqrt[3]{2}\left(\cos\left(\dfrac{11\pi}{18}\right) + i\sin\left(\dfrac{11\pi}{18}\right)\right)$, $\quad \sqrt[3]{2}\left(\cos\left(\dfrac{23\pi}{18}\right) + i\sin\left(\dfrac{23\pi}{18}\right)\right)$,

$\sqrt[3]{2}\left(\cos\left(\dfrac{35\pi}{18}\right) + i\sin\left(\dfrac{35\pi}{18}\right)\right)$

39. $\cos\left(\dfrac{2\pi}{15}\right) + i\sin\left(\dfrac{2\pi}{15}\right)$, $\quad \cos\left(\dfrac{4\pi}{5}\right) + i\sin\left(\dfrac{4\pi}{5}\right)$,

$\cos\left(\dfrac{22\pi}{15}\right) + i\sin\left(\dfrac{22\pi}{15}\right)$

45. $-7 - i\sqrt{3}$

Chapter 4

1. $[8]_{12}$ **3.** $[1]_4$ **5.** $[2]_{22}$

7. $[6]_7$ **9.** $[3]_7$ **11.** $[5]_7$

13. $[4]_5$ **15.** $[10]_{13}$ **17.** $[1]_7$

19. $2 + 2i + 5j - 4k$ **21.** $\dfrac{1}{3} - \dfrac{1}{3}i + \dfrac{5}{3}j$ **23.** $-3i + 9k$

Chapter 6

1. $\mathbb{Q}^{p(x)} = \mathbb{Q}$

3. $\mathbb{Q}^{p(x)} = \mathbb{Q}\left(\dfrac{-1 \pm \sqrt{-3}}{2}\right)$

5. $\mathbb{Q}^{p(x)} = \mathbb{Q}(\sqrt[3]{7}, \sqrt[3]{7}\zeta_3, \sqrt[3]{7}\zeta_3^2) = \mathbb{Q}(\sqrt[3]{7}, \zeta_3)$

7. $\mathbb{Q}^{p(x)} = \mathbb{Q}\left(-\frac{5}{2} \pm \frac{\sqrt{13}}{2}, \frac{1}{2} \pm \frac{\sqrt{3}}{2}i\right)$

25. $r_1 = \sqrt[3]{11}, k_1 = 3, r_2 = \zeta_3, k_2 = 3$

27. $r_1 = \sqrt{29}, k_1 = 2, r_2 = \sqrt[3]{-\frac{5}{2} + \frac{\sqrt{29}}{2}}, k_2 = 3,$

$r_3 = \sqrt[3]{-\frac{5}{2} - \frac{\sqrt{29}}{2}}, k_3 = 3, r_4 = \zeta_3, k_4 = 3$

29. $r_1 = \zeta_5, k_1 = 5$ **31.** $r_1 = \zeta_n, k_1 = n$

Chapter 8

1. irreducible **3.** reducible **5.** irreducible

7. irreducible **9.** irreducible **11.** reducible

13. $q(x) = x^4 - 2x^3 + 3x^2 - 6x + 15, r(x) = -35$

15. $q(x) = x^3 - 8x, r(x) = 59x - 5$

17. $q(x) = x^2 - 2, r(x) = x^2 + 5x - 7$

19. $q(x) = x^4 + [9]_{11}x^3 + [3]_{11}x^2 + [5]_{11}x + [4]_{11}, r(x) = [9]_{11}$

21. $q(x) = x^3 + [3]_{11}x, r(x) = [4]_{11}x + [6]_{11}$

23. $q(x) = [3]_{11}x^3 + [4]_{11}x^2 + x + [2]_{11}, r(x) = [7]_{11}x$

25. 1 **27.** 1 **29.** 1

31. $x^2 - 2$

Chapter 9

15. ax^2, bx^2, cx^2, where $a, b,$ and c are three distinct nonzero elements of F

33. no

Chapter 10

1. one polynomial is $x^2 - 7$ **3.** one polynomial is $x^3 - 5$

5. one polynomial is $(x - (1 + 2i))(x - (1 - 2i)) = x^2 - 4x + 5$

7. minimum polynomial is $x^2 + 1$

9. minimum polynomial is $x^2 + x + 1$

11. minimum polynomial is $x^2 + 2x + 3$

13. minimum polynomial is $x^2 + \frac{4}{3}x + \frac{13}{3}$

15. $\mathbb{Q}(\sqrt{13}) = \{a_0 + a_1\sqrt{13} \mid a_0, a_1 \in \mathbb{Q}\}$

17. $\mathbb{Q}(4 + i) = \{a_0 + a_1(4 + i) \mid a_0, a_1 \in \mathbb{Q}\}$

Chapter 10 (continued)

19. $\mathbb{Q}(2 - 3i) = \{a_0 + a_1(2 - 3i) \,|\, a_0, a_1 \in \mathbb{Q}\}$

21. $\mathbb{Q}(\sqrt{2}, i) = \{(a_0 + a_1\sqrt{2}) + (b_0 + b_1\sqrt{2})i \,|\, a_0, a_1, b_0, b_1 \in \mathbb{Q}\}$

23. $\mathbb{Q}(\sqrt{2}, \zeta_3) = \{(a_0 + a_1\sqrt{2}) + (b_0 + b_1\sqrt{2})\zeta_3 \,|\, a_0, a_1, b_0, b_1 \in \mathbb{Q}\}$

25. $\mathbb{Q}(\sqrt[3]{11}, \zeta_3) = \{(a_0 + a_1\sqrt[3]{11} + a_2\sqrt[3]{11^2})$
$\qquad\qquad + (b_0 + b_1\sqrt[3]{11} + b_2\sqrt[3]{11^2})\zeta_3 \,|\, a_0, a_1, a_2, b_0, b_1, b_2 \in \mathbb{Q}\}$

33. $20 + 6r + 2r^3$ **35.** $200 + 285r + 107r^2 + 140r^3$

39. $4s - 2s^2 + 2s^3 + 3s^4$ **41.** $14s + 30s^2 + 30s^3 - 29s^4$

Chapter 11

1. $(x^2 - 5)(2x^2 + 3)$ **3.** $3(x^5 - 2x)(5x^4 + 4x - 5)$

5. irreducible with prime 5   **7.** irreducible with prime 2

9. Eisenstein's criterion does not apply.

11. $\mathbb{Q}(\sqrt{5}, i) = \{(a_0 + a_1\sqrt{5}) + (b_0 + b_1\sqrt{5})i \,|\, a_0, a_1, b_0, b_1 \in \mathbb{Q}\}$

13. $\mathbb{Q}(\sqrt[3]{12}, \zeta_3) = \{(a_0 + a_1\sqrt[3]{12} + a_2\sqrt[3]{12^2})$
$\qquad\qquad + (b_0 + b_1\sqrt[3]{12} + b_2\sqrt[3]{12^2})\zeta_3 \,|\, a_i, b_j \in \mathbb{Q}\}$

15. $\mathbb{Q}(\sqrt[5]{11}, i) = \{(a_0 + a_1\sqrt[5]{11} + a_2\sqrt[5]{11^2} + a_3\sqrt[5]{11^3} + a_4\sqrt[5]{11^4})$
$\qquad\qquad + (b_0 + b_1\sqrt[5]{11} + b_2\sqrt[5]{11^2} + b_3\sqrt[5]{11^3} + b_4\sqrt[5]{11^4})i \,|\, a_j, b_i \in \mathbb{Q}\}$

Chapter 12

1. $(r^2)^{-1} = 1 - \frac{1}{3}r^5$

3. $(2r^3 - 1)^{-1} = \frac{9}{12}r - \frac{4}{13}r^3 - \frac{11}{39}r^5,$
$(r^5 - 2r^4 + r - 3)(2r^3 - 1)^{-1} = \frac{17}{13} - \frac{51}{12}r + \frac{4}{13}r^2 + \frac{14}{13}r^3 - \frac{5}{13}r^4 + \frac{15}{13}r^5$

5. $[\mathbb{Q}(i):\mathbb{Q}] = 2$, minimum polynomial is $x^2 + 1$

7. $[\mathbb{Q}(3 - i\sqrt{7}):\mathbb{Q}] = 2$, minimum polynomial is $x^2 - 6x + 16$

9. $[\mathbb{Q}(\zeta_5):\mathbb{Q}] = 4$, minimum polynomial is $x^4 + x^3 + x^2 + 1$

11. $[\mathbb{Q}(a):\mathbb{Q}] = 6$, minimum polynomial is $x^6 - 4x^5 + 2x^3 + 6x - 2$

13. $[\mathbb{Q}(\sqrt[5]{13}):\mathbb{Q}] = 5, \{1, \sqrt[5]{13}, \sqrt[5]{13^2}, \sqrt[5]{13^3}, \sqrt[5]{13^4}\}$

15. $[\mathbb{Q}(\sqrt[3]{6}, i):\mathbb{Q}] = 6, \{1, \sqrt[3]{6}, \sqrt[3]{6^2}, i, i\sqrt[3]{6}, i\sqrt[3]{6^2}\}$

17. $[\mathbb{Q}(\sqrt[7]{12}, i):\mathbb{Q}] = 14,$
$\qquad \{1, \sqrt[7]{12}, \sqrt[7]{12^2}, \sqrt[7]{12^3}, \sqrt[7]{12^4}, \sqrt[7]{12^5}, \sqrt[7]{12^6},$
$\qquad\quad i, i\sqrt[7]{12}, i\sqrt[7]{12^2}, i\sqrt[7]{12^3}, i\sqrt[7]{12^4}, i\sqrt[7]{12^5}, i\sqrt[7]{12^6}\}$

25. $[\mathbb{Q}(\sqrt[3]{5}, \sqrt{3}, i):\mathbb{Q}] = 12$

27. $[\mathbb{Q}(1 + 5i, 1 - 4\sqrt[3]{11}, 7 + 3\sqrt{13}):\mathbb{Q}] = [\mathbb{Q}(i, \sqrt[3]{11}, \sqrt{13}):\mathbb{Q}] = 12$

29. $[\mathbb{Q}(\sqrt{11}, i):\mathbb{Q}(i)] = 2$ **31.** $[\mathbb{Q}(\sqrt{6}, \zeta_3):\mathbb{Q}(\zeta_3)] = 2$

33. $[\mathbb{Q}(\sqrt[5]{11}, \zeta_5):\mathbb{Q}] = 20$ **35.** $[\mathbb{Q}(\sqrt[5]{7}, \sqrt{5}):\mathbb{Q}] = 35$

37. $[\mathbb{Q}(\sqrt[5]{7}, \sqrt[3]{5}, \zeta_5):\mathbb{Q}] = 60$ **39.** $[\mathbb{Q}(\sqrt[5]{7}, \sqrt[3]{5}, \zeta_5):\mathbb{Q}(\zeta_5)] = 15$

49. $x^7 - 5$ **51.** $x^5 - 7$

Chapter 13

1. not one-to-one, not onto   **3.** one-to-one, not onto

5. not one-to-one, not onto   **7.** one-to-one, not onto

9. not one-to-one, not onto   **11.** one-to-one, onto

13. not a homomorphism   **15.** homomorphism, not isomorphism

17. not a homomorphism   **19.** not a homomorphism

21. not a homomorphism   **23.** not a homomorphism

25. automorphism

27. $f(a_0 + a_1\zeta_5 + a_2\zeta_5^2 + a_3\zeta_5^3 + a_4\zeta_5^4)$
$$= a_0 + a_1\zeta_5^2 + a_2(\zeta_5^2)^2 + a_3(\zeta_5^2)^3 + a_4(\zeta_5^2)^4$$

29. $f(a_0 + a_1\sqrt[3]{11} + a_2\sqrt[3]{11^2}) = a_0 + a_1\sqrt[3]{11}\zeta_3 + a_2(\sqrt[3]{11}\zeta_3)^2$

31. $f(a_0 + a_1\sqrt[4]{7} + a_2\sqrt[4]{7^2} + a_3\sqrt[4]{7^3}) = a_0 + a_1 i\sqrt[4]{7} + a_2(i\sqrt[4]{7})^2 + a_3(i\sqrt[4]{7})^3$

41. no   **43.** $s: \mathbb{Q} \to \mathbb{Q}, s(q) = 2q$

Chapter 14

1. $(x + 2)^2$ **3.** $5(x - \sqrt[3]{5})(x - \sqrt[3]{5}\zeta_3)(x - \sqrt[3]{5}\zeta_3^2)$

5. $2x^3(x - 3)^2$ **7.** 8 functions

9. 2 functions   **11.** 20 functions

13. 4 functions   **15.** 12 functions

17. 6 functions

19. Gal $\mathbb{Q}^{p(x)}/\mathbb{Q} = \{f_1, f_2\}$, where $f_k(a_0 + a_1\sqrt{-11}) = a_0 + a_1 f_k(\sqrt{-11})$, and

$$\begin{array}{ccc} & f_1 & f_2 \\ \sqrt{-11} \to & \sqrt{-11} & -\sqrt{-11} \end{array}$$

Chapter 14 (continued)

21. Gal $\mathbb{Q}^{p(x)}/\mathbb{Q} = \{f_1, f_2\}$, where $f_k(a_0 + a_1 r) = a_0 + a_1 f_k(r)$, and

$$
\begin{array}{ccc}
 & f_1 & f_2 \\
r \rightarrow & r & -r
\end{array}
$$

23. Gal $\mathbb{Q}^{p(x)}/\mathbb{Q} = \{f_1, f_2, f_3, f_4, f_5, f_6\}$, where

$$f_i((a_0 + a_1\sqrt[3]{11} + a_2\sqrt[3]{11^2}) + (b_0 + b_1\sqrt[3]{11} + b_2\sqrt[3]{11^2})\zeta_3)$$
$$= (a_0 + a_1 f_k(\sqrt[3]{11}) + a_2 f_k(\sqrt[3]{11})^2) + (b_0 + b_1 f_k(\sqrt[3]{11}) + b_2 f_k(\sqrt[3]{11})^2)f_k(\zeta_3),$$

and

$$
\begin{array}{ccccccc}
 & f_1 & f_2 & f_3 & f_4 & f_5 & f_6 \\
\sqrt[3]{11} \rightarrow & \sqrt[3]{11} & \sqrt[3]{11}\zeta_3 & \sqrt[3]{11}\zeta_3^2 & \sqrt[3]{11} & \sqrt[3]{11}\zeta_3 & \sqrt[3]{11}\zeta_3^2 \\
\zeta_3 \rightarrow & \zeta_3 & \zeta_3 & \zeta_3 & \zeta_3^2 & \zeta_3^2 & \zeta_3^2
\end{array}
$$

25. Gal $\mathbb{Q}(\sqrt{11})/\mathbb{Q} = \{f_1, f_2\}$, where $f_k(a_0 + a_1\sqrt{11}) = a_0 + a_1 f_k(\sqrt{11})$, and

$$
\begin{array}{ccc}
 & f_1 & f_2 \\
\sqrt{-11} \rightarrow & \sqrt{11} & -\sqrt{11}
\end{array}
$$

27. Gal $\mathbb{Q}(\sqrt[4]{3}, i)/\mathbb{Q} = \{f_1, f_2, f_3, f_4, f_5, f_6, f_7, f_8\}$, where

$$f_k((a_0 + a_1\sqrt[4]{3} + a_2\sqrt[4]{3}^2 + a_3\sqrt[4]{3}^3) + (b_0 + b_1\sqrt[4]{3} + b_2\sqrt[4]{3}^2 + b_3\sqrt[4]{3}^3)i)$$
$$= (a_0 + a_1 f_k(\sqrt[4]{3}) + a_2 f_k(\sqrt[4]{3})^2 + a_3 f_k(\sqrt[4]{3})^3)$$
$$+ (b_0 + b_1 f_k(\sqrt[4]{3}) + b_2 f_k(\sqrt[4]{3})^2 + b_3 f_k(\sqrt[4]{3})^3)f_k(i),$$

and

$$
\begin{array}{ccccccccc}
 & f_1 & f_2 & f_3 & f_4 & f_5 & f_6 & f_7 & f_8 \\
\sqrt[4]{3} \rightarrow & \sqrt[4]{3} & i\sqrt[4]{3} & -\sqrt[4]{3} & -i\sqrt[4]{3} & \sqrt[4]{3} & i\sqrt[4]{3} & -\sqrt[4]{3} & -i\sqrt[4]{3} \\
i \rightarrow & i & i & i & i & -i & -i & -i & -i
\end{array}
$$

29. Gal $\mathbb{Q}(\sqrt[3]{11})/\mathbb{Q} = \{f_1\}$, where
$$f_k(a_0 + a_1\sqrt[3]{11} + a_2\sqrt[3]{11^2}) = a_0 + a_1\sqrt[3]{11} + a_2\sqrt[3]{11^2}$$

31. Gal $\mathbb{Q}(\sqrt[5]{7}, \zeta_5)/\mathbb{Q} = \{f_1, \ldots, f_{20}\}$, where

$$f_k((a_0 + \cdots + a_6\sqrt[7]{13^6}) + (b_0 + \cdots + b_6\sqrt[7]{13^6})\zeta_5$$
$$+ (c_0 + \cdots + c_4\sqrt[5]{7^4})\zeta_5^2 + (d_0 + \cdots + d_4\sqrt[5]{7^4})\zeta_5^3)$$
$$= (a_0 + \cdots + a_4 f_k(\sqrt[5]{7})^4) + (b_0 + \cdots + b_4 f_k(\sqrt[5]{7})^4)f_k(\zeta_5)$$
$$+ (c_0 + \cdots + c_4 f_k(\sqrt[5]{7})^4)f_k(\zeta_5)^2 + (d_0 + \cdots + d_4 f_k(\sqrt[5]{7})^4)f_k(\zeta_5)^3,$$

and

	f_1	f_2	f_3	f_4	f_5	f_6	f_7	f_8	f_9	f_{10}
$\sqrt[5]{7} \to$	$\sqrt[5]{7}$	$\sqrt[5]{7}\zeta_5$	$\sqrt[5]{7}\zeta_5^2$	$\sqrt[5]{7}\zeta_5^3$	$\sqrt[5]{7}_5^4$	$\sqrt[5]{7}$	$\sqrt[5]{7}\zeta_5$	$\sqrt[5]{7}\zeta_5^2$	$\sqrt[5]{7}\zeta_5^3$	$\sqrt[5]{7}\zeta_5^4$
$\zeta_5 \to$	ζ_5	ζ_5	ζ_5	ζ_5	ζ_5	ζ_5^2	ζ_5^2	ζ_5^2	ζ_5^2	ζ_5^2

	f_{11}	f_{12}	f_{13}	f_{14}	f_{15}	f_{16}	f_{17}	f_{18}	f_{19}	f_{20}
$\sqrt[5]{7} \to$	$\sqrt[5]{7}$	$\sqrt[5]{7}\zeta_5$	$\sqrt[5]{7}\zeta_5^2$	$\sqrt[5]{7}\zeta_5^3$	$\sqrt[5]{7}\zeta_5^4$	$\sqrt[5]{7}$	$\sqrt[5]{7}\zeta_5$	$\sqrt[5]{7}\zeta_5^2$	$\sqrt[5]{7}\zeta_5^3$	$\sqrt[5]{7}\zeta_5^4$
$\zeta_5 \to$	ζ_5^3	ζ_5^3	ζ_5^3	ζ_5^3	ζ_5^3	ζ_5^4	ζ_5^4	ζ_5^4	ζ_5^4	ζ_5^4

33. Gal $\mathbb{Q}(\sqrt[7]{13}, \zeta_7)/\mathbb{Q} = \{f_1, \ldots, f_{42}\}$, where

$$f_k((r_0 + \cdots + r_6\sqrt[7]{13}^6) + \cdots + (w_0 + \cdots + w_6\sqrt[7]{13}^6)\zeta_7^5)$$
$$= (r_0 + \cdots + r_6 f_k(\sqrt[7]{13})^6) + \cdots + (w_0 + \cdots + w_6 f_k(\sqrt[7]{13})^6)f_k(\zeta_7)^5$$

and

	f_1	$\cdots$	f_7	f_8	$\cdots$	f_{14}	f_{15}	$\cdots$	f_{21}
$\sqrt[7]{13} \to$	$\sqrt[7]{13}$	$\cdots$	$\sqrt[7]{13}\zeta_7^6$	$\sqrt[7]{13}$	$\cdots$	$\sqrt[7]{13}\zeta_7^6$	$\sqrt[7]{13}$	$\cdots$	$\sqrt[7]{13}\zeta_7^6$
$\zeta_7 \to$	ζ_7	$\cdots$	ζ_7	ζ_7^2	$\cdots$	ζ_7^2	ζ_7^3	$\cdots$	ζ_7^3

	f_{22}	$\cdots$	f_{28}	f_{29}	$\cdots$	f_{35}	f_{36}	$\cdots$	f_{42}
$\sqrt[7]{13} \to$	$\sqrt[7]{13}$	$\cdots$	$\sqrt[7]{13}\zeta_7^6$	$\sqrt[7]{13}$	$\cdots$	$\sqrt[7]{13}\zeta_7^6$	$\sqrt[7]{13}$	$\cdots$	$\sqrt[7]{13}\zeta_7^6$
$\zeta_7 \to$	ζ_7^4	$\cdots$	ζ_7^4	ζ_7^5	$\cdots$	ζ_7^5	ζ_7^6	$\cdots$	ζ_7^6

37. $w = \sqrt{7} + \zeta_3$ **39.** $w = \sqrt{5} + \sqrt{3} + i$

41. $w = \zeta_3 + \sqrt{7} + i$

Chapter 15

3. not a group   **5.** not a group

7. not a group   **9.** a group

11. not a group   **13.** a group

15. a group   **17.** a group

31. not a group   **35.** $I, R, R^2, R^3, Y, YR, YR^2, YR^3$

Chapter 16

1. (1362) **3.** $(152)(36)$ **5.** $(14)(25)(36)$

7. (15432) **9.** (153) **11.** $(12)(354)$

Chapter 16 (continued)

13. (13)(265) **15.** (126534) **17.** (153)(246)

19. (1346)(25) **21.** (1246)(35) **23.** (1632)

25. (123), (132) **27.** $\sigma(a + bi) = a - bi$, $\tau(a + bi) = b + ai$

31. (12) $\in S_3$ **33.** not Abelian

Chapter 17

1. one-to-one, not onto, a homomorphism

3. one-to-one, not onto, a homomorphism

5. one-to-one, not onto, a homomorphism

7. not one-to-one, onto, not a homomorphism

9. not one-to-one, onto, not a homomorphism

11. not one-to-one, onto, a homomorphism

13. not one-to-one, not onto, a homomorphism

15. a homomorphism **17.** not a homomorphism

21. $\ln x$ **23.** not one-to-one

25. $T([n]_6) = f_{[n]_6}$, where $f_{[n]_6}([k]_6) = [n]_6 + [k]_6$

27. $T(\sigma) = f_\sigma$, where $f_\sigma(\tau) = \sigma\tau$ **29.** T is onto

31. T is onto **33.** T is not onto

37. $T(f) = (1254)$, where $r \sim 1$, $r\zeta_5 \sim 2$, $r\zeta_5{}^2 \sim 3$, $r\zeta_5{}^3 \sim 4$, $r\zeta_5{}^4 \sim 5$

41. $f: \mathbb{Q}[x] \to \mathbb{Q}[x]$, $f(a_0 + a_1x + \cdots + a_nx^n) = a_1 + a_2x + \cdots + a_nx^{n-1}$

Chapter 18

1. a subgroup **3.** a subgroup **5.** not a subgroup

7. a subgroup **9.** not a subgroup **11.** not a subgroup

13. a subgroup **15.** not a subgroup **17.** a subgroup

19. a subgroup **21.** not a subgroup **31.** $\{(n, n) \mid n \in \mathbb{Z}\}$

Chapter 19

1. (16)(56)(35)(37)(27) **3.** (15)(45)(46)(23)(35)(57)

5. (17)(37)(36)(56)(25)(45) **7.** $\mathbb{Z}_7$

9. $(3) = \{3n \mid n \in \mathbb{Z}\}$ **11.** $\{2m + n\sqrt{3} \mid n, m \in \mathbb{Z}\}$

13. $\{\sigma \in S_{11} \mid \sigma(k) = k \text{ for all } k \in \{5, 6, 7, 8, 9, 10, 11\}\}$

15. S_5 **17.** $\left\{ \begin{pmatrix} k\pi & 0 \\ 0 & k\pi \end{pmatrix} \middle| k \in \mathbb{Z} \right\}$

19. $\{\pm 1\}$ **21.** Q

Chapter 20

1. index $= 3$, $([a]_6 + S) + ([b]_6 + S) = [a + b]_6 + S$

3. index $= 3$, $[a]_7 S[b]_7 S = [ab]_7 S$

5. index $= 12$, $(34)S(234)S$ is not a left coset

7. index $= 6$, $(23)S(243)S$ is not a left coset

9. index $= 2$, $\sigma S \tau S = \sigma \tau S$

13. $G = S_3$, $S = \{\iota, (12)\}$, $(13)S = (123)S$, but $(13)(123)^{-1} \notin S$

15. $G = S_3$, $S = \{\iota, (12)\}$, $(13)S \neq S(13)$ and $(13)S \cap S(13) \neq \varnothing$

Chapter 21

1. 3	**3.** 3	**5.** 6
7. 6	**9.** 12	**11.** infinite
13. order $= 6$	**15.** order $= 6$	**17.** index $= 9$
19. index $= 180$	**31.** order $= 3$	**33.** order $= 4$
35. 4 distinct subgroups	**37.** 4 distinct subgroups	**39.** 6 distinct subgroups
45. $f: \mathbb{Z} \to \mathbb{Z}, f(n) = 0$		**47.** 1

Chapter 22

7. for all $0 \le n < 64$, $[[n]_{64}] = \{[n]_{64}\}$

9. $[\iota] = \{\iota\}$, $[(12)] = \{(12), (13), (14), (23), (24), (34)\}$,

$[(123)] = \{(123), (124), (132), (134), (142), (143), (234), (243)\}$,

$[(1234)] = \{(1234), (1243), (1324), (1342), (1423), (1432)\}$,

$[(12)(34)] = \{(12)(34), (13)(24), (14)(23)\}$

11. $[[0]_9] = \{[0]_9\}$, $[[1]_9] = \{[1]_9, [2]_9, [4]_9, [5]_9, [7]_9, [8]_9\}$, $[3]_9 = \{[3]_9, [6]_9\}$

Chapter 22 (continued)

13. $[[0]_{36}] = \{[0]_{36}\}$,

$[[1]_{36}] = \{[1]_{36}, [5]_{36}, [7]_{36}, [11]_{36}, [13]_{36}, [17]_{36}, [19]_{36}, [23]_{36}, [25]_{36},$
$\qquad [29]_{36}, [31]_{36}, [35]_{36}\}$,

$[[2]_{36}] = \{[2]_{36}, [10]_{36}, [14]_{36}, [22]_{36}, [26]_{36}, [34]_{36}\}$,

$[[3]_{36}] = \{[3]_{36}, [15]_{36}, [21]_{36}, [33]_{36}\}$,

$[[4]_{36}] = \{[4]_{36}, [8]_{36}, [16]_{36}, [20]_{36}, [28]_{36}, [32]_{36}\}$,

$[[6]_{36}] = \{[6]_{36}, [12]_{36}, [24]_{36}, [30]_{36}\}$,

$[[9]_{36}] = \{[9]_{36}, [27]_{36}\}$,

$[[18]_{36}] = \{[18]_{36}\}$

15. $[1] = \{1\}, [-1] = \{-1\}, [i] = \{\pm i\}, [j] = \{\pm j\}, [k] = \{\pm k\}$

17. $[[1]_7] = \{[1]_7, [2]_7, [3]_7, [4]_7, [5]_7, [6]_7\}$

19. $[[1]_8] = \{[1]_8, [3]_8, [5]_8, [7]_8\}$,

$[[2]_8] = \{[2]_8, [6]_8\}$,

$[[4]_8] = \{[4]_8\}$

21. $[1] = \{q \in \mathbb{Q} \mid q \neq 0\}$

Chapter 23

1. normal subgroup

3. subgroup, not normal

5. subgroup, not normal

7. $(13)N$

11. $\dfrac{1087}{20} + \mathbb{Z}$

13. $\{1\}, \{\pm 1\}, \{\pm 1, \pm i\}, \{\pm 1, \pm j\}, \{\pm 1, \pm k\}, Q$

Chapter 24

1. $f(n) = n + (5), \ker f = (5)$

3. $f(x) = x\{\pm 1, \pm i\}, \ker f = \{\pm 1, \pm i\}$

5. $|H| = 4$

7. one-to-one

9. not one-to-one

11. not one-to-one

Chapter 25

1. $r_1 = \sqrt[3]{5}, k_1 = 3; r_2 = \sqrt[3]{7}, k_2 = 3$

3. $r_1 = \zeta_3, k_1 = 3;\quad r_2 = \zeta_5, k_2 = 5;\quad r_3 = \sqrt[3]{5}, k_3 = 3;$
$\quad r_4 = \sqrt[3]{6}, k_4 = 3;\quad r_5 = \sqrt[5]{-10}, k_5 = 5$

5. $r_1 = \zeta_3, k_1 = 3; r_2 = \alpha, k_2 = 3; r_3 = \beta, k_3 = 3$

7. $r_1 = \sqrt[4]{3}, k_1 = 4; r_2 = i, k_2 = 2$

9. $R(\sigma) = \sigma$ restricted to $\mathbb{Q}(\zeta_3)$

11. $R(\sigma) = \sigma$ restricted to $\mathbb{Q}(i, \sqrt{2})$

15. $x^3 - 7 \in \mathbb{Q}[x]$

Chapter 26

1. $[0]_4$ **3.** ι **5.** 1

7. $(13)(24)$ **15.** $\{[0]_{169}\} \subseteq \mathbb{Z}$

17. $\{(\iota, [0]_3)\} \subseteq \{\iota, (123), (132)\} \times \mathbb{Z}_3 \subseteq S_3 \times \mathbb{Z}_3$

19. $\{(\iota, \iota)\} \subseteq \{\iota, (123), (132)\} \times S_2 \subseteq S_3 \times S_2$

21. $\{\iota\} \subseteq \text{Gal } \mathbb{Q}(\zeta_3)/\mathbb{Q}$

23. $\{\iota\} \subseteq \text{Gal } \mathbb{Q}(\sqrt{5}, i)/\mathbb{Q}(i) \subseteq \text{Gal } \mathbb{Q}(\sqrt{5}, i)/\mathbb{Q}$

25. $\{\iota\} \subseteq \text{Gal } \mathbb{Q}(\sqrt[5]{5}, \zeta_5)/\mathbb{Q}(\zeta_5) \subseteq \text{Gal } \mathbb{Q}(\sqrt[5]{5}, \zeta_5)/\mathbb{Q}$

27. $(\mathbb{Z}_{169})' = \{[0]_{169}\}$

29. $(S_3 \times \mathbb{Z}_3)' = \{\iota, (123), (132)\} \times \{[0]_3\}, (S_3 \times \mathbb{Z}_3)^{(2)} = \{(\iota, [0]_3)\}$

31. $(S_3 \times S_2)' = \{\iota, (123), (132)\} \times \{\iota\}, (S_3 \times S_2)^{(2)} = \{(\iota, \iota)\}$

35. true   **37.** true   **41.** $M^{(2)} = \left\{ \begin{pmatrix} 1 & 0 \\ 0 & 1 \end{pmatrix} \right\}$

Chapter 27

7. $\mathbb{Q} = \{k \in \mathbb{Q}(i) \mid \sigma(k) = k$ for all $\sigma \in \text{Gal } \mathbb{Q}(i)/\mathbb{Q}\}$

9. $\mathbb{Q} = \{k \in \mathbb{Q}(\zeta_3) \mid \sigma(k) = k$ for all $\sigma \in \text{Gal } \mathbb{Q}(\zeta_3)/\mathbb{Q}\}$

11. $\mathbb{Q} = \{k \in \mathbb{Q}(\sqrt[4]{3}, \sqrt{13}, i) \mid \sigma(k) = k$ for all $\sigma \in \text{Gal } \mathbb{Q}(\sqrt[4]{3}, \sqrt{13}, i)/\mathbb{Q}\}$

17. $p(x)^\sigma = (4 - i)x^3 + ix^2 + 11$

19. $p(x)^\sigma = (2\sqrt[3]{7} + (\sqrt[3]{7})^2 + \zeta_3^2)x + ((1 - 5(\sqrt[3]{7})^2 + (3 + 3\sqrt[3]{7})\zeta_3^2)$

21. $p(x)^\sigma = x^4 - [(2 + (\sqrt[3]{7}\zeta_3^2)^2) + (\sqrt[3]{7}\zeta_3^2)^2\zeta_3^2]x^3$
$\qquad\qquad + 13(\sqrt[3]{7}\zeta_3^2)^2 x^2 - x + [4\sqrt[3]{7}\zeta_3^2 + 8(\sqrt[3]{7}\zeta_3^2)^2\zeta_3^2]$

Chapter 28

1. solvable by radicals   **3.** solvable by radicals

5. not solvable by radicals   **7.** solvable by radicals

Chapter 28 (continued)

9. not solvable by radicals

15. $p(x) = 3x^{11} - 110x^9 + 1285x^7 - 5410x^5 + 6335x^3 + 5$

Chapter 29

1. 144   **3.** 192   **5.** 5760

7. $\{([0]_2, [0]_4), ([0]_2, [1]_4), ([0]_2, [2]_4), ([0]_2, [3]_4),$
    $([1]_2, [0]_4), ([1]_2, [1]_4), ([1]_2, [2]_4), ([1]_2, [3]_4)\}$

9. $\{([0]_3, \iota), ([0]_3, (12)), ([0]_3, (13)), ([0]_3, (23)), ([0]_3, (123)), ([0]_3, (132)),$
    $([1]_3, \iota), ([1]_3, (12)), ([1]_3, (13)), ([1]_3, (23)), ([1]_3, (123)), ([1]_3, (132)),$
    $([2]_3, \iota), ([2]_3, (12)), ([2]_3, (13)), ([2]_3, (23)), ([2]_3, (123)), ([2]_3, (132))\}$

11. $([2]_5, (123))$ **13.** $([4]_5, (123))$ **15.** $([1]_5, \iota)$

17. $([1]_4, (1243), (12))$ **19.** $([2]_4, (132), (132))$ **21.** $([4]_7, (34), (123), [0]_3)$

23. $([2]_7, (1324), \iota, [2]_3)$ **25.** $([2]_7, (1243), (132), [2]_3)$

27. $\mathbb{Z}_2 \times \mathbb{Z}_2 \times \mathbb{Z}_2$ **43.** $A = \mathbb{Z}_3, B = \mathbb{Z}_7$

Chapter 31

1. $2\mathbb{Z}_{10} = \{[2]_{10}, [4]_{10}, [6]_{10}, [8]_{10}, [0]_{10}\},$
    $S_2(\mathbb{Z}_{10}) = \{[0]_{10}, [5]_{10}\},$
    $S_2(\mathbb{Z}_{10}) = S([5]_{10})$

3. $2\mathbb{Z}_8 = \{[2]_8, [4]_8, [6]_8, [0]_8\},$
    $S_2(\mathbb{Z}_8) = \{[0]_8, [4]_8\},$
    $S_2(\mathbb{Z}_8) = S([4]_8)$

5. $2\mathbb{Z}_{24} = \{[2]_{24}, [4]_{24}, [6]_{24}, [8]_{24}, [10]_{24}, [12]_{24},$
        $[14]_{24}, [16]_{24}, [18]_{24}, [20]_{24}, [22]_{24}, [0]_{24}\},$
    $S_2(\mathbb{Z}_{24}) = S([0]_{24}, [12]_{24}\},$
    $S_2(\mathbb{Z}_{24}) = S([12]_{24})$

7. $3A = \{([0]_3, [0]_{12}), ([0]_3, [3]_{12}), ([0]_3, [6]_{12}), ([0]_3, [9]_{12})\};$
    $S_3(A) = \{([0]_3, [0]_{12}), ([0]_3, [4]_{12}), ([0]_3, [8]_{12}),$
        $([1]_3, [0]_{12}), ([1]_3, [4]_{12}), ([1]_3, [8]_{12}),$
        $([2]_3, [0]_{12}), ([2]_3, [4]_{12}), ([2]_3, [8]_{12})\};$
    $S_3(A) = S(([1]_3, [0]_{12})) \oplus S(([0]_3, [4]_{12})).$

9. $2(\mathbb{Z}_2 \times \mathbb{Z}_3 \times \mathbb{Z}_8) = \{([0]_2, [0]_3, [2]_8), ([0]_2, [0]_3, [4]_8), ([0]_2, [0]_3, [6]_8),$
$([0]_2, [0]_3, [0]_8), ([0]_2, [2]_3, [2]_8), ([0]_2, [2]_3, [4]_8),$
$([0]_2, [2]_3, [6]_8), ([0]_2, [2]_3, [0]_8), ([0]_2, [1]_3, [2]_8),$
$([0]_2, [1]_3, [4]_8), ([0]_2, [1]_3, [6]_8), ([0]_2, [1]_3, [0]_8)\}$

$S_2(\mathbb{Z}_2 \times \mathbb{Z}_3 \times \mathbb{Z}_8) = \{([0]_2, [0]_3, [0]_8), ([0]_2, [0]_3, [4]_8),$
$([1]_2, [0]_3, [0]_8), ([1]_2, [0]_3, [4]_8)\}$

$S_2(\mathbb{Z}_2 \times \mathbb{Z}_3 \times \mathbb{Z}_8) = S(([1]_2, [0]_3, [0]_8)) \oplus S(([0]_2, [0]_3, [4]_8))$

11. $2(\mathbb{Z}_2 \times \mathbb{Z}_2 \times \mathbb{Z}_8 \times \mathbb{Z}_8)$
$= \{([0]_2, [0]_2, [0]_8, [2]_8), ([0]_2, [0]_2, [0]_8, [4]_8), ([0]_2, [0]_2, [0]_8, [6]_8),$
$([0]_2, [0]_2, [0]_8, [0]_8), ([0]_2, [0]_2, [2]_8, [2]_8), ([0]_2, [0]_2, [2]_8, [4]_8),$
$([0]_2, [0]_2, [2]_8, [6]_8), ([0]_2, [0]_2, [2]_8, [0]_8), ([0]_2, [0]_2, [4]_8, [2]_8),$
$([0]_2, [0]_2, [4]_8, [4]_8), ([0]_2, [0]_2, [4]_8, [6]_8), ([0]_2, [0]_2, [4]_8, [0]_8),$
$([0]_2, [0]_2, [6]_8, [2]_8), ([0]_2, [0]_2, [6]_8, [4]_8), ([0]_2, [0]_2, [6]_8, [6]_8),$
$([0]_2, [0]_2, [6]_8, [0]_8)\}$

$S_2(\mathbb{Z}_2 \times \mathbb{Z}_2 \times \mathbb{Z}_8 \times \mathbb{Z}_8)$
$= \{([0]_2, [0]_2, [0]_8, [0]_8), ([0]_2, [0]_2, [0]_8, [4]_8), ([0]_2, [0]_2, [4]_8, [0]_8),$
$([0]_2, [0]_2, [4]_8, [4]_8), ([0]_2, [1]_2, [0]_8, [0]_8), ([0]_2, [1]_2, [0]_8, [4]_8),$
$([0]_2, [1]_2, [4]_8, [0]_8), ([0]_2, [1]_2, [4]_8, [4]_8), ([1]_2, [0]_2, [0]_8, [0]_8),$
$([1]_2, [0]_2, [0]_8, [4]_8), ([1]_2, [0]_2, [4]_8, [0]_8), ([1]_2, [0]_2, [4]_8, [4]_8),$
$([1]_2, [1]_2, [0]_8, [0]_8), ([1]_2, [1]_2, [0]_8, [4]_8), ([1]_2, [1]_2, [4]_8, [0]_8),$
$([1]_2, [1]_2, [4]_8, [4]_8)\}$

$S_2(\mathbb{Z}_2 \times \mathbb{Z}_2 \times \mathbb{Z}_8 \times \mathbb{Z}_8) = S(([1]_2, [0]_2, [0]_8, [0]_8)) \oplus S(([0]_2, [1]_2, [0]_8, [0]_8))$
$\oplus S(([0]_2, [0]_2, [4]_8, [0]_8)) \oplus S(([0]_2, [0]_2, [0]_8, [4]_8))$

17. $\mathbb{Z}_{10} = H \oplus S([2]_{10})$

19. $\mathbb{Z}_2 \times \mathbb{Z}_6 = H \oplus S(([1]_2, [0]_6)) \oplus S(([0]_2, [3]_6))$

21. $\mathbb{Z}_2 \times \mathbb{Z}_2 \times \mathbb{Z}_4 \times \mathbb{Z}_8$
$= H \oplus S(([1]_2, [0]_2, [0]_4, [0]_8)) \oplus S(([0]_2, [1]_2, [0]_4, [0]_8))$

23. $2\mathbb{Z}_8 = S(2[1]_8), \mathbb{Z}_8 = S([1]_8)$

25. $3(\mathbb{Z}_3 \times \mathbb{Z}_9) = S(3([0]_3, [1]_9)), \mathbb{Z}_3 \times \mathbb{Z}_9 = S(([1]_3, [0]_9)) \oplus S(([0]_3, [1]_9))$

27. $2(\mathbb{Z}_2 \times \mathbb{Z}_2 \times \mathbb{Z}_{16}) = S(2([0]_2, [0]_2, [1]_{16})),$
$\mathbb{Z}_2 \times \mathbb{Z}_2 \times \mathbb{Z}_{16} = S(([1]_2, [0]_2, [0]_{16}))$
$\oplus S(([0]_2, [1]_2, [0]_{16})) \oplus S(([0]_2, [0]_2, [1]_{16}))$

29. $\mathbb{Z}_6$

Chapter 32

1. $(\mathbb{Z}_{10})_5 = \{[0]_{10}, [2]_{10}, [4]_{10}, [6]_{10}, [8]_{10}\}$

3. $(\mathbb{Z}_6 \times \mathbb{Z}_6)_2 = \{([0]_6, [0]_6), ([0]_6, [3]_6), ([3]_6, [0]_6), ([3]_6, [3]_6)\}$

5. $(\mathbb{Z}_{12} \times \mathbb{Z}_{18})_2 = \{([0]_{12}, [0]_{18}), ([0]_{12}, [9]_{18}), ([6]_{12}, [0]_{18}), ([6]_{12}, [9]_{18})\}$

7. $\mathbb{Z}_{16}, \ \mathbb{Z}_2 \times \mathbb{Z}_8, \ \mathbb{Z}_2 \times \mathbb{Z}_2 \times \mathbb{Z}_4, \ \mathbb{Z}_2 \times \mathbb{Z}_2 \times \mathbb{Z}_2 \times \mathbb{Z}_2, \ \mathbb{Z}_4 \times \mathbb{Z}_4$

9. $\mathbb{Z}_{18}, \ \mathbb{Z}_2 \times \mathbb{Z}_9, \ \mathbb{Z}_3 \times \mathbb{Z}_3 \times \mathbb{Z}_2$

11. $\mathbb{Z}_{22}$

13. $\mathbb{Z}_{36}, \ \mathbb{Z}_4 \times \mathbb{Z}_9, \ \mathbb{Z}_4 \times \mathbb{Z}_3 \times \mathbb{Z}_3, \ \mathbb{Z}_2 \times \mathbb{Z}_2 \times \mathbb{Z}_3 \times \mathbb{Z}_3, \ \mathbb{Z}_2 \times \mathbb{Z}_2 \times \mathbb{Z}_9$

Chapter 33

1. $D_3 = \{I, X, X^2, Y, YX, YX^2\}$

3. $D_6 = \{I, X, X^2, X^3, X^4, X^5, Y, YX, YX^2, YX^3, YX^4, YX^5\}$

5. $D_4, \ \{I\}, \ \{I, X^2\}, \ \{I, Y\}, \ \{I, YX\}, \ \{I, YX^2\}, \ \{I, YX^3\},$
   $\{I, X, X^2, X^3\}, \ \{I, X^2, Y, YX^2\}, \ \{I, X^2, YX, YX^3\}$

7. $\{I\}, \ \{I, Y\}, \ \{I, YX\}, \ \{I, YX^2\}$

9. $D_4, \ \{I\}, \ \{I, X^2\}, \ \{I, Y\}, \ \{I, YX\}, \ \{I, YX^2\}, \ \{I, YX^3\},$
   $\{I, X, X^2, X^3\}, \ \{I, X^2, Y, YX^2\}, \ \{I, X^2, YX, YX^3\}$

11. $\{I\}, \ \{I, X, X^2, X^3, X^4\}$

13. $\{I, X, X^2\}$

15. $\{I, Y\}, \ \{I, YX\}, \ \{I, YX^2\}, \ \{I, YX^3\}, \ \{I, YX^4\}$

17. $\mathbb{Z}_8, \ \mathbb{Z}_2 \times \mathbb{Z}_4, \ \mathbb{Z}_2 \times \mathbb{Z}_2 \times \mathbb{Z}_2, \ D_4, \ Q$

19. $G = \{\iota, (1324), (12)(34), (1423), (13), (24)(23), (34), (14)(23)\}$

Chapter 34

1. $1[2]_{10}$ 3. $2[3]_{10}$ 5. $4[5]_{12}$

7. i^2 9. $(1234)^2$ 11. $((132)(46))^2$

13. $((12654), [5]_{12})^4$ 15. $Z(S_3) = \{\iota\}$

17. $Z(S_3 \times Q) = \{(\iota, 1), (\iota, -1)\}$ 19. $C([7]_{10}) = \mathbb{Z}_{10}$

21. $C((12)) = \{\iota, (12)\}$ 23. $C(((12), [3]_{10})) = \{\iota, (12)\} \times \mathbb{Z}_{10}$

25. $C(((132), -i)) = \{\iota, (123), (132)\} \times \{\pm 1, \pm i\}$

27. $C((12)(34)) = \{\iota, (12), (34), (12)(34), (1324), (1423), (13)(24), (14)(23)\}$

29. $\{[3]_{10}\}$ **31.** $\{(12), (13), (23)\}$

33. $\{(12)(34), (14)(23), (13)(24)\}$

35. $\varnothing$ **37.** $\{(12), (123)\}$

39. $\{(\iota, i), (\iota, j), (\iota, k), ((12), 1), ((12), i), ((12), j), ((12), k),$
$\quad ((123), 1), ((123), i), ((123), j), ((123, k)\}$

Chapter 35

7. $\{\iota\}, \{\iota, (12)\}, \{\iota, (13)\}, [\iota, (23)\}$

9. $\{\iota, (12)\}, \{\iota, (13)\}, \{\iota, (23)\}$, none are normal

11. $N(\{\iota, (12)\}) = \{\iota, (12)\}$

13. $\{\iota, (12)\}^{S_3} = \{\{\iota, (12)\}, \{\iota, (13)\}, \{\iota, (23)\}\}$

15. $\{\iota, (123), (132)\}^{\{\iota, (12)\}} = \{\{\iota, (123), (132)\}\}$

17. $\{\iota, (12)\}^{S_3} = \{\{\iota, (23)\}, \{\iota, (13)\}, \{\iota, (12)\}\}$,
$\quad [\{\iota, (23)\}]_{\{\iota, (23)\}} = \{\{\iota, (23)\}\}$,
$\quad [\{\iota, (13)\}]_{\{\iota, (23)\}} = \{\{\iota, (13)\}, \{\iota, (12)\}\}$

19. $Q \lhd Q$

21. $\{\pm 1, \pm k\}^{Q} = \{\{\pm 1, \pm k\}\}$

23. $\{\pm 1, \pm k\}^{Q} = \{\{\pm 1, \pm k\}\}$,
$\quad [\{\pm 1, \pm k\}]_{\{\pm 1, \pm i\}} = \{\{\pm 1, \pm k\}\}$

25. $\{e\}, \{e, x, x^2, x^3, x^4\}$

27. $\{e, x, x^2, x^3, x^4\} \lhd D_5$

29. $N(\{e, x, x^2, x^3, x^4\}) = D_5$

31. $\{e, x, x^2, x^3, x^4\}^{D_5} = \{\{e, x, x^2, x^3, x^4\}\}$

33. $\{\iota, yx\}^{D_5} = \{\{\iota, y\}, \{\iota, yx\}, \{\iota, yx^2\}, \{\iota, yx^3\}, \{\iota, yx^4\}\}$,
$\quad [\{\iota, y\}]_{\{\iota, yx^4\}} = \{\{\iota, y\}, \{\iota, yx^3\}\}$,
$\quad [\{\iota, yx\}]_{\{\iota, yx^4\}} = \{\{\iota, yx\}, \{\iota, yx^2\}\}$,
$\quad [\{\iota, yx^4\}]_{\{\iota, yx^4\}} = \{\{\iota, yx^3\}\}$

35. $\{\iota, (123), (132)\}, \{\iota, (124), (142)\}, \{\iota, (134), (143)\}, \{\iota, (234), (243)\}$

37. $\{\iota, (123), (132)\}, \{\iota, (124), (142)\}, \{\iota, (134), (143)\}, \{\iota, (234), (243)\}$,
none are normal

39. $N(\{\iota, (134), (143)\}) = \{\iota, (134), (143), (13), (14), (34)\}$

41. $\{\iota, (1234), (13)(24), (1432)\}, \{\iota, (1324), (12)(34), (1423)\}$,
$\quad \{\iota, (1243), (14)(23), (1342)\}$

Chapter 35 (continued)

43. $\{\iota, (1234), (13)(24), (1432)\}^{\{\iota, (123), (132)\}}$

$= \{\{\iota, (1234), (13)(24), (1432)\}, \{\iota, (1324), (12)(34), (1423)\},$

$\{\iota, (1243), (14)(23), (1342)\}\}$

$N(\{\iota, (1234), (13)(24), (1432)\})$

$= \{\iota, (1234), (13)(24), (1432), (13), (24), (14)(23), (12)(34)\}$

45. $N(S) = \{(e, [0]_3), (x^3, [0]_3), (yx, [0]_3), (yx^4, [0]_3), (e, [1]_3), (x^3, [1]_3),$

$(yx, [1]_3), (yx^4, [1]_3), (e, [2]_3), (x^3, [2]_3), (yx, [2]_3), (yx^4, [2]_3)\}$

63. $G = S_3 \times S_3, p = 2,$

$X = \{(\iota, \iota), \quad (\iota, (12)), \quad ((12), \iota), \quad ((12), (12))\},$

$Y = \{(\iota, \iota), \quad (\iota, (12)), \quad ((13), \iota), \quad ((13), (12))\},$

$Z = \{(\iota, \iota), \quad (\iota, (23)), \quad ((23), \iota), \quad ((23), (23))\}$

Chapter 36

1. $\mathbb{Z}_{77}$ **3.** $\mathbb{Z}_{169}, \quad \mathbb{Z}_{13} \times \mathbb{Z}_{13}$

17. $1 \sim S(i), 2 \sim jS(i), F(x) = \begin{cases} \iota & \text{if } x \in S(i) \\ (12) & \text{if } x \notin S(i) \end{cases}$

19. $1 \sim S((e, b)), 2 \sim (a, b^2)S((e, b)), 3 \sim (a^2, e)S((e, b)),$

$F((x, y)) = \begin{cases} \iota & \text{if } (x, y)S((e, b)) = S((e, b)) \\ (123) & \text{if } (x, y)S((e, b)) = (a, b^2)S((e, b)) \\ (132) & \text{if } (x, y)S((e, b)) = (a^2, e)S((e, b)) \end{cases}$

Chapter 37

1. $\iota, (123), (132)$

3. n odd

Chapter 38

3. $\mathbb{Z}_2 \times_\theta \mathbb{Z}_1, \Theta_{[0]_1}([k]_2) = [k]_2$

5. $\mathbb{Z}_2 \times_\theta \mathbb{Z}_3, \Theta_{[i]_3}([k]_2) = [k]_2$

7. $\mathbb{Z}_3 \times_\theta \mathbb{Z}_3, \Theta_{[i]_3}([k]_3) = [k]_3$

11. $\mathbb{Z}_4 \times_\theta \mathbb{Z}_4; f_1([k]_4) = [k]_4, f_2([k]_4) = 3[k]_4; \Theta_y \begin{cases} f_1 & \text{if } y \in \{[0]_4, [2]_4\} \\ f_2 & \text{if } y \in \{[1]_4, [3]_4\} \end{cases}$

13. $\mathbb{Z}_4 \times_\Theta \mathbb{Z}_{10}; f_1([k]_4) = [k]_4, f_2([k]_4) = 3[k]_4;$

$$\Theta_y = \begin{cases} f_1 & \text{if } y \in \{[0]_{10}, [2]_{10}, [4]_{10}, [6]_{10}, [8]_{10}\} \\ f_2 & \text{if } y \in \{[1]_{10}, [3]_{10}, [5]_{10}, [7]_{10}, [9]_{10}\} \end{cases}$$

15. $Q \times \mathbb{Z}_2, D_4 \times \mathbb{Z}_2$

17. $Q \times \mathbb{Z}_5, D_4 \times \mathbb{Z}_5$

Bibliography

General

While the presentation of the preceding material is my own, its original discoverers and subsequent expositors are legion. I am especially indebted to the following works.

Curtis, Morton L. *Abstract Linear Algebra.* New York: Springer-Verlag, 1990.
Jacobson, Nathan. *Basic Algebra I.* 2d ed. New York: W. H. Freeman, 1985.
———. *Basic Algebra II.* New York: W. H. Freeman, 1980.
———. *Lectures in Abstract Algebra, Volume III—Fields and Galois Theory.* Princeton: D. van Nostrand Co. Inc., 1964.
MacLane, Saunders, and Garrett Birkhoff. *Algebra,* New York: Macmillan, 1967.
McCoy, Neal H. *The Theory of Rings.* New York: Macmillan, 1964.
Rotman, Joseph J. *An Introduction to the Theory of Groups.* 3d ed. Dubuque, Iowa: Wm. C. Brown, 1988.

The following were mentioned in the text:

Aleksandrov et al., eds. *Mathematics: Its Content, Methods, and Meaning.* Cambridge, Mass.: M.I.T. Press, 1963. (Original Russian edition 1956.)
Dudley, Underwood. *Mathematical Cranks.* Washington, D.C.: Mathematical Association of America, 1992.
McKay, James H. "Another Proof of Cauchy's Theorem." In *American Mathematical Monthly* 66 (1959): 119.

Historical Notes

For the most part, I drew the basic information in the historical notes from

Gillispie, Charles Coulston, ed. *Dictionary of Scientific Biography,* 16 vols. New York: Charles Scribner's Sons, 1970.

The accounts given in the dictionary supplied at best only the bare bones; I added flesh from the following sources.

Albers, Donald J., and Gerald L. Alexanderson, eds. *Mathematical People.* Boston: Birkhäuser, 1985.

Albers, Donald J., Gerald L. Alexanderson, and Constance Reid, eds. *More Mathematical People.* Boston: Harcourt Brace Jovanovich, 1990.

Belhoste, Bruno. *Augustin-Louis Cauchy: A Biography.* New York: Springer-Verlag, 1991.

Birkeland, Bent. "Ludvig Sylow's Lectures on Algebraic Equations and Substitutions, Christiana (Oslo), 1862: An Introduction and a Summary." In *Historia Mathematica* 23 (1996): 182–99.

Boole, George. *The Mathematical Analysis of Logic.* Oxford: Basil Blackwell, 1965. A reprint of the edition published in 1847 by Macmillan, Barclay & Macmillan, Cambridge.

Bourbaki, Nicolas. *Eléments d'Histoire des Mathématiques.* Paris: Hermann, 1969.

Brewer, James K., and Martha K. Smith, eds., *Emmy Noether: A Tribute to Her Life and Work.* New York and Basel: Marcel Dekker, 1981.

Browder, Felix, ed. *Mathematical Developments Arising from Hilbert Problems.* In *Proceedings of Symposia* 27. Providence: American Mathematical Society, 1976.

Bühler, W. K. *Gauss: A Biographical Study.* New York: Springer-Verlag, 1981.

Dick, Auguste. *Emmy Noether, 1882–1935.* Translated by H. I. Blocher. Boston: Birkhäuser, 1981.

Dieudonné, Jean. *Mathematics: The Music of Reason.* New York: Springer-Verlag, 1992.

Dold-Samplonius, Yvonne. "Bartel Leendert van der Waerden (1903–1996)." In *Historia Mathematica* 24 (1997a): 125–30.

———. "Interview with Bartel Leendert van der Waerden." In *Notices of the American Mathematical Society* 44 (1997b): 313–20.

Gindikin, Semyon Grigorevich. *Tales of Physicists and Mathematicians.* 2d ed. Boston: Birkhäuser, 1985.

Gowing, Ronald. *Roger Cotes: Natural Philosopher.* Cambridge, Mass.: Cambridge University Press, 1983.

Grinstein, Louise S., and Paul J. Campbell, eds. *Women of Mathematics.* New York: Greenwood Press, 1987.

H. H. "Camille Jordan." Obituary in the *Proceedings of the London Mathematical Society* 21 (1923): xliii–xlv.

Henkin, Leon et al., eds. *Proceedings of the Tarski Symposium.* In *Proceedings of Symposia in Pure Mathematics* 28. Providence: American Mathematical Society, 1974.

Kline, Morris. *Mathematics: The Loss of Certainty.* New York: Oxford University Press, 1980.

———. *Mathematical Thought from Ancient to Modern Times.* New York: Oxford University Press, 1972.

Kramer, Edna E. *The Nature and Growth of Modern Mathematics.* Princeton: Princeton University Press, 1981.

MacFarlane, Alexander. *Ten British Mathematicians.* New York: Wiley and Sons, 1916.

MacHale, Desmond. *George Boole.* Dublin: Boole Press, 1985.

MacLane, Saunders. "Van der Waerden's Modern Algebra." In *Notices of the American Mathematical Society* 44 (1997): 321–22.

Muir, Jane. *Of Men and Numbers: The Story of the Great Mathematicians.* New York: Dover Publications, 1996. (Originally published by Dodd Mead & Company, New York, 1961.)

Nicholson, Julia. "The Development and Understanding of the Concept of Quotient Group." In *Historia Mathematica* 20 (1993): 68–88.

O'Donnell, Sèan. *William Rowan Hamilton.* Dublin: Boole Press, 1983.

Ore, Oyestein. *Niels Henrik Abel.* New York: Chelsea, 1957.

Rothman, Tony. "Genius & Biographers: The Fictionalization of Evariste Galois." In *American Mathematical Monthly* 89 (1982): 84–106.

Smith, David Eugene. *History of Mathematics, I.* New York: Dover Publications, 1958. (Originally published by Ginn and Company, 1923).

Struik, D. J. *A Source Book in Mathematics 1200–1800.* Princeton: Princeton University Press, 1969 and 1986.

van der Waerden, B. L. *Geometry and Algebra in Ancient Civilizations.* New York: Springer-Verlag, 1983.

———. *A History of Algebra.* New York: Springer-Verlag, 1985.

Weaver, J. R. H., ed. *The Dictionary of National Biography.* London: Oxford University Press, 1937.

Young, Laurence. *Mathematicians and Their Times.* Amsterdam: North-Holland, 1981.

Photo Credits

Notation Index

A_4, 373, 376, 378, 381
A_5, 384–385
A_n, 373, 376, 380, 382, 384–385, 387
A_p, 329, 330, 331
$|a + bi|$, 39, 40
$\arg(a + bi)$, 39, 40
$\text{Aut}(G)$, 389, 392
$\mathbb{C}$, 35
$\mathbb{C}^\times$, 186
$C(x)$, 350, 354
∇, 102
(d), 120, 121, 122–123, 132
D_n, 194, 225, 339–342, 376, 383
$F[x]$, 83, 103–113, 171
$F^{p(x)}$, 85, 175, 292, 296
$F(r)$, 135, 136, 151, 173
$F(r_1, \ldots, r_n)$, 82, 136–137, 143–144
$G \times H$, 192, 208, 218, 219, 266, 303–306
$G \times_\theta H$, 392, 393
$G_1 \oplus G_2$, 312–316
G/N, 256, 257, 264, 395
$\text{Gal } F^{p(x)}/F$, 175, 205–207, 292, 296
$\text{Gal } K/F$, 170, 175, 184, 186, 196, 205, 217, 272–278, 280, 292, 296
  construction, 176–178
$\mathbb{H}$ (See Q), 63, 68, 187
$\mathbb{H}^\times$, 187
i, 35, 37, 63
$[i]_n$, 58–59, 187
j, 63
k, 63

$[K{:}F]$, 149, 151, 152, 175
$M_2(\mathbb{R})$, 99, 187, 188
$M_2(\mathbb{R})^{ns}$, 188
(n), 121, 187, 228
nx, 213
$N_G(S)$, 361
$N(S)$, 361
pA, 320
π_n, 262
Q, 191, 225, 258, 342, 343, 395
$\mathbb{Q}$, 34, 250
$\mathbb{Q}^\times$, 186
$[r{:}F]$, 133, 151
$\mathbb{R}$, 34
$\mathbb{R}^+$, 208
$\mathbb{R}^\times$, 186
S^H, 363
S_n, 197–200, 201, 205, 223, 224, 235, 285, 373, 375
$S_p(A)$, 320, 323
$S(X)$, 221, 313
T, 374, 376, 388–389
U_n, 214
x^G, 353
x^n, 213
$\mathbb{Z}$, 34
$\mathbb{Z}[x]$, 127
$\mathbb{Z}^\times$, 186
$\mathbb{Z}_n$, 59–61, 62, 187, 238, 334
$\mathbb{Z}_p^{\;\times}$, 187
$Z(G)$, 219, 258, 350, 354, 356
ζ_n, 42, 187, 234, 240

Subject Index

A1, 35, 60, 72, 79, 97
A2, 35, 60, 72, 79, 97
A3, 35, 60, 72, 79, 97
A4, 35, 60, 72, 79, 97
A5, 35, 60, 72, 79, 80, 97
Abel, Niels Henrik, 91–94, 146, 253, 300, 370
Abelian group, 186, 216, 256, 331, 333, 335
   free, 309–310
   of prime power order, 325
Académie des Sciences, 50, 167, 168, 169, 251, 252
Académie Francaise, 51, 252
Algebra, fundamental theorem of, 46, 49–56
Algebraic, 132, 135
  degree of an element, 133, 135, 151
Alternating group, 380, 382
Amsterdam, University of, 270
And,
  meaning, 398
  truth table, 399
Argand, Jean-Robert, 49, 54–55
Argument, 39, 40
Arithmetic, fundamental theorem of, 12
Artin, Emil, 262, 268, 269
Associates in a ring, 249
Associative laws, 35, 60, 72, 75, 79, 80, 96, 97, 186, 305, 315
Automorphism, 160, 170, 389
Axioms, 35, 72, 79, 97, 98, 99, 186, 197, 200
  (*See* **A1**, **A2**, **A3**, **A4**, **A5**, **D1**, **D2**, **DM**, **G1**, **G2**, **G3**, **G4**, **G5**, **M1**, **M2**, **M3**, **M4**, **M5.**)

Basel, University of, 51
Basis theorem, 331

Basis, vector space, 149, 151, 414–417
Berkeley, University of California at, 116, 117
Berlin Academy, 147, 337
Berlin Society of Sciences, 52
Berlin, University of, 310, 337
Bernoulli, Daniel, 51
Bernoulli, Jean, 53
Bernoulli, Johann, 51
Bernoulli, Nikolaus, 51
Bertrand, Joseph, 252
Binary operation, 72
Binomial coefficients, 15
Binomial theorem, 15
Bombelli, Raffael, 48
Bonn, University of, 345
Boole, George, 17, 410–412
Boolean ring, 102
Bounded below, 3, 4
Breslau,
  Technical Institute at, 128
  University of, 147, 337
Brouwer, L. E. J., 338
Brunswick,
  Duke of, 55, 56
  Polytechnikum in, 180
Bryn Mawr College, 128
Burnside, William, 287

Cambridge,
  St. John's College, 287
  University of, 209, 210
Cancellation law, 101
Canonical homomorphism, 262

Cantor, Georg, 338
Cardano's formula, 20, 21, 48–49
Cardano, Gerolamo, 31–33, 48, 49
Category theory, 309
Catherine I, 51
Catherine the Great, 50, 52
Cauchy, Augustin-Louis, 50, 93, 167, 168, 251–253, 300
    Cauchy's theorem, 243, 247, 251, 348, 354, 370
    Cauchy's theorem for Abelian groups, 349
Cayley, Arthur, 195, 209–210
    Cayley's theorem, 204–205, 209
Center of a group, 219, 258, 350, 354, 356
Chicago, University of, 310
Class equation, 354
Class,
    conjugacy, 249, 358
    equivalence, 245, 354
Closed, 35, 60, 72, 75, 79, 96, 97, 119, 186, 212, 213
Codomain of a function, 159, 197
Coefficients, method of undetermined, 150–151
Commutative,
    diagram, 264
    group, 186, 216, 333, 335
    laws, 35, 60, 72, 75, 79, 80, 96, 97, 98, 186, 197, 200, 305, 315
    ring, 98
Commutator, 281
    subgroup, 282, 283
Complex,
    conjugate, 37, 45
    numbers, 35–46
    plane, 38
Conclusion, 402
Congruence relation, 249–250, 259
Conjugacy classes, 249, 358
Conjugate,
    complex, 37, 45
    of a group element, 249, 353
    subgroup, 363, 364, 365
Constructed by straightedge and compass, 418
Construction of Gal *K/F,* 176–178
Contradiction, proof by, 406–407
Contrapositive, 401, 405–406
Converse, 401
Copenhagen, University of, 54
Correspondence theorem, 266
Coset, 226–230, 243–244
Cotes, Roger, 49, 52–53, 54

Counterexamples, 403–404
Courant, Richard, 128
Crelle, Leopold, 92, 93, 147
Cubic equations, 19–25
Cycle, 198–200, 235
    disjoint, 200, 235
Cyclic group, 221, 237, 241, 325, 331, 333, 355, 372

d'Alembert, Jean le Rond, 49–51, 52, 55
D1, 35, 60, 72, 79, 80, 97
D2, 35, 60, 72, 79, 80, 97
Danish Academy of Sciences, 54
Darboux, Gaston, 345
Davis, Martin, 117
De Moivre, Abraham, 49, 53–54
    De Moivre's theorem, 41, 53, 54
De Morgan, Augustus, 16–17, 67, 102, 411
    De Morgan's laws, 102
Dedekind, Richard, 180–181, 260
Degen, Ferdinand, 91, 92
Degree,
    of an algebraic element, 133, 135, 151
    of a polynomial, 105, 133
Del Ferro, Scipione, 31, 33
Della Nave, Annibale, 31, 33
Derivative of polynomial, 172, 179, 180
Descartes, René, 48–49
Diagram commutes, 264
Diderot, Denis, 50
Dihedral groups, 194, 225, 287, 339–342, 347, 355, 376, 383, 387
Dimension, 149, 152, 154, 155, 175
Direct,
    product, 192, 208, 218, 219, 266, 303–306, 309, 312, 318, 325, 331, 333, 335
    proof, 402–403
    sum, 303, 312, 318, 325, 331, 333, 335
Dirichlet, P. G. LeJeune, 147, 180, 337,
Disjoint,
    cycles, 200
    union, 230, 246, 354
Distributive laws, 35, 60, 72, 75, 79, 80, 96, 97, 305, 315
Divides, 103, 106, 125
Division algorithm,
    for polynomials, 105, 110
    for the integers, 4

Division ring, 102
Divisor, polynomial, 103, 106
DM, 99
Domain,
    integral, 99, 100, 158, 250–251
    of a function, 159, 197
    principal ideal, 122, 127
Duels,
    mathematical, 31–33
    with weapons, 169
Dumas, Alexandre, 168, 169
Duplicating a cube, 425
Dyck, Walther von, 195, 260

École,
    Normale, 168
    Polytechnique, 167, 251, 252, 356, 386
Edinburgh, University of, 310
Eidgenössische Technische Hochschule, 180
Eisenstein, Gotthold, 146–147
Eisenstein's irreducibility criterion, 142
Elliptic integrals, 91
Equations,
    cubic, 19–25
    quadratic, 18–19
    quartic, 25–29
Equivalence,
    class, 245, 246, 364
    relation, 244, 246, 364
Equivalent statements, 405, 407–408
Erlangen, University of, 127, 346
Erlanger programm, 346
Euclidean algorithm,
    for the integers, 6–7
    for polynomials, 110
Euler, Leonhard, 49, 50, 51–52, 55, 241
Even permutation, 380, 381
Extension,
    field, 74, 75, 82, 173
    of one group by another, 395
External direct product, 192, 208, 218, 219, 266,
    303–306, 309, 325, 331, 333, 335

Factor, 303
Feit, Walter, 285
Fermat's theorem, 66

Ferrari, Ludovico, 32–33
Fibonacci numbers, 13–14, 15
Field, 73
    extension, 74, 75, 82, 173
    fixed, 289
    homomorphism, 160, 161
    primitive element of, 173–175
    of quotients, 250–251
    sub-, 74, 75, 81, 82
Finite dimensional, 149, 152, 154, 155, 175
Fiore, Antonio Maria, 31, 33
First isomorphism theorem, 260, 267
First principle of mathematical induction, 9, 10, 11, 16,
    173, 325, 381
First Sylow theorem, 359
Fixed field, 289
Fontana, Niccolò, 31–33
Formula,
    Cardano's, 20, 21, 48–49
    quadratic, 18–19
Fourier, J. B. J., 168
Francais, J. F., 54, 55
Frederick the Great, 50, 52
Free Abelian group, 309–310
Frobenius, Georg, 260
Fruedenthal, Hans, 269
Fundamental theorem,
    of algebra, 46, 49–56
    of arithmetic, 12
    of finite Abelian groups, 333

G1, 186
G2, 186
G3, 186
G4, 186
G5, 186
Galois,
    Evariste, 29, 167–169, 260, 370
    group, 184, 186, 196, 205, 217, 272–278, 280, 292,
    296
    theory, 180
Gauss, Carl Friedrich, 49, 55–56, 92, 147, 180, 337
    Gauss's Lemma, 141
Generated subgroup, 221, 223, 224
Generator,
    of an ideal, 121
    of a subfield, 82

Generators and relations, 341, 342, 374, 376–377
Geometric series, 14
Girard, Albert, 49
Gödel, Kurt, 261
Göttingen, University of, 55, 56, 127, 128, 180, 260, 268, 337, 345
Grassman, Hermann Günther, 68
Greatest common divisor,
  in a principal ideal domain, 126
  of polynomials, 107, 108, 109
  of two integers, 5, 6
Greenwich, Royal Naval College at, 287
group, 186
  Abelian, 186, 216, 333, 335
  Abelian of prime power order, 325
  alternating, 380
  center, 219, 258, 350, 354, 356
  commutative, 186, 216, 333, 335
  conjugate elements, 249, 353
  cyclic, 221, 237, 238, 241, 325, 331, 333, 355, 372
  dihedral, 193–194, 225, 287, 339–342, 347, 355, 376, 383, 387
  external direct product, 192, 208, 218, 219, 266, 303–306, 309, 325, 331, 333
  free Abelian, 309–310
  Galois, 184, 186, 196, 205, 217, 272–278, 280, 292, 296
  homomorphism, 203, 205, 317
  isomorphism, 203, 238, 305, 317
  metabelian, 287
  noncommutative, 186, 197
  operation, 186–187
  order, 201, 233, 236, 237, 238, 257, 304, 325, 331, 333, 349, 354, 355–356, 359, 365, 373, 376, 382
  permutation, 197–200, 201, 205, 206, 223, 224, 235, 373, 375, 380, 381
  prime order, 238
  quaternion, 191, 225, 258, 342, 343, 395
  quotient, 256, 261, 277, 281, 282
  simple, 382, 384, 387
  solvable, 280, 283, 285, 292, 297
  sub-, 211–213, 215, 216, 221, 226–230, 236, 243, 255, 263, 360, 361, 363, 382
  table, 189–190

Halley, Edmund, 53
Hamilton, William Rowan, 62, 66–68, 146, 195

Hankel, Herman, 68
Hecke, Erich, 268
Heisenburg, Werner, 269
Hensel, Kurt, 338
Hilbert,
  David, 117, 127, 128, 268
  problems, 117
Hölder, Otto, 259–261
Holmboe, Berndt, 91, 92
Homomorphic image, 215, 216, 283
Homomorphism, 160, 161, 162, 164, 170, 203, 205, 216, 236, 262, 305–306, 312, 317, 334, 342, 374, 376, 393
  one-to-one, 263
  restriction, 276
  theorem, 260
  theorem for groups, 260, 264, 292
Humboldt, Alexander von, 147
Hypotheses *S,* 274
Hypothesis, 402

Ideal, 119, 120, 132, 187
  principal, 120, 121, 123, 126, 127, 132
Identity, 35, 60, 72, 79, 96, 97, 98, 186
If and only if,
  meaning, 398
  proofs involving, 405, 407–408
Implies,
  logical equivalents, 400, 401
  meaning, 398–399
  truth table, 399
Index, 226, 236, 352, 364, 375, 447, 448
Induction
  first principle, 9, 10, 11, 16, 173, 325, 381
  second principle, 12, 113, 325–326, 359
Infinite order, 233
Inquisition, 33
Integral domain, 99, 100, 158, 250–251
Internal,
  direct product, 312, 318
  direct sum, 312–316, 318, 325, 331, 333, 335
Inverse, 35, 61, 62, 72, 79, 80, 96, 97, 102, 186
Irreducible polynomial, 111, 112, 113, 132, 133, 142, 171, 296, 297
Isomorphic groups, 203, 264, 305–306
Isomorphism, 160, 164, 170, 203, 264, 305–306, 317, 334, 342, 374, 376, 393

Jacobi, Karl, 93, 147
Johns Hopkins University, 270, 346
Jordan, Camille, 260, 345, 386–387

Kant, Immanuel, 56
Kernel, 215, 216, 256, 262, 263, 264, 375
Klein, Felix, 195, 260, 345–347
Kristiana University, 370
Kronecker, Leopold, 260, 337–338, 370
Kummer, Ernst Eduard, 337

Lagrange, Joseph Louis, 30–31, 241–242, 251
    Lagrange's theorem, 236, 375, 376
Laplace, Pierre, 251
Lattice, subgroup, 214–215, 236
Least common multiple of polynomials, 115
Left coset, 226, 229–230, 343
Legendre, Adrien-Marie, 54, 93
Leibniz, Gottfried Wilhelm, 53
Leipzig, University of, 346
Leopoldovna, Anna, 52
Lie, Sophus, 260, 345, 370
Linear combination, 149, 414
Linearly independent, 149, 414
Liouville, Joseph, 252
Logical connectives, 398–399
Logically equivalent, 401, 405, 407–408
London University, 16, 17
Louis XIV, 53
Louis XVIII, 252

M1, 35, 60, 72, 79, 97
M2, 35, 60, 72, 79, 97
M3, 35, 60, 72, 79, 98
M4, 35, 61, 72, 79, 98
M5, 35, 60, 62, 72, 79, 98, 102
MacLane, Saunders, 269
Mathematical Analysis of Logic, The, 17, 397, 411
Matijasevic, Yuri, 117
Matrix ring, 99, 187, 188
McKay, James H., 247
Metabelian group, 287
Method of undetermined coefficients, 150–151
Minimal element, 3, 4
Minimum polynomial, 133, 135, 162, 164
Minkowski, Hermann, 260

Mod, 61, 256
Modular arithmetic, 141–143
Modulus, 39, 40
Monge, Gaspard, 251
Monic, 107, 132
Munich Technische Hochschule, 195, 346

Napoleon, 55, 56, 242
Neugebauer, Otto, 270
Newton, Isaac, 50, 53, 411
Noether, Emmy, 127–128, 181, 262, 268, 311
Noncommutative group, 186, 197
Nontrivial,
    subgroup, 236, 286
    vector space, 148, 413
Normal subgroup, 255–257, 263, 343, 382
Normalizer, 361, 362
Not,
    meaning, 398
    truth table, 399
nth root, 41, 42, 43, 70, 187

Odd permutation, 380, 381
One-to-one, 159–160, 170, 203, 205, 230, 263
Onto, 159–160, 170, 203, 230
Operation,
    binary, 72
    group, 186–187, 189–190
Or,
    meaning, 398
    proofs involving, 408–409
    truth table, 399
Order,
    of an element, 233, 235, 236, 237, 238, 320, 349, 354
    of a group, 201, 233, 236, 237, 238, 257, 304, 325, 331, 333, 349, 354, 355–356, 359, 365, 373, 376, 382
Ozanam, Jacques, 53

p-Subgroup, 360
Pacioli, Fra Luca, 31
Padua, University of, 32
Permutation, 196, 205
    even, 380, 381
    group, 197–200, 201, 205, 206, 223, 224, 235, 373, 375, 380, 381
    odd, 380, 381
Peter the Great, 51
Poinsot, Louis, 251, 252

Poisson, Siméon Denis, 168
Polynomial,
    degree, 105, 133
    derivative of, 172, 179, 180
    division algorithm, 105, 110
    divisor, 103, 106
    greatest common divisor, 107, 108, 109
    irreducible, 111, 112, 113, 132, 133, 142, 171, 296, 297
    least common multiple, 115
    minimum, 133, 135, 162, 164
    monic, 107, 132
    reducible, 111, 112
    ring, 98, 100, 122, 123, 127
    root of, 83, 84, 85, 106, 162, 171
Preserves,
    addition, 160, 170, 203
    multiplication, 160, 170, 203
    the operation, 203
Primary decomposition theorem, 331
Prime number, 7, 8, 9, 11, 12, 13, 235, 238, 329, 330, 331, 333, 355, 356, 359, 365, 383
Prime order,
    elements, 235, 329
    groups, 238
Primitive,
    element, 173–175
    nth root of unity, 48
Princeton University, 128, 311
Principal,
    ideal, 120, 121, 123, 126, 127, 132
    ideal domain, 122, 127
Principle,
    first induction, 9, 10, 11, 16, 173, 325, 381
    second induction, 12, 113, 325–326, 359
    well-ordering, 3, 123
Product,
    direct, 192, 208, 218, 219, 266, 303–306, 309, 312, 318, 325, 331, 333, 335
    semidirect, 388–394
Proof by contradiction, 406–407
Proper subgroup, 219, 236, 286
Prussian Academy, 50
Putnam, Hilary, 117

Quadratic formula, 18–19
Quantifiers, 401–402
Quartic equations, 25–29

Quaternions, 61–64, 68, 187, 210
    group, 191, 225, 258, 342, 343, 395
Quotient,
    field of an integral domain, 250–251
    group, 256, 261, 277, 281, 282
    ring, 259, 267

Radicals, 43, 85–88, 272–278, 290–293, 297, 298
Rameau, Jean-Philippe, 50
Rational numbers, construction of, 250
Reducible polynomial, 111, 112
Reflexive, 243, 244
Reid, Constance, 116
Relation,
    congruence, 249–250, 259
    equivalence, 244, 246, 364
Remainder, 7, 105, 110
Restriction homomorphism, 276
Riemann, G. F. B., 180, 337
Right coset, 226, 232, 343
Ring, 97, 187
    Boolean, 102
    commutative, 98
    division, 102
    homomorphism, 160, 161, 258
    ideal, 119, 120, 132, 187
    matrix, 99, 187, 188
    polynomial, 98, 100, 122, 123, 127
    principal ideal, 120, 121, 123, 126, 127, 132
    principal ideal domain, 122, 127
    quotient, 259, 267
    sub-, 118
    with unit element, 98
Robinson,
    Julia, 116–117
    Raphael, 116
Romanova, Anna, 51
Root,
    nth, 41, 42, 43, 70, 187
    of a polynomial, 83, 84, 85, 106, 162, 171
Royal Irish Academy, 67
Royal Society, 53
Ruffini, Paolo, 29, 92, 299–300

Saint Petersburg Academy of Sciences, 51, 55
Scalar, 148, 413
Schreier, Otto, 260, 268
Second isomorphism theorem, 267

Second principle of mathematical induction, 12, 113, 325–326, 359
Semidirect product, 388–394
Simple group, 382, 384, 387
Smallest,
  ideal, 120, 121, 132
  subfield, 82
  subgroup, 220, 221
Solution of a polynomial equation, 83
Solvable,
  group, 280, 283, 285, 292, 297
  by radicals, 85, 292, 297
Spans, 149, 414
Stendhal, 252
Subfield, 74, 75, 81, 82
  generated by, 82
Subgroup, 211–213, 216, 226
  commutator, 282, 283
  conjugate, 363, 364, 365
  coset, 226–230, 243–244, 343
  generated by a subset, 221–224
  lattice, 214–215, 236
  nontrivial, 236, 286
  normal, 255–257, 263, 343, 382
  p-, 360
  proper, 219, 236, 286
  Sylow *p-*, 360, 362, 365, 366
Subring, 118
Subspace, 148, 413
Sum, direct, 303, 312, 318, 325, 331, 333, 335
Summand, 303
Sylow,
  p-subgroup, 360, 362, 366
  Peter Ludvig, 370–371
  theorems, 359, 365
Symmetric, 243, 244
  difference, 102

Table, multiplication or addition, 189–190
Tait, Peter Guthrie, 68
Tarski, Alfred, 117
Tartaglia, 31–33
Thompson, John, 285
Transitive, 243–244
Transposition, 223
Triads, 68
Trinity College,
  Cambridge, 16, 53
  Dublin, 67
Trisecting an angle, 424
Trisection problem, 155, 424
Truth tables, 399–401
Tübingen, 260

Undetermined coefficients, method of, 150–151
Unit element, 98

Van der Waerden, B. L., 128, 260, 268–270
Vector space, 148, 210, 413
Vectors, 148, 210, 413

Wedderburn, J. H. M., 310–311
Weierstrass, Karl, 259, 260, 337, 338, 370
Well defined, 60
Well-ordering principle, 3
Wessel, Caspar, 49, 54
Wiener, Norbert, 128
Wilson, Woodrow, 310

Zero divisor, 99